D1708965

Handbook of Research on Urban Informatics:
The Practice and Promise of the Real-Time City

Marcus Foth
Queensland University of Technology, Australia

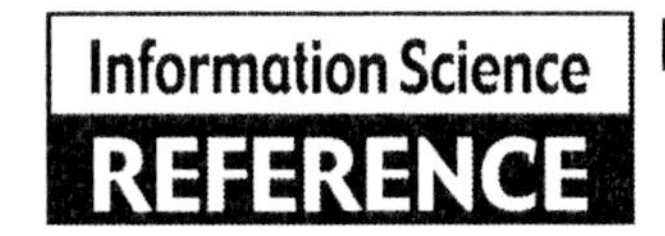

INFORMATION SCIENCE REFERENCE
Hershey · New York

Director of Editorial Content: Kristin Klinger
Director of Production: Jennifer Neidig
Managing Editor: Jamie Snavely
Assistant Managing Editor: Carole Coulson
Typesetter: Michael Brehm
Cover Design: Lisa Tosheff
Printed at: Yurchak Printing Inc.

Published in the United States of America by
Information Science Reference (an imprint of IGI Global)
701 E. Chocolate Avenue, Suite 200
Hershey PA 17033
Tel: 717-533-8845
Fax: 717-533-8661
E-mail: cust@igi-global.com
Web site: http://www.igi-global.com

and in the United Kingdom by
Information Science Reference (an imprint of IGI Global)
3 Henrietta Street
Covent Garden
London WC2E 8LU
Tel: 44 20 7240 0856
Fax: 44 20 7379 0609
Web site: http://www.eurospanbookstore.com

Library of Congress Cataloging-in-Publication Data

Handbook of research on urban informatics : the practice and promise of the real-time city / Marcus Foth, editor.

p. cm.

Includes bibliographical references and index.

Summary: "This book exposes research accounts which seek to convey an appreciation for local differences, for the empowerment of people and for the human-centred design of urban technology"--Provided by publisher.

ISBN 978-1-60566-152-0 (hardcover) -- ISBN 978-1-60566-153-7 (ebook)

1. Cities and towns--Effect of technological innovations on--Case studies. 2. Sociology, Urban--Case studies. 3. Computer networks--Social aspects. 4. Information technology--Social aspects. 5. Telecommunication--Social aspects. 6. Information society. I. Foth, Marcus.

HT166.H364 2009

307.76--dc22

2008020498

British Cataloguing in Publication Data
A Cataloguing in Publication record for this book is available from the British Library.

List of Reviewers

Udo Averweg
Information Services, eThekwini Municipality, Durban, South Africa

Bhishna Bajracharya
Mirvac School of Sustainable Development, Bond University, Australia

Mark Bilandzic
Media Informatics Group, Ludwig-Maximilians-Universität München, Germany

Jean Burgess
ARC Centre of Excellence for Creative Industries and Innovation, Queensland University of Technology, Australia

Bharat Dave
Faculty of Architecture, Building and Planning, University of Melbourne, Australia

Ernest Edmonds
Faculty of Information Technology, University of Technology Sydney, Australia

Alexia Fry
Faculty of Arts, Deakin University, Australia

Mark Gaved
Knowledge Media Institute, The Open University, UK

Mariann Hardey
Department of Sociology, University of York, UK

Greg Hearn
Creative Industries Faculty, Queensland University of Technology, Australia

Dan Hill
Arup, Sydney, Australia

Vikki Katz
Annenberg School for Communication, University of Southern California, USA

Mark Latonero
Department of Communications, California State University, Fullerton, USA

Marcus Leaning
Trinity College, University of Wales, UK

Ian MacColl
Australasian Cooperative Research Centre for Interaction Design, Australia

Tikva Morowati
Interactive Telecommunications Program, New York University, USA

Kristina Rauschan
Institute for Media Research, Braunschweig University of Arts, Germany

Gavin Sade
Creative Industries Faculty, Queensland University of Technology, Australia

Barry Saunders
Creative Industries Faculty, Queensland University of Technology, Australia

Giandomenico Sica
Polimetrica Publishers, Italy

Matt Ward
Goldsmiths College, University of London, UK

Michele Willson
Faculty of Media, Culture and Society, Curtin University of Technology, Australia

List of Contributors

Table of Contents

Section V:
Wireless and Mobile Culture

Section VI:
The Not So Distant Future

Detailed Table of Contents

This chapter critically examines the notion of "the city" within urban informatics. Arguing that there is an overarching tendency to construe the city as an economically and spatially distinct social form, it reviews a series of system designs manifesting this assumption. Systematically characterizing the city as a dense ecology of impersonal social interactions occurring within recognizably public places, this construction can be traced to turn-of-the-century scholarship about the metropolis. An alternative perspective which foregrounds the experience rather than the form of the metropolis is advocated. Users become actors embedded in global networks of mobile people, goods, and information, positioned in a fundamentally heterogeneous and splintered milieu. Grounding this approach in a preliminary study of mobility practices in Bangkok, Thailand, the chapter illustrates how urban informatics might refine its subject, accounting for local particularities between cities as well as the broader global networks of connection between these sites.

This chapter describes the design and installation of a new kind of public opinion forum—*TexTales*, a public, large-scale interactive projection screen—to demonstrate how public city spaces can become sites for collective expression and public opinions can be considered social constructions. The design and implementation of *TexTales* installations is analyzed and a number of interaction design elements critical for designing expressive urban spaces are identified such as, starting "intermodal" conversations; authoring for nomadic, unfamiliar audiences; distributing public discourse across mediated and physical space; and editing and censoring dialog to ensure that it reflects the norms and values of forum designers.

This chapter describes a small networked community in which residents of an apartment building in Washington, D.C., USA supplement their face-to-face social interactions with a Yahoo email listserver. Analysis of over 460 messages that have been archived since July 2000 when the list began reveals that the same issues that drive participation on the list also drive participation off the list. Threats to safety, high rent increases, and changes in management practices, such as parking regulations and access to facilities, motivate communication on and offline. Furthermore, those who are most active online are typically most active offline. Activity on the list is strongly fuelled by interest and discussion around local events, hence the term event-driven, and is promoted by activist tenants. Friendly notes about new restaurants, bird observations and other niceties may help a little to create a sense of overall community, but they do little to motivate online participation.

After more than a decade of e-participation initiatives at the urban level, what remains obscure is the alchemy—i.e., the "arcane" combination of elements—that triggers and keeps citizens' involvement in major decisions that affect the local community alive. The Community Informatics Lab's experience with the Milan Community Network since 1994 and its two more recent spin-off initiatives enable us to provide a tentative answer to this question. This chapter presents these experiments and looks at election campaigns and protests as triggers for (e-)participation. It also discusses these events as opportunities to engender more sustained participation aided by appropriate technology tools such as software that is deliberately conceived and designed to support participation and managed with the required expertise.

Section IV:
Location, Navigation and Space

This chapter describes a platform that enables the systematical study of online social networks alongside their real-world counterparts. The system, Cityware, merges users' online social data, made available through Facebook, with mobility traces captured via Bluetooth scanning. Furthermore, the system enables users to contribute their own mobility traces, thus allowing users to form and participate in a community. In addition to describing Cityware's architecture, the chapter discusses the type of data that is being collected, and the analyses the platform enables, as well as users' reactions and thoughts.

In our everyday lives, we are surrounded by information which weaves itself silently into the very fabric of our existence. Much of the time we act in the world based on recognising qualities of information which are relevant to us in the particular situation we are in. These qualities are very often spatial in nature and, in addition to information in the environment itself, we also access representations of space, such as maps and guides. Increasingly, such forms of spatial information are delivered on mobile devices, which enable a different relationship with our spatial world. This chapter discusses an empirical study which attempts to understand how people acquire and act on digital spatial information. In conclusion, it draws on the outcomes of a study to discuss how we might better embed and integrate information in place so that it enables a more relational and shared experience in the interaction between people and their spatial setting.

This chapter proposes that locative urban information necessitates the use of visual instruments, such as maps integrated into spatial annotation systems. The thesis is that the dynamics of the movement and behavior of messages appearing, disappearing, and spreading on the urban maps provide clues as to what extent a specific type of information is dependent on urban space for context, that is, its level of location-sensitivity. A parallel is drawn between the interpretation of dynamic patterns appearing on urban maps and of scientific discovery supported by the use of visual instruments. In order to illustrate how the question of locativity arises when developing technologies for urban life, an examination of BlueSpot, a locative media project in Budapest, is provided.

Car navigation systems, based on "augmented reality," no longer direct the driver through traffic by simply using arrows, but represent the environment true to reality. The constitutional moment of this medium is the constant oscillation between environmental space and two-dimensional projection space. Temporal information in addition to spatial information is becoming increasingly important with features such as real time gridlock reports aided by highway sensors or guidance to the nearest event. Does the future lie in the fusion of travel guides and navigation systems? This chapter argues that future developments in urban informatics resulting from the convergence in cartographic, media and communication technologies can be inferred based on the increasing phenomenon of mobile augmented reality applications.

A lot of street view services, which present views of urban landscapes, have recently appeared. The conventional method for creating street views requires on-vehicle cameras. This chapter proposes a new method, which relies on people who voluntarily take photos of an urban landscape using a system called QyoroView. The system receives photos from users, adjusts the photos' position and orientation, and finally synthesizes them to generate a street view. The chapter reports on two experiments in which the subjects generated a street view using our system.

Many research projects have studied various aspects of smart environments including rooms, homes, and offices. Few projects, however, have studied smart urban public spaces such as smart railway stations and airports due to the lack of an experimental environment. This chapter proposes virtual cities as a testbed for examining the design of smart urban public spaces. An intelligent emergency guidance system for subway stations is presented. A virtual subway station platform is used to analyze the effects of the system. The chapter argues that simulations in virtual cities are useful to pre-test the design of smart urban public spaces and estimate the possible outcome of real-life scenarios.

As cities have become more "computable," capable of manipulation through their digital content, large areas of social life are migrating to the web. This chapter focuses on the virtual city in software, presenting speculations about how such cities are moving beyond the desktop to the point where they are rapidly becoming the desktop itself. But what emerges is a desktop with a difference, a desktop that is part of the web, characterized by a new generation of interactivity between users located at any time in any place. This chapter first outlines the state of the art in virtual city building drawing on the concept of mirror worlds and then comments on the emergence of Web 2.0 and the interactivity that it presumes. It characterizes these developments in terms of virtual cities through the virtual world of Second Life, showing how such worlds are moving to the point where serious scientific content and dialogue is characterizing their use often through the metaphor of the city itself.

Section V:
Wireless and Mobile Culture

This chapter introduces the role of community wireless networks (CWNs) in reconfiguring people, places and information in cities. CWNs are important for leading users and innovators of mobile and wireless technologies in their communities. Their identities are geographically-bounded and their networks that they imbued with social, political and economic values. While there has been much discussion of the networked, virtual and online implications of the Internet, the material implications in physical spaces have been overlooked. By analyzing the work of CWNs in New York and Berlin, this chapter reconceptualizes the interaction between technologies, spaces and forms of organizing. This chapter introduces the concept of codespaces in order to capture the integration of digital information, networks and interfaces with physical space.

From WiFi (802.11b) with its fixed and mobile high-speed wireless broadband Internet connectivity to WiMAX (802.16e), the newest wireless protocol, extending the reach of WiFi across longer distances and more difficult terrain, new wireless technologies are increasingly thought to impact the ways in which we encounter social spaces in public, civic and commercial sites within large urban centers. This chapter explores how and to what extent these new wireless technologies might also be reconfiguring and reorganizing domestic practice and social relations. Drawing on a year-long ethnographic study of WiFi and WiMax provisioned homes in a major Australian metropolitan center, the chapter argues that new wireless infrastructures are impacting how people imagine and use mobile devices, computers and the Internet in and around the home but not in ways wholly anticipated by commercial Internet service providers.

Section VI: The Not So Distant Future

u-City is South Korea's answer to urban community challenges leveraging ubiquitous computing technology to deliver state-of-the-art urban services. Korea's experience designing and constructing u-City may be a useful benchmark for other countries. This chapter defines the concept of u-City and analyzes the needs that led Korea to embark on the u-City project ahead of others. It examines the opportunities and challenges that the nation faces in the transition stage. What has enabled Korea to pioneer the u-City concept is the development of IT infrastructure and the saturation of the IT market on the one hand, and the balanced national development strategy on the other hand. Success of u-City requires a national capability of designing forward-looking institutions to enable better cooperation among stakeholders, the establishment of a supportive legal framework and promotion of technology standardization.

This chapter examines the development of information and communication technologies in urban China, focusing mainly on their impact on social life. The key question raised by this study is how the Internet and mobile technologies are affecting the way people make use of urban space. The chapter begins with some background to China's emergence as a connected nation. It then looks at common uses of web-based and mobile phone technologies, particularly bulletin boards, SMS and instant messaging. The chapter then presents findings of recent research that illustrates communitarian relationships that are enabled by mobility and the use of technologies. Finally, these findings are contextualized in the idea of the City 2.0 in China.

The real-time city is now real! The increasing deployment of sensors and handheld electronic devices in recent years allows for a new approach to the study and exploration of the built environment. The WikiCity project deals with the development of real-time location-sensitive tools for the city and is concerned with the real-time mapping of city dynamics. This mapping, however, is not limited to representing the city, but becomes also instantly an instrument for city inhabitants to base their actions and decisions upon in a better informed manner, leading to an overall increased efficiency and sustainability in making use of the city environment. This chapter discusses the WikiCity Rome project, which was the first occasion for implementing some of WikiCity's elements in a public interface—it was presented on a large screen in a public square in Rome.

This chapter presents an important new shift in mobile phone usage—from communication tool to "networked mobile personal measurement instrument." It explores how these new "personal instruments" enable an entirely novel and empowering genre of mobile computing usage called citizen science. It investigates how such citizen science can be used collectively across neighborhoods and communities to enable individuals to become active participants and stakeholders as they publicly collect, share, and remix measurements of their city that matter most to them. It further demonstrates the impact of this new participatory urbanism by detailing its usage within the scope of environmental awareness. Inspired by a series of field studies, user driven environmental measurements, and interviews, the chapter present the design of a working hardware system that integrates air quality sensing into an existing mobile phone and exposes the citizen authored measurements to the community—empowering people to become true change agents.

What happens to urban space given a hypothetical future where all information loses its body, that is, when it is offloaded from the material substrate of the physical city to the personal, portable, or ambient displays of tomorrow's urban information systems? This chapter explores the spatial, technological and social implications of an extreme urban informatics regime. It investigates the total virtualization of the marks, signage, signaling and display systems by which we locate, orient ourselves, and navigate through the city. Taking as a vehicle a series of digitally manipulated photographs of specific locations in New York, this study analyzes the environmental impact of a pervasive evacuation of information—at various sites and scales—from the sidewalks, buildings, streets, intersections, infrastructures and public spaces of a fictional future De-saturated City.

Foreword

Unlike most scholarly texts, let us begin with some definitions. Since we are in relatively uncharted territory, it is worth a short divergence.

Webster's online dictionary defines *urban* as "of, relating to, characteristic of, or constituting a city". *Informatics*, a bit more obtuse, is "the collection, classification, storage, retrieval, and dissemination of recorded knowledge treated both as a pure and as an applied science". So how would we synthesize these terms to define *urban informatics*? One version might read "the collection, classification, storage, retrieval, and dissemination of recorded knowledge of, relating to, characteristic of, or constituting a city". This definition emphasizes information as the dominant structural aspect or reason for being—that information is literally what constitutes a city. What is left of Carthage, after all, but legends? Another possible definition for *urban informatics* is "the collection, classification, storage, retrieval, and dissemination of recorded knowledge *in* a city". This definition highlights the physical city's persistent role as a container for information-based human activity. Either way, both definitions illustrate that information processing is an age-old function of cities—as Mumford (1961) noted, writing and urbanization were more or less simultaneous historical developments.

Taking a long view of urban informatics, the simultaneous urbanization and global economic integration we are currently experiencing can best be seen as a refinement of the city as a system for information processing. In the pre-electronic era, face-to-face proximity and the clustering of functions was the most efficient means of replicating, transmitting and searching for information in social and economic networks. Over time, new tools augmented this function, but in a sense the city itself is our original and greatest information technology.

Over the last two decades, urban scholars have built a powerful case in support of the hypothesis that advances in informatics have been a powerful force driving urbanization. The literature on "global cities", most notably, argues that the centralization of high-level decision-making in a handful of global cities, augmented by information and communications technology, is how globalization is enacted. Indeed, urban informatics does play a critical role in accelerating every step of this process—from information retrieval ("what is going on in Nigeria?") to analysis and decision-making ("should we invest in more oil production?") to dissemination ("drill that well!").

But while a historical view tells us that urban informatics is not a new thing—it is as innate a part of urbanity as anything else we study, and has been used since the dawn of cities to reinforce political and economic control—there is something unique about this moment in history. It seems that after 50 years of incubating digital information technologies on the desktop, we are now at a point where they are to become inextricably woven into the everyday social and economic life of dwellers in every city on the planet. On top of the centralized informatics infrastructure of the 20th century, we are juxtaposing layers of tools for material sensing and broad, decentralized cooperation among groups.

The first big shift, the pervasive spread of sensing in urban environments is already reshaping both the day-to-day and long-term processes of urbanization. While humans still set the boundaries, more and more of the critical life support systems of the city are instrumented to both sense and make sense of the world around them. Like Frankenstein's monster, the physical fabric of cities is waking up and becoming aware of itself.

Much of the sensing of urban settings today comes from the top-down. Congestion pricing, dynamic power grids and biometric surveillance are all examples of informatics systems that manage the status quo. But as many examples in this book describe, as sensing technology becomes more broadly diffused, it can be leveraged for disruptive uses that challenge established views of the city. For instance, inexpensive networked sensing embedded in mobile devices recasts urban dwellers as participants in an agile, dense swarm of pollution and traffic probes.

A parallel shift is the lightening-up of the informatics infrastructure, in favor of more decentralized, bottom-up frameworks. A popular metaphor for the industrial city was the machine. But if the city as machine has given way to the city as computer, what we are living in more closely resembles the messy collective capability of the Internet than a mainframe. In an ironic twist, information systems have evolved to become much like our most successful cities—open and modularized platforms on which human activity can take place—and less like master-designed utopias. In analogy form: Brasilia is to Big Blue, as Los Angeles is to Web 2.0. In a sense, by becoming more "cosmopolitan", tolerant of differences and inter-connected information systems are thriving.

This volume, then, comes along at an opportune moment—to reflect on this historic moment, and to chart both directions of change and specific principles and techniques for how to proceed into unknown territory. It conveys, I believe for the first time, the sense that we are starting to actually "see" informatics transforming cities before our eyes. When he was tearing apart New York City to make room for the automobile, power broker Robert Moses was reputed to announce, "When you operate in an overbuilt metropolis, you have to hack your way with a meat ax." But today's urban informatics effect change at the other end of the spectrum, through persuasion, surveillance, personalization and contextualization. Instead of rewriting space with a few large-scale strokes, they allow us to re-engineer an infinite number of small-scale relationships. But ubiquitous sensing is giving us the ability to sense, map and visualize these previously invisible processes.

What truly mark this volume more than any other on the topic, however, are the clear signs that scholars working in this area are developing transdisciplinary approaches to their research. In 2005, the Institute for the Future conducted a 50-year scan of future trends in science and technology for the UK government's Department of Trade and Industry (now the Department of Trade and Innovation). One of the eight high-level forecasts to emerge from this year-long effort, the idea of trandisciplinarity essentially meant that rather than putting together teams of specialists from established fields, we would see ever more young scholars seek training in multiple disciplines to develop new approaches to particularly messy or difficult problems. As author Howard Rheingold described it, "transdisciplinarity goes beyond bringing together researchers from different disciplines to work in multidisciplinary teams. It means educating researchers who can speak languages of multiple disciplines—biologists who have an understanding of mathematics, mathematicians who understand biology" (IFTF, 2006, p. 31).

The ability to easily form new communities around topics is a key driver of transdisciplinarity—historically, "*disciplines have been social as well as intellectual institutions. They've helped define what research problems and areas are important; identified who is worth knowing; rewarded innovative work; and helped allocate financial and human resources. Now though, an emerging cluster of online services offer scientists the means to find colleagues working on similar problems, irrespective of geography and institutional affiliation. Social software tools allow individuals to self-organize around common interests. Digital preprint services, wikis, and Web logs offer a spectrum of means to rapidly publish new research.*" (IFTF, 2006, p. 31).

This book is the result of just such a process unfolding, through a dialogue that without electronic sustenance would have been unlikely just a decade ago. Ten years ago, this would have been two or more separate books—one for social scientists, one for information scientists, yet another for architects and urban designers. As I sat down to write this foreword, I considered the traditional task of trying to draw out common themes from the diversity of manuscripts—until I realized that to do so would completely miss the point of what the authors have accomplished. This group of authors has stepped outside their disciplinary silos to engage in a dialogue that I suspect many will be loathe to return from.

Not having been around to experience it, as best I can from the literature, the last great burst of interest in urban informatics seems to have occurred in the late 1960s and early 1970s, about the time that computing began to be introduced on a large scale in government and business. The excitement about using computers to improve data analysis and even do predictive modeling of urban systems was widespread. In fact, my own organization, the Institute for the Future, was established in 1968 at Wesleyan College with a grant from the Ford Foundation to do computer-based urban modeling. These efforts soon fell apart, as the underestimated complexity of the effort became apparent. In a sense, the undelivered promises of that era's technocracy have driven much of the urban informatics research agenda since.

Transdisciplinarity is a beacon—instead of ignoring the input of colleagues across the table, it will be our own minds offering alternate hypotheses. Instead of blindly pursuing technological possibilities, social research is informing the definition of computational problems. The future research agenda both stated and implied by the contributions in this volume suggests decades of future work, and that is why I believe this new wave of urban informatics research will be sustained.

In conclusion, I like to think that this collaboration between urban scholars and information scientists will be more nuanced and more productive, building on the mistakes of the past collaboration. I also like to think that this will in the end result in better cities—more energy efficient, more fun, more just. The timing is certainly right. Urban planning is well into an undeclared crisis of thought leadership—despite it being one of the best avenues for dealing with global challenges like climate change and migration. Information science is poking its head out of the burrow and seeing the enormous intellectual challenge of expanding what worked on the desktop of the elites, to a diverse and mobile urban population.

It's worth speculating on the long-term impact of this developing body of research. Unlike the 20th century's urbanization, the big story by mid-century won't be the changes to the hardware. The mega-buildings of Shanghai and Dubai are just that—over-sized versions of the familiar 20th century forms borrowed from Chicago and Manhattan. Coal-fired or solar-powered, we will still be living in an energy-intensive civilization powered by electricity—billions more of us than ever.

But where we will see lots of change is in the software that shapes cities. Embedded sensing will replace a lot of human watchers and they will watch things on a frequency and scale we can barely imagine. But what will be important is how these abundant data streams provide a new ability to model and simulate very complex urban systems in real-time. Whereas today urban managers and planners react on the time cycle of a census, by mid-century real-time dashboard and predictive models will rule the trade. Already today, firms like Inrix provide fine-grained traffic forecasts for dozens of metropolitan areas in the United States.

Advances in the tools we have for "seeing" cities, from the first maps to the latest in satellite imagery, have always had major impacts on how we define problems, opportunities and aspirations. Sherman Fairchild, the father of aerial photography, described the impact of his invention (1924):

[It] shows the city with the minutest detail. It shows every structure from contractor's temporary tool-shed to skyscraper; back-yards, gardens and parks with every tree and bush visible; avenues and alleys, streets and unrecorded foot-paths; big league ball parks; water-front clubs, with their yachts and motor boats; the boardwalk of Coney Island, and crowds of people appearing like small black dots.

As Campanella (2001) describes in his history of the Fairchild Aerial Survey Company, the aerial perspective unleashed a wave of re-thinking urbanism:

If Le Corbusier was rhapsodic about the airplane's possibilities, he was shocked by what it revealed of the city. More than anything, it was the aerial view that ratified his conviction about the bankruptcy of the urban-architectural past. Rather than unfold in new light the wondrous legacy of urban civilization, the airplane supplied Corbusier with damning proof that the city of man was deeply pathological. The airplane peeled back the shrouds of the city, revealed its wrinkles and warts to Corbusier's unforgiving eye.

In essence, this new perspective granted by aerial photography rendered cities as abstract expressions of steel and concrete—malleable designs to be reworked from above by technocratic deities. The legacies of that fantasy are the planning disasters of the post-War period, costly lessons browbeaten into the minds of young urban planners today.

To use a crude analogy, if aerial photography showed us the muscular and skeletal structure of the city, the revolution in urban informatics is likely to reveal its circulatory and nervous systems. I like to call this vision the "real-time city", because for the first time we'll see cities as a whole the way biologists see a cell—instantaneously and in excruciating detail, but also alive. This is in contrast to the way astronomers see a heavenly body—as it was, some time ago, light-years in the past. And as these capabilities become more widespread, the real-time city could become a place where everyone is an amateur urban planner, using urban informatics to understand the larger impacts of their everyday decisions. That, so fundamental a shift in our perception of our own civilization, seems to be something worth working towards.

Anthony Townsend
Research Director, Technology Horizons Program
Institute for the Future

REFERENCES

Campanella T. (2001). *Cities from the Sky: An Aerial Portrait of America.* Princeton, New Jersey: Princeton Architectural Press.

Fairchild S. M. (1924). Aerial Mapping of New York City. *The American City*, January.

IFTF. (2006). *Science & Technology Perspectives: 2005-2055* (Report No. SR-967). Palo Alto, CA: Institute for the Future.

Mumford L. (1961). *The city in history: its origins, its transformations, and its prospects*. New York: Harcourt, Brace & World.

Anthony Townsend recently joined the Institute for the Future, an independent non-profit research group based in Palo Alto, California. As a research director, he will contribute to the Institute's long-range technological forecasting programs. Prior to joining the Institute, Anthony enjoyed a brief but productive career in academia, where his research focused on the role of telecommunications in urban development and design. Between 2000 and 2004 he taught courses in geographic information systems, telematics, and urban design in two graduate schools at New York University: the Interactive Telecommunications Program in the Tisch School of the Arts, and the Urban Planning Program in the Wagner Graduate School of Public Service. During this period, he directed several major research projects funded by the National Science Foundation and Department of Homeland Security. Townsend has been a key organizer in the wireless community networking movement since 2001. He is a co-founder and advisory board member of NYCwireless, a non-profit organization that promotes community broadband initiatives using unlicensed wireless spectrum. From 2002 to 2004 he was a principal of Emenity, a successful startup company that built and manages public local wireless networks in public spaces in Lower Manhattan. Anthony's work continues to develop an international focus. He has lectured and consulted throughout Asia, Europe and North America. He lived in Korea during the summer of 2004 as a Fulbright scholar, investigating that nation's rapid development of broadband technology. Anthony holds a PhD in urban and regional planning from Massachusetts Institute of Technology, a master's in urban planning from New York University, and a BA from Rutgers University. More information at http://urban.blogs.com/

Preface

Reading Anthony's foreword tempted me to change the title of this volume from *Urban Informatics* to *Urban Anatomy*—studies of the structure of the living city. Cities are indeed living organisms. They are alive with movement. A rapid flow of exchange is facilitated by a meshwork of infrastructure connections. Transport grids, building complexes, information and communication technology, social networks and people form the bones, organs, muscles, nerves and cell tissue of a city. Studying the organisation and structure of these systems may seem straightforward at first, since there are visible appearances and tangible objects that we can observe and examine. We can count the number of cars on the road, the number of apartments in a building, the number of emails on our computer screens and the number of profiles on social networking sites. We could also qualify these observations by recording the make and model of cars, the size and price of apartments, the sender and recipient of emails and the content and popularity of online profiles. This approach would potentially produce a large amount of data and render a detailed map of various levels of a city's infrastructure, but a large quantity of detail does not necessarily result in a great quality (and clarity) of meaning. How do we analyse this data to better understand the "city" organism? How do the cells of the city cluster to form tissue and organs, and how do various systems communicate and interact with each other? And, recognising that we ourselves are cells living in cities as active agents, how do we evaluate the effectiveness and efficiency of the processes we observe in order to plan, design and develop more liveable cities?

A macroscopic perspective of urban anatomy does not easily reveal those meticulous details which are necessary to help us understand and appreciate what Anthony calls the *urban metabolism* (Townsend, 2000), that is, the nutrients, capacities, processes and pace which nurture the city to keep it alive. Some of the fascination with human anatomy stems from the fact that a living body is more than the sum of its parts. Similarly, the city is more than the sum of its physical elements. Trying to get to the bottom of a city's existence, urban anatomists have to become dissectors of urban infrastructure by trying to microscopically uncover the connections and interrelationships of city elements. Yet, this is anything but trivial for at least three reasons. First, time is a crucial factor. Many events that trigger urban processes involving multiple systems result in a timely, interrelated response. A dissection by isolating one system from another would cut the communication link between them and jeopardise the study of the wider process. The city comprises many of these real-time systems and requires approaches and tools to conduct real-time examinations. Second, the physical city is increasingly complemented with a virtual mirror that digitally augments and enhances urban infrastructures by means of information and communication technology, including mobile and wireless networks. This world, which Mitchell (1995) called the "city of bits," is invisible to the human eye, and we require instruments for live surgery to render the invisible visible. Third, and most importantly, the "cells" of the urban body, the lifeblood of cities, are the city dwellers who have a life of their own and who introduce human fuzziness and socio-cultural variables to the study of the city. The toolbox of what could be termed anthropological urban anatomy thus calls for research approaches that can differentiate (and break apart) a universally applicable model of "The

City" by being sensitive to individual circumstances, local characteristics and socio-cultural contexts.

Fulfilling these three challenges, urban informatics offer research methods and instruments that become the microscope of urban anatomy. Urban informatics provides real-time tools for examining the real-time city, to picture the invisible and to zoom into a fine-grained resolution of urban environments to reveal the depth and contextual nuances of urban metabolism processes at work. Although I contemplated, for a minute, following the fame and glory of Henry Gray's *Anatomy of the Human Body*, I decided that *Urban Informatics* would be a more fitting title. Employing Anthony's portrait of the "real-time city" and following a suggestion by Paul Dourish, who co-authored the first chapter of this volume, the collection of chapters in this book is now—aptly, I think—titled, *Urban Informatics: The Practice and Promise of the Real-Time City*.

At this stage, the term "informatics" requires further elaboration. Why not call this field of research and development urban technology, urban infrastructure, or urban computing? Valid terms as they may be, I feel that they are too focused on the technology and that there is an important element missing, which they do not capture adequately, and that is the human element: people, citizens, urban residents, city dwellers, urbanites. Informatics, with its implied reference to information systems and information studies, slightly shifts the attention away from the hardware and more towards the softer aspects of information exchange, communication and interaction, social networks, and human knowledge. Similar thinking probably guided Michael Gurstein (2000) who coined the term "community informatics"—rather than, say, community technology—to underline the attention scholars and practitioners in this field pay to the impact of using information and communication technology on the socio-cultural and economic development of communities. Likewise, urban informatics research and development is concerned with the impact of technology, systems and infrastructure on *people* in urban environments.

The invention of the term "urban informatics" is not mine. The earliest prominent public occurrence I could find is from September 2003. Back then, Howard Rheingold of *Smart Mob*'s fame (Rheingold, 2002) wrote an article for the now discontinued *TheFeature.com* entitled *Cities, Swarms, Cell Phones: The Birth of Urban Informatics* in which he introduced his interviewee Anthony Townsend as an "urban informatician and wireless activist". I'm honoured that Anthony, surely one of the original urban informaticians, accepted my invitation to write the foreword for this book! Since 2003, Stephen Graham's (2004) *Cybercities Reader*, the late Patrick Purcell's (2006) *Networked Neighbourhoods*, as well as a number of special issues of journals (e.g., Shklovski & Chang, 2006; Kindberg, Chalmers & Paulos, 2007; Dave, 2007; Ellison, Burrows & Parker, 2007) are some of the hallmarks of urban informatics research. These works give rise to an emerging field populated by researchers and practitioners at the intersection of people, place and technology with a focus on cities, locative media and mobile technology. It is interdisciplinary in that it combines members of three broad academic communities: the social (media studies, communication studies, cultural studies, etc.), the urban (urban studies, urban planning, architecture, etc.), and the technical (computer science, software design, human-computer interaction, etc.), as well as the three linking cross sections of urban sociology, urban computing, and social computing. Furthermore, as Anthony explained in his foreword, the field's increasing transdisciplinarity is dissolving the rigid boundaries between disciplinary silos. "Nomadic" researchers, who enjoyed more than one higher education and traverse seamlessly between academic schools, enter the stage. The contributors to this book are prime examples: architects with degrees in media studies, software engineers with expertise in urban sociology, human-computer interaction designers grounded in cultural studies, and urban planners with an appetite for digital media and social network research.

A nucleus within this broad ecology of urban informatics is particularly worth tracing back, and that is the development of the *digital cities* notion. Toru Ishida and Peter van den Besselaar are arguably two of the most noteworthy scholars in the digital cities field of research. They initiated and supported the digital cities series of workshops that began in Kyoto, Japan in 2000 (Ishida & Isbister, 2000) and

2002 (Tanabe, van den Besselaar & Ishida, 2002). The series then continued in conjunction with the *International Conference on Communities and Technologies* (C&T) with workshops held in Amsterdam, The Netherlands in 2003 (van den Besselaar & Koizumi, 2005), and Milan, Italy in 2005 (Aurigi & De Cindio, 2008). The collection of studies published in the workshop proceedings can be roughly categorised into three distinct but related understandings of the term. First, social scientists teamed up with software designers to simulate urban environments and provided two and three dimensional visual interfaces resembling features and qualities of a physical city. These virtual cities would offer a post office to collect your electronic mail, a shopping mall to shop online and conduct e-business transactions, a town hall to pay your parking fines, and a market square to chat and socialise. Second, the online public sphere of these digital cities captured the imagination of city officials and public servants to assist in the delivery of local government services (e-government) and in the civic engagement and participation of residents in matters of urban planning (e-participation and e-democracy). And third, digital cities also refer to the attempt to digitally augment the physical urban infrastructure with ubiquitous technology and pervasive computing. This development has now culminated in South Korea's ambitious national *u-City* strategy, which Jong-Sung Hwang of Korea's National Information Society Agency discusses in his chapter in the future section of this book.

The latest instalment of the workshop series took place on 28th June 2007 at Michigan State University in East Lansing, USA, as part of the third C&T conference. The key research questions informing the presentations and discussions at *Digital Cities 5: Urban Informatics, Locative Media and Mobile Technology in Inner-City Developments* were as follows:

- How can a balance be achieved between the opportunities of locative media and mobile technology on the one side and issues of access, trust and privacy on the other?
- What is the role of locally relevant content, such as personal and community images and narratives, in the establishment of sustainable social networks as well as in the context of civic participation?
- What can we learn from the communication models of global social networking sites such as myspace.com and facebook.com in order to animate local interaction and civic participation of residents and friends locally?
- What is the role of location, (geo)graphical representations such as maps of various kinds, in supporting people to understand and navigate the augmented urban landscape?
- What is the impact of these new technologies on the challenges in moving from e-government to e-governance and e-participation to e-democracy at the urban level? Will these technological developments help increase or decrease the opportunities for citizens to play a role in shaping sustainable cities?
- What are the implications for the architecture and urban design of cities and public spaces?

Ten chapters in this book (IV, V, VIII, X, XII, XV, XVII, XXII, XXVII, XXVIII) are based on presentations given at the *Digital Cities 5* workshop. The workshop series is interesting insofar as its conceptual development reflects a broadening of both disciplinary input and academic scope. The workshops have always been a friendly meeting place for computer engineers to exchange ideas and findings with social scientists, but other disciplinary voices from urban planning, communication studies and the arts have also been welcome to join in the discussion over the years. Furthermore, the three main streams outlined above continue to play a key role in the discussions, but they, too, are being complemented by studies examining the impact of significant new technical developments such as mobile telephony, urban screens, location-based games, as well as social developments such as participatory culture, online activism and cultural citizenship. Ensuring this research is situated in real life and real time contexts as well as addressing the individual needs and realising the opportunities of urban residents living in a variety of situations is pivotal in not only keeping the research momentum alive and kicking, but also in yielding maximum impact.

According to United Nations (2008) estimates, not just the majority of people in developed countries, but now the majority of the world's entire population resides in cities, and this share is growing. More than every second human being lives, works, sleeps, eats and socialises in cities. Yet, about a third of these urban residents live in slums and squatter cities (Neuwirth, 2005). The magnitude of population dynamics and pressures on existing urban environments resulting in urban sprawl, pollution, crime, and an accelerated depletion and destruction of natural resources draw stark attention to the significance of urban research. Notwithstanding these trends, rural areas grow in importance, too since cities rely on non-urban, that is, rural and natural areas, to maintain their fast-paced metabolism which feeds on water, air and agricultural produce, and requires space for its "excrements" such as waste and CO_2. Establishing and maintaining a balanced and sustainable ecology of urban and rural areas, as well as the environment, in the face of global challenges such as population growth and climate change is imperative in order to safeguard the health and well-being of humankind.

I hope that this book will stimulate your mental metabolism with a rich and multi-faceted degustation menu. Sampling the "dishes" prepared for this urban smorgasbord will take you on a Grand Tour covering a great range of timely and significant topics and issues such as sustainability, digital identity, surveillance, privacy, access, environmental impact, activism, participatory planning, and community engagement. The book exposes research accounts which seek to convey an appreciation for local differences, for the empowerment of people and for the human-centred design of urban technology. Both contributors and coverage are international. They are not limited to cases based in Europe and America only. Rather, I purposefully sourced chapters covering Asia, Africa and Australia by a most engaging and prolific group of authors not afraid of presenting challenging and controversial ideas. The book starts with some introductory examinations that situate urban informatics research in the field and critique some of the assumptions behind urban informatics, as well as propose new ways of thinking. The second section focuses on ways people use technology to participate in urban planning scenarios and online deliberations. The engagement of urban communities is the central theme of the third section of the book and brings together examples from Germany, Mexico, Australia, and Canada dealing with multiculturalism, user-led innovation, creative expression and social sustainability. The fourth section comprises examples of studies investigating the link between the physical and digital city in the context of location, navigation and space. Wireless and mobile technology and its socio-cultural impact on urban communities and environments is the topic of the chapters in section five. And for dessert, the book concludes with a selection of speculative chapters, which examine trends in Korea and China, socio-technical innovation that supports location-sensitive tools for the real-time city and citizen science, and commentaries exploring the digital desaturation of the city and—in the afterword—the relation of urban informatics to social ontology.

Guten Appetit!

Marcus Foth, PhD
Australian Postdoctoral Fellow
Institute for Creative Industries and Innovation
Queensland University of Technology
Brisbane, Australia

April 2008

REFERENCES

Aurigi, A., & De Cindio, F. (Eds.). (2008, in press). *Augmented Urban Spaces: Articulating the Physical and Electronic City*. Aldershot, UK: Ashgate.

Ellison, N., Burrows, R., & Parker, S. (Eds.). (2007). Urban Informatics: Software, Cities and the New Cartographies of Knowing Capitalism. Guest editors of a special issue of *Information, Communication & Society*, 10(6). London: Routledge.

Dave, B. (Ed.). (2007). Space, sociality, and pervasive computing. Guest editor of a special issue of *Environment and Planning B: Planning and Design*, 34(3). London: Pion.

Graham, S. (Ed.). (2004). *The Cybercities Reader*. London: Routledge.

Gurstein, M. (Ed.). (2000). *Community Informatics: Enabling Communities with Information and Communication Technologies*. Hershey, PA: IGI.

Ishida, T., & Isbister, K. (Eds.). (2000). *Digital Cities: Technologies, Experiences, and Future Perspectives* (Lecture Notes in Computer Science No. 1765). Heidelberg, Germany: Springer.

Kindberg, T., Chalmers, M., & Paulos, E. (Eds.). (2007). Urban Computing. Guest editors of a special issue of *Pervasive Computing*, 6(3). Washington, DC: IEEE.

Mitchell, W. J. (1995). *City of Bits: Space, Place, and the Infobahn*. Cambridge, MA: MIT Press.

Neuwirth, R. (2005). *Shadow Cities: A billion squatters, a new urban world*. New York: Routledge.

Purcell, P. (Ed.). (2006). *Networked Neighbourhoods: The Connected Community in Context*. Berlin: Springer.

Rheingold, H. (2002). *Smart Mobs: The Next Social Revolution*. Cambridge, MA: Perseus.

Shklovski, I., & Chang, M. F. (Eds.). (2006). Urban Computing: Navigating Space and Context. Guest editors of a special issue of *Computer*, 39(9). Washington, DC: IEEE.

Tanabe, M., van den Besselaar, P., & Ishida, T. (Eds.). (2002). *Digital Cities 2: Computational and Sociological Approaches* (Lecture Notes in Computer Science No. 2362). Heidelberg, Germany: Springer.

Townsend, A. M. (2000). Life in the Real-Time City: Mobile Telephones and Urban Metabolism. *Journal of Urban Technology*, 7(2), 85-104.

United Nations. (2008). *World Urbanization Prospects: The 2007 Revision*. New York, NY: Department of Economic and Social Affairs, United Nations Secretariat.

van den Besselaar, P., & Koizumi, S. (Eds.). (2005). *Digital Cities 3: Information Technologies for Social Capital: Cross-cultural Perspectives*. (Lecture Notes in Computer Science No. 3081). Heidelberg, Germany: Springer.

Acknowledgment

This project which began in early 2007 has been a tremendous experience. Being strategically positioned as the central interface between a great number of outstanding authors and an equally great number of terrific reviewers allowed me to be exposed to the bleeding edge of urban informatics research in the making. 1548 emails later and with the publisher's imminent submission deadline and with the publisher's submission deadline imminent, I'm feeling slightly sentimental writing these acknowledgements since it signals having to let go and hand over the final manuscript. It also means that I now have to tame my naturally impatient self in this period of anticipation of the final published book.

I am infinitely indebted to the many colleagues who were involved in the writing, collation and review process of this handbook for their generosity and proactive spirit. With deep appreciation and gratitude I thank the authors for their intellectual contributions, their creativity, time and patience during the review and editing process, and for putting up with my part-time role as the benevolent dictator and my persistent nagging when submission deadlines lapsed or the publisher's rigid formatting requirements called for reworkings. You are a wonderful bunch of people!

Each chapter has been carefully refereed by three to four reviewers plus myself as the editor. Most of the authors of chapters included in this handbook also served as referees for chapters written by other authors. Additionally, I invited a group of external reviewers made up of highly acclaimed senior academics who provided excellent feedback and advice. Their names are listed below. I am grateful to these colleagues for their comprehensive commentary and quick turnaround of reviews. Thank you.

Special thanks also go to the publishing team at IGI Global whose support throughout the whole process from inception of the initial idea to final publication have been invaluable. I particularly thank Jessica Thompson and Julia Mosemann who continuously prodded with e-mails for keeping the project on schedule, and Mehdi Khosrow-Pour, IGI's Executive Editor, who recognised the significance of this book and selected it for publication as a *Handbook of Research* in IGI's most prestigious imprint, *Information Science Reference*.

Special thanks go to Fiorella De Cindio for co-chairing the Digital Cities 5 workshop with me at the *Third International Conference on Communities and Technologies* in East Lansing, USA, on 28th June 2007, and to Eric Paulos for delivering the keynote presentation on that day. Many workshop participants followed my call to submit chapters based on their workshop presentations. I also thank Anthony Townsend (and Laura Forlano for introducing us) and Roger Burrows (and Bill Dutton for introducing us) who accepted my invitation to write the foreword and afterword for the book and read semi-final drafts of the entire book manuscript.

Finally, I acknowledge the Creative Industries Faculty at Queensland University of Technology for providing a level of support and an intellectual environment conducive to develop this book and see it through to completion. Finally, I am grateful to be the recipient of an Australian Postdoctoral Fellowship supported under the Australian Research Council's Discovery Projects funding scheme (project number DP0663854) and I thank the team of this Discovery project, *New Media in the Urban Village: Mapping Communicative Ecologies and Socio-Economic Innovation in Emerging Inner-City Residential Developments*, for the support and strength they have given me in the past two years.

Marcus Foth, PhD
Australian Postdoctoral Fellow
Institute for Creative Industries and Innovation
Queensland University of Technology
Brisbane, Australia

April 2008

About the Editor

Marcus Foth is a senior research fellow at the Institute for Creative Industries and Innovation, Queensland University of Technology (QUT), Brisbane, Australia. He received a BCompSc(Hon) from Furtwangen University, Germany, a BMultimedia from Griffith University, Australia and an MA and PhD in digital media and urban sociology from QUT. Marcus is the recipient of an Australian Postdoctoral Fellowship supported under the Australian Research Council's Discovery funding scheme and a 2007 Visiting Fellowship from the Oxford Internet Institute, University of Oxford, UK. Marcus' work is positioned at the intersection of people, place and technology with a focus on urban informatics, locative media and mobile applications. His research has significantly shaped the social strategies of the Kelvin Grove Urban Village, the Queensland Government's flagship urban renewal project. Employing human-centred and participatory design methods, Marcus and his team pioneer new interactive social networking systems informed by community, social and urban studies.

Since 2003, Marcus has (co-)authored over 50 publications. The high quality of his research output has attracted over $1.16M in national competitive grants from the Australian Research Council and industry in 2006 and 2007. He is a chief investigator on the projects New Media in the Urban Village: Mapping Communicative Ecologies & Socio-Economic Innovation in Emerging Inner-City Residential Developments, and Remembering the Past, Imagining the Future: Embedding Narrative and New Media in Urban Planning. He is lead chief investigator of Opportunities of Media and Communication Technology to Support Social Networks of Urban Residents in Mexico, South Africa, UK and Australia, and Swarms in Urban Villages: New Media Design to Augment Social Networks of Residents in Inner-City Developments. He is a member of the Australian Computer Society and the Executive Committee of the Association of Internet Researchers. More information at http://www.urbaninformatics.net/

Section I
Introductory Examinations

Chapter I
Urbane-ing the City:
Examining and Refining the Assumptions Behind Urban Informatics

Amanda Williams
University of California, Irvine, USA

Erica Robles
Stanford University, USA

Paul Dourish
University of California, Irvine, USA

ABSTRACT

This chapter critically examines the notion of "the city" within urban informatics. Arguing that there is an overarching tendency to construe the city as an economically and spatially distinct social form, we review a series of system designs manifesting this assumption. Systematically characterizing the city as a dense ecology of impersonal social interactions occurring within recognizably public places, this construction can be traced to turn-of-the-century scholarship about the metropolis. The idealized dweller of these spaces, the flâneur, functions as the prototypical user for urban computing technologies. This assumption constrains the domain of application for emergent technologies by narrowing our conception of the urban experience. Drawing on contemporary urban scholarship, we advocate an alternative perspective which foregrounds the experience rather than the form of the metropolis. Users become actors embedded in global networks of mobile people, goods, and information, positioned in a fundamentally heterogeneous and splintered milieu. Grounding this approach in a preliminary study of mobility practices in Bangkok, Thailand, we illustrate how urban informatics might refine its subject, accounting for local particularities between cities as well as the broader global networks of connection between these sites.

INTRODUCTION

Over the past several years, "urban informatics" has emerged as a significant research area, drawing together researchers from various disciplines to focus on problems and opportunities at the intersection of computer science, design, urban studies, and new media art. This volume, for example, attests to the richness and diversity of this program.

Relying on the city as a unique and important context for investigation and design, the endeavor remains, in many ways, marked by contradictions. On the one hand, a city can be a specific setting for technologies that might otherwise be dubbed "ubiquitous" and, in their ubiquity, be located nowhere in particular. On the other hand, "the city" is a highly generalized site; to speak of "the city" is to strip away the specificities of particular cities. Indeed, urban informatics is marked by a focus on world cities at once globally similar but locally specific: New York rather than Boise, Paris rather than Arles, Sydney rather than Wagga Wagga. Similarly, while the city is a social and cultural phenomenon that speaks to complex ensembles of economic, technological, spatial, and social production, it also allows itself to be reduced to problems of scale and navigation.

As a focus for the development of ubiquitous computing technologies, the city is framed as a source of problems to be resolved: problems of location, resource identification, and access. Wayfinding applications, which might operate in terms of cartographic navigation (e.g. Sohn et al., 2005), or particular forms of commodity or cultural consumption (e.g. Axup et al., 2006; Brown & Chalmers, 2003), instead draw upon research in mobile and positioning technologies. The urban environment becomes no more than an appealing design resource. Providing rich and familiar social settings, they are environments already thick with information technologies and infrastructures, full of mobile people using mobile technologies.

We have three goals in this chapter. The first is to critically examine a series of assumptions about the nature of "the city" underlying many efforts in urban informatics. Locating the specific historical, geographical and cultural circumstances of early 20th century urban scholarship that still colors our design efforts, we begin to understand the characteristic perspective of urban computing systems. Scrutinizing the types of user experiences favored by this paradigm, we then question designers' role in their construction.

Second, we advocate a perspective shift, replacing emphasis on the urban *form* with emphasis on the urban *experience*. Focusing on what it is like to move through and live in contemporary cities brings the multitude of experiences co-existing within even the same urban space into plain view; this heterogeneity is critical to our approach. A focus on experience informs both analysis and for design.

Third, we ground our perspective through ongoing ethnographic work in Bangkok, Thailand. This choice of globally connected non-Western city not only expands the corpus of field engagements in urban informatics, but more importantly, provides a specific standpoint from which we can critically examine assumptions about "world cities" as generic, usually (culturally) Western, and indistinguishable from one another. The ethnographic treatment strongly counteracts any elision of the designer's subjectivity as part of fieldwork. Engagement with Bangkok speaks simultaneously about the places we are studying and the places from which we come (Marcus & Fischer, 1986).

Across these three goals, then, we advocate a reconstruction of urban informatics that emphasizes how cities operate in more particular, divergent relations. We hope our work usefully complements more technically oriented scholarship on urban informatics, rendering contemporary discourse on the topic not just more urban but also more urbane.

THE CITY AND ITS DWELLERS: THEMES IN URBAN INFORMATICS

There is perhaps no psychic phenomenon so unconditionally reserved to the city as the blasé outlook. It is at first the consequence of those rapidly shifting stimulations of the nerves which are thrown together in all their contrasts and from which it seems to us the intensification of metropolitan intellectuality seems to be derived... The essence of the blasé attitude is an indifference toward the distinctions between things.

Georg Simmel, 1903

Design efforts in urban informatics might best be understood as technological responses to the conditions of city life. Focused on enabling connections between mobile individuals within public spaces, the role of urban technologies is to help individuals, isolated within the teeming crowds, successfully locate and engage opportunities for expression and interaction. These design values systematically favor an interpretation of the city that, consciously or unconsciously, constrains how we think about both the urban experience and the urban inhabitant.

By tracing the notion of "the city" to its origins in modernization, we can critique its capacity to account for social life in a networked world. In contemporary use, it tends to uncritically favor users that are mobile, young, affluent, cosmopolitan and technologically savvy. By constructing users as latter-day *flâneurs*, this strategy seriously under-representing the diversity of social practices endemic to the city. Ultimately, opportunities for technological engagement within the urban sphere remain circumscribed.

The City is A Place Full of Strangers

From the earliest writings on urbanism, the city has been understood as a dense ecology of strangers. A social condition both liberating and alienating, metropolitan life offers countless opportunities for personal exploration, reinvention, and encounter, while simultaneously threatening to overwhelm the individual with a sense of isolation.

Writing at the turn of the century, sociologist Georg Simmel described the city-dweller's psyche as shaped in response to conditions of contrasting sensory impressions, mobility, a dense built environment, and constant encounters with diverse crowds of people. Urbanites negotiate these demands by affecting an attitude of indifference, or a *blasé outlook*. Simmel's contemporary, Émile Durkheim, characterized social interactions in the city as impersonal, individualistic, and formalized (Durkheim, 1933). Paradoxically, it is proximity to others that produces a lack of personal connection and shared sentiment (Durkheim, 1933; Riesman & Glazer, 1950; Simmel, 1971).

Both Simmel and Durkheim wrote in the context of modernization. A widespread shift from agrarian to industrial modes of production—from traditional to modern society—was reconfiguring spatial practices within Europe and the United States. "The city" referred to the ensemble of technological, economic, spatial and political practices whereby centers of rationalistic production coordinated the flow of goods and information (Harvey, 1989; Lefebvre, 1974). Their writings provide insight about how individuals, enmeshed within an emergent cultural form, negotiated its consequences.

The dominance of this economic-spatial organization endured well into mid 20th century, ensuring that these early theorizations migrated into subsequent scholarship. Acquiring an empirical dimension, the social conditions of city life became a favorite bugbear for theorists and social critics. Chicago School sociologists Robert Park and Louis Wirth, used the ideas of *social distance* to explicate why people might feel psychically far from each other despite their spatial proximity (Park, 1924; Wirth, 1938). Similarly, sociologist Erving Goffman (1963) quoted Simmel at length in describing the practice of *civil inattention*, in which people in close proximity (as in elevators or subway cars) nonetheless politely pretend that others are not there.

Inheriting this conceptualization of city life as simultaneously dense and isolating, designers produce systems with a certain characteristic logic: a focus on connection. For example, the *LoveGety* (Yukari, 1998), a popular in Tokyo during the late 1990's, promotes meetings between proximate strangers. The devices search within five meters for *LoveGety* holders of the opposite sex. By allowing the pair to find one another, the technology implies that love can be "gotten" on the streets of Tokyo, and that the city is full of strangers who might become friends, if only you had an introduction.

Dodgeball, a service available to residents of many United States metropolitan areas, promotes introductions by exploiting local social networks. Users identify or invite friends on the *Dodgeball* website. The service then "introduces" friends of friends located within ten blocks of one another. The "buddy finder" is often the first application, and certainly one of the most familiar tropes, associated with location-based technologies.

Mobile urban friend-finders presume that the chief design problem posed by the city is how to connect to people in a landscape teeming with impersonal others; the city is a static social condition requiring technological solutions in response. Articulating the city as a dense ecology of strangers, veined throughout with invisible networks of friendship and acquaintance, urban computing takes on the function of "curing" anomie. Not necessarily by turning *every* stranger into a friend, these systems emphasize latent social networks against a backdrop of anonymous public space, while recognizing the boundaries through which identity is enacted (Satchell, 2006; Satchell et al., 2006).

An exception to the tradition of highlighting the dichotomy of friends and strangers, Eric Paulos and Elizabeth Goodman (2004) conduct a nuanced examination of the concept of the *urban stranger*. Based on the notion of the *familiar stranger* (Milgram, 1977), these are persons repeatedly encountered in public but not interacted with for example, a fellow commuter on your daily bus route. Paulos and Goodman explore the contribution of these recognizable, but unknown, figures to a sense of place, and even a sense of safety. Designing a system called *Jabberwocky*, they play with relationships of mobility, legibility, and relation to the city's inhabitants. Hinting at positive aspects of urban anonymity, they highlight how the indifference of strangers might foster unprecedented autonomy. City dwellers are free to move about, developing and presenting identities in public urban spaces.

Turning now towards an explicit examination of what urban computing makes of public space, we point to a characterization weighted towards anonymous individuals' presentation and imagination in urban publics.

The City is a Space for Public Interaction

Concomitant with the dichotomization of friends and strangers is the tendency for urban computing systems to presume the city as a place of public (rather than domestic) and anonymous interactions. Augmenting the pleasure of being 'out and about', designers promote a particular historical imagination about lifestyle in the modern city.

Mobile and embedded social software services employ the trope of *ubiquity*, "reminding" users that in a networked world, they are "more connected than ever". Information technologies constitute the infrastructure for mutual visibility; networked communications are the architectures of public interaction. Eagle and Pentland (2004) point towards this shift, discussing the contrast between personal computers and mobile phones as application platforms:

Today's social software is not very social. From standard CRM systems to Friendster.com, these services require users to be in front of a computer in order to make new acquaintances. Serendipity embeds these applications directly into everyday social settings: on the bus, around the water cooler, in a bar, at a conference.

Systems like *Serendipity* (Eagle & Pentland, 2004) and *Digidress* (Persson et al., 2005) enable personal profile sharing via Bluetooth-enabled mobile phones. Profile sharing functions not only as a way to make friends, but also as a way of being peripherally aware of people sharing a space. These personal representations are springboard for imagination about otherwise anonymous individuals.

Similarly, mobile music-sharing applications like *tunA* and *BluetunA* (Bassoli et al., 2004; Baumann et al., 2007) allow users to push songs to one another or to find that a stranger's song has spontaneously migrated onto their player. Listeners tune in to each other, participating in a shared but distinct listening experience without necessarily breaching anonymity.

In each of these cases, computing serves as an analogue to public space. Personal digital devices, networked together, create an information space as public as the physical space in which they appear. Supporting fleeting, low-obligation interactions, these mobile social applications invite users to participate in collective and imaginative experience of a "public" setting. Twinning the public plaza or boulevard in digital space, they enable the sort of malleable identity work that echoes Walter Benjamin's imagination of the Paris Arcades (Benjamin, 2001).[1]

The City Presents Mobile Opportunities

Echoing Wirth's (1938) assertion that the heterogeneity and anonymity of urban populations is fundamentally related to mobility, urban computing systems respond are preoccupied with mapping the bodies of their users. Location information provides new opportunities for interaction, place-based content, or openings in the journey itself. Urban games (Benford et al., 2004) and pervasive location-based entertainment systems, like *Mogi Mogi* (Joffe, 2005), take advantage of a convergence of mobile and positioning technologies. Layered on top of the "real" world, participants engage in opportunistic gameplay while sitting at home, waiting for a train, or otherwise going about their daily lives.

Grafting new opportunities for interaction on pre-existing urban infrastructures, *Undersound* is a music-sharing system through which tracks may be downloaded in London Underground stations. These download points gather metadata on the movement of songs throughout the Underground, fueling public displays of the music's migration. *Undersound* treats urban mobility not merely as a way to get from origin to destination, but as an aesthetic experience rich with memory, imagination, and brief encounters, one of the "aspects of daily life that can make the urban experience a pleasurable one" (Bassoli et al., 2007).

Systems like *Undersound* inhabit complex design tensions. Situating unique movements within larger flows, they must simultaneously represent individual experiences and the collective structuring of urban space. The resulting approach is well aware of the complex spatialities of city-life while nevertheless remaining tied to a particular category of urban experience, and a particular class of mobile users.

These systems are situated within the particular mobile practices of middle and upper-middle class consumers, existing within a space of flows and a rhythm of travel, commute, work, and leisure (Castells, 2000). By attending to one sort of discretionary mobility (versus the sort practiced by migrant populations), urban computing constructs the user as a latter-day flâneur. It is to this construction of the user (and ultimately, the designer) that we now turn.

Intersecting Themes: The Urbanite as Mediated Flâneur

At the intersection of mobility, imagination, and urban visual culture stands the *flâneur*. Literally translated as "stroller", the term "mall rat" might better capture the spirit of the word. Native to modern commercial spaces such as arcades, the flâneur emerged on the wide boulevards and sidewalks of Paris after Baron von Haussmann's "modernization" of the city during the 1850's (Berman, 1988). These reforms replaced many of the mazelike streets of Paris's poorer neighborhoods with wider roads, facilitating the movement of both traffic and troops, and connecting the city as a whole entity rather than a collection of individual (and unnavigable) neighborhoods. The new boulevards, offering mutual visibility and space to stroll, along with electric lighting and glass storefronts, transformed the city into a space of movement and spectacle. The flâneur moves through space (slowly and aimlessly, yet still moving), looks (the space may be saturated with commodities on view), and imagines.

While the built environment is a significant part of the urban spectacle, it is the crowds that provide fodder for the flâneur's imagination. Where Simmel discusses with shock the multitudes of personal impressions, Baudelaire, a prototypical flâneur, turns the experience into enriched imaginings that about everyday life in the city:

The solitary and thoughtful stroller finds a singular intoxication in this universal communion... He adopts as his own all the occupations, all the joys and all the sorrows that chance offers. (Baudelaire, 1988)

This sentiment is echoed over a century later in the urban computing literature:

More than just problem solvers, we are creatures of boundless curiosity. Mixed within our moments of productivity are brief instances of daydreaming. We find ourselves astonished and in awe of not just the extraordinary, but the ordinary. We marvel at mundane everyday experiences and objects that evoke mystery, doubt, and uncertainty. How many newspapers has that person sold today? When was that bus last repaired? How far have I walked today? How many people have ever sat on that bench? Does that woman own a cat? Did a child or adult spit that gum onto the sidewalk? ... How can we design technology to support such wonderment? (Paulos & Beckman, 2006)

Authors, designers, and users alike are positioned as flâneurs. While the imagination surely plays an important role in good design, the fleeting nature of these interactions raises questions. Do these experiences of imagination, wonderment, or flânerie involve a deep understanding of strangers, or are they merely voyeuristic? For whom are we designing, and for what sorts of engagements with the city? Susan Buck-Morss (1986) points out that native to the regulated, relatively sanitized commercial spaces of the covered arcade:

...the flâneur, now jostled by crowds and in full view of the urban poverty which inhabited public streets, could maintain a rhapsodic view of modern existence only with the aid of illusion... If at the beginning, the flâneur as a private subject dreamed himself out into the world, at the end, flânerie was an ideological attempt to reprivatize social space, and to give assurance that the individual's passive observation was adequate for knowledge of social reality.

Exemplifying contemporary flânerie, magazine articles from the early 1980's describe the "poverty, punk rock, drugs and arson" of New York's Lower East Side as "ambience" even as gentrification drove out the original inhabitants of the neighborhood (Smith, 1992).

Our concern is that these descriptions, equating serious social ills and human suffering with "ambience" or "local color", echo in design techniques employed in urban computing. Paulos and Jenkins (2005) describe their use of *urban probes* to inspire the design of public interfaces. They record people's interactions with a public trashcan for an afternoon, and notice an "interesting" anecdote:

...a single bottle of Sprite soda was first observed being dropped vertically into the can, almost full, only minutes later to be picked out, finished and thrown back. The bottle resurfaced only a few minutes later in the same stalking as it was collected for recycling by a third individual.

The resulting design object was an "augmented trashcan" that displayed its "archeological layers" of trash as a sidewalk projection. Designed to be engaging and fun for passers-by, it may even assist those populations who rely on trashcans as a source of food or income (but who were not part of the evaluation or design scenarios). The protuct itself is not the object of our critique so much as certain aspects of the design process, which exhibits a tendency to favor a particular brand of engagement with the city, that of flânerie, which implicates users and designer alike.

The flâneur is a problematic figure. Produced by unique historic and economic circumstances, his practices are enabled by particular (commercial, public, inhabited, safe, visible) spaces. As Susan Buck-Morss (1986) argues, flânerie is "a form of perception... preserved in the characteristic fungibility of people and things in mass society". The flâneur has become more than a specific character; rather it is a privileged subject position manifesting in "a myriad of forms" that "continue to bear his traces" and replicate the conditions of his existence (see Friedberg, 2006 on the migration of the flâneur into the digital era).

A mobile consumer moving effortlessly through *spaces of flow* (Castells, 2000), the contemporary flâneur can afford to elide the distinctions between charismatic "world cities" (Dourish et al., 2007). A relatively empowered subject position, the flâneur is easy and profitable to design for. His habitats—covered walkways and commercial spaces like shopping centers—are widely replicated. Thus certain areas in distinct cities—New York, Tokyo, London—take on similar character, as though materializing an abstract entity called "the city" (Lefebvre, 1974).

As designers, we often enable flânerie and, more alarmingly, design with it as our methodology. As a mode of engagement, flânerie is diametrically opposed to "thick description" (Geertz, 1973) and deep understanding of others' lived experiences. Instead we privilege passive voyeurism and imagination tending towards illusion. The alternate mobilities, inhabitations, and appropriations alive in the city (homelessness and immigration, among other things) are left for examination by someone else.

The flâneur as contemporary user often blithely ignores the arrangement of social, cultural, and economic conditions enabling his existence. Designing for flânerie involves reifying these conditions through new technological forms. Thus, despite our best intentions, we replicate anachronistic patterns that are best undone.

Design practices need not be generalized for universality, nor need they incorporate all "other" populations. Rather, by attending to *positionality*, we can begin to account explicitly for the ways in which the imagined user relates to the complex urban environment. These relationships, encountered, constitute the essence of the urban experience. This perspective distinguishes technologies that take the urban seriously from those that merely consider the city as a place where technologies are used.

REFINING URBAN INFORMATICS: NETWORKS, CONTINGENCIES, SPLINTERINGS AND SITUATIONS

The "city" dominating urban informatics is based on early 20th century responses to modernization in Europe and North America (Harvey, 1985; Jacobs, 1961). Characteristically distinct from the suburban or rural, it is considered uniquely suited to mobile computing applications. The concept itself, however, often fails to reflect many of 21st century urban practices, worldwide. Refining this idea requires considering the relationship between urban informatics and what geographer Saskia Sassen (2001) calls "global cities." *Global cities*, advantageously positioned within networks of people, ideas, goods and capital, transcend their geographical specificity and regional role. It is perhaps not surprising that when urban informatics reaches beyond culturally Western cities, it looks towards other global cities, only to discover that they offer more of the same (e.g. Mainwaring et al., 2005).

By instead accounting for the post-modern cultural conditions of a networked global economy (Castells, 2000; Harvey, 1989), urban informatics can embrace opportunities to blend technology, art, activism, and design. In this section we discuss qualities of the postmodern city. Moving from "the city" to a networked of particular cities, we describe particular social actors *positioned* within flows of capital that structure these spaces, negotiate their circumstances via independent processes of mobility. We then consider how technologies mediate relationships between residents and their cities, pointing towards the diversity of lived experiences that results from these interactions. Finally, we move to questions of methodology. By situating the actions and expressions of particular people within their particular cities, we advocate a view of the local connected to the global rather than mere flânerie.

From "The City" To A Network of Particular Cities

By dismantling a series of presuppositions about the urban space, informatics gains an opportunity to participate in the ensemble of connections, distinctions, and spatial logics composing the contemporary metropolis. Leaving behind "the city" we enter the "networked world".

Cities are not isolated entities. Generally framed as bounded entities featuring heterogeneity (Wirth, 1938), mobility, and centralized calculation (Lefebvre, 1974), cities are fundamentally *interconnected.* The network of other cities and surrounding hinterlands plays a key dialectical role, crafting the notion of any distinctiveness at all. Moreover, contemporary cities participate in a global economy driven by the flow of capital, information, and goods and services (Castells, 2000; Harvey, 1989; LeFebvre, 1974), which shapes everyday urban and technological experiences (Mainwaring et al., 2005). As such, they are embedded in, subject to, and sometimes excluded from widespread networks.

Cities are not identical. Each city possesses its own unique cultural heritage. Even when spatial or architectural forms do bear resemblance, they "mean" differently. For example, grids exist in both American and Japanese cities but their origins and meanings are entirely distinct (LeFebvre, 1974). A preoccupation with urban forms—grids, piazzas, malls, parks, downtowns—risks collapsing the distinctions to a universal, but thin, set of commonalities.

Cities are more than "centers". "The city" has taken on a quality of hyper-specification, referring to the public spaces of a certain type of democratic industrial population center in dialectical relationship with suburban and rural surroundings (LeFebvre, 1974; Weber, 1969). This model fails to account for the exurban expanses of Orange County, the massive apartment blocks of Seoul, or the rapidly growing "desakota" regions of Southeast Asia (McGee, 1995).

In networked urban spaces, individual experience becomes a powerful analytical tool. The movements, actions, practices, and experiences of the urban inhabitant represent a series of tactics that negotiate between global regimes of production and particular local conditions (de Certeau, 1984). Where individual agencies meet social and technological infrastructures, the characters of the modern city are revealed.

Flows and Contingent Mobilities

Contemporary social life and its accompanying spatial forms can be described in terms of post-modern practices and networks of production. Social theorists like Manuel Castells, David Harvey, and Henri Lefebvre argue that the spatial and economic conditions are mutually constitutive. For example, the *flows* enabled by information technologies (of capital, technology, information, symbols, goods, people, etc.) simultaneously shape the physical and social world. Flows give rise to and rely upon spaces that function as nodes within networks. *Position* within these networks enables privileged access, allowing hubs to develop that, in turn, shape the network. Powerfully accounting for certain aspects of "global cities" like New York or Tokyo, as well as cities prominent within specialized global networks like Milan for fashion or Austin for indie rock, this model provides a viable starting point for theorizing within urban informatics.

Just as nodes acquire relative weight, so too flow favors particular social actors and institutions. Power tends to be maintained by elites possessing organizational capacity. The resulting spaces reproduce social conditions in which populations that are numerically superior, but segmented and disorganized, remain politically and economically disempowered:

Space plays a fundamental role in this mechanism. In short: elites are cosmopolitan, people are local. The space of power and wealth is projected throughout the world, while people's life and experience is rooted in places, in their culture, in their history. (Castells, 2000)

If flow describes the formation of cities and systematically produced spaces favoring elites, then power and social class become necessary categories for examining the role informatics plays in shaping the urban milieu.

Castells' assertion, however, is perhaps overly simplistic. Local residents are not powerless to modify their circumstances, nor are they uniformly pinned in place. Mobile labor plays a critical role within global networks of production. Doreen Massey (1993) describes the relationship between spatial manifestations of power and mobility as defining a *power geometry*:

...different social groups and different individuals are placed in very distinct ways in relation to these flows and interconnections. The point concerns not merely the issue of who moves and who doesn't, although that is an important element of it; it is also about power in relation to the flows and the movement. Different social groups have distinct relationships to this anyway-differentiated mobility: some are more in charge of it than others; some initiate flows and movement, others don't; some are more on the receiving end of it than others; some are effectively imprisoned by it.

Social actors, embedded within spaces of flow, negotiate their conditions using experience and heritage as resources. So too, technologies and infrastructures provide key materials for mediating individual response.

Infrastructures and Splintering

A central concern for both technologists and social theorists, *infrastructures* configure social interactions in material, spatial, and institutional ways (Bowker & Star, 1999). Technological systems depend on existing infrastructures—electricity, wireless Internet, cellular positioning, etc.—that may prove unreliable or problematic (Benford et al., 2004). These properties shape subsequent adoption, design, and use (Kline, 2000).

Given the expense of constructing infrastructures, we can expect that they generally reflect entrenched social interests. "Premium" (read profitable) spaces receive better and cheaper infrastructural services. Less profitable areas receive poorer, more expensive service, or none at all (Graham & Marvin, 2001). Premium spaces connect globally to each other. Underserved regions are systematically cut off from accessing privileged spaces nearby (Castells, 2000; Harvey, 1989). More recently, digital technologies enable finely targeted provision of services, what Graham has called "software-sorted geographies" (Graham, 2005).

Privileged connection is certainly present in "Western" cities. Overhead walkways in Minneapolis and Calgary connect downtown businesses, allowing (certain) people to move between them without encountering the street, homeless people, or bad weather. Selectively connecting downtown Los Angeles to the highway while disconnecting it from walkways and bus routes creates *de facto* spatial segregation between those who can and cannot afford to own cars, proof that infrastructures do, in fact, have politics (Bowker & Star, 1999; Winner, 1986). Cities do not correspond to the Modernist ideal of a unified, coherent public infrastructure. Rather, many inhabitants experience the fractured and disconnected infrastructures that Graham and Marvin (2001) call "splintering urbanism."

Turning to non-Western examples, these splintering urbanisms become glaringly apparent, highlighting the tenacity of historical relationships between infrastructure and power. Many cities inherit colonialist histories of artificially slowed growth prior to World War II followed by explosive growth atop inadequate infrastructures. Apartheid laws and "homelands" are stark examples of elitist attempts to advantageously deploy infrastructures in order to circumscribe the movements of native populations (Mbembe & Nuttall, 2004). Further reinforcing the politics of spaces, urban in-migration laws against encroachment were often coupled with refusals to provide basic services to native populations. This infrastructural assertion precluded the disenfranchised from belonging in or owning their cities:

> *In India, Burma, and Ceylon, their [British] refusal to improve sanitation or provide even the most minimal infrastructure to native neighborhoods ensured huge death tolls from early-twentieth-century epidemics (plague, cholera, influenza) and created immense problems of urban squalor that were inherited by national elites after independence. (Davis, 2006)*

In such a context, the German proverb "Stadtluft macht frei" ("city air makes you free") takes on a bitter irony.

Situated Urbanisms

We cannot assume that cities offer a uniform brand of social life, mobility, public space, or lived experience. Breaking down this essentialist notion is crucial for urban computing researchers and designers.

Increasingly, scholars in anthropology, urban studies, geography, and development studies problematize "mobility", "identity" and "space", favoring instead *situated* analyses of the urban experience. Their studies, generally ethnographic, articulate relationships between local and global mobilities, information technologies, and everyday practice.

le Marcis (2004), studying AIDS patients in Johannesburg, highlights the extent to which concepts of urban practice are predicated on healthy, unproblematic bodies. Mobility for the "suffering body" requires moving, despite its ill health, through an ever-expanding network of clinics, hospitals, support groups and hospices, resting finally in the graveyard. Local social, political and cultural factors further mediate this travel; hospital location and quality reflect remnants of Apartheid, social stigma prompts people to travel several hours to support groups rather than risk recognition in their own neighborhood. One wonders what latent networks a mobile social system might reveal for these patients.

Profoundly local, these experiences also connect to international flows of capital and technology. The expense of conventional treatments prompts many AIDS sufferers to turn to alternatives like Ayurvedic medicine or natural products. Instead, treatment is sought through drug trials both from international drug companies and local searches for a specifically "African" solution.

The issue of *Public Culture* in which le Marcis' ethnography appears demonstrates the necessity of writing both Africa and the world from a local point of view (Mbembe & Nuttall, 2004). The emerging spatial structures of São Paulo—fortified and surveilled private enclaves coupled to the disappearance of democratic public spaces—are tied to individual choices rooted in stories of violent crime and experiences of fear (Caldeira, 2000). "Intimate economies" in Bangkok (Wilson, 2004) based in local traditions of work, commerce, gender, and social support, intersect (via clientele) with the larger global economy. The urban landscape takes shape from an ensemble of individual knowledge workers, tomboys, bargirls, and telecom tycoons opportunistically exploring intersections. Many studies of these sites come not from Western anthropologists working abroad but from local scholars. Local scholarship and "halfie ethnographies" (Abu-Lughod, 1991) inherently question constructions of the less powerful as "other" and the authority presumed in speaking for them (Sittirak, 1996).

GROUNDING THE DISCUSSION: BANGKOK, THAILAND

Urban ethnographies examine individual experiences within larger configurations of power, contingent mobilities, and globalized and localized configurations of space, and may reveal valuable opportunities for designers concerned with experiences of technology. At this time one author (Williams) is conducting an investigation of mobility and technology use in and around Bangkok. The city, chosen for its role as a central focus for rural hinterlands and as a node in a transnational network provides a context for exploring how local meanings mediate the use of global technologies. Moreover, the setting contains a complex cast of identities, residents that locate themselves within global assemblages of mobile goods and differential infrastructures. Finally, the setting demands consideration of the urban experience through a historical lens.

Bangkok's role as a center for government, culture and commerce predates significant European colonial presence in Southeast Asia. The city operated as a port well before becoming the capital, in 1782. Its cultural and economic centrality derives largely from its position within a trans-national network. Many residents of Bangkok practice *oscillating mobilities*, with residences in multiple locales both within and outside of Thailand. So too, neighborhood *sois* challenge dichotomies of private and public, formal and informal, global and local.

Oscillating Mobilities and Temporality

Participant-observation over the course of several months in Bangkok highlights the "oscillating mobilities" (Askew, 2002) composing the city. Workers in the informal sector, including construction workers, sex workers, food vendors, domestic help, and even higher-paid professional positions, are crucial participants in city life. Not easily classified as permanent residents or temporary migrants, they periodically leave jobs and return to rural villages to visit family, help with important harvests, or when they have sufficient money to stop working for a while. This pattern, observed by ethnographers (Askew, 2002), informants (Williams et al., 2008), and the government, was recently acknowledged by a national holiday allowing migrant workers to vote on the 2007 constitutional referendum in their home provinces (Bangkok Post, 2007).

Oscillation also occurs at the trans-national scale. Interviews and home stays with twenty-three Thai-American retirees in New York, Seattle, St. Louis, Bangkok, and the agricultural province of Chantaburi indicated oscillations on another amplitude and frequency (Williams et al., 2008). Participants lived both in Thailand (mostly Bangkok) and metropolitan areas in the United States, traveling between them once or twice a year. One family also migrated between Bangkok and Chantaburi to accommodate the needs of aging siblings and to maintain their home and orchard. Participants' rhythmic mobilities provided a counterpoint to the linear life phase transitions that modulated their participation in the social life of their cities, communities and families.

The networked city is typically portrayed as a recent phenomenon, a city emerging with post-industrial production practices (Castells, 2000; Graham & Marvin, 2001). However, cities like Bangkok were never unified modern cities and thus did not become post-modern with the rise of the information economy. Historically, Bangkok has long been where Thailand encounters and appropriates the rest of the world. The oscillating mobilities of Bangkok's residents continue economic and spatial practices that have always characterized that city; to paraphrase Bruno Latour, they have *never* been modern.

Soi Ecologies and Urban Public Spaces

A *soi* (translation: alley) can range from a full two-way street to an alleyway in which a restaurant has placed its tables, to an intimate walkway appropriated by families as a place to set up washing machines or kitchen tables:

Near Petchaburi Road, December 9, 2007: *We cut off the main street into some smaller sois.. which just get smaller and smaller and narrower... you really feel you are walking through someone's yard or across their porch (though not as pronounced as in Khlong Toei, where I felt like I was in people's living rooms)... A few places we can see right into someone's open kitchen or laundry room.*[2]

Tracing the paths of waterways that once subdivided local, privately owned rice-fields, Bangkok's sois and the social and economic activities that take place in them form a vital part of the city's character. While informal economic activities and food vending occur across many cities, Bangkok is arguably unique in the extraordinarily high rate at which people rely on informal food vending systems (with more than half the food expenditures in the metropolitan area spent on food prepared outside the home), and the particular prominence of women both as vendors and customers (Yasmeen, 2006). Informants that were asked to describe their journeys through the city reported foraging for breakfast in their soi during their commutes, and consistently articulated the transition from soi to main thoroughfare as a notable point in their journeys, a place where the pace and sensory or emotional experience of travel would change noticeably. In comparison to main thoroughfares, sois are not merely smaller, quieter streets; rather histories of property ownership often mean that they are loosely regulated and flexibly defined in ways that impact commerce and local entrepreneurship, and ultimately social life on the street level. Vendors in sois typically must acquire the landowner's permission to sell, or may set up in front of their own residence. Living space and working space thus blur together, though customers and foot traffic through a busy soi might give it a "public" feel. Yet in dealing with local police and municipal inspectors, vendors can question their authority by appealing to the fact that they are on "private" property (ibid). In her study of Bangkok food vendors, Yasmeen notes that "similar leeway is not granted to those on the city's major arteries", and informal interviews with vendors confirm that there are significant financial barriers to vending on a main road:

Silom Rd., November 22, 2007: *Quality fake watches in Patpong, K's friend. Has 3 spots. Each is 5000 baht [about 165 U.S. dollars] per night. Who does she pay rent to?—the landlord. The one who owns the building that she's in front of?—no no different person. So someone else owns the sidewalk?—yeah kind of. Officially?—no more like the mafia. Also bribes to the police which are variable, but may sometimes be even more than the "rent".*

Though soi entrepreneurship is often part of the informal economy, its activities are not necessarily displaced by growth in the neighborhood's formal economy (Askew, 2002). Commuting office workers support food vendors and motorcycle taxi drivers. Formal storefront establishments rent space and electricity to street vendors. Privileged spaces such as shopping centers and high-rise condominiums anchor them. These observations are not meant to dismiss the power differential between more and less privileged residents of cities, rather they emphasize the blurrings between private and public spaces and the bridges between formal and informal economies. The local residents of Bangkok's sois actively engage with and appropriate the proliferating urban forms of Bangkok the global city.

Designing for a Networked City

Bangkok's particular spatial practices raise considerations for the kinds of mobile and urban applications that might be deployed there.

First, Bangkok's oscillating mobilities challenge binary ideas of stability and mobility. On the one hand ubicomp researchers have typically (though with some exceptions) treated built structures as fixed settings into which sensing and computational technology might be deployed (Beckmann et al., 2004) rather than changeable environments (Brand, 1994) through which people might move or linger in ways relevant to technology design (Crabtree et al., 2003);

Figure 1. Commerce in a soi in downtown Bangkok near Ploenchit Rd., literally in the shadows of high-rise condominiums

"home" might be taken to represent an absolute stability. On the other hand, an assumption of anywhere-anytime mobility underlies the design of mobile technologies ranging from ubicomp games to international SIM cards. Oscillation, between Bangkok and either a rural village or a metropolitan area in another country troubles assumptions about users who are either at rest or roving unpredictably. This form of urban (and inter-urban) mobility may also provide different challenges and opportunities for the design of new sorts of systems. For example:

- Delay-tolerant networking applications currently might depend on social as well as communications networks, exploiting the fact that people who know each other meet and pass information between each other. Imagine instead an approach that exploits the fact that people return to the same few places over and over.
- Cell phone contact lists could have different capabilities supporting two or more home locations.
- Urban games could be designed for networks of cities, rather than a single metropolis.

Second, Bangkok's sois, as spaces that are typically defined flexibly and situationally, trouble dichotomies of private and public. While ubicomp research currently treats "the home" and "the city" as separate private and public domains for investigation and design, one might instead envision technologies that would allow users to claim a flexible space as their own, or to invite others into it.

More broadly, these considerations point to ways in which Bangkok itself oscillates. The character of a space can change drastically over the course of hours, days, seasons, or according to various conflicting needs of the people who occupy and constitute it. Food vendors descend on a spot near a shrine and shopping center for a few hours around 5pm, leaving the space calm for the rest of the day; a market at noon is the parking lot of an expensive Italian restaurant at 7pm. Residents understand the city as a place in flux, changing rhythmically and linearly; perhaps not coincidentally, maps are not typically considered useful representations of the space. In designing technologies to engage with such a city, we have an opportunity to employ other representations, orienting ourselves towards temporal, kinetic, auditory, olfactory, embodied or performative experiences of cities.

While situating practice in Bangkok facilitates designing systems targeted to Bangkok, these design considerations provide the broader benefit of de-familiarizing and reframing the cities we think we already know. In the process, we accumulate insights, surface themes, and articulate diverse responses to related conditions. In short, changing frameworks means changing methods and re-imagining what it means to design systems for urban use.

CONCLUSIONS: IMPLICATIONS FOR DESIGNERS

If we accept that the domain of urban informatics is not "the city" but rather some complex of more-and-less connected cities and parts of cities then we must re-examine our own methodologies and theoretical frameworks. Multi-sited ethnographies, for example, take on increased importance (Marcus, 1995). Entailing more than visits to multiple research locations, multi-sited ethnography acknowledges that ethnographic informants *already* consider their relationships to global structures. Their considerations are important forms of local knowledge. Tracing the lived experiences of particular social actors enables patterns of negotiation between the local and global to emerge.

Particularly where local individual experience meets global structures, urban computing offers a valuable lens by foregrounding ways in which users understand their interactions with technological and material infrastructures (Bowker & Star, 1999; Dourish & Bell, 2007). A focus on infrastructure can allow a flexible understanding of cities both as centers of regional hinterlands (Harvey, 1985; Harvey, 1987), and nodes in a network of flows (Castells, 2000); the density of layered infrastructures can be seen as differentiating urban from non-urban areas, and yet it can also be seen as a factor that connects rather than just differentiates; and examining differences and breaks in infrastructure can expose local difference as well as global exchange (Graham & Marvin, 2001).

While it is true that different places are different, and that different people might use technology in different ways, a taxonomic understanding of urban difference does not capture the full picture. Working with an unquestioned notion of the city may bias our urban designs in certain ways, emphasizing, for example, discretionary mobility, visual representations and maps, or making friends out of strangers. Rather, a generative understanding of cities and urban practices as produced by people living within particular cultural and historical contexts can open up new opportunities for analysis and design. While we are not typically historians, cities are temporal entities, and "historical context" is an ongoing process, evident in the rhythms and transformations that characterize the experience of many cities.

Finally, implicit in our discussion is the belief that urban informatics not only responds to but also shapes the conditions for social life. Information technologies—from systems of addressing (Smail, 1999) to those of representation (Scott, 1988)—constitute the infrastructures for urban growth, and thus play critical roles in organizing urban experience (Philips & Curry, 2003; Goss, 1995). Urban informatics not only conceives the city, it creates it.

ACKNOWLEDGMENT

Support for this research has been provided by Intel Corp. and by the U.S. National Science Foundation under a Graduate Research Fellowship and grant awards 0133749, 0205724, 0527729, 0524033, and 0712890.

REFERENCES

Abu-Lughod, L. (1991). Writing against culture. In *Recapturing anthropology: Working in the present*. 137-162.

Askew, M. (2002). *Bangkok: Place, practice and representation*. New York, NY: Routledge.

Axelsson, F., & Östergren, M. (2002). SoundPryer: Joint music listening on the road. *Adjunct Proceedings of the 2002 International Conference on Ubiquitous Computing (UbiComp).*

Axup, J., Viller, S., MacColl, I., & Cooper, R. (2006). Lo-Fi matchmaking: A study of social pairing for backpackers. *Proceedings of the 2006 International Conference on Ubiquitous Computing (UbiComp)*, 351-368.

Bassoli, A., Brewer, J., Martin, K., Dourish, P. & Mainwaring, S. (2007). Aesthetic journeys: Rethinking urban computing. *IEEE Pervasive Computing*, *6*(3), 39-45.

Bassoli, A., Moore, J. and Agamanolis, S. (2004). tunA: Synchronised music-sharing on handheld devices. *Adjunct Proceedings of the 2004 International Conference on Ubiquitous Computing (UbiComp)*, 171-172.

Baudelaire, C.P. (1988). Crowds. In *Paris spleen*. New York, NY: New Directions Publishing Corporation.

Baumann, S., Bassoli, A., Jung, B. & Wisniowski, M. (2007, April) BluetunA: Let your neighbor know what music you like. Extended abstracts for the *2007 Conference on Computer Human Interaction (CHI)*, San Jose, CA.

Beckmann, C., Consolvo, S. and LaMarca, A. (2004) Some assembly required: Supporting end-user sensor installation in domestic ubiquitous computing environments. *Proceedings of the 2004 International Conference on Ubiquitous Computing* (*UbiComp*), 107-124.

Benford, S., Crabtree, A., Flintham, M., Drozd, A., Anastasi, R., Paxton, M., Tandavanitj, N., Adams, M. & Row-Farr, J. (2004). Can you see me now? *Transactions on Computer Human Interaction (TOCHI), 13* (1), 100-133.

Benjamin, W. (2001). *Arcades project.* New York, NY: Belknap Press.

Berman, M.(1988). *All that is solid melts into air: The experience of modernity.* New York, NY: Penguin Books.

Boddy, T. (1992). Underground and overhead: Building the analogous city. In M. Sorkin (Ed.), *Variations on a theme park: The new American city and the end of public space* (pp. 123-153). New York, NY: Hill & Wang.

Bowker, G. & Star, S.L. (1999). *Sorting things out: Classification and its consequences.* Cambridge, MA: MIT Press.

Brand, S. (1994). *How buildings learn: What happens after they're built.* Viking.

Brown, B., and Chalmers, M. (2003). Tourism and mobile technology. *Proceedings of the 2003 European Conference on Computer Supported Cooperative Work (ECSCW)*, 335-355.

Buck-Morss, S. (1986). The flaneur, the sandwichman and the whore: The politics of loitering. *New German Critique, Second Special Issue on Walter Benjamin, 39*, 99-140.

Caldeira, T.P.R. (2000). *City of walls: Crime, segregation, and citizenship in São Paulo.* Berkeley, CA: University of California Press.

Castells, M. (2000). *The rise of the network society.* New York, NY: Blackwell.

de Certeau, M. (1984). *The Practice of Everyday Life.* Berkeley, CA: University of California Press.

Chang, M., Jungnickel, K., Orloff, C. & Shklovski, I. (2005, April). Engaging the city: Public interfaces as civic intermediary. Extended abstracts for the *2005 Conference on Computer Human Interaction (CHI)*, Portland, OR.

Crabtree, A., Rodden, T., Hemmings, T. and Benford, S. (2003). Finding a place for UbiComp in the home. *Proceedings of the 2003 International Conference on Ubiquitous Computing* (*UbiComp*), 208-226.

Davis, M. (2006). *Planet of slums.* New York, NY: Verso.

DiSalvo, C., & Vertesi, J. (2007, April). Imaging the city: Exploring the practices and technologies of representing the urban environment. Extended abstracts for the *2007 Conference on Computer Human Interaction (CHI)*, San Jose, CA.

Drozd, A., Benford, S., Tandavanitj, N., Wright, M. & Chamberlain, A. (2006). Hitchers: Designing for cellular positioning. *Proceedings of the 2006 International Conference on Ubiquitous Computing* (*UbiComp*), 279-296.

Dourish, P., & Bell, G. (2007). The infrastructure of experience and the experience of infrastructure: Meaning and structure in everyday encounters with space. *Environment and Planning B: Planning and Design, 34* (3), 414-430.

Durkheim, É. (1933). *The division of labor in society.* New York, NY: Free Press.

Eagle, N. & Pentland, A. (2004). Social serendipity: Proximity sensing and cueing. *MIT Media Lab Technical Note 580.* Cambridge, MA: MIT Press.

Friedberg, A. (2006). *The virtual window: From Alberti to Microsoft*. Cambridge, MA: MIT Press.

Gaver, B., Dunne, T. and Pacenti, E. Cultural Probes. *Interactions*, *6* (1). 21-29.

Geertz, C. (1973). Thick Description: Toward and Interpretive Theory of Culture. In *The Interpretation of Cultures*. New York, NY: Basic Books.

Goffman, E. (1963). *Behavior in public places: notes on the social organization of gatherings*. New York, NY: Free Press.

Goss, J. (1995). We know who you are and we know where you live: The instrumental rationality of geodemographic systems. *Economic Geography, 71*(2), 171-198.

Graham, S. (2005). Software-sorted geographies. *Progress in Human Geography, 29*, 562-580.

Graham, S. & Marvin, S. (2001). *Splintering urbanism: Networked infrastructures, technological mobilities and the urban condition*. New York, NY: Routledge.

Håkansson, M., Rost, M., Jacobsson, M. & Holmquist, L.E. (2007). Facilitating mobile music sharing and social interaction with Push! Music. *Proceedings of the 2007 Hawaii International Conference on System Sciences (HICSS)*, 87-88.

Harvey, D. (1985). *Consciousness and the Urban Experience*. New York, NY: Blackwell.

Harvey, D. (1989). *The condition of postmodernity: An enquiry into the origins of cultural change*. New York, NY: Blackwell Press.

Jacobs, J. (1961). *The death and life of great American cities*. New York, NY: Vintage Books.

Joffe, B. (2005, September). Mogi: Location and presence in a pervasive community game. Presented at the *2006 Conference on Ubiquitous Computing Workshop on Ubiquitous Gaming and Entertainment*, Tokyo, Japan.

Kline, R. (2000). *Consumers in the country: Technology and social change in rural America*. Baltimore, MD: Johns Hopkins University Press.

LeFebvre, H. (1974). *The production of space*. New York, NY: Blackwell Publishing.

LeFebvre, H. (1984). *Everyday Life and the Modern World*. London: Continuum.

le Marcis, F. (2004). The suffering body of the city. *Public Culture*, *16*(3), 453-477.

Mainwaring, S., K. Anderson, et al. (2005). Living for the global city: Mobile kits, urban interfaces, and ubicomp. *Proceedings of the 2005 International Conference on Ubiquitous Computing (UbiComp)*.

Marcus, G.E. (1995). Ethnography in/of the world system: The emergence of multi-sited ethnography. *Annual Review of Anthropology, 24* (1), 95-117.

Marcus, G. & Fischer, M. 1986. *Anthropology as cultural critique: An experimental moment in the human sciences*. Chicago, IL: University of Chicago Press.

Massey, D. (1993). Power-geometry and a progressive sense of place. In *Mapping the futures: Local cultures, global change* (pp. 59-69). New York, NY: Routledge.

Mbembe, A. & Nuttall, S. (2004).Writing the world from an African metropolis. *Public Culture, 16* (3), 347-372.

McGee, T.G., and Robinson, I.M., eds. (1995). *The Mega-Urban Regions of Southeast Asia*. Vancouver, BC: UBC Press.

Milgram, S. (1977). *The individual in a social world: Essays and experiments*. Reading, MA: Addison Weseley.

Mumford, L. (1938). *The culture of cities*, New York, NY: Secker & Warburg.

O'Neill, E., Kindberg, T., gen Schieck, A.F., Jones, T., Penn, A. & Fraser, D.S. (2006). Instrumenting the city: Developing methods for observing and understanding the digital cityscape. *Proceedings of the 2006 International Conference on Ubiquitous Computing (UbiComp)*.

Park, R. E. (1924). The concept of social distance. *Journal of Applied Sociology 8*, 339-344.

Paulos, E., Anderson, K., Chang, M., Burke, A. & Jenkins, T. (2005, September). *Metapolis and urban life*. Presented at the *2005 Conference on Ubiquitous Computing*, Tokyo, Japan.

Paulos, E., Anderson, K. & Townsend, A. (2004, September). *UbiComp in the urban frontier. Presented at the 2004 Conference on Ubiquitous Computing,* Nottingham, England.

Paulos, E. & Beckmann, C. (2006). Sashay: Designing for wonderment. *Proceedings of the 2006 Conference on Computer Human Interaction (CHI)*.

Paulos, E. & Jenkins, T. (2005). Urban probes: Encountering our emerging urban atmospheres. *Proceedings of the 2005 Conference on Computer Human Interaction (CHI)*, 341-350.

Paulos, E. & E. Goodman. (2004). The familiar stranger: Anxiety, comfort, and play in public places. *Proceedings of the 2004 Conference on Computer Human Interaction*, 223-230.

Persson, P., Blom, J. & Jung, Y. (2005). DigiDress: A field trial of an expressive social proximity application. *Proceedings of the 2005 International Conference on Ubiquitous Computing (UbiComp)*,195-212.

Philips, D. & Curry, M. (2002). Privacy and the phenetic urge. In D. Lyons (Ed.), *Surveillance as social sorting: Privacy, risk, and digital discrimination* (pp. 137-152). New York, NY: Routledge.

Sassen, S. (2001). *The global city: New York, London, Tokyo*. Princeton, NJ: Princeton University Press.

Satchell, C. (2006). Contextualising Mobile Presence with Digital Images. Paper presented at the 2nd International Workshop on Pervasive Image Capturing and Sharing, UbiComp 2006, Orange County, CA.

Satchell, C., Shanks, G., Howard, S., & Murphy, J. (2006). Knowing Me—Knowing You. End User Perceptions of Digital Identity Management Systems. Paper presented at the 14th European Conference on Information Systems (ECIS), Göteborg, Sweden.

Scott, J. (1998). *Seeing like a state: How certain schemes to improve the human condition have failed*. New Haven, CT: Yale University Press.

Simmel, G. (1903). The metropolis and mental life. In D. Levine (Ed.), *On individuality and social forms: Selected writings* (324-339). Chicago, IL: University of Chicago Press.

Sittirak, S. (1996). *Daughters of development: The stories of women and the changing environment in Thailand*. Bangkok, Thailand: Women and Environment Research Network in Thailand.

Smail, D. (1999). *Imaginary cartographies: Possession and identity in late medieval Marseille*. Ithaca, NY: Cornell University Press.

Smith, N. (1992). New City, New frontier: The lower east side as wild, wild west. In M. Sorkin (Ed.), *Variations on a theme park: The new American city and the end of public space*. New York, NY: Noonday Press.

Sohn, T., Li, K., Lee, G., Smith, I., Scott, J., & Griswold, W. (2005). Place-its: A study of location-based reminders on mobile phones. *Proceedings of the 2005 International Conference on Ubiquitous Computing (UbiComp)*, 232-250.

Vaneigem, R. (1983). *The Revolution of Everyday Life*. London: Aldgate Press.

Weber, M. (1921) The nature of the city. In R. Sennett (Ed.), *Classic essays on the Culture of Cities*. New York: Appleton-Century-Crofts.

Williams, A., Anderson, K., & Dourish, P. (2008). Anchored mobilities: Mobile technology and transnational migration. *Proceedings of the 2008 Conference on Designing Interactive Systems (DIS)*.

Wilson, A. (2004). *The intimate economies of Bangkok: Tomboys, tycoons, and Avon ladies in the global city*. Berkeley, CA: University of California Press.

Wirth, L. (1938). Urbanism as a way of life. *The American Journal of Sociology 44* (1),1-24.

Winner, L. (1986). Do artifacts have politics? In *The whale and the reactor: A search for limits in an age of high technology* (pp. 19-39). Chicago, IL: University of Chicago Press.

Yasmeen, G. (2006). *Bangkok's foodscape: Public eating, gender relations, and urban change.* Bangkok, Thailand: White Lotus.

Yukari, I. (1998, June). Love: Japanese style [Electronic Version]. *Wired Magazine*.

Aug 20 Declared a Public Holiday. (2007, August 8) *Bangkok Post*, p. A1.

KEY TERMS

Anomie: Emile Durkheim used the term "anomie" to refer to the experience of an absence of social norms. Various writers have employed it to characterise the social isolation and alienation from communitarian life and social ties associated with the scale and anonymity of urban environments.

Civil Inattention: Sociologist Erving Goffman coined the term "civil inattention" to refer to the ways in which people maintain a comfortable social order in public spaces by explicitly disattending to one another and their actions (for instance, the minimal social interaction amongst people packed into an elevator).

Flâneur: Critical theorist Walter Benjamin draws the term "flâneur" from the writings of French poet Charles Baudelaire. To Baudelaire, the flâneur is a figure unique to the city, one who wanders through urban space in order to consume and revel in the images that it offers. Flânerie, then, is an experience of urban space. Benjamin notes the historical and economic specificities of the flâneur, arguing that the kinds of narrative afforded by flânerie depend upon forms of leisure and mobility associated with wealth and power. Critically, the flâneur may be in the crowd, but is not of the crowd.

Positionality: In cultural accounts of experience, positionality refers to both the fact of and the specific conditions of a given social situation. So, where one might talk about the "position" of an individual in a social structure, "positionality" draws attention to the conditions under which such a position arises, the factors that stabilize that position, and the particular implications of that position with reference to the forces that maintain it.In urban informatics, positionality is relevant in the ways in which information systems create and sustain particular networks of positions, spatially and socially.

Power Geometry: Feminist geographer Doreen Massey introduced the term "power geometry" to point to the ways in which spatiality and mobility are both shaped by and reproduce power differentials in society. Examples might include the control over distribution of goods and services, or the different circuits enabled by transportation systems.

Situationism: The Situationists were a group of avant-garde artists, radicals, and intellectuals active in Europe particularly in the late 1950s and 1960s. Situationism argued that the conditions of contemporary capitalism had rendered people passive subjects whose relationship to their own experience was one of the consumption of daily life as spectacle. Urban life was a particular example of a domain in which they sought to revolutionize the experience of everyday life by encouraging people to become conscious, active participants in the reality that their everyday actions produced.

Space of Flows: Urban sociologist Manuel Castells uses the term "space of flows" to reimagine urban space as a nexus of flows of people, capital, goods, and information. This helps us understand the city as a component in broader social and economic processes, and draws attention to the dynamics of those processes.

Splintering Urbanism: A term coined by geographers Steven Graham and Simon Marvin to refer to the ways in which infrastructures, including information and communication technologies, can fragment the experience of the city.

ENDNOTES

1. In his posthumously published *Arcades Project* (2001) Walter Benjamin examines the emergence of spaces for leisurely consumption in 19th century Paris. Using these sites as a lens, he magnifies the nexus of social, technological, aesthetic, and architectural practices involved in modernization. His work prefigures the contemporary interest in relationships between visual culture, consumption, and the spatial ordering of everyday life, as discussed by Lefebvre (1984), de Certeau (1984), and Vaneigem (1983).
2. Excerpted from field notes.

Chapter II
To Connect and Flow in Seoul:
Ubiquitous Technologies, Urban Infrastructure and Everyday Life in the Contemporary Korean City

Jaz Hee-Jeong Choi
Queensland University of Technology, Australia

Adam Greenfield
Independent Scholar, New York, USA

ABSTRACT

Once a city shaped by the boundary conditions of heavy industrialisation and cheap labour, within a few years Seoul has transformed itself to one of the most connected and creative metropolises in the world, under the influence of a new set of postindustrial prerogatives: consumer choice, instantaneous access to information, and new demands for leisure, luxury, and ecological wholeness. The Korean capital stands out for its spatiotemporally compressed infrastructural development, particularly in the domain of urban informatics. This chapter explores some implications of this compression in relation to Seoulites' strong desire for perpetual connection—a desire that is realised and reproduced through ubiquitous technologies connecting individuals both with one another and with the urban environment itself. We use the heavily managed urban creek Cheonggyecheon as a metaphor for the technosocial milieu of contemporary Seoul, paying particular attention to what its development might signify for Seoulites both as a constituent node of the city and as an outcropping of networked information technology. We first describe some of the historic, social and economic contexts in which the Cheonggyecheon project is embedded, then proceed to discuss the most pertinent facets of Korean-style everyday informatics engaged by it: ubiquity; control and overspill; government-industry collaboration; lifestyle choice; and condensed development timelines.

HISTORY AND CONTEXT

A stream of fresh water. Shoals of fish orbit in a leisurely manner; curious children point them out, all the while being photographed by their delighted parents. Through the sound of the running water, surrounded by laughter and the little shutter-clicks from cameras and camera phones, a young couple are crossing evenly-spaced stepping stones, hand in hand. The air feels lush, fragrant, alive.

Standing on the many bridges arching over the stream, you realise you are at the centre of one of the most populous, polluted, quickly-developing, and densely interconnected metropolises on the planet. You are at Cheonggyecheon, in the very heart of Seoul.

Originally stretching ten kilometres from its origin to the point at which it eventually meets the Han River, Cheonggyecheon's history as an urban feature dates to the Joseon Dynasty's selection of Seoul as its new capital, at the beginning of fifteenth century CE. As a restored and managed stream, it now runs for almost six kilometres across the central city.

Recognition of Cheonggyecheon's potential benefits for Seoul residents was initially realised in simple forms: as 'a sewage system, a laundry and playground for children' (Park, 2007, p. 9) and adults alike (Seoul Development Institute, 2004, p. 1). Its use as an open sewage system evidently became unsustainable sometime during the Japanese occupation, leading to a first attempt at dredging and partial covering, with the aim of safeguarding Japanese citizens from disease and crime (*ibid.*). However, with the intense national focus on economic reconstruction in the post-liberation (1945) and post-war (1953) periods, and a corresponding slide into social and environmental negligence on the part of a preoccupied government, attempts at improvement fell by the wayside. Cheonggyecheon remained—and was generally perceived as—a perilous seam in the fabric of Seoul.

The stream's natural flow finally came to an end during Park Chung-hee's authoritarian administration (1961-79), a period in which the thrust toward national greatness was heavily predicated on, and identified with, export-oriented industrialisation. During this period, the government's need to make its authority and legitimacy visually manifest in modernisation—amidst a broad concomitant suppression of nature, history, and human rights—began to shape the city in ways that are still visible today. The result of this approach was evidenced in a contemporary statement of Kim Hyeong-ok, then mayor of Seoul: 'The city is lines.' Straight wires and streets started to replace traditional winding roads.

As part of this rapid national modernisation process, Cheonggyecheon was filled with cement, and was used as the foundation for both local streets and a high-capacity roadway transporting products and people in and out of the city centre. This was the height of the period often called the 'miracle on the Han (한강의 기적)'—approximately three decades from the mid-1960s to the Asian financial crisis of 1997 (Kleiner, 2001, p. 254) although the term is generally used to refer to the first two decades—in conscious emulation of the postwar West German *Wirtschaftwunder* (economic miracle), or 'miracle on the Rhine.' The stream was effectively ploughed under, literally subducted beneath the infrastructural development perceived as necessary to the advance of one of Asia's surging 'tiger' or 'little dragon' economies.

During this "Miracle" phase, a large-scale national effort—both the iconography and the subjectivity of which frequently involved themes of heroic sacrifice—was directed toward the end of rapid economic development. The predominant institutional structure which South Korea relied upon to accomplish this breakneck industrialisation was the *chaebol*, a huge and highly centralised, but heavily diversified, family-owned form of business conglomerate with no direct comparison in the Western world.

Chaebol are often compared with the Japanese *zaibatsu*, written with the same Chinese characters. But although they were originally modelled after the Japanese exemplar, during Park Chung-hee's administration, the distinctions between the two institutional forms are more than simply a matter of different pronunciation: zaibatsu have the organic means to manage their financial sustenance, typically through a network of wholly-owned banks and financial institutions, whereas chaebol lack these structures and are thus largely dependent upon the state's tight control over the mobilisation and distribution of financial resources (Chang, 1992, p. 46). During the Miracle period, comparatively few chaebol, working in close coordination with national government, established Korean competitiveness in shipbuilding, automotive manufacture, consumer electronics, and especially construction - first for the domestic and then, eventually, for the global market.

Beneath this ostensibly monolithic surface, things were far from quiescent. It was perhaps inevitable that a rapidly-developing nation would experience multiple and major social, political, and cultural shifts in the wake of any such breakneck economic expansion, and this is in fact precisely what happened: South Korean society experienced simultaneous shifts away from the more overt forms of authoritarianism politically, and from a largely agrarian population base towards an intense degree of urbanisation demographically and economically (Choe, 2005). There were important psychological shifts as well: to some degree, the readiness of the Miracle generation to sublimate their personal hopes and dreams to the national good was predicated on the belief that their sacrifice would purchase all the fruits of choice (both democratic and consumerist) for their children.

Major democratic reforms were launched in the yearlong run-up to the Seoul Olympics of 1988, and were pursued alongside a national agenda of globalisation (*segyehwa*), which persisted for a decade until, in 1997, the Asian economic crisis enveloped the nation. This ultimately led to a painful socioeconomic reconfiguration, under a restructuring mandate imposed by the International Monetary Fund that was designed to bring Korea into line with the prevailing neoliberal international framework (Crotty & Lee, 2006) (or, as one World Bank report calls it, the 'standard Anglo-Saxon blueprint')(Lee, Kim, Lee, & Yee, 2005, p. 4).

In the years since the IMF intervention and subsequent renewal of economic growth, Korea has gone through yet another dramatic shift, this one flowing outward from the technological and cultural industries. This is the so-called 'Korean Wave' (Choi, 2008, forthcoming), driven at least in part by the widespread local adoption of network technologies, including mobile telephony and broadband Internet.

Contemporary media-cultural and digital communication developments have occurred in a co-evolutionary spiral; the Korean Wave itself has been the result of 'non-static exogenous and endogenous convergence processes in an evolving system' (*ibid.*) in which the individual user—of the city and of informatic apparatuses—plays a crucial role in sustaining and expanding the network as a whole. This is a complex and organic infrastructural development linking micro- and macro-networks, rather than one that is hierarchically controlled and ordained from the top down. It is in this framework that the recuperation of Cheonggyecheon can best be understood as symbolic of the paradigmatic shifts now taking place in contemporary Korean society, with its new emphasis on individual desire, choice, amenity, and lifestyle.

The Cheonggyecheon restoration project was launched in 2003 by the then-mayor of Seoul (and recently-elected President of South Korea), Lee Myung-bak, as part of a comprehensive 'public betterment' initiative aimed at improving transportation safety, cultural understanding, and industrial, economic, and ecological conditions in areas surrounding the capital (Seoul Metropolitan Government, 2002). Within a comparatively short two years, disputes amongst various commercial, residential, and political parties were negotiated and resolved through an official body consisting of representatives from these and other sectors (known as the "Citizen's Committee"), and the construction of the waterway was completed, at an estimated cost of KRW 900 billion (approximately EUR 667M / USD 900M).

The result of this effort was 5.84 kilometres of cleanwater stream, sited between two parallel walking paths (See Image 1a) leading from residential suburbs in the east, through industrial and commercial districts, into the City Hall (While the neighbourhood where the pathways end is undoubtedly the civic and business centre of northern Seoul, it remains a contrast to the younger, more fully-developed and more privileged Seoul on the south side of the Han river [See Image 1b]).

The new Cheonggyecheon serves Seoulites as an open and accessible multi-functional place for leisure in and of itself. At the same time, it clearly functions as a space of mobility and flow, a conduit connecting multiple sectors of Seoul. This multiplicity of readings, meanings, and uses is one of the main characteristics of contemporary urban development in Seoul, a typological obscurity that tends to confound simple classification. It is in this respect that Cheonggyecheon can be understood to epitomise four factors shaping the technosocial contours of contemporary Korean life, four onrushing streams so intricately interbraided that it can be difficult to disentangle them:

- As we shall see, Cheonggyecheon captures in its very essence the **complicated negotiations between flow, control, and more-than-occasional overspill** that seem to inhere in everyday Korean spatial practice.
- The **institutional framework** within which the creek was developed demonstrates the way in which, compared to Western democracies particularly, South Korean society depends on a high degree of coordination between government and industry (chaebol in particular) in determining urban, industrial, and technological policy.
- Finally, the **compressed timeline** of the creek's redevelopment project reflects the prevailing *ppali-ppali* (hurry hurry / 빨리

Image 1.

a. the new Cheonggyecheon

b. Map of Seoul

빨리) ethos, signalling both 'hastiness' and 'dynamism' (Kang, 2006, p. 47) in adopting and adapting to technological and social change.

- In the wake of the transition to a postindustrial economy, Cheonggyecheon's role as a symbol and manifestation of leisure space/time epitomises the **broad public endorsement of a hedonic agenda**, dedicated to consumer choice, the pursuit of the 'noble' life, and perhaps even a greater awareness of and respect for the natural environment.

We argue that these factors are likely to shape the experience of ubiquitous and ambient informatics not merely within Korea, but—owing to Korea's emerging status as a leading exporter of technical products, components, and frameworks—globally as well. We begin our discussion by examining the concept of computational ubiquity in the context of contemporary Seoul, and then proceed to a discussion of each of these four factors in detail.

UBIQUITY

In order to situate the contemporary desire for ambient informatics correctly, it is necessary to first understand that the everyday Korean experience of information technology is *already* one of ubiquity: in his article, 'Seoul: birth of a broadband metropolis,' Townsend (2007) cites 2004 government figures claiming 80% household broadband penetration, one of the highest rates in the world, while the International Telecommunication Union has placed Korea at the top of its Digital Opportunity Index for the two years 2005 and 2006 (International Telecommunication Union, 2006).

In such an environment, Internet-derived conventions become part of the daily *lingua franca,* with manifestations such as emoticons—for example, (^_^)—rendered without explanation in newspaper headlines, or in branding intended for the mass audience. Technical terms and jargon infiltrate everyday life, in a way that is clearly beneficial to those institutions with something at stake in the mass adoption of technology. One result of this is that the single Koreanized-English word "ubiquitous," and the *u-* prefix derived from it (e.g. "u-City," see Hwang in this volume), is now commonly understood by the general public to refer to a technological regime positioned as desirable; an example epitomizing two of the aspects discussed in this paper is a current slogan used by the Ministry of Information and Communication (MIC), 'Happy U-life that do with U-Korea realization.'

This is nowhere more visible than at Cheonggyecheon, where government policy, aspirational ubiquity, and public space have become fused to a degree that is hard to convey to those unfamiliar with the Korean way of doing and making things. The official government portal *Korea.net* invites residents and visitors to 'Experience Ubiquitous Seoul at Cheonggyecheon Event,' where they might enjoy a 'rush of high-tech cyberspace and nature in one central-Seoul spot' (Korea.net, 2007).

By re-designating the stream 'U-Cheonggyecheon,' and touting the deployment of high-tech assemblies to 'monitor its purity and water and pollution levels,' the notion of technological *testbed* is collapsed against that of *riverbed.* Once thus embedded, visitors can indulge themselves in technologically-enhanced leisure with a frisson of ecological responsibility, interacting with features such as 'Free Board' (a digital bulletin board on which the user can create their own multimedia content, and email it for free); a touch-screen based 'Interactive Media Board' providing a variety of information about Cheonggyecheon; and LED-equipped street lamps, which are also Internet hotspots in disguise (*ibid.*).

If, as an unnamed Seoul Metropolitan official explains, this overcoded space is explicitly a 'standard model and guide for other ubiquitous projects,' it is not the only one. Although sponsored by the private Korea Home Network Industries Association, the Ubiquitous Dream Hall displays cutting-edge domestic technologies in a privileged home on the grounds of the Ministry of Information and Communication, literally across the street from historic Gyeongbokgung Palace, while visions of domestic Weiseriana are presented to consumers in seasonal exhibitions like the popular Daelim Model House, in the fashionable, upper-class Apgujeong neighbourhood.

All of these manifestations enthusiastically embrace, to a degree that tends no longer to be the case among Western technology vendors, the traditional Weiserian vision of ubiquitous computing as heavily-instrumented space (cf. Bell & Dourish, 2007; Weiser, 1995), and are in turn welcomed with equal gusto by a nation of consumers increasingly primed to regard such technological interventions as *de rigeur* appurtenances of the good life. Indeed, during the exhibition's season, long lines of would-be residents file through the dazzling, spacious dream apartments; Image 2 shows the files of bowing, elegantly-dressed models deployed to greet their tour buses on arrival.

Here visions of computational ubiquity are closely coupled to notions of ease, leisure, and luxury, which may go some way toward explaining why MIC's vision for rolling out next-generation ubiquity, the so-called 'IT 839 Strategy,' enunciates a (somewhat peculiar for a highly technical infrastructure-development program) success metric of GDP USD 30,000 per capita. There is no doubt that a vision of robust domestic ubiquity is latent in the 'eight services, three infrastructures [and] nine growth engines' enumerated in IT 839 (see Table 1), which also claims the emergent WiBro (wireless broadband) and DMB (digital multimedia broadcasting) standards as Korean innovations. But is it a particularly Korean one?

The implied seamless model of spatiality—with no differentials or gradients of access within it—embraced by the IT 839 strategy is somewhat at variance with the way ubiquitous urban informatics are currently experienced in the everyday life of Seoul, an experience whose features are largely shaped by the unique '*bang* (room)' culture of Korea. Translating 'bang' as 'room,' however, threatens to obliterate a variety of meaningful distinctions between the typologies, distinctions that clearly condition the type, timing, and intensity of social activities that take place within.

In contrast to the general understanding of a 'room' embedded in Western spatial practice—a single-purpose space, designed, designated, and provisioned for a specific function—bangs in traditional Korean culture are generally required to support a multiplicity of functions, and are

Image 2. Tour bus greeters

Table 1. Adapted from MIC (2005)

8 Services	3 Infras	9 Engines
HSDPA/W-CDMA	Broadband convergence network (BcN)	Mobile communication/ Telematics devices
WiBro	u-sensor network (USN)	Broadband/ Home network devices
Broadband convergence service	Soft infraware	Digital TV/Broadcasting
DMB/DTV service		Next-generation computing/ Peripherals
u-Home service		Intelligent robot
Telematics/ Location-based service		RFID/USN devices
RFID/USN application service		IT SoC/Convergence parts
IT service		Embedded software
		Digital content/ Software solution

provisioned accordingly. Take, for example, the custom of serving food on low tables that can be folded away. It was a common practice (and still is, in small residential spaces) for the living room to be metamorphosed into the dining room when the table is set up, for it then to be converted for study during the evening, and finally into a bedroom at night, when *yo* (Korean futon) are unfolded on the floor. This inherent reconfigurability of domestic space has become commercialised in the contemporary Seoul urbanscape (in fact, by the chaebol themselves, in the appointments of their large-scale developments), blurring the border between what is imagined and lived as private and as public.

The following excerpt from Sung Hong Kim's curatorial statement (Kim, 2004) for the Korean Pavilion at the 9th Architecture Biennale of Venice—themed *City of the Bang*—aptly captures the essence of this culture:

[B]ang has infiltrated the Korean urban landscape of commercialized space with enterprises such as the PC bang, Video bang, Norae bang, Jjimjil bang, Soju bang, and others. The Norae bang, a scaled-down version of the Karaoke bar, is the primeval cave festival in the midst of the contemporary city. Visual, audible, olfactory, tactile, and gustatory sensations are simultaneously experienced in this tiny black box. Meanwhile, the Jjimjil bang, which combines a steam bath, fitness room, lounge, restaurant, and sleeping area, provides space where half-clothed bodies intersperse between a variety of functional areas. The Jjimjil bang blurs the lines between the collective and the individual, normal and deviant behaviour, privacy and voyeurism. The bang is an incarnation of the room, the house and the city, but it does not belong to any of them. The city of the bang oscillates between the domestic realm, institutionalized place, and urban space.

Like the majority of other interior spaces in South Korea, most bangs are now heavily mediated and connected via broadband Internet, providing additional opportunities for instant and spontaneous connection through geo-social mobility (bangs as decentralised connection points), and at the same time, constant and now 'given' connection through immobility (bangs as physically and socially constrained spaces) (Choi, 2007b; cf. Hjorth, 2007 on mobility/immobility). Ubiquitous computing, in this sense, is socially established and experienced through multiple overlaying infrastructural arrangements, both tangible and intangible, giving a sense of what Bell and Dourish call 'messiness' (2007, pp. 139-141) and what we have here termed overspill.

CONTROL AND OVERSPILL

A sense of boundaries being overrun is also inherent to Cheonggyecheon, albeit primarily through its absence or negation ("control"). Over time, the primary aim of the various governmental efforts at managing the creek have concerned the relationship between useful, life-giving *flow* and an undesirable *overspill*, with more or less heavy-handed interventions aimed at limiting the latter. The first known project on the site was one devoted to the control of seasonal flooding, ordered by King Taejong in 1406. Such flooding continued to occur despite near-continual efforts at intervention, with the most recent taking place in 2001 (though it was of a minor class, and water only flowed into the underground level).

If Castells (1989) characterised the modern world as a 'space of flows,' Seoul embodies this in several respects. Some days everything in the city seems to have slipped its bounds: the stuff of apartment lives tumbles onto balconies—'verandas displaying racks of drying laundry, children's bicycles, and brown pottery jars for staple condiments and kimchi,' as noted by Nelson (2004, p. 5); clubs, parties, conversations, arguments and even commercial services extend heedless into the street (Lee, 2004, pp. 74-75); restaurants seem to store half their crockery on the sidewalk. Meanwhile, omnipresent Columbia and UCLA sweatshirts stand as synecdoche for the continual flux of traffic both cargo and passenger between Incheon Airport and New York and Los Angeles, the latter city enjoying the second-largest Korean population on the planet. The cultural roots of Korean identity are certainly felt, but what is visible is a hybrid, an overcoded overspill of cultural eclectics and constant negotiation.

Just as subject to change, perhaps surprisingly, is the visual envelope and appearance of local buildings. Amongst arrays of identical, matchbox-like apartment buildings in the Apgujeongdong district is the Galleria department store (see Image 3), the entire façade of which is made up of networked, programmable display elements capable of generating 16 million colours (Arup, 2004). Here, spatial demarcation becomes obscured, and space itself thus becomes a fundamentally subjective experience, conceptually and sensorially. Through grids of such connected lights, Seoul becomes the 'circuit city' (Vanderbilt, 2005) where individual narratives flow together to create a common history.

Seoul is found in the flux hinging between control and overspill: the circuit city of bangs, of screens, and in flux (Choi, 2007a). How is such a city managed politically and economically today? We point to the concept of chaebol to answer this question.

Image 3. Galleria department store

CHAEBOL NATION

As a high-profile, high-prestige development in the very core of the capital city, Cheonggyecheon's recent evolution would be hard to imagine without the involvement of the chaebol, as contractors and executors of the national will. It is almost impossible to overstate the influence these massive business combines have on Korean life: the commercial hegemony established by the chaebol early on in the post-Korean War reconstruction effort, and consolidated in the Park Chung-hee years, continues to be manifested in the Korean landscape, literally in concrete—not least in the arrays of identical housing blocks that cover the city and the countryside beyond in endless domino ranks, each proudly emblazoned with its corporate logo (see Image 4).

A drive from central Seoul to one of its outlying newtowns simultaneously epitomises how far into daily life the chaebol reach, and captures something of the current national mood. Accessible via Cheonggyecheon walking paths, Doota, or Doosan Tower, is one of the biggest and most well-known shopping malls in the popular fashion district of Namdaemun, while the same conglomerate has recently (2005) developed an entire edge-city newtown near Seoul. This scale of private development is by no means considered particularly excessive by local standards; as their primary slogan—*We've*—boasts, Doosan is a

Image 4. Apartment blocks

conglomerate whose business interests reach from wine production and ownership of a baseball team to the design and construction of surface-to-air missile systems.

Indeed, nothing is more ubiquitous in Korea than the chaebol, and none of the chaebol is more ubiquitous than Samsung: as the Seoul-based net artists YOUNG-HAE CHANG HEAVY INDUSTRIES (http://www.yhchang.com), themselves named in parodic imitation of chaebol, point out, one can be born in a Samsung hospital, attend a Samsung school, marry in a Samsung chapel, live in a Samsung home, and be buried in a Samsung casket ("Samsung will help me get over being dead...and being alive").

It is only natural, therefore, that the chaebol loom large in any Korean discussion of ubiquitous development, both at the level of infrastructure (Doosan has a business unit dedicated to "ubiquitous framework standardisation") and consumer-grade interfaces, nor that governmental specification of the relevant technical standards is pursued in close cooperation with them. The emphasis on institutional coordination has clear implications for the prospective development of ambient informatics, not always those that an onlooker might be tempted to imagine. More specifically, programs undertaken in the light of close chaebol-governmental cooperation would appear to benefit from:

- a markedly accelerated speed of development, in accord with the ppali-ppali mentality;
- a certain consistency of aspiration and execution, especially as concerns the production of physical space, with the associated desires reproduced and diffused via the single, centralized national media market; and
- increased interoperability among and between communication devices and platforms, resulting in a smoother and more fluid user experience.

Not all of these things are necessarily true. However tempting it may be—however often their policies seem to evolve in tune with the personal desires of one or another charismatic chairman, even when that desire contravenes general business ethics or an obvious profit motive—the chaebol cannot be thought of as monolithic organisations. The chaebol, by and large, are in fact internally heterogeneous, with sub-companies, divisions and business units run by other members of the family (siblings or cousins, for example) each of which will retain a certain level of autonomy.

Particularly, the fact that chaebol are not simple monoliths can lead to user experiences that are sporadic and disconnected. From this perspective, the chaebols' internal heterogeneity means that devices bearing the Samsung or the LG brand were likely developed by entirely discrete design organisations, with no overlap of personnel, process, or practice. Despite the Weiserian promise to 'encalm as well as inform' (Greenfield, 2006, p. 29), very few of the current generation of Korean-designed digital tools have the same interfaces, very few of them work well together or have been designed with seemingly any recognition that their actual environment of use would likely be one of saturation and synchrony. (In fairness, the same situation is true of both Japanese and Western commercial competitors and "free" or open-source alternatives, with few exceptions, however, the design organisation is internally organised. The point is merely that the apparent homogeneity of the chaebol confers no evident benefit to users of products and services they design and bring to market.)

There is also a clear question as to whether what is good for the chaebol is good for Korea. In criticising excessive urban renewal occurring in South Korea, Yim (2006, pp. 116-117) argues that an average of 50-60 percent of the total profit generated by Korean construction industry comes from apartment construction, which amounts to approximately 8% of the national GDP. This massive incentive, he further asserts, leads to chaebol promoting the physically higher, high-tech, and highly self-contained (and thus isolated) living environment of the apartment complex as an aspirational image for 'good living.' Clearly, chaebol produce space and fill it with their increasingly networked products and services. This environment, people are constantly persuaded, is one in which the resident can remain safe, secure and comfortable, while managing every outward-facing aspect of their lives via fluid high-speed connections: a utopia of privilege, reserved for those who are worthy of the 'noble life.'

Image 5. The Polus development

NOBLESSE OBLIGE

Across the hoardings where a new apartment development will rise in trendy Apgujeongdong are emblazoned the words *The Noble Community*; nearby, a bus-shelter ad urges waiting passengers to consume, *For your nobility life*. There's even a (high)life-style magazine, positioned somewhere between *Vogue* and *Architectural Digest*, called *Noblesse* (노블레스). This is nothing if not a culture that wears its aspirations—like its brand names—on its sleeve.

So what counts as noble here? Judging from the images splashed across those same hoardings, three times as large as life: bowls, handkerchiefs, and umbrellas monogrammed with Italianate names. Tumi wheelaways, bottles of Johnny Walker Black clanking dully in the duty-free bag. Enough time to play golf and to visit the wonders of the world. (Another development, Polus, modestly compares itself to the Taj Mahal, the Sagrada Familia, the Arc de Triomphe, and the Empire State Building. Its slogan: "Over the borderline & Over the luxury"—see Image 5)

The trappings of nobility even extend to the architecture of the body. In this context, 'noble' means smaller faces - sleeker, but retaining enough space for the bigger, rounder, doll-like eyes and higher nose. This is the new popular Korean-Western aesthetic, smooth, synthetic, and purchased. Although this vision of what constitutes the noble is not universally held, it is a broadly popular one, and to a significant degree it informs what happened at Cheonggyecheon: a perfected stream forged for the pleasure of a perfected population.

This new 'skin job' aesthetic, with its unachievable limit-case dream of nanometer-smooth surfaces, can contrast with the reality of Seoul's urban fabric, in places starkly. Across the river from Cheonggyecheon, a pedestrian bridge runs south from Banpodong's Express Bus Terminal across six lanes of traffic; with its dramatic uplighting, sharply-raked struts and pointless blue Ming the Merciless flanges, it looks like an escapee from William Gibson's "Gernsback Continuum." To gaze upon this bridge from far away, or to drive under it at speed, especially at night, is to enter a city that still lives mostly in the renderings favoured by Seoul's developers and civic boosters (and in subway-ad versions of which a 're-gooded' Cheonggyecheon figured heavily, for years before its unveiling). Urban furniture like this bridge, along with new subway lines and new building complexes, and enough English signage to tie them all together, were trotted out for the 1988 Olympics (which resulted in top-down democratic reforms), and then again for the 2002 World Cup (during which brigades of Red Devils, a Korean soccer supporters' club which was organised from the bottom up, gained international attention).

These two events, and all the impedimenta introduced in their wake, were intended to usher this proud city at long last through the velvet rope and into the 'world class,' an ambition clearly stated in the Seoul Olympics' very slogan, *Seoul to the World, the World to Seoul* (서울은 세계로, 세계는 서울로). However, the pedestrian crossing the bridge may see something else entirely. At close range, one cannot help but notice that the welds on its steps and stanchions are sloppy, incontinent, gappy, and that the translucent blue wings have long gone dull with wear. The bridge is not an aberration, an outlier. In its haphazardness, the pedestrian bridge joins the rivets of the girders holding up a parking garage at the Yongsan Electronics Market, the unfinished light wells in the ceiling of a high-end hotel's lobby, and the letter "E" that for many years hung ten degrees off true, five stories up, on the side of the Newcore shopping center. The whole city, in fact, can feel to a visitor very much as if it has been assembled so quickly that large parts of it are in imminent danger of falling apart entirely; various observers (e.g. Feffer, 2003) ascribe this to the perceived need to *ppali-ppali* on the way to yet another economic Miracle.

While the ppali-ppali mode has seen South Korea make its entry to the domain of developed nations in impressively short order, this has been won at the cost of serious, and occasionally fatal, consequences—most notably, the 1995 Sampoong Department Store collapse, in which 937 people were injured and a further 501 lost their lives. The implications of a continued reliance on ppali-ppali for any domain of development as sensitively dependent on accurate configuration and ongoing maintenance as the deployment of ubiquitous information technologies are significant. Physical danger from faulty, undertested or undershielded devices is one possibility, especially with base stations and other high-energy retransmission equipment being placed in far closer proximity to living and sleeping quarters than would be considered prudent in the West. Just as evident, however, is a cavalier attitude with regard to less tangible potential hazards of ubiquity, privacy concerns, and so-called 'digital divide' issues chief among them (Bell & Dourish, 2007, p. 138). (The latter is a problem that is already apparent in Korea despite, and paradoxically exacerbated by, the high-broadband penetration rate in urban areas; the Internet is now perceived to be so unremarkably vital to the management of everyday life that those without access are doubly disadvantaged.)

The tension between the reality resulting from *ppali-ppali* development and visions of the noble is acute, but generally addressed only obliquely. One manifestation, however, is that Koreans find it increasingly hard to accept the natural, with all its inescapable imperfections and variations from the statistical norm as evidenced in the rising popularity of plastic surgery; a study shows that approximately 81.5% of women between the ages of 25 and 29 feel that they need cosmetic surgery and 61.5% of them have already had more than one surgery done (Kim, 2007). Particularly in conjunction with the traditional collective mentality, in which deviance is deplored (Triandis & Suh, 2002), reparative cosmetic surgery becomes desirable not only for people, but also for the environment—an aspect for which the Cheonggyecheon restoration project has been heavily criticised as a costly urban facelift. It remains to be seen whether this wide pursuit of perfection and consumerist nobility can do anything other than create everyday predicaments, imperfectly concealed in the compressed timeframe of urgent development. Nevertheless, we can at least be certain of the destination to which all of these efforts are supposed to lead: a life of happiness.

HAPPY FOREVER

At the end of 2007, Samsung Anycall launched an advertising campaign urging people to *Talk, Play, Love*. While this slogan is incessantly animated across the brilliantly-coloured display façade of the aforementioned Galleria department store, Samsung presents another message on a huge video billboard in central Seoul, this one in English: *Happy Forever*. The context in which these slogans—so easily perceived by Western observers as Orwellian, even Stepfordian—are encountered suggests three things: these words accurately capture a mass aspiration in the contemporary Korean soul; they are meant literally; and they are meant seriously.

The word 'happy,' particularly, seems to occupy the place in the popular Korean imaginary that 'dream' does for many Japanese. However, whereas 'dream' generally suggests a state that is aspirational, and thus perhaps eternally deferred, 'happy' connotes a state that a person might reasonably expect to achieve in the course of ordinary existence. 'Happy,' for this audience only a very few decades removed from the most harrowing imaginable experience of wartime suffering and deprivation, is not outside history; it is meant to be realised in the here and now.

The chaebol, with their multipronged need to shift products and sell services, can often lend the impression that quality of life is a matter primarily of infrastructural technical innovation, rather than anything conventionally subsumed under the rubric of urbanism. High-quality urban environments are increasingly perceived by South Koreans as something to be planned, budgeted for, and delivered politically, commercially, and quickly, and not something forged in the contestation and negotiation of uses for public space, let alone in the active participation of residents *qua* citizens (cf. Gelézeau, 2007; Yim, 2006; Yang, 2005).

At this point, we ask, largely but admittedly not entirely in the spirit of devil's advocacy, if this city-as-lifestyle-as-service—where happiness itself is constructed as something consumed (cf. Luke, 2005) rather than participated in (cf. Rheingold, 2008, forthcoming)—is perceived by the overwhelming majority of its users as delivering value and satisfaction reliably and consistently, what benefit would be served by the minority, non-participant observers, advancing claims to the contrary? What would raising such claims do but complicate the swift and smooth delivery of services to the people who have freely engaged them? (Such an assertion would certainly seem to be among the many messages laminated into the recent victory of Lee Myung-bak.)

Given such conditions, democracy itself—defined as a process that attempts to balance interests through a satisficing churn of discourse, deliberation, and disputation on the part of nearly all of the members of a community—may come to be deprecated locally, yielding perhaps to a softer strain of that authoritarian/consumerist fusion which finds its fullest contemporary expression in the People's Republic of China.

Those children whose democratic and economic entitlement had to be fought for by the Miracle generation are now parents of their own children—perhaps those we last saw curiously pointing at the shoals of fish at Cheonggyecheon. Looking at these children play, one might almost imagine that the balance of their lives will unfold as serenely and as generously as a late-spring day, in both material and spiritual registers; that happiness is truly something to which one might subscribe; that most if not all will be able to do so; and that it will all last.

In the late Joseon period, a noted poet and scholar, Lee Duk-moo, wrote a poem called *Song of the Full Moon* depicting a joyous scene of people waiting for the full moon on a bridge over Cheonggyecheon (Box 1.)

For the Korea(n)s, this imagined and long-deferred happiness may finally be realised in this city conceived—however rightly or wrongly—as a machine capable of delivering the good life to the majority of the people who live in it. Whether

Box 1. Song of the Full Moon

雪色澄明惟此宵　　유독 오늘밤은 달빛이 맑고 밝아

Tonight snow is auspiciously bright and clear

人人候月廣通橋　　사람마다 광통교에서 달을 기다린다

Everyone waits for the moon on the Gwangtong Bridge

歌童一隊聯群袂　　노래하는 아이들 한무리가 여러 옷깃을 이어

A group of children bridge their collars together

齊唱東方行樂調　　동방의 행락조를 함께 부르네

Singing the song of joy from the East

any such state can be lived as and in reality remains open to question, but we see an undeniable promise in the optimism so abundantly evident at Cheonggyecheon, and elsewhere in the streets of contemporary Seoul. This optimism inheres not least, perhaps, in the rise of technologies able to support collaborative efforts amongst individuals, who are together able to reimagine their once-guarded, isolar and literally defended city as a connected space, to and within which access is open and flexible.

For our own part, we believe that the ongoing convergence—imagined, in process, and lived—between what Seoul is and what it could be must occur organically, true to the desires and intentions of the Korean people. We remain hopeful that appropriately-designed ubiquitous technology will indeed empower people towards this end, and that we will once again witness (though in a vastly different technosocial context) Seoulites from diverse backgrounds coming together as joyously and hopefully as those depicted in the *Song of the Full Moon*.

ACKNOWLEGEMENTS

We would like to thank all the reviewers for their valuable comments. We would also note that the following images were externally sourced under Creative Commons licence (http://www.creativecommons.org):

- **Image 1a:** This image was sourced from Kai Henry's flickr page (http://www.flickr.com/photos/hendry/154550723), and is licensed under Creative Commons Attribution 2.0 (cc-by- 2.0). See this page for further information about the license: http://creativecommons.org/licenses/by/2.0/deed.en
- **Image 1b:** This image was modified from the original version on Wikimedia Commons, under the terms of the GNU Free Documentation Licence, Version 1.2 or any later version published by the Free Software Foundation. See this page for further information about the image and licence: http://commons.wikimedia.org/wiki/Image:Image-Map_Seoul-teukbyeolsi-big.png

REFERENCES

Arup. (2004). The biggest pixels in the world clad the Galleria West shopping centre, Seoul. Retrieved May 16, 2007, from http://www.arup.com/newsitem.cfm?pageid=5323

Bell, G., & Dourish, P. (2007). Yesterdays tomorrows: notes on ubiquitous computings dominant vision. *Personal and Ubiquitous Computing, 11*(2), 133-143.

Castells, M. (1989). *The Informational City*.

Chang, L. (1992). Financial mobilization and allocation: The South Korean case. *Studies in*

Comparative International Development, 27(4), 41.

Choe, S.C. (2005). "The impact of globalization on the urban spatial-economic system in Korea." In H.W. Richardson & C.-H. C. Bae (Eds.), *Advances in Spatial Science: Globalization and Urban Development*. Heidelberg: Springer.

Choi, J. H.-j. (2007a, July 5-6). *All Things Big and Small: Rising Importance of Mobile Media in South Korea*. Paper presented at the China East Asia Media New Media Conference, QUT, Brisbane.

Choi, J. H.-j. (2007b, December 3-5). *Ready, Unsteady, Go: trans-youth mobile media and spatial experience in seoul*. Paper presented at the 2007 Australasia Interactive Entertainment Conference,, RMIT.

Choi, J. H.-j. (2008, forthcoming). The Korean Wave of U. In H. K. Anheier & Y. R. Isar (Eds.), The Cultures and Globalization Series: The Cultural Economy. London: Sage.

Crotty, J., & Lee, K.-K. (2006). The Effects of Neoliberal "Reforms" on the Post-Crisis Korean Economy *Review of Radical Political Economics, 38*(4), 669-675.

Feffer, J. (2003). *North Korea/South Korea: US Policy at a Time of Crisis*. New York: Seven Stories Press.

Gelézeau, V. (2007). *The Republic of Apartments (아파트 공화국)*. Seoul: Humanitas (후마니타스).

Gibson, W. (1997). The Gernsback Continuum.

Greenfield, A. (2006). *Everyware : the dawning age of ubiquitous computing*. Berkeley, Calif.: New Riders.

Hjorth, L. (2007, Oct 25-26). *Mobile and immobile imaging communities*. Paper presented at the Rikkyo-Yonsei International Conference, Tokyo.

International Telecommunication Union. (2006). *World Information Society Report*. Geneva: International Telecommunication Union.

Kang, J.-m. (2006). *Korean Code (한국인 코드)*. Seoul: Inmulgwasasang (인물과 사상).

Kim, H.-A. (2004). Korea's Development Under Park Chung Hee: Rapid Industrialization, 1961-1979. London: RoutledgeCurzon.

Kim, S. H. (2004). City of the Bang. *9th International Architecture Exhibition 2004 Venice Biennale* Retrieved May 8, 2007, from http://www.korean-pavilion.or.kr/04pavilion/e_2004_02.htm

Kim, Y. S. (2007). 70% of Women 'Under Stress from Appearence". Retrieved Jan 2, 2008, from http://www.medical-tribune.co.kr/news/sub_main_view.php?subm=2&number=13902&uplink=&code=ME31

Kleiner, J. (2001). Korea: A Century of Change. Singapore: World Scientific Publishing Company.

Ko, H. B. (2007). *Mobile Phone, Calling Philosophy (휴대전화, 철학과 통화하다)*. Seoul: Chaeksesang (책세상).

Korea.net. (2007). "Experience Ubiquitous Seoul at Cheonggyecheon Event." Retrieved from *korea.net/news/news/newsView.asp?serial_no=20071126025* on 02 Dec 2007.

Lee, K., Kim, B.-K., Lee, C. H., & Yee, J. (2005). *Visible Success and Invisible Failure in Post-Crisis Reform in the Republic of Korea: Interplay of Global Standards, Agents, and Local Specificity* New York: World Bank.

Lee, M.-Y. (2004). The Landscape of Club Culture and Identity Politics. *Korea Journal, 44*(3), 65-107.

Luke, R. (2005). he Phoneur: Mobile Commerce and the Digital Pedagogies of the Wireless Web. In P. P. Trifonas (Ed.), *Communities of Difference:*

Culture, Language, Technology (pp. 185-204). New York: Palgrave Macmillan.

MIC. (2005). A Leap to Advanced Korea based on IT: IT 389 Strategy. Retrieved Jan 2, 2008, from http://eng.mic.go.kr/eng/secureDN.tdf?seq=6&idx=2&board_id=E_04_03

Nelson, L. C. (2004). *Measured Excess: Status, Gender and Consumer Nationalism in South Korea*. New York: Columbia Univeresity Press.

Park, K.-d. (2007). Cheonggyecheon Restoration Project. Paper presented at the JFES-WFEO Joint International Symposium on River Restoration, Hiroshima University.

Republic of Korea Ministry of Information and Communications. Slogan. Retrieved from *eng.mic.go.kr/eng/index.jsp* on 02 Dec 2007.

Republic of Korea Ministry of Information and Communications. Ubiquitous Dream Home. Retrieved from *ubiquitousdream.or.kr/main.html* on 02 Dec 2007.

Rheingold, H. (2008, forthcoming). Using Participatory Media and Public Voice to Encourage Civic Engagement. In W. L. Bennett (Ed.), *Civic Life Online: Learning How Digital Media Can Engage Youth* (pp. 97–118). Cambridge, MA: The MIT Press.

Ryu, J.-h. (2004). Naturalizing Landscapes and the Politics of Hybridity. Korea Journal, 44(3), 8-34.

Seoul Development Institute. (2004, Jul 26, 2004). The current state of Cheonggyecheong-related Folk Play and Its Contemporary Enactment (청계천에 얽힌 민속놀이 현황과 현대적 재현 방안). *Seoul Research Focus* Retrieved Dec 24, 2007, from http://www.sdi.re.kr/nfile/zcom_focus/14-1.pdf

Seoul Metropolitan Government. (2002). Cheonggyecheon Restoration Project: Needs & Influences. Retrieved Oct 24, 2007, from http://www.metro.seoul.kr/kor2000/chungaehome/en/seoul/sub_htm/2sub_02.htm

Triandis, H. C., & Suh, E. M. (2002). Cultural Influences on Personality. *Annual Review of Psychology, 1*(53), 133-160.

Vanderbilt, T. (2005). Circuit City. *Artforum, 44*(3), 65-66.

Weiser, M. (1995). The computer for the 21st century. In R. M. Baecker, J. Grudin, W. A. S. Buxton & S. Greenberg (Eds.), *Human-computer interaction: toward the year 2000* (pp. 933-940). San Francisco: Morgan Kaufmann Publishers Inc.

Yim, S. J. (2006). *Architecture, Our Self-Portrait (건축, 우리들의 자화상)*. Seoul: inmulgwasa-sangsa (인물과사상사).

KEY TERMS

Bang: Although the literal translation is 'room,' the designation *-bang* connotes a social and multipurpose space. *Dabang*, for example, means tea-house, but is literally translated as 'tea room.' There are many commercialised bangs today, such as PC-bang, DVD-bang, and *noraebang* (karaoke).

Chaebol: Large, family-owned conglomerates in Korea, such as Samsung, Hyundai, and LG, with multiple child-companies in diverse fields belonging to one parent company.

Korean Wave: Refers to the sharp increase in popularity of Korean popular culture—and subsequent increase in Korean media export—especially in the Asian region, starting in the late 1990s.

"Miracle on the Han": the catchphrase used to refer to the post-Korean-War period economic expansion, from mid-1960's to 1997, during which Korea's economy grew at an extraordinarily fast rate through industrialisation and modernisation.

Ppali-Ppali: Literally, 'hurry-hurry.' The ppali-ppali ethos refers to the general tendency to rush through any given activity, widely accepted as a common characteristic of contemporary Korean culture.

Ubiquitous Computing: A milieu in which computation resources are distributed through the objects and surfaces of everyday life.

Segyehwa: Literally, 'globalisation.' A public slogan used by the Kim Young-sam administration (1993-1998), representing the top-down reform of the Korean political economy in response to global pressure for market liberalisation

Chapter III
Creating an Analytical Lens for Understanding Digital Networks in Urban South Africa

Nancy Odendaal
University of KwaZulu-Natal, South Africa

ABSTRACT

Recent literature on African cities examines the way in which social networks function as critical livelihood arteries in the ongoing survival strategies of the poor. An understanding of livelihood strategies is not new, but these transactions cannot be defined in space or frozen in time. This terrain comprises a divergent range of intentions, communications and movements exchanged between a multiplicity of actors making sense of their life worlds; negotiating, scheming and bargaining. Urban life continues to be reinvented at the margins, despite prevailing exclusionary economic and social forces. The potential exists for harnessing these strategies for developmental aims—building on the social capital created despite the absence of, or in addition to, the usual resources available for survival. One of these resources is Information and Communication Technology (ICT). Clearly the "real-time" communication, information transfer and exchange functions facilitated by mobile phones, e-mail and the Internet create the potential for informed decision making around the use and distribution of scarce resources. However, this chapter begins with the premise that ICT can only be considered a meaningful development tool if it is appropriated as ongoing input into the day to day decision-making of the poor. It is at this scale—the local, the individual, the social—that the appropriation of digital technologies is examined. The social appropriation of technology is considered in tandem with the network strategies people employ to manage and access resources. A conceptual bridge between the theoretical foundations of actor-network theory and the more contemporary writings on the African city is constructed to posit a theoretical lens for understanding digital networks in South African cities. The chapter concludes with a number of methodological implications with regards to future research into ICT and social networks in developmental contexts.

INTRODUCTION

Urban Informatics as a field of study sits between various disciplines. Its multi-disciplinarity is partially its strength. Drawing on the fields of urban studies, sociology and information technology creates opportunities for conceptual 'borrowing' across disciplines. However, as with any endeavor to integrate, there are conceptual blind spots when applied to certain contexts.

One such gap has emerged in research currently being undertaken by this author that entails the development of web sites for selected community groups within South Africa's primary port city, Durban. The aim is to understand the relationship between urban dynamics and digital technologies in South African cities at a local level. The study draws from various bodies of literature such as ICT and cities (S. Graham, 2004; S. Graham & Marvin, 2001; Mitchell, 2000), community informatics (such as Day, 2005; Liff, 2005; Williams, 2006) as well as selected writings about South African cities and African urban spaces (for example Pieterse, 2005; Simone, 2004). This is not enough. What is revealed in this research is the need for a theoretical lens that examines the power relations, institutional dynamics and livelihoods specific to (South) African urban spaces. The threads are there, but they require the conceptual mesh to encapsulate them.

The aim of this paper is to make suggestions on what may constitute such a framework. It begins by sketching the research project mentioned and its preliminary findings. The conceptual difficulties in gaining insight into finer dimensions of the relationship between technology and community within this context are defined. These are then addressed through a consideration of literature on associational life in (South) African cities and the tenets of Actor-network-theory (ANT). The argument here is that in combination, these two sets of debates provides more appropriate theoretical tools for understanding digital communities in African spaces. The emphasis on South Africa is by virtue of the fact that the study is conducted in Durban, considered an example of an African metropolis with its own distinguishing characteristics.

DESIGNING COMMUNITY WEB SITES IN DURBAN

Two areas within Durban were selected as cases. KwaMashu is a township on the outskirts of Durban whilst Wentworth / Lamont contains a number of residential areas on the edges of Durban's primary industrial area, the Southern Basin that essentially forms an extension of the port. Both study areas could be considered marginal in terms of income and economic opportunities. Selected community groups worked with undergraduate Internet Studies students from the University of KwaZulu-Natal in developing web sites for a network of craft producers, a group of female home-based care workers (health workers that primarily provide support to HIV/AIDS victims in their homes), a Primary Health Care network (collective of clinics and Health Care workers) and a network of schools engaged in a market gardening and tree planting project in KwaMashu. The second generation of sites was developed for a Soccer team and clinic in Lamont as well as a school voluntary association and school environmental pressure group in Wentworth. The project has evolved over two years but is not yet complete. The author has been a participant-observer throughout this process and has conducted focus groups with members of the web design teams. Maintenance training is currently planned for representatives from all groups and follow-up interviews are to be conducted.

Preliminary findings indicate that the relationship between community networking and ICT manifests at a number of scales and in various dimensions. The first issue relates to the digital divide. Early studies of digital divides have tended to focus on physical and educational barriers that

Figure 1. Developing web sites with community members in KwaMashu (Nancy Odendaal, 2006)

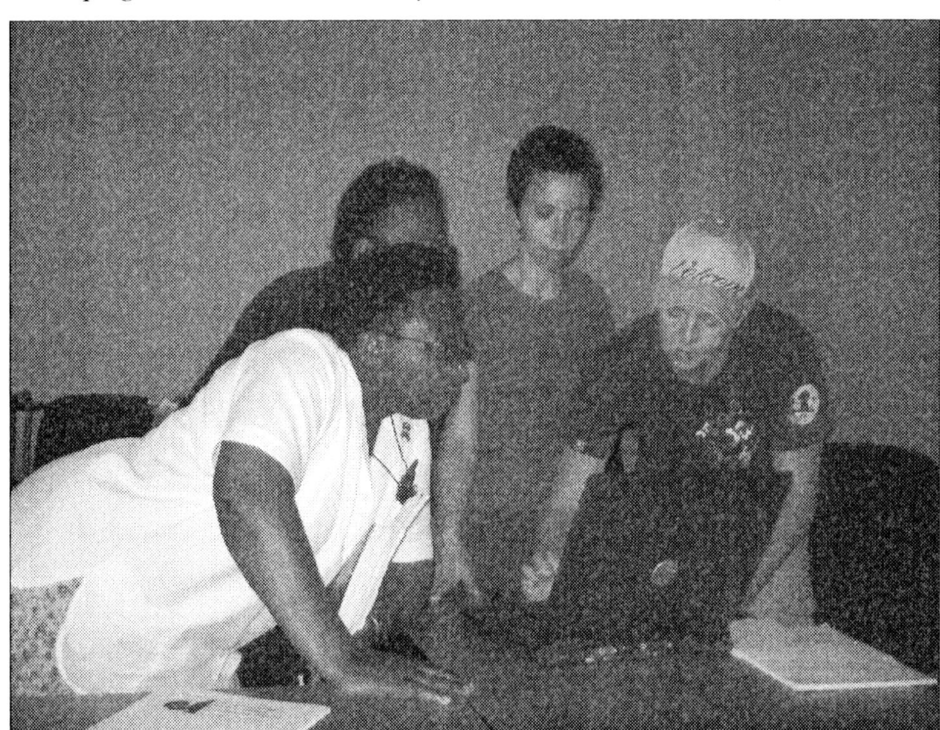

Figure 2. ICT in Lamont: phone shops and Internet cafe (Nancy Odendaal, 2007)

separate the information 'haves' from the 'have-nots'. Later work takes a more subtle approach that is critical of the determinist notion that technology is enough and capable of addressing complex social issues (Servon, 2002). Furthermore, demographic and socio-economic factors influence the choice and use of ICTs as well as how they are used in conjunction with other resources (Crang, 2006 et al; Selwyn & Facer, 2007). Differential access is influenced by perceptions of the usefulness of ICT, fears and suspicions (Bridges.org., 2002). The Durban cases reveal important dimensions of the digital divide. Barriers to ICT access are present at varying scales; in the urban, within the more immediate neighbourhood, within a group and at a more personal level. The Foucauldian notion (Philp, 1985) that knowledge equals power together with self-imposed barriers that reflect insecurities surrounding digital technologies are evident. Individuals are constrained by perceptions, definitions of what technology is, whose interests it represents and limited confidence. For example, the female members of the home-based care workers groups stubbornly refused to participate in the training necessary to maintain their web site. The only male member, significantly younger than the other constituents, was selected to participate; technology was considered young and male. The perceptions of digital technology as Western, representing progress and growth emerged as a powerful discourse. The impact on participation in the research is profound.

Nevertheless, "IT is deeply rooted in geography" (Servon, 2002: 226). Lack of investment in ICT infrastructure in poor areas represents market failure that result in inequitable access (S. Graham & Marvin, 2001). The larger context that impacts on Durban entails a deregulated telecommunications environment that yields broadband access to isolated pockets of prosperity and economic specialization. Access to ICT is constrained by limited availability of computers and dated telecommunications infrastructure. Mobile phone usage is high but expensive. Whilst infrastructural capacity is clearly limited, human capacity to adopt digital technologies is layered and complex.

Given that social networks provide the lifeblood for many South Africans (Simone, 2004; Pieterse, 2005), an emphasis was placed on community relations and network dynamics at a local level. Participation was impacted upon by local-level power relations. The craft network, dominated by one particular member, became polarized with the resulting web site reflecting the market interests of her particular group. Internal dynamics within groups became more pronounced, or different dimensions of the relations between members became apparent during the web design process. This also raises questions with regards to the nature of community networks: the actors implicated in the relations that contribute to the functioning of those networks and the precarious nature of those relations. Upon the introduction of web technology, observation reveals that the dynamic between actors could profoundly shift in the subtle interplay of power relations and bargaining.

Legitimacy issues have emerged with regards to the content of web sites. One group sought to use the site to propagate the views of a controversial (National) AIDS dissident for example. The importance of content management through collaborative design became increasingly important.

Finally, the hopes and aspirations associated with digital technologies are important also. Participants viewed the web sites as enabling them to access funding for projects, publicize their group goals and generally market themselves. Potentially the scope of these networks could stretch beyond the spatial limits of our study areas as enabled through the web sites; the potential of which became clear as participants understood the potential of digital technologies. Interestingly, participants did not raise the issue of inter-community networking as a need or see it as particularly important.

These preliminary research findings reveal the need for a lens that engages with context, recognizes the essentials of not just the urban, but the African urban as represented by Durban. Engagement with the digital relates to how resources are accessed, information exchanged and terms of access negotiated. Power relations impact, as do the nature of associational networks. Macro factors that relate to the broader urban environment are profound. Conceptual understanding needs to include context, as well as scale-and movement/relations between scales. Normative concerns around technological inequalities cannot be separated from the relational aspects of digital networking. The dynamic between these two is deeply contextual.

The rest of this chapter posits a conceptual lens that can facilitate such an examination. It is informed by literature on social capital, digital networks and recent work on African urbanity. The review of this literature is contrived to reveal the overlaps in considering these various debates in tandem.

UNDERSTANDING (SOUTH) AFRICAN URBAN SPACES: ON HUMAN AGENCY, "BELONGING" AND "BECOMING"[1]

Urban poverty is a global phenomenon. Polarization, exclusion, fragmentation and marginalization have become dominant discourses in understanding the spatial distributions and socio-economic functioning of an increasing global urban population. The image of the African city has, however, become synonymous with visions of disorder and discontent: high degrees of informality, a breakdown in infrastructure and increasing chasms between local elites and the poor. South African cities, in particular, display the legacy of Apartheid planning where segregated spaces are no longer ordered according to race, but economic class (Harrison, Huchzermeyer, & Mayekiso, 2003). Cities such as Johannesburg, Durban and Cape Town each have their charms and international profiles but nevertheless shock with their chasms of inequality. Social exclusion is a concern (Khan, 2004; Swilling, Simone, & Khan, 2002)concerns of social justice remain imperative (Watson, 2002) and spatial inequalities persist (Harrison et al., 2003; Tomlinson, Beauregard, Bremner, & Mangcu, 2003). The future of cities will be contingent upon how well social exclusion is combated. To enable this Khan (2004) suggests the poor be considered as part of the city development equation—as contributors to, and active participants in urban life, rather than passive recipients of limited benefits.

The focus on urban dwellers as co-contributors to the play of city life is a compelling idea: it is a perspective that emphasizes empowerment, dignity and creativity. Poor urban inhabitants are seen to be in possession of capital, of resources that enable them to become part of the urban equation. The emphases on associational networks and social capital have become important contributors to African urban studies. Social capital implies both action and outcome: like financial capital, "using creates more of it" (Williams, 2006: 594). In the absence of the usual money, skills and educational resources available, much is made of the nature of associational life in African cities as a means towards resource exchange. Social capital also refers to the associational networks that facilitate the exchange of resources, the norms, and rules of behavior, traditions and relationships that underpin these exchanges. In the African context of extremely poor urban conditions, high degrees of economic marginalization and high degrees of impermanence associated with cross-border migration due to war and famine, these exchanges are often elusive, dynamic, and unpredictable, manifested in informal and temporary economic and social arrangements. South African cities, such as Johannesburg in particular, but also Durban, host large immigrant communities and others reliant on informal livelihoods through engagement in the informal economic sector, or housed in informal housing. Harrison (2006) borrows from Mignolo (2000) in defining these strategies as 'subaltern', below the surface, not immediately discernable. It is dangerous to romanticize these livelihood strategies as rebellious responses to the constraints of contemporary urban living. The networks that facilitate them are not stable, often impermanent and contradictory. Putman's definition of social capital as "features of social organization such as networks, norms, and social trust that facilitate coordination and cooperation for mutual benefit" (1995: 67) implies stable terms of engagement. It is a problematic term that seen as being potentially nebulous, simplistic perhaps in its equation with other forms of capital, as well as conceptually problematic in its application across scales (Fischer, 2005; Portes, 1998). The term 'associational life' is more useful. Defined as the formal groups (church groups, community based organisations for example) as well as the less defined social groupings and familial networks that people use to contribute to livelihoods, it acknowledges impermanence and informality. Tostensten, Tvedten and Vaa (2001) outline the following characteristics:

- Groups include adherents or followers, not formal members
- Activity relies on participation, not formal enrolment
- Leaders are often self-proclaimed, not appointed or elected
- Principles of accountability and transparency are not always present
- These forms deviate from the usual Weberian principles that inform institutions.

A romantic notion of associational life as providing the lifeblood for urban Africans is problematic. Often these groups can be exclusionary in terms of gender (Ibid. p. 23), they are also often unstable and under-resourced. These "localized constellations of interests and urban practices" (Simone, 2001: 46) produce a "complex topography of intersecting social networks, which simultaneously "dissolve" into each other but also often maintain rigid operational hierarchies, norms and criteria for participation" i.e. formally agreed norms and expedient arrangements co-exist.

Approaching any form of intervention in African spaces, should therefore take cognizance of the workings of associational networks. There are no absolutes, no models, and no fixes. Considering the digital networks as potentially enabling inclusion and empowerment clearly requires a perspective that acknowledges this. This may be difficult since networks are not necessarily stable and tied to particular places. The transactional spaces created to bargain and negotiate do not necessarily correspond with the physical spaces impacted upon since they often assume permanence and a sense of 'belonging'. "African identities also display a remarkable capacity not to need fixed places". (Simone & Gotz, 2003: 125). Activities in cities may relate to survival in specific spaces, but enabling the engagement with the specifics of the local often entails the negotiation of social spaces across boundaries, markets and immediate spaces. African urban areas can therefore be linked in simultaneously different ways to "national, regional, and global markets as well as different modes of production and spatial organization." (Simone, 2004: 239). The local and the global co-exist in negotiations around surviving in particular spaces but do not necessarily lead to connection to those spaces. "Becoming" does not necessarily lead to a "belonging". Thus associational networks are not necessarily employed as 'capital' in a 'bank' of strategies intended to reinforce attachment to place; they are often fleeting exchanges of temporary alignments of interests. Despite their lack of guaranteed permanence, these social networks nevertheless provide the vehicles for information sharing, resource negotiations and support in precarious living conditions.

Globalization and market liberalization have required cities to become nodal points in international market networks and trade intentions that may be more adept at excluding, rather than including (Swilling, Simone et al. 2002). Much is made of the prominence of networks in the knowledge economy but these do not have even benefits (Castells, 2000). 'Subaltern' networks are necessary as they constitute the less permanent relationships that are reconfigured and renegotiated within a context of exclusion. Like the formal economic networks that inform city functioning, they are often global. Yet they are subject to rules of behavior that are less definable and sometimes normatively problematic, i.e. not the associations policy makers would like to encourage (Watson, 2003).

Human agency, in this context, often occurs in conditions of political crisis and instability (Simone, 2005) yet Watson's work shows the questionable alliances that inform community networks in relatively stable Cape Town (2003). The subtleties of human endeavor are often informed by more than the linearity of resource need and access. Contextual factors, histories, intentions, perceptions and the agreements that underpin associational ties contribute to a layered human intention. Culture (in the broader sense of the term, beyond ethnicity) is a dimension explored by Pieterse in his argument for a 'transgressive' urban politics that uncovers the relationship between agency and culture; "language, discourse and symbolic meanings" are extensions of agency (2005: 140). Whilst ethnic prescriptions may potentially exclude, drawing on cultural resources, argues Pieterse, potentially provide the means whereby the disciplinary boundaries imposed by the 'governmentality' framework posed by the state and its allies may be challenged. Here he uses the Foucauldian definition of the term 'governmentality' (Philp, 1985) to denote relations of power between policy and the local. Yet, the very same cultural discourses that constitute what Pieterse refers to as 'subaltern practices' can potentially discipline to create spaces of inequality and disempowerment in social networks. Agency is informed by the energies that inform the ability to act, to contribute. In cultural settings where rules and norms are not necessarily determined through consensus, variables such as gender, age and religion may lead to exclusion rather than empowerment. These limitations are not necessarily external to the actor since:

Discourses provide a lens on the world, our everyday spaces and us. ... we internalize discourses about what is appropriate to think about how to think (or believe) about the issues we should think about, and how to act in consistent ways with what we believe. All this comes to us as unquestionable truths and that is the core of the power of discourse." (Pieterse, 2005, p. 158)

Thus, Pieterse's work departs from the exclude/include dichotomy raised in earlier work on associational networks. Human agency is informed by how we internalize cultural norms and values—yet cultural practice can also emancipate.

A relational perspective provides a useful vehicle for understanding human agency in relation to larger structural issues. Maintaining multiple networks, building relational understandings across multiple scales and maintaining a sense of "belonging" whilst at the same time constantly "becoming" in multiple social networks implies a dynamic interplay between human action and urban space. The potential role for technologies that transcend the usual limitations of time and space is therefore important. The following section explores this by considering the interface between networks and digital technologies.

THE VIEW FROM EVERYWHERE[2]

The use of ICT implies a multi-layered presence where one can be present in different places at once; communicate with several others simultaneously and preserve a presence whilst engaged with other tasks. Understanding this 'view from everywhere' in conjunction with urban living comprises a literature that spans disciplinary boundaries, most notably between urban studies, sociology and information technology. There is considerable overlap where the one discipline has informed the other on the character and functioning of this interface. Essentially a practice-based body of literature, social informatics has emerged as "research that examines the design, use, and consequences of information and communication technologies in ways that take into account their interaction with institutional and cultural contexts" (Kling 2000: 217—218 cited in Pigg & Crank, 2004). In addition to that, community informatics (CI) focuses specifically on 'the community' within these institutional and cultural contexts and is particularly concerned with 'communities of place' that are implicated in the interplay between technology and society. Day (2005) contends that CI emerged as a response to network society phenomena. Defined as "the application of ICT to enable community processes and the achievement of community objectives" (Gurstein, in Day, 2005) CI is described as academic interest as well as community practice. The primary intention of CI is to enhance and sustain the social capital of local communities (Ibid.). It is, therefore, by definition, concerned with the local, and with the networks that sustain places.

Little direct investigation of the interface between African cities and ICTs exists in the literature. While this may be, to some degree, due to the fairly recent large-scale urbanization of African countries, it is nevertheless disconcerting given the strong emphasis placed on ICT for development by bi-lateral development agencies. Little comprehensive analysis has been done of how the intersection between African urban life and advanced technologies pans out. The emphasis is often on developmental objectives (Donner, 2004; Moyi, 2003; Van Belle & Trusler, 2005), where networks are assumed to be contributors to development.

A number of studies in Europe and North America examine the many dimensions of virtual capital (K.N. Hampton, 2003; Licoppe & Smoreda, 2005; Liff, 2005; Shah, Kwak, & Holbert, 2001; Weare, Loges, & Oztas, 2005; Zhao, 2006). Whether such a phenomenon exists was the first question to be answered. Initially anxiety existed on the potential for digital technologies to undermine physical networks, but research indicates that on-line and mobile activity deepens rather than undermines social networks (DiMaggio, Hargittai, Russel Neuman, & Robinson, 2001; Licoppe & Smoreda, 2005) Drawing from the Netville case in North America, Hampton (2002) argues that Internet relationships are, indeed, relationships, if only in forms unaccustomed to. They should therefore be treated as extensions of existing relationships, "not as entities in themselves as if existing social networks and existing means of communication did not exist." (p. 229). Internet use actually deepened community involvement by encouraging activity at the neighborhood level. ICT potentially increases place-based community despite the opportunities to not be place-bound and build on "interest-based" networks alone (Ibid.).

The nature of on-line networks and their sustainability are issues explored in more recent work on digital networks and virtual capital. The key to a more subtle view of technology and society is in understanding the nature of social capital whilst uncovering the dimensions of on-line and mobile communication. The nature of networks and the functions to which technologies are employed are important. Putman distinguished between 'bridging' and 'bonding' types of social capital (cited in Williams 2006). 'Bridging' capital refers to looser relationships with broader reach whereby opportunities are created for the expansion of networks and resources. 'Bonding' capital refers to the more narrow kinship networks that provide more permanent ongoing emotional support. Weak ties with broader reach are more conducive to the expansion of existing networks whilst a deepening of local networks can deepen 'belonging' and strengthen ties. The nature of on-line networks is informed by what digital technologies are used for, what functions dominate and the actual nature of those uses. Technology allows for simultaneous and broad networks across space which implies that they tend to favor a bridging function, yet that would depend on the nature of the technologies used and existing networks (Liff, 2005; Weare et al., 2005).

The looser social arrangements facilitated by the Internet are not necessarily of less value than the more intimate, stronger ties that bond. Initial opinions on the relationship between social capital and virtual networks tended to underplay the 'cross cutting nature of community' (Hampton 2003: 418), the supportive qualities of which should not be underestimated. Looser network constellations are not necessarily without purpose as shown by Hampton in his work on Netville residents' ability to unify and mobilize around a common cause. Not only does the actual reach of communication broaden but Hampton found a denser web of interaction also. Online communication tends to deepen existing bonding relations whilst increasing the breath as well as frequency of usual communications. In the Netville example, connectivity was found to be associated with more recognition of neighbours, increased communications and participation in private and public spheres (K.N Hampton & Wellman, 2003)

Bridging capital loosens, broadens and expands; depth and commitment are more typical of bonding capital but require strong social norms, 'thick' trust and shared values. Bridging capital emphasizes the looser connections largely concerned with increasing agency density and reach (Pigg and Crank 2004). There may, therefore, be a convenient 'fit' between what is emerging as a looser arrangement of social ties in African cities and the tendency for virtual capital to be more of the bridging type. Can on-line activity provide the means whereby increased agency can assist individuals in overcoming the disempowering implications of 'bonding' capital formations such as repressive kinship networks and institutional constraints? Could these network formations empower? Research on this in the African context is limited.

Seeing ICT as an enabler of empowerment assumes that there are advantages to ICT-enabled networking. Recognizing that the line between virtual and 'real' social capital is blurred—that digitally enabled communication is an extension of face-to-face contact and communication through other means, does pose the 'so-what?' question. Advantages to using new technologies may only deepen what exists, whilst uneven access may worsen inequalities. The interplay between technology and society requires more careful consideration of the nature of technologies and how they are put to use. Individuals use technologies very differently. This use is informed by purpose, context as well as education and access—not just the 'what'—but also the 'why' and 'how'. The motives for ICT use are as important as the attributes of the technology and impressions of the potentials of computers, mobile phones and the Internet. These perceptions are closely tied to other socio-economic indicators such as education, income and age (Crang, 2006).

Whatever use is determined, the Internet and e-mail are communication media that avail a wide range of options with regards to information dissemination and exchange. Different modes of communication are combined with a variety of content representations in one medium (DiMaggio et al. in Pigg and Crank 2004). Two-way communication can deepen relations whilst one-to-many communication functions could broaden information dissemination. Information functions are enhanced by the breadth of representational forms. Moyi (2003) investigates ICT use amongst informal traders in Kenya and notes that the need for financial and market information as well as credit related information amongst informal traders presupposes a high priority for ICT. Yet, his research finds that it is not as critical as factors like high costs, bureaucracy and poor infrastructure. Internet technology is not considered useful because it is not accessible due to poor access to telephones, electricity as well as computers. An interpretive approach to ICT may allow for a more nuanced understanding of social/technical relations but structural constraints remain an issue.

The many technologies available to the urban dweller, therefore, avails a range of opportunities that extend current networks. A glance at the relationship between urban spaces and ICTs reveals a picture of inequalities but also possibilities whilst the way that people make sense of these spaces may be enhanced by electronic networking. The latter is not clear however, ICT-enabled communication does not replace face-to-face contact, yet there is uncertainty on how it may contribute to associational life. Throughout the discussion there emerges the need for an understanding of urban networks that move between scales, acknowledges agency in terms of individual aspirations as well as social interaction and the structural context that constraints or encourages it. Digital technologies do not sit outside the social, the individual, and the institutional. ICT, from Internet and e-mail to mobile phones are woven into the fabric of social networking and functioning. This requires an analytical lens that allows for a nuanced understanding of urban networking where technology is considered a part of the dynamic rather than a resource or contextual element.

Our personal, social and economic life worlds are informed by improved means of communication and more immediate means of connecting. However, it is evident that technology does not sit 'outside' us. Technologies get appropriated in different ways, which in itself is informed by context and agenda. ICT inform and enrich networks (sometimes it may undermine them) but the technology itself is also part of an economic network relation: a web of ownership patterns, regulation and corporate relations. Thus, an analytical frame needs to take cognizance of the various scales of networks yet be malleable enough to allow for insights into the local.

The shaping of technology and social appropriation of ICT is a multi dimensional dynamic. Material inputs, sociological processes and psychological barriers have bearing on the relationship between the social and technological. Structural forces that determine access are important but the more intricate and reciprocal relations between machines and humans determine ongoing function and use. Furthermore, the latter ties into the network equation. Material resources and symbols contribute to constituting society; technology is part of society but also part of its social networks. ICT contributes to the broadening of the 'kaleidoscope of network formations' (Linde, Linderoth, & Raisanen, 2003). Epistemologically, the lens through which social/technical relations are understood needs to reveal the intricacies of these formations, the relational underpinnings of their functioning.

The View From Everywhere and Nowhere: The Analytical Contribution of Actor-Network Theory

The incorporation of technology into social networks may just be considered a deepening process where ICT is an additional resource factored into the equation. The review of empirically based literature above (and the Durban case) reveals a need to consider technology an actor, a contributor to networks. Furthermore, the functioning of social networks is revealed as an unstable process that can be highly volatile yet also reassuringly supportive of livelihood strategies. Considering ICT as part of these networks requires deeper insights into how these networks stabilize and evolve and how digital technologies may alter these dynamics.

Actor-network-theory (ANT) contributes through a deeper understanding of the interplay between actor, agency and structure (Latour, 2005). The incorporation of ICT as an actor in a network is useful analytically. For example, when examining mobile phone use amongst informal street traders one would include considering the mobile phones they use, the cell phone operators that enable this use and the airtime vendors. The consideration of ICT as actor opens up network possibilities and may reveal relations previously not considered.

The ANT notion of symmetry, however, is contentious in that it argues for analytical equality in the treatment of human and non-human actors. It may also be misunderstood. Networks are considered heterogeneous, comprising human and non-human actors or 'actants'. The distinction between 'actor' and 'actant' is not without consequence. Human and non-human agencies are not equivalent in terms of character and intention (Rose and Jones 2005) but deserve equivalent interrogation with regards to understanding networks that contain technological elements. The symmetry does not constitute 'equality' and does not assume similar character, argue Rose and Jones (2005), it considers non- and human agency in equal measures. Equality is measured in terms of the 'power to act' due to its position in a network (Rose and Jones 2005), as a component of a network; stabilizing and maintaining those networks. Thus, machines do not have human characteristics, but in the process of acting, in the process of network formation and maintenance, human and non-human agency become entangled in ongoing interaction.

The perceived autonomy of machines is misleading. It is the interpretative process that accompanies the perceptions that machines are autonomous as they become more advanced, more embedded in networks, and the intentions that evolve from this interpretation that matters. Intention informs action, which in turn constitutes agency.

The focus on agency is paramount. ANT however sees non-human actors as equal contributors to the agency dynamic in that they have 'transformative capacity' (hence the term 'actants'—to move away from the association of agency with humans only) (Rose & Jones, 2005). A distinction is drawn between the attributes of human and non-human agency and the contribution that agency makes to network formation:

> *....humans and machines can both be understood to demonstrate agency, in the sense of performing actions that have consequences, but the character of that agency should not be understood as equivalent. Human agents have purposes and forms of awareness that machines do not. The two kinds of agency are not separate, but intertwined, and their consequences emergent. Those consequences are also the subject of human interpretations which provide part of the context for future actions.* (Rose and Jones 2005: 27)

Rose and Jones develop a 'double dance of agency' model that consider the properties, processes and conditions that underpin networks (2005). This is useful since it reveals the way in which machines and humans act, as well as the relations that distinguish them. Machines are used as tools, proxies and automata; they are not reflexive. Yet, the awareness and interpretation of their value, the innovation that accompanies their design are human qualities. Discourses and the embedded nature of meaning through attribution of value are human endeavors. Machines contribute to this dynamic. Interactions with machines are not neutral; they are imbued with disciplinary power, discourses and aspirations.

The processes by which machines act in networks as explored by Rose and Jones (2005) go beyond the traditional functions of enabling and constraining. Machines posit a situation through which humans *may be compelled* to act. The act of network formation—network making—would most likely be informed by not only what is possible, but what is perceived to be possible by individuals. Intentions as well as aspirations are informed by the perceptions and prior experiences of technologies. The 'fit' between machine and human would also determine the outcome of such processes; "....some human intentions fit more easily than others with the design trajectories of the machines..." (Rose and Jones 2005: 30). But what informs those design trajectories? Markets, research networks, profit motives, policy—elements of structure. Outcomes are seldom informed by one type of agency, are unpredictable and multi-layered.

Personal histories and broader social structures contribute to agency and evolve (emerge) through the process of agency creation (Rose and Jones 2005: 31). The 'dance' metaphor implies action and interaction; fleeting unions of artistic expression are imprinted on the minds of the audience and carried through into the next performance. Relations perform and it is in their performance that agency is created (Cordella & Shaikh, 2006). It is the interplay, the energies that are generated, that should be considered as essential in the shaping of relations and networks: "...the movement, the dynamic interaction of actors in the circularity of the interactions between human and non-human entities, and in their 'mutual constituency' in the process..." (Cordella and Shaikh 2006: 7). Thus, the processes through which socio-networks are created are as important as the actual outcomes of these processes. The open-ended nature of these processes through which these relationships are defined and redefined is important; 'the actor is generated in and by these relationships" (Cordella and Shaikh 2006: 9). This implies that nothing can be intrinsically assumed as a priori and that actor-networks are potentially unstable. Furthermore, actors are contributors to networks, but they also emerge as outcomes of those relationships. More important (ontologically) is that the actual relationship is an actor/actant.

Ongoing maintenance of networks entails 'translation' through definition, interpretation and negotiation. This process redefines and reasserts the position and functions of the actants in the network on an ongoing basis. The emergent process redefines reality which is "achieved through the interplay between different actors, both human and non-human, with equal constitutive characteristics." (Cordella and Shaikh 2006: 14). 'Becoming' a part of a network is ongoing, 'belonging' is not a static condition but one that is continuously negotiated and redefined; 'being' refers to the continuously dynamic state that constitutes human experience. The energies created by the interaction between technology and society constitute a layer of inputs, contributions and outputs that sometimes deepen, sometimes constrain and potentially broaden human experience.

METHODOLOGICAL CONCLUSIONS: REFLECTING ON COMMUNITY WEB SITE DEVELOPMENT IN DURBAN

Urban change as mediated through ICT is experienced at many scales but not necessarily as progressive, developmental or even resourceful (although that may be the aspiration). Examining a case on its own terms; i.e. whether it provides a service or not, whether web sites are used or not, is important. It is not enough however. Successful outcomes (such as the completion of the craft network web site in KwaMashu for example) are not necessarily experienced as such by all participants (those that felt marginalized). The addition of a particular actant (the web site) facilitated the emergence of a dominant actor (one group within the craft network). Urban interaction is such that a methodological lens needs to allow for an interactive perspective whilst also allowing for the nature of network energies to be examined. The addition of ICT to the equation necessitates a consideration of non-human agency since the technologically determinist stance that technology sits 'outside' us is simply not useful.

Uneven distribution of technologies is not necessary solved through technology provision—access constraints are more subtle and linked to individual constraints / perceptions / backgrounds and network relations. ICT is embedded in structural relations of production and ownership that requires a consideration of context and institutional regulation. Simply put: urban informatics needs to be considered at varying scales and in its many dimensions. Given the intricacy of urban networks and the complex underpinning of technology ownership, uneven access and rapid development, this could be an onerous task. Structure and agency need to be considered simultaneously. Network relations matter: at metropolitan and local scales. Individual access to technology is affected by structural economic relations yet can also be challenged by mobilizing networks at a local level. Those networks are often linked. In the Wentworth case, web development and training of the environmental pressure group took place in a well equipped computer room donated by a large petroleum company situated in close proximity to the school and blamed for many of the community's environmental woes. Web developers were frustrated by lack of participation by school children and staff involved in the project. The two issues may be connected (this would need to be probed further).

The definition of what technology is-what it can do and what it should do—is meaningful in itself. It is clear that these perceptions had a profound impact on the Durban case. A gendered perspective emerged in some cases, in another it was clear that age was a factor whilst the presence of computers inhibited some. Clearly community members were not only constrained by these perceptions but this sometimes changed and evolved as the web design projects progressed. ANT contributes by assigning equality to non-human actors in network interactions. It is the nature of these interactions that are important, not the actual distinguishing characteristics of machines and humans. Those interactions are embedded with interpretations, expectations, aspirations and definitions. They are also influenced by cultural frames, self-imposed disciplinary boundaries, perceptions and normative constraints. Action is not taken at face-value, but as the outcome of various factors. Similarly, action is agency in itself in that it has consequence—it is open ended. The introduction of technology into the network formations in Durban had an impact in all cases. This does not relate to the web site only, but to the process and continued engagement with digital technology.

An epistemological definition of knowledge as that which is constituted and achieved through interaction and intersubjective exchange allows for a more nuanced understanding of urban informatics. "Becoming" is as important as "belonging" as urban dwellers make sense of their circumstances and address livelihoods. The innovation employed can often be overlooked if structure is overemphasized; on the other hand focusing only on local survival may underplay the very real contextual elements that constrain and inhibit. The layering of ICT onto this dynamic further necessitates the need for a relational perspective: the appropriation of ICT is not only locally informed but it has consequence. It is also contingent; informed by the larger networks of capital accumulation and ownership that can constrain but can also be challenged and undermined locally. The emphasis on agency is useful as a conceptual anchor that considers action as well as the roles of various actors in networks.

The ANT ontological position allows the actions to speak for themselves. The researcher does not become an actor in the network through 'putting words into mouths' of respondents but is interested in what emerges from the interplay; "... so in a sense reality becomes 'real' when actors interact." (Cordella and Shaikh 2006: 18). The Durban case shows that experience of technology does not necessarily lead to a substantive outcome (many of the web sites are not used) but it can be meaningful within itself in generating new networks. For example, a strong relationship has been established between some of the actors involved in developing the web sites and some of the community members particularly interested in ICT.

The conceptual elasticity enabled through the notions of agency and actor-networks allows for analysis at varying scales and incorporates human experience as well as institutional agency. The contextual understanding provided by literature on associational networks (particularly from an African perspective) combined with the exactitude of ANT in analyzing network interactions allows for an engagement with substance and process. Common to both is a relational understanding of Urban Informatics. The structural underpinnings of digital access are understood, whilst the interactive nature of network creation is rigorously examined. It creates the conceptual space for an engagement with the complexity, messiness and layered nature of what it is to be human…..and machine.

ACKNOWLEDGMENT

The author thanks the reviewers, and particularly the editor of this volume, for constructive comments and helpful suggestions. This research is enabled through a South African National Research Foundation Thuthuka Grant for emerging researchers. This research project would not be possible without this funding, nor would it be possible without the participation of Dr Marijke Du Toit and her Internet Studies students. The author is extremely grateful to both parties.

REFERENCES

Bridges.org. (2002). *Taking Stock and Looking Ahead: Digital Divide Assessment of the City of Cape Town*: City of Cape Town.

Castells, M. (2000). The Rise of the Network Society: The Information Age: Economy, Society and Culture (Vol. 1).

Cordella, A., & Shaikh, M. (2006). From Epistemology to Ontology: Challenging the Constructed "truth" of ANT, *Working Paper Series*. London: London School of Economics and Political Science.

Crang, M. C., T & Graham, S. . (2006). Variable Geographies of Connection: Urban Digital Divides and the Uses of Information Technology. *Urban Studies, 43*(13), 2551-2570.

Day, P. (2005). Community Research in a Knowledge Democracy: Practice, Policy and Participation, *Community Informatics Research Network Conference*. Cape Town.

DiMaggio, P., Hargittai, E., Russel Neuman, W., & Robinson, J. P. (2001). Social Implications of the Internet. *Annual Review of Sociology, 27*, 307-336.

Donner, J. (2004). Microentrepreneurs and Mobiles: An Exploration of the Uses of Mobile Phones by Small Business Owners in Rwanda. *Information Technologies and International Development, 2*(1), 1-21.

Fischer, C. S. (2005). Bowling Alone: What's the Score? (Book Review). *Social Networks, 27*, 155-167.

Graham, S. (Ed.). (2004). *The Cybercities Reader*. London: Routledge.

Graham, S., & Marvin, S. (2001). *Splintering Urbanism: Networked Infrastructures, Technological Mobilities and the Urban Condition*. London: Routledge.

Hampton, K. (2002). Place-based and IT Mediated 'Community' *Planning Theory and Practice, 3*(2).

Hampton, K. N. (2003). Grieving for a Lost Network: Collective Action in a Wired Suburb. *The Information Society, 19*, 417-428.

Hampton, K. N., & Wellman, B. (2003). Neighboring in Netville: How the Internet supports Community and Social Capital in a Wired Suburb. *City and Community, 2*(4), 277-311.

Harrison, P. (2006). On the Edge of Reason: Planning and Urban Futures in Africa. *Urban Studies, 43*(2), 319-335.

Harrison, P., Huchzermeyer, M., & Mayekiso, M. (Eds.). (2003). *Confronting Fragmentation: Housing and Urban Development in a Democratising Society*. Cape Town: UCT Press.

Khan, F. (2004). The City and its Future? The Eternal Question. *Development Update, 5*(1).

Latour, B. (2005). *Reassembling the Social: An Introduction to Actor-Network-Theory*. Oxford: Oxford University Press.

Licoppe, C., & Smoreda, Z. (2005). Are Social Networks technologically embedded? How Networks are changing today with changes in communication technology. *Social Networks, 27*, 317-335.

Liff, S. (2005). *Local Communities: Relationships between 'real' and 'virtual' Social Capital*. Paper presented at the Communities and Technologies Milano, Italy.

Linde, A., Linderoth, H. C. J., & Raisanen, C. (2003). An actor-network theory perspective on IT-projects: A battle of wills. In G. Golkuhl, M. Lind & P. J. Agerfalk (Eds.), *Action in Language, Organizations and Information Systems* (pp. 237-250). Linkoping: University of Linkoping.

Mignolo, W. (2000). *Local Histories/Global Designs: Coloniality, Subaltern Knowledges, and Border Thinking*. Princeton: Princeton University Press.

Mitchell, W. (2000). *E-topia: "Urban Life, Jim—But not as we know it"*. Cambridge: MIT Press.

Moyi, E. D. (2003). Networks, Information and Small Enterprises: New Technogies and the ambiguity of Empowerment. *Information Technology for Development, 10*, 221-232.

Philp, M. (1985). Michel Foucault. In Q. Skinner (Ed.), *The Return of Grand Theory in the Human Sciences*. Cambridge: Cambridge University Press.

Pieterse, E. (2005). At the Limits of Possibility: working notes on a Relational Model of Urban Politics. In A. Simone & A. Abouhani (Eds.), *Urban Africa: Changing Contours of Survival in the City*. London: Zed Books.

Pigg, K. E., & Crank, L. D. (2004). Building Community Social Capital: The Potentials and Promise of Information and Communication Technologies. *The Journal of Community Informatics, 1*(1), 58-73.

Portes, A. (1998). Social Capital: Its Origins and Applications in Modern Sociology. *Annual Review of Sociology, 24*, 1-24.

Putman, R. (1995). Bowling Alone: America's Declining Capital. *Journal of Democracy, 6.1*, 65-78.

Rose, J., & Jones, M. (2005). The Double Dance of Agency: A Socio-Theoretic Account of How Machines and Humans Interact. *Systems, Signs & Actions: An International Journal on Communication, Information, Technology and Work, 1*(1), 19-37.

Selwyn, N., & Facer, K. (2007). Beyond the Digital Divide: rethinking digital inclusion for the 21st century, *Opening Education* Bristol: Futurelab.

Servon, L. (2002). Four Myths about the Digital Divide. *Planning Theory and Practice, 3*(2).

Shah, D. V., Kwak, N., & Holbert, R. L. (2001). "Connecting" and "Disconnecting" with Civic Life: Patterns of Internet Use and the Production of Social Capital. *Political Communication, 18*(2), 141-162.

Simone, A. (2004). *For the City Yet to Come*. London: Duke University Press.

Simone, A. (2005). Introduction: Urban Processes and Change. In A. Simone & A. Abouhani (Eds.), *Urban Africa: Changing Contours of Survival in the City* (pp. 1-26). London: Zed Books.

Simone, A., & Gotz, G. (2003). On Belonging and Becoming in African Cities. In T. R, R. A. Beauregard, L. Bremner & X. Mangcu (Eds.), *Emerging Johannesburg: Perspectives on the Postapartheid City*. New York: Taylor and Francis.

Swilling, M., Simone, A.-M., & Khan, F. (2002). 'My Soul I can See': The Limits of Governing African Cities in a Context of Globalisation and Complexity. In S. Parnell, E. Pieterse, M. Swilling & D. Wooldridge (Eds.), *Democratising Local Government: the South African Experiment*. Cape Town: UCT Press.

Tomlinson, R., Beauregard, R. A., Bremner, L., & Mangcu, X. (Eds.). (2003). *Emerging Johannesburg: Perspectives on the Postapartheid City*. New York: Routledge.

Tostensten, A., Tvedten, I., & Vaa, M. (Eds.). (2001). *Associational Life in African Cities*. . Stockholm Nordiska Afrikainstitutet.

Van Belle, J., & Trusler, J. (2005). An Interpretivist Case Study of a South African Rural Multi-Purpose Centre. *The Journal of Community Informatics, 1*(2), 140-157.

Watson, V. (2002). The usefulness of normative planning theories in the context of sub-Saharan Africa. *Planning Theory, 1*(1), 27-52.

Watson, V. (2003). Conflicting Rationalities: Implications for Planning Theory and Ethics. *Planning Theory and Practice, 4*(4), 395-408.

Weare, C., Loges, W. E., & Oztas, N. (2005). Does the Internet enhance the Capacity of Community Associations? . In P. Van Den Besselaar, G. De Michelis, J. Preece & C. Simone (Eds.), *Communities and Technologies 2005*. Dordrecht: Springer.

Williams, D. (2006). On and Off the 'Net: Scales for Social Capital in an Online Era. *Journal of Computer-Mediated Communication, 11*, 593-628.

Zhao. (2006). Do Internet Users have more Social Ties? A call for DIfferentiated Analyses of Internet Use. *Journal of Computer-Mediated Communication, 11*, 844-862.

KEY TERMS

Actants: A term used in Actor-Network Theory to describe human and non-human actors engaged in an ongoing network relationship.

Associational Life: Term that describes the conditions that define a context where individuals rely on social networks and kinship relationships to survive and access resources. The term has become popular recently in urban studies, particularly with regards to African cities.

Digital Networks: Social networks enabled through the use of digital technologies such as mobile phones, the Internet and e-mail.

Digital Communities: Communities of interest or place that rely on digital technologies such as mobile phones, the Internet and e-mail to communicate, network and disseminate information.

Discourse: Arguments, opinions and statements that are represented as facts ('truths') supported by definitions, theories and contentions that are part of a particular discipline. This term was developed by the social theorist Michel Foucault and is often used to provide a deeper understanding of the power relations that often underpin representation of knowledge and the imposition thereof.

Livelihoods: Used in Development Studies and Development Sociology, livelihoods refer to the survival strategies as well as the human, natural, social and financial capital people employ to function and survive.

Social Capital: Assets used by individuals in the absence of financial and monetary capital to assist in accessing resources. Examples include familial relationships, social networks and clubs/societies.

Townships: Predominantly residential areas on the outskirts of South African cities designed during the Apartheid era to accommodate 'non-White' residents. These areas were racially segregated and enabled through the notorious Group Areas Act (abolished in South African in the early 1990s), legislation intended to reinforce spatial segregation through the designation of racially exclusive areas.

ENDNOTES

1. The notions of "becoming" and "belonging', developed by Simone and Gotz (2003) are used in this work to encapsulate the identity formation that takes place when urban space is negotiated.
2. The term is taken from Foth, M., Odendaal, N., & Hearn, G. (2007, Oct 15-16). The View from Everywhere: Towards an Epistemology for Urbanites. In D. Remenyi (Ed.), *Proceedings 4th International Conference on Intellectual Capital, Knowledge Management and Organisational Learning (ICICKM)* (pp. 127-133). Cape Town, South Africa.

Section II
Participation and Deliberation

Chapter IV
Place Making Through Participatory Planning

Wayne Beyea
Michigan State University, USA

Christine Geith
Michigan State University, USA

Charles McKeown
Michigan State University, USA

ABSTRACT

Community planning is facing many challenges around the world, such as the rapid growth of megacities as well as urban sprawl. The State of Michigan in the United States is attempting to re-invent itself through place making by using participatory planning supported by new information tools, models and online training. The Michigan State University Land Policy Institute framework for place making includes Picture Michigan Tomorrow, an informatics initiative to democratize data and incorporate it into scenario planning methodologies and tools, and Citizen Planner, an on-ground and online training program for local planning officials. Still in the early phases of implementation, these initiatives provide promising models for use in other regions of the world that seek consensus among citizens, developers and government on the vision and plan for their communities.

INTRODUCTION

Community planning is the art of envisioning and implementing a common goal for a community's future physical development. More than simply regulating land-use, planning is a process that gives communities a forum to build a common vision, a process for engagement and a way to build consensus about land-use policy issues. The community planning process varies globally depending on legal frameworks, local traditions and priorities.

In the United States (U.S.), the planning process reflects a strong tradition of individual rights and local government, and nowhere more so than in the State of Michigan, where strong land-use planning rights are granted to the smallest units of government. Michigan has a total of 1,815 local units of government with planning authority. This presents a unique challenge relating to educating local officials, providing planning and informatics tools, and elevating the level of knowledge so that wise decisions are made. In addition, Michigan, once a global industrial giant as the home of the auto industry, has been impacted negatively by globalization more than any other state in the U.S..

The Land Policy Institute (LPI) at Michigan State University (MSU) is developing and implementing tools, methods of engagement and a framework for moving the state forward. Reliance on traditional tools will not be sufficient to meet the challenges in Michigan. A new strategy of engagement and building understanding has been developed. This strategy includes in-depth training, information structures, and technology tools so communities have a fundamental understanding of their critical assets in order to enable wise land-use decisions.

In this chapter, we describe a strategic framework that is being implemented in Michigan for engaging communities in addressing the state's challenge. This chapter presents a review of urban planning challenges and how those challenges manifest in Michigan. It also provides an overview of the importance of participatory planning. The chapter features the strategies that MSU's LPI has developed in response to Michigan's needs. These include the Picture Michigan Tomorrow data democratization and community engagement project and the Michigan Citizen Planner program that provides on-ground and online training to citizens in planning roles. The solutions being developed in Michigan, can be informative to other regions of the world that are addressing similar issues of population and land-use change.

BACKGROUND

Urban Planning

The growing population of the world continues to gravitate towards population centers to meet basic human needs such as housing, employment, food and water. Demographers with the United Nations have noted "the world is in the midst of a massive urban transition" (Hill, Wolfson & Targ, 2004). According to statistics prepared by the United Nations Population Division and Environment programs:

In 1975, just over one-third of the world's population lived in urban areas. Currently, [n]early half of the world's population (47 percent) lives in urban areas, a figure which is expected to grow by 2 percent per year during 2000 - 15. By the year 2025, the United Nations estimates that almost two-thirds of the world's population will live in densely populated metropolitan areas. Much of this growth will continue to occur in megacities like Tokyo, Japan; Cairo, Egypt; Sao Paulo, Brazil; Mexico City, Mexico; Bombay, India; and Buenos Aires, Argentina. (Hill, Wolfson & Targ, 2004)

The growth of megacities will have profound impacts on the quality of life on vast numbers of the world's people. Methods for responding to growth through city planning vary considerably among nations around the world. Much of the variation is deeply rooted in each country's value system. The system imposed by each nation provides unique opportunities and constraints for addressing human settlement planning concerns within each nation. In Europe, for example, there is a strong tradition of top-down land-use planning in many countries. However, recent trends have been to transfer some of this authority to local municipalities. In Brazil, statutes emphasize a citizens' right to sustainable cities and the participation of municipalities and citizens in development decisions. In China, the law sets targets and tasks for improving the environment through urban planning. However, in reality there is not yet a functional national law to address land-use concerns (Nolan, 2006).

U.S. Context

While the rest of the world is in the midst of an urban transition, the U.S. has been growing in population under a decentralization scheme. Fueled by cheaper land in the rural areas, easy access to interstate highways with private automobiles, and comparatively low gas prices, most urban population centers, particularly in the northeastern U.S., have experienced urban out-migration.

A study by the Connecticut-based Center for Environment and Population (CEP) paints a troubling picture of how decentralization and out-migration from U.S. cities impacts the environment (Hearn, 2006). According to the CEP study, "each American currently occupies 20 percent more land for housing, school, shopping, and other uses than the average American did two decades ago." (Markham, 2006, p.5). In addition, some experts put the average American's "ecological footprint," the amount of land and water needed to support an individual and absorb his or her waste, at 24 acres (Knickerbocker, 2006). "By that calculation, the long-term 'carrying capacity' of the U.S. would sustain less than half of the nation's current population" (Hearn, 2006).

Many Americans have become dissatisfied with the pattern of development taking place in their communities. Cities and regions have come to look the same no matter where you travel in the U.S. with national franchise operations increasingly dominating the landscape. Yet, Americans yearn for places that are unique and they spend considerable time and money trying to locate, and travel to, these places (Bunnell, 2002). Planning officials daily face citizens who oppose the onslaught of new development in their community and lack an understanding of what alternatives exist.

Noted city planning author Kevin Lynch illustrated in his landmark book on the subject, *The Image of the City* (1960), why some places evoke a more positive response than others. Through studying social behavior in urban places and talking with citizens about what makes a place special, certain place-making motivations have emerged since Lynch began his important research. The special qualities that citizens consistently identify as making a place unique are:

Local identity and a sense of place; escape from the "suburban paradigm" (the ability to come and go without using the automobile); a strong, vital, densely developed, pedestrian-oriented downtown with a mixture of uses and activities that makes it the focal point of the community; good neighborhoods and well-kept older housing adjacent to, and surrounding, the downtown; gathering places (so-called "third places") where people can interact with one another and have the sense of being part of a community; and environmental, scenic and open space resources integrated within the city and/or close at hand (Bunnell, 2002, p. 34).

In addition, nearly five decades after the publication of *The Death and Life of Great American Cities* by Jane Jacobs *(1961)*, the evils of post-World War II planning approaches that Jacobs criticized (particularly the seperation of uses into residential, commercial, industrial, etc.) are as prevelant today in city planning as they were when Jacobs first wrote her historic book. The principles of dense, mixed uses and unique places advocated by Jacobs are at the forefront of the place making city planning movement today.

While some states have begun to experiment with Smart Growth strategies no national system of land-use planning exists. States such as Florida and Oregon have had laws on the books for several decades to encourage land-use planning but not until recently have more states begun to require planning official training or guidelines for growth strategies. Most implementation of these strategies is done at the municipal level within the U.S.. Individual property rights, deeply rooted in the U.S. Constitution, prevail as the over-arching value and legal mechanism that fosters a fragmented approach to city planning.

Planning is largely done in the city and town halls throughout the U.S.. It is one of the most contentious democratic processes involving citizens and government officials. The concept of "home rule," which emphasis local autonomy regarding land-use decisions, dominates discussions in the U.S. regarding how and where human settlements should develop. This land-use decision-making reality in the U.S. creates several challenges towards achieving sustainable human settlements. According to Schultink, Memon and Thomas (2005, p. 36), land-use decision making in the U.S. typically "fails to recognize comparative advantages of land-use options, including land productivity, transportation cost and environmental impacts, occurs in isolation of the decision-making process by contiguous units of government, incurs conflicts in cross-boundary designation of land-use plans, zoning ordinances, taxation policies and watershed management, and is not addressed in long-term state planning by identifying regional and sector-specific economic growth preferences or consideration of their aggregate benefits, costs or environmental consequences."

The isolation of local decision making is perhaps the greatest short-coming of the U.S. system of City Planning. One of the negative consequences of this is sprawl. The sprawl model of city planning has prevailed within the U.S. over the past 50 years since World War II. Generally speaking, sprawl can be defined as development dependant on private modes of transportation, miles from shopping and employment centers, and segregated rather than integrated land uses by zones (Schultink, Memon & Thomas, 2005).

While the sprawl model of development is primarily an American experience, many regions of the world are experiencing unsustainable levels of growth in new towns and existing population centers. The critical need for new human settlements to house the poor, and a growing population in general, have overshadowed sustainability concerns regarding the long-term consequences of poor place-making attributes and livable places. The number of land-use conflicts will continue to grow with the urban transition taking place. Yet, few national or local participatory planning techniques exist to resolve these issues, particularly within the U.S..

The cumulative impact of these factors highlight the problem of unchecked human settlement patterns and their impact on the natural environment, not only within the U.S. but around the world. The process of making places special and livable has been a lost art within the U.S. system of city planning and is beginning to be repeated by some growing mega cities around the world.

Participatory Approaches to Planning

Most failed planning efforts are due to the fact that members of the community: (1) do not have access to usable information that will inform them of the land-use trends in their communities; (2) do not have good tools to support their learning about the issues; and (3) are dispersed and fragmented—groups of citizens in different communities likely face similar problems and have similar needs but it is currently difficult for them to share.

Participatory planning techniques are important tools for achieving success. As defined here, participatory planning involves the systematic effort to envision a community's desired future, plan for that future, and involve and harness the specific competencies and inputs of community residents, leaders, and stakeholders in the process. Participatory planning attempts to use a common goal for a community to bring together stakeholders in a constructive, consensus-building process to develop a plan to meet that goal. Four critical elements of the participatory planning process are: (1) a common goal; (2) involved stakeholders with the desire to work toward the goal; (3) common democratized information to inform the process; and (4) a means of efficient productive engagement.

Participatory planning requires mutual understanding between citizens, planning officials, developers and government policymakers (Figure 1). The intersection of these perspectives forms a common ground for not only policy formation, but also for a common vision of a community with a pathway to achieving goals. Engaging citizens and empowering them with the 'whys' and 'hows' of influencing the planning in their communities is a complex problem. This also can present significant challenges when visions, priorities and cultures are in conflict (Godschalk, 2004).

Too often, one of the four critical elements of participatory planning is lacking and the planning process is short circuited and "left to the professionals." The reasons for this are numerous: lack of a well articulated goal and a clearly communicated welcome to the public; lack of information and critical knowledge that pushes certain groups away from the table; and the fact that managing a process where all views are invited can be a daunting logistical challenge that can stall progress. Technology offers solutions to some of these barriers that were not available as recently as a decade ago. Today, information can flow quickly, people can gather in large numbers and exchange ideas, and communication can be more current and relevant than ever before. According to Dhavan et. al (2005), "...online media complement traditional media to foster political discussion and civic messaging. These two forms of political expression, in turn, influence civic participation." Bringing these technologies and processes to bear at the local level has taken some time, but momentum is building.

Figure 1. Participatory planning

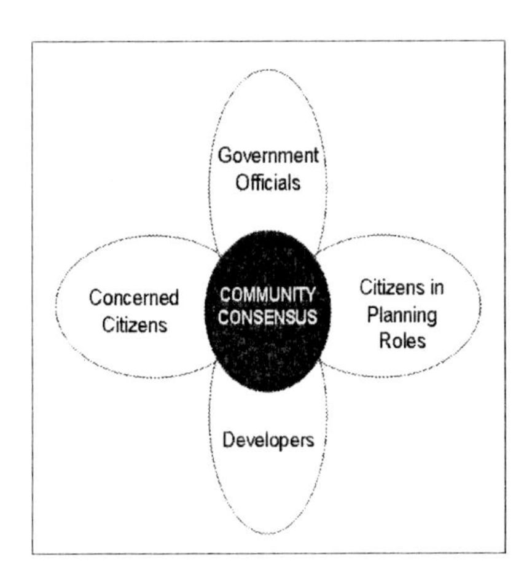

There are well-established tools for participatory planning. Historically they have revolved around public meetings and the traditional media. Today's technology expands the scope as well as the depth of participation. A review of today's tools can be found in Hanzl (2007). These tools include web-based information clearinghouses, visual preference surveys, and scenario planning using simulations. Initiatives in Michigan include web-based tools that allow for maximum public input and can also reach respondents who don't normally attend public meetings.

MAIN FOCUS

Michigan Solutions

Michigan is in a transition phase economically. The state economy has been dominated by traditional manufacturing and economic development strategies for the better part of a century. The last two decades have created a substantial migration of manufacturing to overseas economies. Michigan has been characterized as a "single state recession" (Kaza, 2006). Michigan's recent economic issues related to the decline of manufacturing due to globalization (Glazer & Grimes, 2004), have provided a rare pause in development and sprawl pressure; time that is being seized to understand how policy choices at the state and local levels affect Michigan's future and how to plan for both prosperity and sustainability.

One way that the MSU LPI is responding to the needs of Michigan is through its Picture Michigan Tomorrow (PMT) initiative and its training programs for elected and appointed planning officials, called the Citizen Planner program. Both PMT and Citizen Planner share the common goal of creating community genesis (consensus on a vision), as illustrated in the LPI Model for Strategic Engagement (Figure 2).

PMT democratizes information; making it available to all citizens, by aggregating data and creating understandable, visual models. PMT provides additional means of engagement for community planning, by using PMT data in participatory planning such as online surveys and scenario planning. The Citizen Planner program involves formal and informal stakeholders in the planning process by training citizens to be effective in planning roles.

The Picture Michigan Tomorrow Project

The Picture Michigan Tomorrow (PMT) project (http://www.PMT.MSU.edu/), is building a programmatic framework to enable stakeholders in three ways: it provides access to a common comprehensive database on change, and the drivers of it, in Michigan; it provides technology tools to help build community consensus; and it uses a unique land-use modeling framework to project land-use into the future as well as provide for scenario planning. All of these elements are focused on helping communities understand how they relate to the larger region, state and nation. The core objective of Picture Michigan Tomorrow is to develop land-use forecasts using computer modeling, then, more importantly, to translate those forecasts into things people care about.

Picture Michigan Tomorrow's core vision has been to serve as a tool for local and state policy makers' understanding of the impact of land-use decisions. In contrast to the traditional planning process the PMT framework allows Michigan communities to inventory, identify, and understand the key infrastructure that will lead them to prosperity and sustainability, and build community vision out of fragmented voices. This is accomplished by actively educating the community on the planning process, increasing the base level on knowledge among community members, actively engaging the general public

Figure 2. LPI model for strategic engagement in the planning process

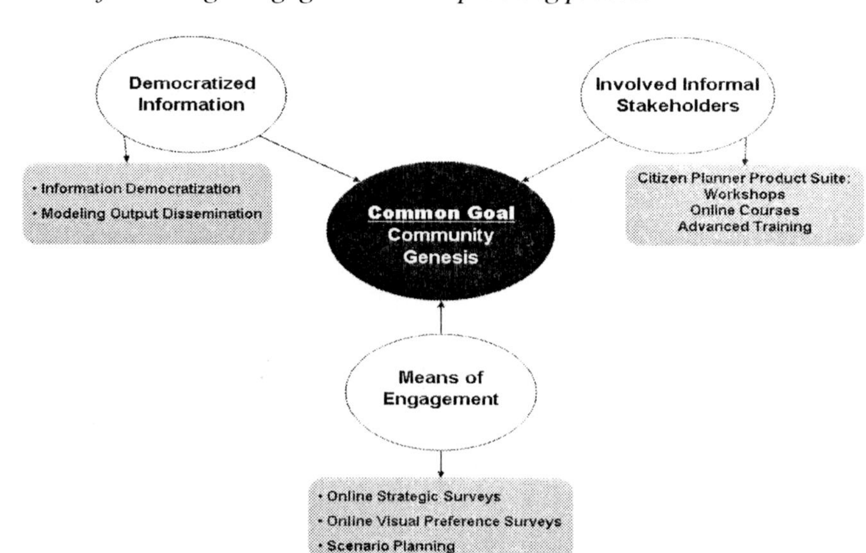

in the planning process, and integrating all of this information to build community consensus. The key building blocks for the PMT initiative are: (a) informatics and data democratization; (b) quantitative modeling and scenario planning; and (c) digital tools for communities.

PMT Informatics and Data Democratization

Figure 3 represents the LPI and PMT vision for informatics as a tool in decision making articulated by Zeleny (1987). The first level of the pyramid is data. It is the foundation of good decision making but its uses are limited. Raw data is usually collected for a singular purpose, or in answer to a limited question, truncating its usefulness in broader integrative processes. In addition, there are copious amounts of data available, making it confusing and difficult to understand. To move from data to information, it is necessary to first collect data into a common repository, filter it down to the critical pieces, standardize the framework, and then democratize it. The last is the most critical component. In the planning sphere, data is a tool for a common vision and as such it must be available freely and at a low transaction cost to all the stakeholders in the process.

Once these first steps have been taken, the result is no longer data but information which encompasses facts and concepts. Here is the point where data can be put to work. To move further, information must be processed through research and analysis, and the results of these must be clearly articulated.

After information undergoes its analytical transformation, it can be referred to as knowledge. This is the critical point where something is understood. Again, knowledge is only useful when it is widely disseminated and understood. Therefore, to move forward, it is necessary to communicate effectively with stakeholders which make education a critical component. Wisdom is arrived at when these preceding steps are taken and there is a common ground of knowledge for groups and individuals to use. At this stage, good decision making is the critical output.

PMT Quantitative Analysis and Scenario planning

PMT is currently building a spatially-explicit econometric model of land-use change in Michi-

Figure 3. Informatics pyramid and processes necessary to build good decision making

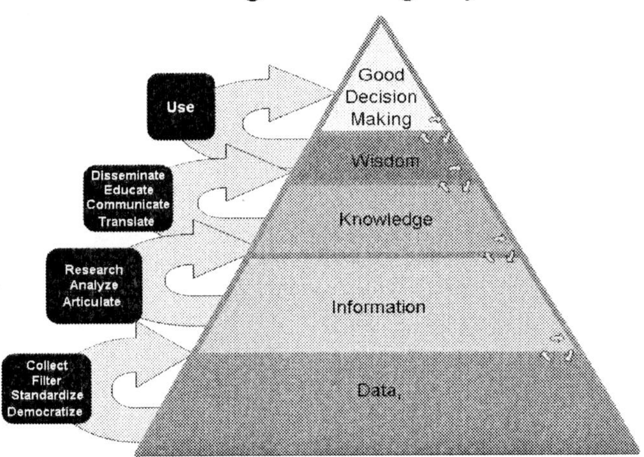

gan. This type of effort is not new. For example, '1000 Friends of Florida' recently completed the Florida 2060 population and land-use modeling effort (Zwick, 2006). A review of land-use modeling is available by Agarwal et. al. (2002). The structure of Michigan, and its prevailing data issues, required PMT to approach this effort from scratch. PMT is building a robust, scalable, econometric model of future land consumption in Michigan. This model will incorporate the history of land-use change, the primary drivers of land-use change in Michigan, and new aerial imagery to forecast future land-use patterns. The outputs will be both spatial and statistical forecasts, scaleable to the municipal level. Those forecasts will then be used to develop detailed reports on the impacts of land-use change to federal, state, and local decision makers. This model is the core engine of the scenario planning process planned for PMT. Communities will be able to evaluate different development and growth scenarios as well as the effects of policy on land-use.

PMT Digital Tools for Communities

Expanding participation in planning requires the use of technologies such as online strategic surveys. Strategic surveys are designed to assess how a community wants to grow. Typical assessments include targeted questions on how a community should grow, where and in what direction growth should be channeled, and how a community views its strengths and assets. It can also be used to effectively assess critical public knowledge gaps, growth priorities and the impact of environmental concerns. A well designed, web-deployed, and marketed survey can provide a wealth of information directly related to community planning. In addition to information gathering, strategic surveys serve as a handshake with the public, inviting their opinions, logging their input, and welcoming them into the planning process. Online strategic surveys, while not new, provide an input mechanism for community members who normally would not be part of the process. See Al-Kodomay (2001) for an early application and its evaluation.

The Citizen Planner Program

To foster engagement of formal and informal stakeholders in the planning process, the MSU Extension division developed the Citizen Planner program in 2002 (http://citizenplanner.MSU.edu/). Simply put, "planning may be considered the conscious organization of human activities to serve human needs" (Saarinen, 1976). As a dynamic process there are many players and stakeholders involved in land-use decision-making including developers, bankers, realtors, architects, engineers, surveyors, planner, lawyers, real estate investors and economic developers. Citizens take the form of neighborhood associations, downtown business owners, and numerous non-profit groups (environmental councils, community development groups, etc.).

In the U.S. and Michigan, local officials are charged with making important local land-use decisions. They are either elected in to their positions by their local community or they are appointed by the current leader of the region, such as a town mayor. Appointed planning officials are residents of the community who give guidance to the land-use, zoning, and planning process in the community. They are appointed by elected city officials for specified terms of office (typically three years) depending on the States' Planning Enabling Legislation.

While planning officials are usually volunteer positions, it is common for planning officials to be paid a small meeting per diem to attend city planning commission meetings. Planning officials often meet at least once per month as a full commission (typically seven or more members) to conduct their business. Primary duties include reviewing and recommending, or approving, development proposals, developing future land-use plans, and recommending zoning changes (e.g. change of use from residential to commercial). Officials have varied backgrounds and personalities, but they are expected to act for the overall good of the community. The appointed planning officials often make major land decisions in conjunction with the local elected officials.

Designed to improve planning decisions and prevent financial and environmental errors, the Citizen planner program provides resources and knowledge to elected and appointed planning officials, community members and local government officials in Michigan. The program provides a basic understanding of the legal framework of planning and zoning, technical planning knowledge, and leadership skills to build a volunteer core to advance good land-use planning and to foster land-use education within communities. To date, the program has reached approximately 20% of the planning officials in Michigan using on-ground and online delivery.

Before designing the online version of Citizen Planner, which launched in 2007, MSU's MSUglobal in conjunction with MSU Extension and LPI, conducted a feasibility study. The 2004 feasibility study included a state-wide mail survey and focus group research targeted to the state's planning officials (Cullen et al., 2006). The survey was designed to discover issues that impact the adoption of online learning such as technology-related issues, cultural context, and metacognitive skills (Brace-Govan & Gabbott, 2004; Clark & Mayer, 2003; Thurmond & Wambach, 2004). The survey asked about computer experience, technology perceptions and technology access, level of prior knowledge, sources and processes used to learn new things, engagement and interaction preferences, and current learning context. Focus groups served to provide additional information about survey responses and to probe more deeply into learning style preferences.

Findings determined that the average age of a planning official in Michigan was 55 and that two-thirds were age 50 or above (Cullen et al., 2006, p. 7). Over half of respondents had an associate's degree or higher (p. 7). Among respondents, only 5% reported no access to a computer or to the Internet (mobile wireless devices were not included in the survey) (p. 7). Respondents were most interested in (1) opportunities to obtain useful answers to urgent questions; (2) opportunities to solve real-world problems; and (3) books, references and peers for double-checking their understanding of the field (p. 11). Of greatest interest, was obtaining information updates from reliable sources about pertinent developments in the field, followed by a resource library with general information. In terms of training, respondents were interested in training that results in helping them do their job better and to prevent or reduce lawsuits by making more informed decisions (p. 11).

Focus group results reinforced that elected and appointed officials in planning roles wanted practical, hands-on information that would help them make decisions. They were not looking to become planning experts. They wanted practical methods that have been used successfully in other communities. The context in which planning officials work requires timely, information-intensive decision making. A focus group participant, for example, described the typical context in which planners need information to make decisions: "A lot of times they'll give you a packet right before the meeting and want a quick decision. They want you to rush a decision that could affect us 20 years down the road" (Cullen et al., 2006, p. 15).

More than 71 percent of survey respondents felt training should be a requirement for serving as a planning official (p. 15). Though most research participants did not have any experience with learning online, they indicated a strong willingness to try it. They also wanted training in a format that provided them with time and place flexibility.

Using the survey data, Cullen suggested a demand function for participation in land use-related education and training driven by education, perceived benefit, and length in service (Cullen, 2005):

Planning officials with graduate education are more willing to participate in land use-related training than those with lower levels of education (high school, some college, associates degree, undergraduate degree, and some graduate courses). Similarly, those planning officials with higher levels of perceived benefit of participating in land use-related training are more willing to participate than those with lesser levels of perceived benefit. Interestingly, the longer a planning official serves the less interested he/she is in participating in land use-related education or training. (p. 34)

These findings suggested focusing training efforts on new planning officials. This is consistent with findings from a national survey conducted by the American Planning Association that found 61% of training programs are designed for new planning officials (Chandler, 2000).

Based on the feasibility research, MSUglobal designed a suite of information and training programs for citizen land-use decision makers (http://cponline.MSU.edu/). Self-paced online training is organized in 30-60 minute modules. Topics include the 7 core module topics ranging from planning official roles and responsibilities to zoning tools and techniques. A growing number of speciality topics include risk management and farmland protection. Modules can be taken self-study or used in combination with face-to-face training at regional MSU Extension sites. The comprehensive 7-module program is the 40-hour Master Citizen Planner program that includes an exam and MSU non-credit designation which can be completed entirely online, on-ground, or in combinations convenient to learners.

The online Citizen Planner program is also used as a reference in the online Smart Growth Readiness Assessment Tool (http://cponline.MSU.edu/sgrat/index.php?sgrat) that scores how well a community is prepared to develop according to Smart Growth principles, such as walkable communities and compact building designs. In addition to expanding online, the Citizen Planner program expanded its face-to-face offerings to include annual conferences and a Master Citizen Planner Leadership Academy.

Evidence of Impact

Evidence is beginning to surface on how the combination of these innovative face-to-face and electronic decision-making tools are providing Michigan communities with a new means for engagement and a way to build consensus about land-use policy and global competitiveness. One example is the City of Fremont—a small community to the north of Grand Rapids, Michigan. Planning officials that participated in the Citizen Planner course used the online Smart Growth Readiness Assessment Tool to address regional place-making concerns as a means to enhance their global competitiveness. The Mayor of Fremont, a graduate of Citizen Planner, was the driving force behind the creation of a Joint Planning Commission, a multiple community authority with full planning powers for Fremont and two of its surrounding townships, Dayton Township and Sheridan Charter Township.

Fremont was among the first cities in Michigan to achieve this milestone. The Mayor is also helping his community adopt Form-Based Zoning, an innovative zoning code that regulates the function of land use and allows diverse architectural development that can serve multiple and changing functions. "We achieved our goals through collaboration and cooperation. The benefits for the future of our region are well worth the effort," said the Mayor of Fremont. Since the founding of the Joint Planning Commission in fall, 2006, the region has already experienced positive progress in the social, economic, and environmental arenas.

The experience in Fremont has created a model for joint planning commissions across the state that will result in regional place making. The process undertook by the City of Fremont has resulted in consolidated resources and has led the way to comprehensive planning at a regional scale; a vital step as Michigan moves toward building its niche in the global market.

CONCLUSION

The State of Michigan in the U.S. is using place making to re-invent itself in the wake of globalization. Technology is enabling a new breadth and depth of participatory planning in communities that are engaging citizens, government officials and developers.

The LPI Model of Strategic Engagement for place making identifies democratized information, means of engagement, and engaged informal and formal stakeholders as the primary components to achieving a common community goal for land use. MSU's PMT initiative is implementing the LPI Model by providing informatics and data democratization, quantitative modeling and scenario planning, and digital tools. MSU's Citizen Planner product suite is implementing the LPI Model by engaging citizens in learning experiences designed to meet their diverse needs for knowledge as well as their requirements for flexibility in the place and time in which they choose to learn.

Enabled by information technology through PMT and Citizen Planner, the LPI Model for place making through participatory planning is in the early stages of implementation in Michigan. Early results indicate that the combination of these tools is enabling effective community planning. Further research will identify their impact. The approaches being developed in Michigan can provide useful models for others to enable communities to envision and implement a common goal for future physical development.

ACKNOWLEDGMENT

The authors would like to thank the W.K. Kellogg Foundation, the MSU Land Policy Institute, MSU Extension and MSUglobal for their generous support.

REFERENCES

Agarwal, C., Green, G. M., Grove, J. M., Evans, T. P. & Schweik, C. M. (2002). A review and assessment of land-use change models: dynamics of space, time, and human choice. Gen. Tech. Rep. NE-297. Newtown Square, PA: U.S.. Department of Agriculture, Forest Service, Northeastern Research Station.

Al-Kodmany, K. (2001). Online tools for public participation. *Government Information Quarterly*, 18 (4), 329-341.

Brace-Govan, J., & Gabbott, M. (2004). General practitioners and online continuing professional education: projected understandings. *Educational Technology & Society*, 7 (1), 51-62.

Bunnell, G. (2002). *Making places special.* Chicago, Illinois: APA Planners Press.

Chandler, M. (2000). Training programs for citizen planners. American Planning Association. Retrieved November 12, 2007, from http://www.planning.org/thecommissioner/19952003/spring00.htm?project=Print

Clark, R. & Mayer, R.E. (2003). *e-Learning and the science of instruction.* San Francisco, California: Pfeiffer.

Cullen, A. E. (2005). The determinants of participation in land use-related education and training: A case study of the state of Michigan. Unpublished master's thesis, Michigan State University, East Lansing. Retrieved January 24, 2008, from http://agecon.lib.umn.edu/cgi-bin/pdf_view.pl?paperid=17432&ftype=.pdf

Cullen, A., Norris, P., Beyea, W. R., Geith, C. & Rhead, G. (2006). *Expanding education and training opportunities for Michigan local government land-use planning officials* (Michigan Agricultural Experiment Station Research Report 574). East Lansing, Michigan: Michigan State University. Retrieved January 24, 2008, from http://web2.MSUe.MSU.edu/bulletins/Bulletin/PDF/RR574.pdf

Glazer, L. & Grimes, D. (2004). *A new path to prosperity? Manufacturing and knowledge-based industries as drivers of economic growth*, (Working Group Report). Lieutenant Governor's Commission on Higher Education and Economic Growth. Lansing, Michigan: State of Michigan Governor's Office. Retrieved January 24, 2008, from http://www.cherrycommission.org/docs/Resources/Economic_Benefits/NewPathToProsperity.pdf

Godschalk, D. R. (2004). Land-use planning challenges: Coping with conflicts in visions of sustainable development and livable communities. *Journal of the American Planning Association*, Vol. 70.

Hanzl, M. (May, 2007). Information technology as a tool for public participation in urban planning: a review of experiments and potentials. *Design Studies*, 28 (3), 289-307.

Hearn, K. (2006). 300 million Americans will take great environmental toll, report warns, *National Geographic News*, Retrieved October 16, 2006, http://news.nationalgeographic.com/news/2006/10/061016-population.html

Hill, B. E., Wolfson, S. & Targ, N. (2004). Human rights and the environment: A synopsis and some predictions, 16 Geo. Int'l Envtl. L. Rev. 359, 363

Jacob, J. (1961). *The death and life of great american cities*. New York: Random House.

Kaza, G. (2006, October 18). Michigan's Single State Recession. *National Review Online*. Retrieved January 21, 2008, from http://article.nationalreview.com/?q=MzRhYzc2ZmY4NzBhMzZjYzA0NDhhM2I1NmE1YjBlZGQ=

Knickerbocker, B. (2006). The environmental weight of 300 million Americans. *Christian Science Monitor*, Retrieved October 7, 2006, from http://www.alternet.org/story/42511/

Lynch, K. (1960). *The image of the city*. Cambridge, Massachusetts: M.I.T. Press.

Markham, V. (2006). U.S. national report on population and the environment. Center for Environment and Population. New Canaan, Connecticut. Retrieved January 21, 2008, from http://www.cepnet.org/documents/USNatlReptFinal.pdf

Nolan, J. (2006). Comparative Land Use Law: Patterns of Sustainability, 37 Urb. Law. 807, 820

Saarinen, T. F. (1976). *Environmental planning: perception and behavior*. Boston, Massachusetts: Houghton Mifflin Company.

Schultink, G., Memon, A. & Thomas, M. (2005). *Land-use planning and growth management: Comparative policy perspectives on urban sprawl and future open space preservation* (Michigan State University Agricultural Experiment Station Research Report 587). East Lansing, Michigan: Michigan State University.

Shah, D. V., Cho, J., Eveland Jr, W. P. & Kwak, N. (2005). Information and expression in a digital age modeling Internet effects on civic participation. *Communication Research*, School of Journalism and Mass Communication at the University of Wisconsin-Madison, 32 (5), 531-565.

Thurmond, V.A., & Wambach, K. (2004). Understanding interactions in distance education: A

review of the literature. *Journal of Instructional Technology and Distance Learning*, 1 (1), 9-33.

Zeleny, M. (1987). Management support systems: towards integrated knowledge management. *Human Systems Management*, 7 (1), 59–70.

Zwick, P. D. & Carr, M.H. (2006). *Florida 2060: A population distribution scenario for the state of Florida.* Research Report for 1000 Friends of Florida. Gainsville, Florida: Florida Geoplan Center. Retrieved January 21, 2008, from http://1000friendsofflorida.org/PUBS/2060/Florida-2060-Report-Final.pdf

KEY TERMS

Informatics: Informatics is the discipline of science which investigates the structure and properties (not specific content) of scientific information, as well as the regularities of scientific information activity, its theory, history, methodology and organization. Mikhailov, A.I., Chernyl, A.I., and Gilyarevskii, R.S. (1966) "Informatika – novoe nazvanie teorii naučnoj informacii." *Naučno tehničeskaja informacija*, 12, pp. 35–39.

Place Making: The process of designing the physical attributes of a community that make it pleasurable and unique.

Participatory Planning: Participatory planning as defined here involves the systematic effort to envision a community's desired future and planning for that future, while involving and harnessing the specific competencies and input of community residents, leaders, and stakeholders in the process. (Author Definition).

Planning Officials: The local elected or appointed citizen responsible for making planning and zoning decisions. This term is interchangeable with planning commissioner, zoning board member, zoning board of appeals member, council member, or township or county board member as used in this chapter. (Author Definition).

Scenario Planning: A process of forecasting multiple future outcomes to simulate the impact of changes in scenarios. Scenario planning avoids the dangers of single point forecasts by allowing users to explore the implications of several alternative futures. By surfacing, challenging and altering beliefs, planners are able to test their assumptions in a non-threatening environment.

Smart Growth: Generally refers to compact, transit accessible, pedestrian-oriented, mixed-use development patterns. In contrast to prevalent development practices, Smart growth refocuses a larger share of regional growth within central cities, urbanized areas, inner suburbs, and areas that are already served by infrastructure. (see American Planning Association Smart growth Policy 2002 at www.planning.org/policyguides/smartgrowth.htm)

Sprawl: A low-density land use pattern that is automobile dependent, energy and land consumptive, and requires a very high ratio of road surface to development served (Michigan Society of Planning, Patterns on the Land, Trends Future Project, final report, 1995).

Chapter V
TexTales:
Creating Interactive Forums with Urban Publics

Mike Ananny
Stanford University, USA

Carol Strohecker
University of North Carolina, USA

ABSTRACT

In this paper, we describe the design and installation of a new kind of public opinion forum—TexTales, a public, large-scale interactive projection screen—to demonstrate how public city spaces can become sites for collective expression and public opinions can be considered social constructions. Each TexTales installation involved different groups of European young people taking photographs of everyday city events and controversial public issues, and then using custom software to invite general public passers-by in urban spaces to annotate the photos with SMS text messages. We analyze the design and implementation of these installations and identify a number of interaction design elements critical for designing expressive urban spaces: starting "intermodal" conversations; authoring for nomadic, unfamiliar audiences; distributing public discourse across mediated and physical space; and editing and censoring dialog to ensure that it reflects the norms and values of forum designers. TexTales is essentially an experiment in understanding how city spaces can be more than venues in which to take public opinion snapshots; instead they might be places that nurture and reveal collaborative, public expression.

INTRODUCTION

Public forums are spaces where individual perspectives come together to reflect and shape political discourse. Designers of such forums become facilitators whose products can help or hinder different voices, constrain or afford certain kinds of discourse and, ultimately, help people to examine and develop their own opinions and the thinking that gives rise to them. We consider the roles of community members—particularly young people—as co-designers, as citizens who express their views on issues of public concern and as learners who become aware of their own ways of forming opinions.

The question of how to discern peoples' public opinions and civic attitudes has long been a topic of research. Downs (1956) argues that individuals are "rationally ignorant" of current affairs and policy options because they think it is unlikely that their perspectives will influence large-scale civic issues. Converse (1970) suggests that most people have "non-attitudes" and questions opinion polls' ability to identify well-formed thoughts, arguing that people usually offer "top-of-the-head" answers to pollsters' questions to avoid appearing ignorant.

These conditions, if true, would be antithetical to democratic life. Several political scientists and technologists are researching ways to counteract such potential deficiencies. Fishkin *et al.'s* Deliberative Polls argue that deliberation among a random sample of voters can "produce better-reasoned preferences grounded in evidence about the complexities of controversial public issues" (Fishkin *et al.*, 2000, p. 665). In essence, people who better understand difficult issues will give less arbitrary and more reasoned answers to poll questions. Wyatt *et al.* (2000) focus on understanding political deliberations that already occur in everyday conversation. After examining how freely and how often Americans engaged in casual political conversations in common spaces, they proposed a conversational model of democracy, arguing that "informal conversation among people who largely agree with each other plays a more vital role in democratic processes than is usually recognized" (Wyatt *et al.*, 2000, p. 72).

Different models of public opinion underly these approaches. Schoenbach and Becker (1995) review various writers' definitions of public opinion: Habermas (1962) considers it as "public reasoning by those who have the intellectual capabilities to arrive at socially useful beliefs and attitudes and to discuss them publicly" (Schoenbach & Becker, 1995, p. 324). This emphasis on the processes by which people arrive at public opinions is consistent with our view of opinion-forming as a development in thinking and therefore a kind of learning. However we question the presumptions about intellectual abilities and social utilities. Aside from being difficult to enact, identifying and excluding those deemed not to have appropriate intellectual capabilities would raise serious questions about hegemony. Requiring citizens to pass standardized tests that evaluate their intellectual capabilities before admitting them to public forums runs counter to an inclusive and participatory model of democracy (Barber, 1984). Further, those advocating preliminary screening misunderstand the nature of democratic forums: participating in such forums *supports* individual development, serving "educative functions" vital to the construction of an informed and active citizenry (Mansbridge, 1999; Pateman, 1970).

De Sola Pool (1973) sees public opinion as the "opinion held by a majority of citizens," invoking a simplistic model of majority-rule democracy that does not adequately account for the role of dissent in the public exchanges. In contrast, Price (1992) considers public opinion as the result of a kind of *collective epistemology* that helps us to consider our own viewpoints and those of our fellow citizens. In this model, both as individuals and as members of collective forums, we separate judgment from fact but may not explicitly resolve differences between them. Price characterizes public opinion as a pragmatic process of dialogue in which individuals come together to form "issue publics." They do not have to adopt any majority opinion; they simply have to agree about what should be done.

Noelle-Neumann (1984) takes an opposing view, characterizing public opinion as a process of *social conformity* that reflects our identifications and allegiances. Whether we choose to express ourselves in public reflects who we think the majority is, where we think public consensus is heading and who we imagine our community to be. In her "spiral of silence" model, Noelle-Neumann defines public opinion not as a dialogue among competing points of view nor even the negotiation of consensus; rather, public opinion is the result of fear of social isolation.

In the *collective epistemology* model, public opinion is a moment in a process of discussion that is ongoing and pluralistic; in the *social conformity* model, public opinion is a mechanism for achieving cohesion and uniformity. Common among these models, though, is the notion that public opinion emerges when citizens produce public deliberation by engaging in the discursive practices of specifically designed forums that let represent both individual and collective views on public issues (Delli Carpini, et al., 2004).

Our interest in generating systems and cultures for public communication is rooted in better understanding how people develop and express these opinions. In our view, public opinions are best considered as epistemological processes—processes of thinking, of learning and potentially of development—that rely on discursive processes at personal and collective scales. By representing ideas so others can appreciate them, and by joining others in extending the representations, people can express and develop perspectives on issues of public concern.

We are interested in how accessible electronic media make possible new, potentially better ways for people to represent, share and develop their perspectives. In particular, our aim in this work is to understand how such representation, sharing and development can best be supported in *urban environments*. While much research on the design and evaluation of technologies for deliberation focuses on assembling groups in online, screen-based forums for discussing political or electoral issues (e.g., Capella et al., 2002; Min, 2007; Price & Capella, 2002; Price, 2006; Iyengar et al., 2003) such work does not tend to see public opinions as more general phenomena that develop in a variety of cultural contexts. Traditional political science research neglects the *situated* nature of public opinions and the potential for new media support and reveal the kind of cultural expressions that occur in contemporary urban spaces.

Our work is more in the tradition of an emerging literature describes how technologies can mediate between the rituals and actions of everyday life, and the affordances and aesthetics of urban space (Manovich, 2006; Silverstone & Sujon, 2005). More specifically, we focus on how large-scale, architecturally integrated interactive projection screens (Auerbach, 2006; Lester, 2006; Slaatta, 2006; Struppek, 2006) might help collaborating individuals comment on and within particular urban locations (Harris & Lane, 2007; Social Tapestries, 2006; West, 2005), creating new kinds of public "media cities" (McQuire, 2006) that use new technologies to help redesign cities (Hanzl, 2007). This focus on the culture and geography of public opinion helps us see public opinions not as snapshot samples of static venues but as *emerging from interactions between people and their built* environments.

Here we describe our experiences designing and implementing urban public opinion forums with four different groups of people, each using mobile phones and a custom piece of software called *TexTales*. We describe the co-design of each media installation and the kinds of interactions that emerged during each forum. We also discuss implications for further participatory design of public opinion forms and forums.

TEXTALES

TexTales is a large-scale interactive projection designed to support multimodal dialogue among many participants in public places.

A TexTales display (Figure 1) consists of a grid of nine photographs, with three captions under each image. Passers-by create the captions by sending short messaging service (SMS) text messages from their mobile phones. Contributors compose captions by deciding what picture to augment and entering its number along with the text. Moments after sending the message, the display refreshes and shows the new caption. It appears at the bottom of the set of three, bumping the oldest from the top as the other two lines scroll upwards. This dynamic effect engages additional passers-by, who quickly understand themselves to be co-creators of the display. As participants continue to initiate captions and respond to some already there, people linger. Some of their discussion plays out among the crowd and some on the display, for all to see. Meanwhile the TexTales system stores and indexes all texts by time, date, phone number from which the text was sent and display number of the associated image.

TEXTALES INSTALLATIONS

During a six-month period we created four different forums with four different groups for four different audiences. The design groups consisted of researchers, photographers, and/or community leaders working with local teenagers and adults to act as "authors" and create the displays. The "audiences" were people of varied ages and backgrounds who found their way into a nearby public area where an installation was taking place and participated by entering the discussions and texting. Members of the design groups generally included themselves among the audiences. The settings and themes of the forums were varied:

Figure 1. Left: The TexTales interface, showing a grid of nine photos with the three most recent captions scrolling underneath. New captions appear in the largest type at the bottom of each image; the screen refreshes approximately every 30 seconds, displaying any new captions that may have arrived. Right: a Dubliner sending an SMS text to be a photo caption.

- **Fatima Mansions:** Dublin, Ireland: addressing fears of eviction and other tensions related to an urban renewal project within a low-income apartment complex (Figure 2);
- **The Big Smoke:** Dublin, Ireland: debating a ban on smoking in pubs, with young people and their parents, via an installation situated in a public square in the city centre;
- **Smokum:** Amsterdam, The Netherlands: examining attitudes on passive and teenage smoking with a group of Dutch young people in a prominent train station;
- **cText:** Kilkeel, Northern Ireland: considering issues of community identity with a group of young people in a mixed Catholic-Protestant community in Northern Ireland during a divisive election campaign.

The installations differed in content, authors and audience but we employed a general process for designing each:

- Establish a collaboration with a group interested in creating an installation in their neighborhood and give an initial demo of the *TexTales* interface.
- Work with a group of citizens, artists and community leaders (*e.g.* photojournalists, youth workers) to decide the installation's focus, setting and audience.
- During several weeks' collaboration, create images and texts for the installation and plan logistics.
- Advertise the installation and present it in a public venue, encouraging broad participation.
- After the installation, reconvene to reflect upon results and plan future engagements or improvements.

We next present each TexTales installation in turn, describing its physical situation, design collaborators, motivating issue of public concern, public participants and the corpus of opinions that resulted.

Fatima Mansions, Dublin

Our first installation was in an urban apartment complex ironically dubbed Fatima Mansions, then undergoing major refurbishment. The complex houses approximately 700 residents in 14 four-story buildings built by the Irish government in 1951 and subsidized to this day. The entire complex was slated for demolition, to be replaced with new living centres. The community was eager to discuss what kind of new social and physical spaces they would like and to experiment with garnering and representing public opinion in this regard. We collaborated with a women's

Figure 2. Children and young adults of Fatima Mansions, Dublin, playing with and around TexTales

history group interested in creating archives for the complex and a local photographer interested in creating visual histories. During the course of six months, we worked with community members to take and edit more than 700 images and to design and present 10 *TexTales* screens on the outdoor wall of the community centre. We ran the installation during three nights and received approximately 150 SMS text captions from the local community, including many from children and teenagers.

The Big Smoke: Temple Bar, Dublin

Our second installation was also in Dublin but was situated instead in a high-traffic public square in the city's centre. We collaborated with a photographer who specializes in small-scale image juxtaposition, a children's arts centre and 12 young people with their parents. This installation differed from Fatima Mansion's in that we asked collaborators to focus on a specific theme: environmental tobacco smoke (or second-hand smoke) and effects of "passive smoking." The Irish government had announced that, beginning January 2004, smoking would be banned in work places (including pubs and restaurants). The announcement sparked debate about the health effects of passive smoking, the rights of pub owners and, more generally, relationships between personal actions and public health. We saw an opportunity to design a *TexTales* installation on a timely and provocative issue. We asked participants to capture images and write captions that could start conversations about the ban and second-hand smoking (Figure 4).

During the course of four weeks, we worked with participants to create and edit more than 130 images and to design and project two *TexTales* displays (18 images) in the public square. We ran the installation during two nights and received approximately 190 texts from passers-by.

Figure 4. One of two TexTales screens created with a mixed-generation group, artist Michael Durand and Jim Ronan of The Ark, an urban cultural centre for children. The installation appeared for two nights in Dublin's Meeting House Square, a large and centrally located high-traffic public area. ("The Big Smoke" is an old nickname for Dublin.)

Smokum: Station Leylaad, Amsterdam

Our third installation, also on the theme of second-hand smoking, was situated in an Amsterdam train station. We collaborated with a Computer Clubhouse (Resnick *et al.*, 1998) whose coordinator was an actively exhibiting visual artist; one of the mentors was studying visual sociology.

Originally we asked participants to focus on the issue of second-hand smoking, but they interpreted the theme more closely to their local context and instead took images and wrote starter captions that focused on smoking among teenagers. During the course of four weeks, we worked with the organizers across the distance between Amsterdam and Dublin as they photographed, selected and arranged dozens of images. One of us (Ananny) then traveled to Amsterdam to help with final selection of images and installation of two *TexTales* displays (18 images) in Lelylaad train station (Figure 5). We ran the installation during two nights and received approximately 50 texts from passers-by. This installation provided the grist for the discourse analysis described below.

cText: Kilkeel, Northern Ireland

Our most recent installation was in Kilkeel, Northern Ireland, a fishing and farming community on the Irish Sea with a mixed Catholic-Protestant (approximately 60-40% split) population of approximately 5,500 people. The town has a long history of sectarian tensions and our local collaborators were eager to focus on a "mixed-community" project, knowing that sectarian issues might surface for discussion. We collaborated with a youth group that focuses on joint Catholic-Protestant youth activities, the British Broadcasting Corporation (BBC) under its project to engage young people in journalism, and an arts group that organizes a yearly Northern Ireland youth festival and was interested in conducting projects outside of Belfast. We worked with nine participants aged 16-19 years during three weeks to take and edit dozens of images, creating seven *TexTales* screens projected as part of the town's Christmas festival. We ran the installation for one night, receiving approximately 50 texts from the local community.

Figure 5. One of two TexTales screens created with a group of Dutch young people in collaboration with the Amsterdam Computer Clubhouse. The installation appeared for two nights in an Amsterdam train station.

INTERACTION DESIGN ANALYSIS

Each site dealt with four design elements: starting "intermodal" conversations; authoring for nomadic, unfamiliar audiences; interplaying public and private messages; and framing, editing and censoring dialogue.

Intermodal Conversation Starting

We intended *TexTales* to support informal conversation about issues of public concern among individuals who might not normally have a reason or opportunity to talk with each other. "Conversation" thus had two senses: discussions occurring through SMS captioning and represented on the display, and those occurring among participants viewing and participating in the installation—each of which fed the other.

A critical aspect of this goal is to create "starting points"—the images that appear initially on the TexTales display, each with just a single caption to begin. The array of image-text combinations should help to ground and urge conversations and tend to encourage broader discussion beyond the displayed image and text (Figures 6 and 7). In preparing all four installations, we asked the design groups to consider what questions they might ask of passers-by through images, texts, a combination of images and texts or, more subtly, juxtapositions of different image-text combinations within *TexTales'* projected grid.

The goal was to create image-text combinations that were evocative enough to elicit participation from the general public and that would encourage and sustain conversation both via *TexTales* and in the physical, social space surrounding the projection.

Framing the interaction design as an exercise in creating conversation starters meant that we could think carefully about how people manage turn-taking, how they might distribute conversations between the projected displays and the surrounding area, how they create coherence

Figure 6. An example of intermodal conversation starting: three generations of Kilkeel farmers peering over a cattle fence. Captions include "will there be a next generation of farmers?" and "i really hope so". (Note: in the projection, captions scroll right-to-left beneath the images and are wholly readable.)

Figure 7. An example of representing public opinions on controversial topics: the popular unionist slogan rejecting unification with the Republic of Ireland "Kilkeel says no!"

between the virtual and physical places, how they change conversational topics and how the intermodal forms might address an installation's intended theme.

Strohecker and Ananny (2003) discuss the need for systems that support compound representations in complementary media forms, describing how people develop "intermodal literacies" as they create and "read" combined textual and visual representations using systems such as *TexTales*. In creating conversation starters, *TexTales* authors consider how to "distribute" meanings across text and image, and what elements of an image to address in the starter text. Additionally, as the captions scroll upwards, starter texts eventually disappear, making the static image the original author's only persistent contribution (unless he or she stays to moderate the conversation in person).

Over time, it is unclear how much influence the starter text created by the original designer will have (though our analyses of the log files reveal some threads of meaning in given sequences of captions). If an image-text designer only wants to launch a discussion, she may focus on an evocative initial image-text combination; if she wants prolonged discussion, she needs to create an image that can stand on its own, to ground discussion throughout an installation's lifetime.

Authoring: Nomadic and Unfamiliar Audiences

Participants created images and texts to pose specific questions or make provocative comments—but, because the texts were anonymous and passers-by could leave easily, we were never quite sure who in the installation's physical space was reading or writing texts. Indeed, for "The Big Smoke," several people participated from beyond the immediate area, sending messages to the installation from nearby pubs and while riding buses. Since we changed image sets periodically to seed new conversations and people were texting continuously, people not present in the space had no way of ensuring that their captions were coherent with an image or with previous texts. And since they could not see their captions appear, remote participants had no feedback and had to imagine their text on the screen, what image it might be under, what other captions it might be with and who might be reading it. However, through the live dialog in the space, people sometimes discovered who had sent a text. These discoveries sometimes influenced responses and contributed to an emerging sense of community among participants.

TexTales highlights several challenges associated with designing large-scale interactive urban projection screens in which the general public creates messages for audiences that are diverse, constantly changing and composed of an unknown number of people.

Specifically, it is difficult to anticipate and design for the many directions an interactive public conversation may take. Since *TexTales* limits the

Figure 8. An abstract image from the "Big Smoke" installation on a particularly cold night in Dublin. Captions include "getting hot out herrrrrre!", "i wish i could warm my hands on that!" and "or i could get a half bottle of wine!" (Note: in the projection, captions scroll right-to-left beneath the images and are wholly readable.)

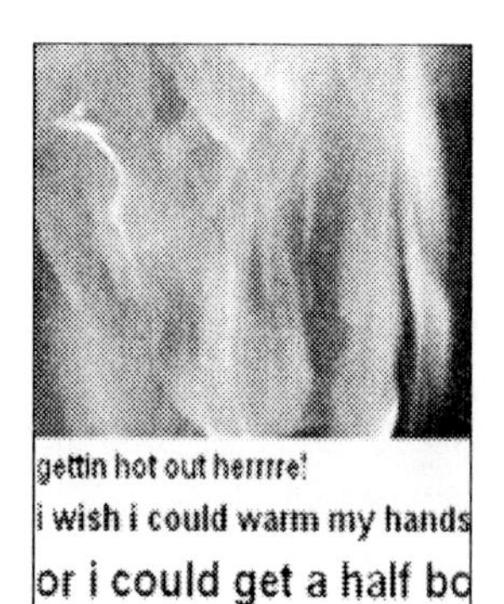

number of lines of text under each picture, texts written by the person who took the picture and created the initial caption—the person who "seeded" that picture's conversation—will disappear when a picture receives its fourth text message; the picture will be the original designer's only persistent contribution. Despite a designer's best efforts to "ground" a conversation in a particular image or text, as the general public participates the conversation may deviate significantly from the designer's intentions. Designers of future similar systems may want to create mechanisms for "re-seeding" the original conversation, branching the public conversation into multiple sub-conversations, or limiting a display's duration or number of comments.

It is also difficult for general public participants to know who is viewing the installation at any particular time. Even in small crowds it is unlikely that passers-by will know everyone at the installation, making it challenging for participants to know whether they creating content for people still in the vicinity or for people who may have long since left. *TexTales* shows no time, date or geographic information on the text messages it displays, making it difficult for people to know how recent the captions are or from where they may have been sent. Designers of similar public urban interactive displays that depend on content from the general public may want to consider how to show who created what content, when and from what location. Such information may help passers-by better understand a public conversation's context and participants.

Participation exists not only in the display but also in the social and physical context surrounding the display. For example, *TexTales* participants were frequently distributing their conversations across both text messages and interpersonal conversations—writing captions for pictures and then elaborating on those captions by talking with those around them. Similarly, because captions were being received continuously, there was no way to anticipate whether your caption would suddenly be juxtaposed to one that referenced a completely different conversation among a different group of people in the same space. Passers-by searching for coherence *only* in the texts displayed at a given moment might be confused since they were not privy to the multiple, concurrent conversations happening among participants in the space. In essence, the content of *TexTales* "conversations"—indeed, perhaps of all discourses among physically situated individuals who create content for shared displays—are distributed in both media and places. Understanding the practices and meanings of such an installation means appreciating two distinct but intersecting kinds of discourse—talk *in* the display and talk *around* the display.

Experimenting: Public and Private Messages

The challenges associated with conversation starting and intermodal authoring for nomadic audiences are closely related to the senses of privacy and publication that are relevant to *TexTales* installations. In one sense, *TexTales* preserves the privacy and anonymity of the installation designers and participants: images are neither "owned" nor moderated by any particular person; no log-in or registration is required to participate; all texts are displayed anonymously; and, since only the three most recent captions are displayed for a given image, captions tend to have an ephemeral quality as their presence in the discursive space gives way to memories of interpreted meanings. The anonymous, open forum and the reliance on personal mobile phones as input devices are intended to lower barriers to participation and encourage casual contributions by as many people as possible (*cf.* Brignull & Rogers, 2003).

Nevertheless, sources of *TexTales* contributions can become known. Some messages were controversial or provocative enough that those around the installation immediately asked everyone assembled who had sent the message and if he or she would elaborate. In this sense, although the projection preserves the texter's privacy and anonymity, the verbal conversations around the projection can make individual opinions public and attributed. This knowledge can, of course, can come to influence the display.

A result of this arrangement is that traces of "public opinion"—ideas in formation—do not just appear in the *TexTales* display. Instead, they are distributed between the displayed text-image combinations and the physical conversations that may occur the installation area or even long after the participants leave the forum.

Another kind of public-private phenomenon were conversations that took place either wholly or partly on the projected screen and that departed from the theme of the forum to broach personal connections among participants. There were numerous instances of people sending personal messages to *TexTales* (sometimes with full names, phone numbers and requests for romantic meetings). These were presumably intended for an audience of only a few people but they appeared in a public form—a kind of reverse, discourse-based voyeurism.

Figure 9. An example of a personal, anonymous—and touching—caption sent to a stark image of a boy with extreme asthma during "The Big Smoke" installation: "MY FATHER DIED ON THE 12 OF APRIL HE SMOKFD [sic]. I STILL DO.IT WORRIES ME.IM HIS SON." An hour and 20 minutes passed before anyone sent another caption to this image. (Note: in the projection, captions scroll right-to-left beneath the images and are wholly readable.)

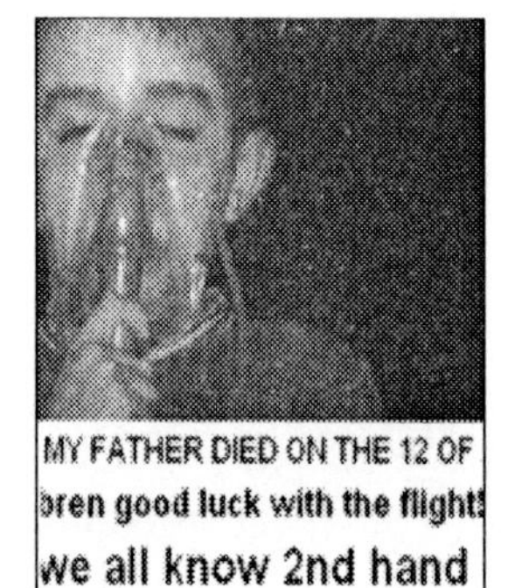

Framing, Editing and Censoring Dialogue

An issue that arose consistently was how to censor or edit the installations to ensure "appropriate" content. There were three aspects to this process: the initial framing of the installation's theme, the real-time moderation of the installation through different editing techniques, and the censorship guidelines and technical architectures designed to "pre-moderate" the installations.

Framing

With the Fatima Mansions group, we framed the installation's theme broadly, asking questions like "show me something you love," "show me something you'd like someone standing here 100 years from now to see," "show me something about your community you'd like to change—or *not* change." Since issues of urban change and reconstruction were foremost in nearly all community discussions, we assumed that they would arise regardless of our framing and that explicit prompting for opinions on urban renewal might make the installation a local cliché.

Similarly, we framed the Kilkeel installation broadly, asking the young participants to describe a day in their life. Like Fatima's focus on urban renewal, issues of Catholic-Protestant sectarian tension are endemic to Kilkeel, with residents almost constantly reflecting or acting upon sectarian issues. Many local leaders want to experiment with new ways of supporting cross-community dialogue and building secular identities.

At the suggestion of our local collaborators, we framed the installation as a way to help issues of religious, nationalist and community identity surface and develop without explicitly asking for opinions on sectarianism. As one local youth worker put it, we "came at sectarianism sideways," addressing it as it arose. Further influencing the decision to let sectarian issues emerge naturally was Northern Ireland's 2003 election. Campaigning was occurring in Kilkeel throughout our preparations, with the final installation coming three days after the election.

In contrast, the "Big Smoke" and "Smokum" installations focused on second-hand smoking, an issue that concerned participants but with which they were not continually engaged. Furthermore they were not part of a close geographic community. In preparing the installations, the collaborators debated far more as they struggled to define the boundaries of the topic and what visuals would capture their interpretations and evoke captioning. Especially since the smoking issue has clear "pro" and "con" sides, we attempted to present a balanced view.

Thus, through the four installations different framing strategies emerged. When the installations were situated within established communities, framing relied on local collaborators, their senses of community issues and what kind of everyday imagery would spark discussion of common issues. When the installations focused on a specific issue and the participants were not members of a geographic community, framing relied on conversations about how to depict issues in balanced ways, what kind of media would support equitable discussion and how to ensure that we were designing for rich and diverse dialog rather than inadvertently seeding a biased debate.

Editing

A consistent question in preparing all four installations was how to ensure that the images and texts would be "appropriate" for public projection. We had more control over the pictures than the texts, as people who cared about the project gathered the images over many weeks and could discuss and re-shoot to achieve different effects or meanings. Indeed, we cropped, lightened, darkened or resized many images to help them communicate particular meanings or be visually effective within the *TexTales* interface.

Ensuring appropriate texts was more difficult. For the Fatima Mansions installation, we agreed that any person whose image appeared in a projection had to give informed consent, understanding that the text messages were not to be filtered or edited before being displayed. Thus, once people had consented to appearing in the projection, they agreed to have their likeness juxtaposed to any text message. We had no clear plan of what to do if someone sent an offensive message and, especially given the nature of the community and the prevalence of casual profanity in everyday speech, we expected controversial captions.

The potentially tense situation was simplified through the insight of a young Fatima resident when his friends texted a playful caption that included his name and a mild joke. Onlookers agreed that the text was a harmless, friendly tease—but in rapid succession, he sent the three captions "`no`", "`i'm`" and "`not`" to the image, re-

Figure 10. The picture that revealed the three-line editing technique (or "bumping"). A Fatima texter sent three texts in rapid succession to remove a teasing message and replace the caption area for this image

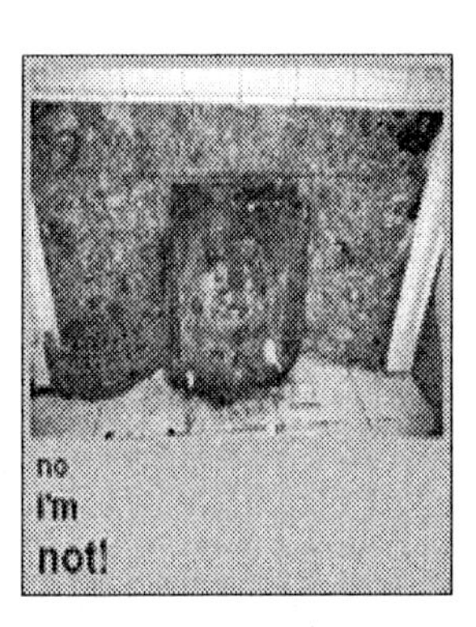

moving the tease and replacing all captions with his own (Figure 10). Thus he enacted a kind of *in situ* editing as he worked with the short, three-line form to compose captions that had nothing to do with the photo's content but took the interface's conversational floor.

A similar but more serious incident occurred during the Fatima installation when a young participant texted "`niggers out`" to the display, prompting community members to send—but only much later in the installation—innocuous messages like "`hi how are you?`," "`good`" and "`i'm fine`" to replace the offensive graffiti (for further discussion see Ananny, Biddick & Strohecker, 2003; Ananny, Strohecker & Biddick, 2004).

Both of these instances became exemplars for how to edit *TexTales* in real-time. It became known as "bumping" messages or the "three-line editing" technique. In planning subsequent installations we referred to these examples, saying that, while we could not censor captions, if participants found any texts to be offensive they could delete the texts from the display by sending three captions in rapid succession.

The critical feature of this editing technique is that it is equitable: *anyone* can be the installation's real-time censor. The editorial control is not the sole responsibility of the designers but is instead distributed among the participants. (It could be argued that we as designers of the original *TexTales* system always have the ultimate responsibility since each installation relied on the core technology we developed. We feel this is a fair criticism and only underscores the need for citizens to design their own public opinion systems with their own technological materials—an option that is becoming increasingly viable.)

In the next installation, "The Big Smoke," we relied on this real-time editing technique to allay the concerns of our principal collaborator: a children's arts organization uncomfortable with co-sponsoring any installation that might have defamatory or offensive material, especially about children. The three-line editing strategy invented at Fatima Mansions initially worked well. (Our collaborators periodically replaced offensive texts with ones they felt were more appropriate). In one instance, however, the "bumping" technique failed completely.

Throughout "The Big Smoke," one group of young people stayed toward the back of the public square, mostly socializing and listening to music but periodically texting captions to the installation. One of these people sent an overtly offensive caption that included the phrase "`kiddie porn`," perhaps testing whether the content was moderated in any way. The text appeared intact and three of us (Ananny, a local collaborating artist and a representative of the children's arts organization) attempted to use the three-line editing technique to "bump" this text from the display.

We began texting but, on the screen's next refresh, the same offensive message appeared under two other images; each of us quickly divided the "bumping" responsibilities and began applying the three-line technique to a particular image. However, on the next refresh the offensive caption appeared on all three lines of all nine images—effectively spamming the projection. (Log files of the installation indicate that the caption spam came from the same phone number and that the 27 spams were sent within about 1 second of each other.)

Our "bumping" technique had failed. In frustration, one of our collaborators texted "`go`" "`away`" "`go`"; another one of us began talking to the spammers, asking them not to send such messages; and, after consultation with the collaborators, another of us (Ananny) suspended the projection for two minutes to delete the offensive texts from the display. The spammers left the area shortly after their texts had been removed.

The incident had little impact on the installation's success but it did show us that the three-line editing technique had limitations. We also learned that, especially when considering the nature of inter-organizational collaborations and public projections, there were limits to the kinds of messages we as designers felt comfortable having associated with the installations.

Censoring

The fallibility of the three-line editing technique, coupled with the sensitive nature of public debates in Northern Ireland and the installation's sponsorship by the British Broadcasting Corporation (BBC) brought several editorial challenges to the next installation in Northern Ireland.

First, "The Big Smoke" experience showed that we could not ensure editorial control using the "bumping" technique alone. Second, sectarian graffiti (some of which is intended to incite hatred against identifiable groups) is common in Kilkeel and we could not ignore the reality that the installation brought serious security concerns. Designing an installation that supported or encouraged violent or offensive messages in any way was antithetical to all individuals involved in the project. Third, since the installation was prepared and presented in the wake of the dissolution of the provincial assembly and during the Northern Ireland election, any potentially sectarian issues needed to be treated with great care. Fourth, the BBC's sponsorship meant that the installation was, in effect, a BBC "broadcast" and was subject to editorial policies designed to ensure objective and balanced reporting, especially with respect to Northern Ireland sectarianism.

These constraints meant that the Kilkeel installation needed a different approach. The informal editing of the previous installations was not adequate. The first step in designing the censorship policy was to discuss different strategies with all participants. We rejected approaches that relied on automatic message filters because we had difficulty defining exactly what constituted an offensive message, even among ourselves. We did agree on three general principles: no one person would have the power to censor messages and all messages would be rejected or accepted in their entirety; we would not delete or substitute any part of a text; any message that mentioned someone by name (whether innocuously or threateningly) would be censored.

A more difficult discussion involved what kind of sectarian messages would be allowed to appear. Our collaborators were adamant that images and messages with sectarian content be included in the installation—the absence of such discussion would have been conspicuous—but were equally insistent that no "offensive" sectarian messages or personal attacks be projected. Further complicating these requirements was the suggestion that we should only censor sectarian messages that were *intended* to be offensive. We failed to agree on a way of determining intent and instead left it as an open problem for the collaborating reviewers (Ananny, a BBC producer and a self-selected subset of collaborators) to address.

Questions about what was technically possible were also present throughout the censorship discussions: Can we block certain phone numbers? (Yes.) Can we block messages from people who may have subverted this blocking by using another phone number? (No.) Can we automatically filter out sectarian messages? (Not with any reliability, given our partial familiarity with the rich set of idioms and slang terms used to convey sectarian messages.) Can we hold all messages in a buffer and discuss them before they are projected? (Yes, but the longer the delay in refreshing the display, the less interactive and engaging the projection.)

Such questions led to a change in the *TexTales* technical architecture. In all previous installations, the software (including the SMS and display servers) resided on a single laptop, the display of which was projected directly to the public screen. Everyone saw all messages when they appeared on the public projection and any kind of editing or moderation occurred after texts had been projected.

For the Kilkeel installation, we split the software in two and added a second laptop. The messages first came into the laptop running the SMS server where they were held for approximately one minute to review a message, discuss our loose editorial policy and decide whether it should remain in the projection's active log file. If the message was acceptable, we would do nothing and let it be projected on the screen's next refresh; if the message was not acceptable, we would move it from an active log file to the master log file for archiving and later review—and it would not be projected. Thus the censoring system's default was to let messages through the filter; we took action only taken if messages were deemed not to be acceptable.

We also ensured that the *process* of review and censorship was public. Declining a suggestion to station ourselves and the two laptops behind the projection screen, we instead situated the two laptops beside the public display. The screens of both the "censorship laptop" and the "projection laptop" were available for anyone in the installation area to see. Thus, although we took care to design editorial guidelines and a technological architecture that supported our values, we chose to make any act of censorship public and transparent. Indeed, people often stood behind us as we reviewed incoming messages, commenting on the raw text messages and offering opinions on their appropriateness.

Censoring interaction in large-scale public projections is a sensitive issue that requires designers to reflect simultaneously upon: the overarching expressive aims of the project; the personal safety of designers and participating publics; the installation's social and political setting and the values and beliefs of the designers and public participants; and the technological and architectural resources available for structuring or censoring discourse. There is no prescription for how navigate these issues—each installation's policies and procedures will emerge from its particular context—but we found that we were best able to address these concerns by making the *process* of censorship public. That is, the limits that we designers—of urban spaces and communication technologies—place on our participants are best understood as opportunities to learn from the communities we work with how and why censorship is sometimes needed. Such decisions may be instantiated in technologies, architectures and policies but they are fundamentally reflections of the community cultures and tensions we ultimately aim to understand and design for.

CONCLUSION

Our broad goal in this paper is to understand determine how citizen-authored media (in this case images and texts) and—more broadly—cities themselves can become "objects to think with" (Papert, 1991) as we understand the socially constructed nature of public opinions. How can cities be seen as cultural learning environments and, more specifically, how can new media forms and forums support and reveal the processes of social discourse out of which collective voices emerge?

By *forms* we mean the representations that personal and public expressions take as people construct opinions. In the case of the *TexTales* installations, the expressions took form in the *photographs* people captured, edited and arranged as they interpreted a particular topic or issue; in the three-by-three *image template* that we as designers provided; in the SMS *captions* people created for the images as they participated in the installations; and in the informal *conversations* people had as they viewed the projected image-text combinations and discussed their particular contributions.

By *forums* we mean the settings and contexts in which people use the forms in particular processes of constructing opinions. In the case of *TexTales*, there were multiple forums:

- the initial *workshops* in which the designers and authors became acquainted with each

other and the *TexTales* technology, and discussed different photojournalistic and communicative techniques that they would use to produce their particular installation;
- the on-going *critiques* of images and issues that arose as people met repeatedly to design their installations;
- the *social spaces* of the projections and the ways in which the participant-designers and the general public came together to experiment with a projections and to explain and discuss the designs and goals of the installations;
- finally—and least explored thus far—the ongoing contexts in which participant-designers and the general public may continue to discuss and think about the issues and opinions they encountered during *TexTales* installations.

Our aim has been to describe one experiment in designing an urban environment for public opinion development. We would need further work, longitudinal case studies and design ethnographies to understand better the exact nature of our participants' public opinions and how they evolve in relation to mediated spaces such as *TexTales*. By understanding how people make and use new media forms and forums for public discourse, we may better appreciate how they envision their roles as expressive city citizens.

ACKNOWLEDGMENT

All the *TexTales* installations emerged from collaborations and friendships that spanned ages, countries, institutions and disciplinary boundaries. Warm thanks go to: the residents of Fatima Mansions, Kieran Doyle O'Brien, Niall O'Baoille, Irene Ward, the women of the Fatima History Project and Kathleen Biddick; Frank O'Connor and his students at Loyalist College Canada, especially Jeff Cooper and Sarah Faulkner; Kodak Canada for donating disposable cameras; Bernadette Larkin, Dara Carroll, Jim Ronan, Eric Fraad and all the support staff at The Ark in Dublin; artist Michael Durand; Marian Koopen, Roelands, Atti, and Kees Bok at the Amsterdam Computer Clubhouse; Grainne Millar and Rob Furey of Temple Bar Properties; Alice Jackson and Suzanne of Young At Art; Mairin Murray at BBC Northern Ireland; Kilkeel photojournalist Ingrid Perry; Tully Kewley of the Kilkeel Bridge Youth Project; all the participant-designers of the Fatima, Big Smoke, Smokum and Kilkeel installations; Erik Blankinship; Helen Doherty; the Everyday Learning group at Media Lab Europe, especially Brendan Donovan, Herve Gomez, Jamie Rasmussen, Matt Karau and Niall Winters; the staff of Media Lab Europe, especially Stephen O'Brien, Martin Pegman, Martin Lynch, Prue Street, Renee Hall and Emma Coyle.

REFERENCES

Ananny, M, Biddick, K. & Strohecker C. (2003). Constructing public discourse with ethnographic/SMS "texts". *Proceedings of Mobile HCI 2003* (pp. 368-373). Berlin: Springer-Verlag Lecture Notes in Computer Science.

Ananny, M., Strohecker, C. & Biddick, K. (2004). Shifting scales on common ground: Developing personal expressions and public opinions. *International Journal of Continuing Engineering Education and Life-Long Learning, 14*(6), 484-505.

Auerbach, A. (2006). Interpreting urban screens. *First Monday, Special Issue #4*. Retrieved January 25, 2008, from http://www.uic.edu/htbin/cgiwrap/bin/ojs/index.php/fm/article/view/1546/1461

Barber, B. (1984). *Strong democracy: Participatory politics for a new age*. Berkeley, CA: University of California Press.

Brignull, H. & Rogers, Y. (2003). Enticing people to interact with large public displays in public spaces. In Rautterberg, M., Menozzi, M., Wesson, J. (Eds.) *Proceedings of INTERACT'03* (pp. 17-24). London, UK: IOS Press.

Cappella J, Price V & Nir L. (2002). Argument repertoire as a reliable and valid measure of opinion quality: electronic dialogue in campaign 2000. *Political Communication, 19,* 73–93.

Converse, P.E. (1970). Attitudes and non-attitudes: Continuation of a dialogue. In Tufte, E.R. (Ed.) *The Quantitative Analysis of Social Problems.* New York, NY: Addison-Wesley.

de Sola Pool, I. (1973). Public opinion. In de Sola Pool, I., F. Frey, W. Schramm, Maccoby, N. & Parker, E.B. (Eds.) *Handbook of communication.* Chicago, IL: Rand McNally College Publishing Company.

Delli Carpini, M.X., Cook, F.L. & Jacobs, L.R. (2004). Public deliberation, discursive participation, and citizen engagement: A review of the empirical literature. *Annual Review of Political Science, 7,* 315-344.

Downs, A. (1956). *An economic theory of democracy.* New York, NY: Harper and Row.

Fishkin, J.S., Luskin, R.C. & Jowell, R. (2000). Deliberative polling and public consultation. *Parliamentary Affairs, 53,* 657-666.

Graber, D.A. (1982). The impact of media research on public opinion studies. In Whitney, D.C., Wartella, E. & Windahl, S. (Eds.) *Mass communication review yearbook 1982.* Beverley Hills, CA: Sage Press.

Habermas, J. (1962). *Structural transformation of the public sphere.* Cambridge, MA: MIT Press.

Hanzl, M. (2007). Information technology as a tool for public participation in urban planning: A review of experiments and potentials. *Design Studies, 28*(3), 289-307.

Harris, K., Lane, G. (2007). Social Tapestries: Conversations and Connections. Evaluation Report for the Ministry of Justice. *Proboscis.* Retrieved January 25, 2008, from http://socialtapestries.net/havelock/ST_Conversations_MoJReport.pdf

Iyengar S, Luskin RC & Fishkin JS. (2003, August). *Facilitating informed public opinion: Evidence from face-to-face and on-line deliberative polls.* Presented at Annual Meeting of the American Political Science Association, Philadelphia, PA.

Lester, P.M. (2006). Urban Screens: the beginning of a universal visual culture. *First Monday, Special Issue #4.* Retrieved January 25, 2008, from http://www.uic.edu/htbin/cgiwrap/bin/ojs/index.php/fm/article/view/1543/1458

Mansbridge, J. (1999). "On the idea that participation makes better citizens." In Elkin, S.L. & Soltan, K.E. (Eds.) *Citizen Competence and Democratic* Institutions (pp. 291-325). University Park, PA: The Pennsylvania State University Press.

Manovich, L. (2006). The poetics of urban media surfaces. *First Monday, Special Issue #4.* Retrieved January 25, 2008, from http://www.uic.edu/htbin/cgiwrap/bin/ojs/index.php/fm/article/view/1545/1460

McCarthy, J. (2002) Using public displays to create conversation opportunities. *Workshop on Public, Community and Situated Displays* at *CSCW'02.* Retrieved January 25, 2008, from http://interrelativity.com/joe/publications/PublicDisplaysForConversations-CSCW2002ws.pdf

McQuire, S. (2006). The politics of public space in the media city. *First Monday, Special Issue #4.* Retrieved January 25, 2008, from http://www.uic.edu/htbin/cgiwrap/bin/ojs/index.php/fm/article/view/1544/1459

Min, S-J. (2007). Online vs. face-to-face deliberation: Effects on civic engagement. *Journal of Computer-Mediated Communication, 12*(4), 1369-1387.

Noelle-Neumann, E. (1984). *The spiral of silence: Public opinion—our social skin*. Chicago, IL: University of Chicago Press.

Papert, S. (1991). Situating Constructionism. In Harel, I. & Papert, S. (Eds.) *Constructionism*. Norwood, NJ: Ablex Publishing Corporation.

Pateman, C. (1970). *Participation and democratic theory*. Cambridge, UK: Cambridge University Press.

Price, V. & D.F. Roberts. (1987). Public opinion process. In Berger, C.R. & Chaffee, S.H. (Eds.) *Handbook of communication science*. Newbury Park, CA: Sage Press.

Price, V. (1992). *Public opinion*. London, UK: Sage Press.

Price V & Cappella J. (2002). Online deliberation and its influence: The electronic dialogue project in campaign 2000. *IT and Society, 1,* 303–29.

Price, V. (2006). Citizens deliberating online: Theory and some evidence. In Davies, T. & Noveck, B.S. (Eds.) *Online deliberation: Design, research, and practice*. Stanford, CA: CSLI Publications.

Resnick, M., Rusk, N. & Cooke, S. (1998). The computer clubhouse: Technological fluency in the inner city. In Schon, D., Sanyal, B. & Mitchell,W. (Eds.) *High Technology and Low-Income Communities*. Cambridge, MA: MIT Press.

Schoenbach, K. & Becker, L.B. (1995). Origins and consequences of mediated public opinion. In Glasser, T. & Salmon, C. (Eds). *Public opinion and the communication of consent*. New York, NY: The Guildford Press.

Silverstone, R., & Sujon, Z. (2005). Urban tapestries: Experimental ethnography, technological identities and place (No. 7). London, UK: Media@lse, Department of Media and Communications, LSE. Retrieved January 25, 2008, from http://www.lse.ac.uk/collections/media@lse/pdf/EWP7.pdf

Slaatta, T. (2006). Urban screens: Towards the convergence of architecture and audiovisual media. *First Monday, Special Issue #4*. Retrieved January 25, 2008, from http://www.uic.edu/htbin/cgiwrap/bin/ojs/index.php/fm/article/view/1549/1464

Social Tapestries (2006). Social Tapestries: St Marks Housing Co-op Project Report. *Proboscis*. Retrieved January 25, 2008, from http://socialtapestries.net/stmarks/StMarks_Report_2006.pdf

Strohecker, C. & M. Ananny (2003). Constructing Intermodal Literacies. *Proceedings of Technology Enhanced Learning*. Milan, Italy. Retrieved January 25, 2008, from http://carolstrohecker.info/PapersByYear/2003/IntermodalLiteracies.pdf

Struppek, M. (2006). The social potential of urban screens. *Visual Communication*, 5(2), 173-188.

West, N. (2005). Urban tapestries: The spatial and social on your mobile (Cultural Snapshot No. 10). London, UK: Proboscis, London School of Economics. Retrieved January 25, 2008, from http://proboscis.org.uk/publications/SNAPSHOTS_spatialandsocial.pdf

Wyatt, R.O., Katz, E. & Kim, J. (2000). Bridging the spheres: Political and personal conversation in public and private spaces. *Journal of Communication, 50*(1), 71-92.

KEY TERMS

Intermodal Forms: Cohesive, expressive units in which meaning is distributed across multiple media (e.g., image, text, video), the combination of which may represent the perspective of a single author or multiple, collaborating authors.

Intermodal Literacies: The social and rhetorical skills associated with: creating and reading intermodal texts that distribute meaning across multiple media (e.g., image, text, video); authoring intermodal forms that start and sustain public dialogues; and negotiating the meanings of such forms with audiences.

Nomadic Audiences: Publics who visit a particular urban public forum for a short period of time to read messages left by past visitors and to write messages for future passers-by.

Situated Public Opinions: Perspectives on social issues created through interactive, public processes in which individuals speak and listen to those around them through both conversation and mediated representations embedded in their shared, built environments.

TexTales: A public, large-scale interactive projection screen with which passers-by in urban spaces develop public opinions by captioning photos with SMS text messages.

Three-line Editing Technique: A TexTales-specific form of post-hoc censorship in which participants send three text messages in rapid succession to replace an offensive text; an example of a more general form of in-context public forum censorship in which participants themselves monitor and edit expressions.

Urban Public Forums: City locations in which individuals create and read the expressions of others; purposefully designed spaces for both synchronous and asynchronous public expression through mediated forms (e.g., image, text, video) and interpersonal conversation.

Chapter VI
An Event-Driven Community in Washington, DC: Forces That Influence Participation

Jenny Preece
University of Maryland, USA

ABSTRACT

This chapter describes a small networked community in which residents of an apartment building in Washington, D.C., USA supplement their face-to-face social interactions with a Yahoo e-mail listserver. Analysis of over 460 messages that have been archived since July 2000, when the list began, reveals that the issues driving participation on the list also drive participation off the list. Threats to safety, high rent increases, and changes in management practices, such as parking regulations and access to facilities, motivate communication on and offline. Furthermore, those who are most active online are typically most active offline. Activity on the list is strongly fuelled by interest and discussion around local events, hence the term event-driven, and is promoted by activist tenants. Friendly notes about new restaurants, bird observations and other niceties may help a little to create a sense of overall community, but they do little to motivate online participation.

INTRODUCTION

People join online communities for many different reasons: they want to meet new people and make friends by social networking; get and exchange information; find support; debate and persuade others to take action or adopt their point of view; work and learn together; explore ideas; take on new personas; avoid being alone; play games; hang-out with like minded people and many more. Networked communities are a particular type of online community which is typically geographically based but which utilizes the Internet to distribute information, coordinate activities and mobilize people. Networked communities therefore operate within both physical and virtual places.

Depending on the purpose of the community and members' personal motivations for joining the community, different kinds of technical infrastructure are needed by these communities. If the community's focus is to provide another medium for communication for people who live locally and share local facilities, as in this study, the motivation of its members and their patterns of usage will be different from those in online communities with only a virtual common place. Foth (2006a, 2006b) and Foth and Hearn (2007) distinguishes between collective interaction which involves many-to-many interactions that tend to be structured and are sometimes formal and associated with community associations and groups that hold discussions about place-based interests such as rent increases and street rejuvenations, and networked interaction that involves peer-to-peer interactions that tend to be transitory and informal, in which the interaction is not limited to place-based interests.

Increasingly, however, many researchers are noting the blurring between place-based community interaction and virtual interaction. (Wellman & Haythornthwaite, 2002; Mesch & Levanon, 2003; Maloney-Krichmar & Preece, 2005; Boase, Horrigan, Wellman, Rainie, 2006). This trend is particularly pronounced with the increased use of cell phones and other mobile devices with Internet capability and the ubiquitous role of the Internet in many people's lives, particularly among teens, college students and young adults under thirty-five years of age in many parts of the world.

Another trend that researchers observe is that many communities use the online space in ways that are unintended by the community software developer or technology owner. Community members tend to take advantage of affordances available through the software design to fulfill their own needs regardless of the purpose for which it was developed (Lefebvre, 1991). For example, the owner of the community discussed in this chapter intended the community to be much more socially-oriented, whereas it turned out to be strongly focused on activism geared towards combating the activities of the apartment building managers.

In addition to participants that actively contribute to online discussions, many people join virtual community spaces and do not post, a concept variously referred to as lurking, visiting, or participating silently (Nonnecke and Preece, 2000). There are several reasons why people fail to participate online, and chief among them are: getting what they needed without having to participate actively (also known as lurking or social loafing); thinking that they were being helpful by not posting because what they were going to say had already been said; wanting to learn more about the community before diving in; not being able to use the software because of poor usability; not liking the dynamics that they observed within the group and feeling that they did not fit in the community (Nonnecke and Preece, 2000; Selwyn, 2003; Preece et al., 2004; Nonnecke et al., 2006; Bishop, 2007).

Even with more knowledge about what motivates people to contribute online, there is still a very poor response rate in some communities and the question of what motivates people to participate in online discussions continues to be a central one (Joyce & Kraut, 2006). Gradually, recent research is starting to tackle this issue in the context of specific communities. For example, in a study of newcomers to newsgroups, it was found that newcomers who received a response to their first post were 12% more likely to post again (Joyce and Kraut, 2006). A variety of other rhetorical strategies have also been tried with some success to elicit responses from online community participants including introductions and requests (Burke et al., 2007). Of course, others are strongly motivated by desire to obtain answers to questions or to offer support, and a study of communication in an online health support community revealed that communication behavior closely mirrored that of established face-to-face patient self-help groups (Maloney-Krichmar and Preece, 2005).

Patterns of posting, and not posting appear, not surprisingly, to be highly contextual, and to depend on the type of community (e.g., Nonnecke and Preece, 2000; Nonnecke et al., 2006) and other internal and external events. Internal events are concerned with the behavior of individuals and groups within the community and may include announcements of events, reports of activities, and political commentaries. There are many such examples, including early accounts about participation in the WELL, one of the first networked communities to be developed in the early 1990s. Comments of an outspoken protagonist enraged others and incited them to respond and argue. When the person died, much of the community interaction died with him (Rheingold, 1993).

For those with access to the Internet, it has often been a tool for coordinating behavior and a focus for activism within local communities (Schuler, 1996) as well as connecting friends across distance. The first type of use is geared towards collective action and tends to employ mailing lists, discussion boards and Yahoo Groups, while the latter is geared more towards interaction around personal connection and tends to employ software such as Facebook, Myspace, and text messaging.

In a study of financial communities peaks of activity correlated strongly with activity in the financial markets (Schoberth et al., 2006), this close coupling of online behavior with offline events or events that occur elsewhere in the digital world is not surprising. Indeed research reveals that individuals who participated in online support communities after the attacks of September 11th, 2001 on the World Trade Center and the Pentagon were more likely to participate in their place-based, often referred to as "real", communities (Dutta-Bergman, 2005). The debate about what actually constitutes a community has been a subject of debate among sociologists for many years, and continues to be discussed within the context of the community supported by the Internet (Bateman and Lyons, 2002; Arnold et al., 2003).

In this chapter, I describe a networked community in which people in the same apartment building communicate via an email list, and whose online activity is strongly driven by events that happen offline in the apartment building. I also discuss how the key activists online were also the key activists offline. Thus, a strong distinction between online and offline community does not exist in this example. The conditions that make this close coupling possible appear to be that everyone who wants access to the Internet has it. Furthermore, community members live in close proximity so they have a strong shared context that creates common ground for understanding the impact of the events that drive their behavior online and offline and their potential consequences on tenants.

THE COMMUNITY

This community consists of the residents of 142 apartments situated in a single building in a residential area in north-west Washington, DC. The community comprises members of all ages, with the majority of members ranging in age from 30 to 45 years old. Many members can be characterized as young professionals who have moved into the Washington, DC area from other cities in the USA to take up jobs in government, and to staff embassies, law firms, not for profit organizations and other companies. There are also a few retired people, a few students and a few children. Many of those working in embassies are from outside of the USA and they typically stay for one or two years only.

The local area in which the building is situated is about three miles north of the Whitehouse on Connecticut Avenue. It is located in the north west of the city and tends to attract professional people and those with reliable and strong incomes. It is close to the Van Ness metro stop on the red metro line and approximately four miles from the Washington, DC beltway around the city. The immediate neighborhood around the apartment building offers quite a few facilities including restaurants, dry cleaners, banks, a supermarket, a pharmacist, a deli, a wine store and other stores. Half a mile north is a terrific bookstore that hosts invited speakers every evening, and half a mile south is a very nice restaurant area with many different kinds of cuisine. These facilities offer ample opportunities for residents, their friends and colleagues to socialize.

The building itself is one of the newest in the area and is popular because of its swimming pool on the roof, business center, library and large lobby that is used for meetings and parties. Apartments range in size from one bedroom and one to two bathrooms and all 141 units are for rental. There is no owner occupied property but some residents have lived in the building for over five years and about five percent have lived there since it opened eight years ago. Despite the good facilities, the community is quite transient with people using the apartment complex as a place to live while they get to know the local area before buying their own house or condo or moving to another rental unit with a lower rent. Others leave to return to their country or state of origin.

The management organizes a few community activities for those living in the building. There are typically two or three parties a year. These include pool-side parties in the summer, a holiday party in December, and one or two other events. There are also residents' breakfasts in the library, which is located on the ground floor of the building, on the first Saturday of each month. Parties tend to be quite well attended. The residents' breakfasts were well attended during the first few years when a core of members got to know each other but in recent years, attendance at breakfasts has diminished. Many residents now collect food from the plates put out for the community breakfast and take it to their own apartments.

Typically community members are busy people with little time to socialize with others who happen to live in the building; and if they do have time, they may lack motivation, preferring to spend time with other friends that they already know. During the first few years, the residents tended to know their neighbors and many knew most of the people living on their floor. Almost seven years later, this desire to belong to and contribute to the community in the apartment building has almost disappeared. There are several reasons to explain the demise of community spirit and most are associated with early key community-builders leaving the community. When a retired member, Gerald, known for his parties, his knowledge of what was going on in the building, and his efforts to look after others, died of cancer in 2003, much of the community spirit died with him. Another reason is due to the transient behavior of the residents, and the fact that many are young professionals who are working to develop a career with little time, energy or motivation for socializing outside of work and their immediate sphere of friends and family. Gradually, over the years, the social capital (Putnam, 1995) that it built up in the early years has been severely eroded.

COMMUNITY ONLINE BEHAVIOR

On July 7, 2000, soon after the building opened a Yahoo listserv was set up by one of the residents to facilitate communication and coordinate activities among the 250 or so people living in the 142 apartments. Online membership was voluntary and somewhat *ad hoc*. The list owner took the initiative to set up the list and, assisted by others, he collected email addresses and contact information from other members at the monthly breakfast meetings. Members also invited others to join or were invited to join when acquaintances were made in elevators, at the pool, in the lobby and at parties. The first email was sent by the list owner and manager on July 7th, 2000 to around 20 people. In this email the list owner stated the purpose of the list as:

...to establish some sense of community among tenants. This would enable all of us to find bridge and tennis partners, invite each other for parties or walks, share extra theater tickets, and organize our suggestions to Management...

People joined the list and others left as they moved away. Traffic on the list was never strong but the list has stood the test of time and still exists over seven years after its inception. A devoted set of residents who moved away have elected to remain on the list, but they no longer submit comments.

WHAT THE LIST ARCHIVES REVEAL

Exact counts of the messages sent to the list is not possible because in managing the list the list owner deleted messages that were concerned with joining or leaving the list and there were many of these due to the transient nature of the community. The current archive covers the period of July 7, 2000 to October 31, 2007 contains 466 messages and the list owner, who is the same person who started the list in 2007 estimates that there were probably around 100 messages that were concerned with registration issues that have been deleted. Fifty of the 466 messages were sent by the list owner and 17 were sent by the management company to the list owner who forwarded them to the list. The remaining 399 messages were sent by 83 members over the seven year period. Many messages addressed more than one issue, which is why the total recorded for each type of communication is greater than the total number of messages. Community activity online was low. However, more detailed analysis of the content of the messages is revealing.

What the community talked about: Two hundred and twenty six of the messages were concerned with discussions about rent increases; this is 57% of the message corpus. This topic periodically caused major concern and heated debates among tenants and with management, particularly when some people appeared to be singled out for unreasonably high rent increases or the algorithm used to determine rent increases was not clear. The three next most talked about topics were: 120 messages about insecure parking, changes in parking fees or regulations (30%); 58 messages about availability and management of the pool and hot tub (15%); and 48 messages about the safety of residents and their property inside the building and the responsibilities and practices of staff on the front desk (12%). These messages are all concerned with living conditions and they dominate discussion among residents on and offline. Other topics discussed on the list include: discussion of restaurants in the vicinity of the building (41 messages, 10%); holiday greetings and welcome messages (31 messages, 8%); invitations to parties (27 messages, 7%); and a discussion about bird observations (3 messages, 1%). Interestingly, discussion of these topics more closely fits the list owner's original intentions when starting the list. However, in practice the list has become a medium for discussing problems faced by tenants, organizing meetings to discuss these problems and keeping each other informed.

Who did the talking?: As has been reported often a large number of people in online communities send just one message (e.g., Burke et al., 2007). In this online community thirty-five (42%) of the members posted just once; twenty seven posted (33%) two to five messages; nine (11%) posted six to ten messages; and eight (10%) posted eleven or more messages. The remaining 4 members included the list owner and apartment managers.

The list owner posted fifty messages. One member posted forty-five messages about a variety of community oriented topics. In addition to messages about rent raises and other concerns, she sent pictures of a group of neighbors walking in the snow and organized a yoga group, which more closely matched the way that the list owner envisaged the list would be used. Another member sent seventy-five messages which is 19% of the total number of messages sent by the community during the seven years of its existence.

Community activism: The main role of the list server has been to support community activism, even though this was not the intention of the list owner. The three people (including the list owner) who posted most frequently (45, 50 and 75 messages) are all strong community activists. Whenever there was a problem in the building these three people immediately became involved. The list owner did not intend to take on this role, but people began to look to him for leadership and he stepped forward. In addition to stimulating discussion on the list he also led face-to-face discussions, and was the spokesperson for the group when there were meetings with apartment managers.

The other two members are both retired so they have more time to search for and find information to support activist causes such as when management convinced one of the older members of the community to move apartments and then also increased her rent. The personalities and behavior of these two members are the same online as offline. The person who sent seventy five messages – tenant75 we will call her – has spent an extraordinary amount of time visiting pro-bono lawyers, checking municipal records, reviewing and comparing rents in comparable apartment buildings and so on, and then communicating her findings to others in the community. Whenever there is a problem or potential problem with apartment management, tenant75 springs into action; she is relentless online and offline.

The on-site manager was not included on the list, and though she probably knew about its existence, she took no action to influence activity on the list in any way.

DISCUSSION

So what can we learn from this review? Some of the issues discussed have been observed and reported by others but this study adds to the growing body of knowledge about the role of communications technology in networked communities. Other observations raise new or less frequently asked questions. Here is what we learned from this study:

It is one interaction space. There is no distinction between online and offline interaction.

Observations from this community strongly support more recent claims that communications technology simply provides another channel of communication. In that respect it is like a telephone. In other words, activity online is closely integrated with activity offline. The online community component does not exist separately from the offline community component as it does in some geographically distributed online communities. The close physical proximity of members, shared context, and comparatively easy access to computing technology enables this community to function as one integrated community in which some people participate and some do not, and whether the action is online or offline is not a distinguishing factor.

Leaders make a difference. Without the community owner, manager and list originator, who manages list membership and frequently initiates communication, and tenant75 and the other activists, this community would not function. A few individuals can make all the difference.

The community evolved and developed its own character. The listserver was developed to support socializing among people living in the apartment building, but it evolved primarily into a community of activists.

The community was event-driven. Much of the time, the community was inactive in that there was little or no message traffic on the list. It only became really active when there was a problem. The main episodes of activity occurred when some members received unreasonable rent increases. Other changes brought about by the apartment managers also stimulated discussion on the list. The role of events in stimulating activity in this and other communities is discussed further in the next section.

Longevity is not always associated with posting activity. Seven years and still going is a highly respectable life for a networked or online community. Indeed it is several years longer than many communities exist. This study strongly illustrates that longevity is not necessarily linked with posting activity. To understand the success of a community we need to understand its purpose from the perspective of the community members themselves.

This last issue is worthy of further discussion, particularly in a research climate that strives to understand what motivates community members to participate after joining an online or networked community (Burke et al., 2007; Joyce et al., 2006; Nonnecke and Preece, 2000; Nonnecke et al., 2006; Preece et al., 2004). Of course part of this obsession is driven by advertisers who want to be sure that their advertisements impact as many people as possible, and software developers who want to make sure that their products are used. But does it matter if some communities have low rates of posting, and is there a correlation with community longevity and success? This study flies in the face of much conventional wisdom by suggesting that "no, it does not matter". What matters is that we researchers need to take a more nuanced approach in which we pay far more attention to the context of the community including the purpose that it serves and how well that purpose meets the motivations and needs of the community's membership. In turn this means that the characteristics of the community members need to be well understood by the community managers, software developers and the community members themselves. This also begs the question of which methods of study we should use and when? Metrics such as number of participants and number of messages posted are useful, but they only tell part of the story; triangulating this data with qualitative data and ethnographic findings promises to throw more light on what is really happening (Hine, 2005).

CONCLUSION

In event-driven communities such as this one, frequent and abundant posting does not seem necessary. What is necessary is that the community responds when needed. This is a characteristic of other activist communities, such as emergency response and recovery communities (Shneiderman and Preece, 2007; Jaeger et al., 2007). The important thing for these communities is that the members respond when appropriate and needed. The community may appear to be dormant or low-functioning the rest of the time. This pattern of activity could be disasterous for other communities, such as health communities (Maloney-Krichmar and Preece, 2005) in which members need and expect a constant flow of new messages. The style of participation adopted depends on the purpose of the community; how it is managed (e.g., its norms, policies and practices); and the people who form the community; these, in turn, impact the choice of technology for supporting the community (Preece, 2000).

The networked community discussed in this case study functions because the listserver pumps email messages into the inboxes of its members. It would not have survived, if the software had been a discussion board that members had to visit because there would have been insufficient messages to keep members sufficiently interested to make accessing the URL worthwhile. Conversely, many high volume communities, like the patient support community that we studied (Maloney-Krichmar and Preece, 2005) would lose members, if they were overwhelmed with email or text messages. Increasingly software designers are paying attention to these kinds of issues and are developing strategies to mitigate them. For example, many discussion boards can be tailored to send email messages to users and many emails lists aggregate messages into a single digest. Even without the aid of designers, some active email lists have adopted FAQ-documents or, more recently, wiki repositories to help mitigate unnecessary redundancies (Hansen et al., 2007).

Interest in Web 2.0 applications and the increase in research on social interaction online are exciting and are producing a wealth of information about what motivates people's behavior. While research and experience are producing some "rules of thumb" for designing and managing networked and online communities, each has its own individual characteristics that have to be considered.

ACKNOWLEDGEMENT

I thank the list owner for giving me access to the archive of list messages and my colleagues, Paul Jaeger, Derek Hansen, Ben Shneiderman, Philip (Fei) Wu, and three anonymous reviewers for their helpful comments on a draft of this chapter. I am especially grateful to Marcus Foth for his thoughtful and extensive comments which have enabled me to enrich the chapter.

REFERENCES

Arnold, M., Gibbs, M.R., Wright, P. (2003) Intranets and Local Community: 'Yes, an intranet is all very well, but do we still get free beer and a barbeque?' In M. Huysman, E. Wenger & V. Wulf (Eds.), *Proceedings of the First International Conference on Communities and Technologies*, 185-204. Amsterdam, NL: Kluwer Academic Publishers.

Bateman, D. R. and Lyon, L. (2002) Are virtual communities true communities? Examining the environments and elements of community. *City and Community*, 1(4), 373-390.

Bishop, J. (2007). Increasing participation in online communities: A framework for human–computer interaction. *Computers in Human Behavior*, 23(4), 1881-1893.

Boase, J., Horrigan, J. B., Wellman, B., & Rainie, L. (2006). *The Strength of Internet Ties.* Washington, DC: Pew Internet & American Life Project.

Burke, M., Joyce, E., Kim, T., Anand, V., Kraut, B. (2007) Introductions and request: Rhetorical strategies that elicit response in online communities. *Proceedings of Communities and Technologies Conference*, 21-40.

Dutta-Bergman, M. J. (2005) The Antecedents of Community-Oriented Internet Use: Community Participation and Community Satisfaction, *Journal of Computer-Mediated Communication* 11(1), 97–113.

Foth, Marcus (2006a) Analyzing the Factors Influencing the Successful Design and Uptake of Interactive Systems to Support Social Networks in Urban Neighborhoods. *International Journal of Technology and Human Interaction* 2(2), 65-79. http://eprints.qut.edu.au/archive/00001912/

Foth, Marcus (2006b) Facilitating Social Networking in Inner-City Neighborhoods. *Computer* 39, 9, 44-50. http://eprints.qut.edu.au/archive/00004750/

Foth, Marcus and Hearn, Greg (2007) Networked Individualism of Urban Residents: Discovering the Communicative Ecology in Inner-City Apartment Complexes. *Information, Communication & Society* 10(5), 749-772 http://eprints.qut.edu.au/archive/00006100/

Hansen, Derek L., Mark S. Ackerman, Paul J. Resnick, Sean Munson (2007). Virtual Community Maintenance with a Repository. *Proceedings of the American Society of Information Science and Technology Annual Conference*, October 19-24. Milwaukee, Wisconsin.

Hine, C. (Ed.). (2005). *Virtual Methods: Issues in Social Research on the Internet*. Oxford: Berg.

Jaeger, P. T., Shneiderman, B., Fleischmann, K. R., Preece, J., Qu, Y., & Wu, F. P. (2007). Community response grids: E-government, social networks, and effective emergency response. *Telecommunications Policy, 31,* 592-604.

Joyce, E., Kraut, R. E., (2006) Predicting Continued Participation in Newsgroups, *Journal of Computer-Mediated Communication* 11(3), 723–747

Lefebvre, H. (1991). *The Production of Space* (D. Nicholson-Smith, Trans.). Oxford: Blackwell.

Maloney-Krichmar, D., Preece, J. (2005) A multilevel analysis of sociability, usability and community dynamics in an online health community *Transactions on Human-Computer Interaction (TOCHI),* 12(2), 1-32.

Mesch, G. S., & Levanon, Y. (2003). Community Networking and Locally-Based Social Ties in Two Suburban Localities. *City and Community, 2*(4), 335-351.

Nonnecke, B. and Preece, J. (2000) Lurker Demographics: Counting the Silent. *Proceedings of CHI'2000*, Hague, The Netherlands, 73-80.

Nonnecke, B., Andrews, D., Preece, J. (2006) Non-public and public online community participation: Needs, attitudes and behavior. *Electronic Commerce Research,* 6(1), 7-20.

Preece, J. (2000) *Online Communities: Designing Usability, Supporting Sociability.* Chichester, UK: John Wiley & Sons.

Preece, J., Nonnecke, B., Andrews, D. (2004) The top 5 reasons for lurking: Improving community experiences for everyone. *Computers in Human Behavior,* 20(2), 201-223.

Putnam, R. D. (1995) Bowling Alone: America's declining social capital. *The Journal of Democracy*, 6(1), 65-78.

Rheingold, H. (1993) *The Virtual Community: Homesteading on the Electronic Frontier.* Reading, MA: Addison Wesley.

Schoberth, T., Heinzl, A., Preece, J. (2006) Exploring Communication Activities in Online Communities: A Longitudinal Analysis in the Financial Services Industry. *Journal of Organizational Computing and Electronic Commerce.* 16(3 /4), 247-265. Lawrence Erlbaum Associates, Inc. Country of Publication: United States - Mahwah, NJ.

Schuler, D. (1996) *New Community Networks: Wired for Change.* Reading, MA: ACM Press and Addison Wesley.

Shneiderman, B., Preece, J. (2007) 911gov. *Science 315*, 944. http://www.sciencemag.org/cgi/content/summary/315/5814/944

Selwyn, N. (2003). Apart from technology: understanding people's non-use of information and communication technologies in everyday life. *Technology in Society*, 25(1), 99-116.

Wellman, B., & Haythornthwaite, C. A. (Eds.). (2002). *The Internet in Everyday Life*. Oxford, UK: Blackwell.

KEY TERMS

Event-Driven Community: A community that becomes active in response to events that typically are initiated by people outside of the community that impact the community.

Motivation: The psychological literature on motivation is extensive with many definitions. In this chapter the word motivation is used to describe an event or desire that encourages activity among the community members.

Networked Community: A local community that is strongly associated with a particular geographically located place but which uses the Internet to coordinate and communicate its membership.

Sociability: In this context sociability refers to the social interaction that occurs via the Internet medium.

Social Capital: Refers to connections among individuals to form social networks and the norms of reciprocity and trustworthiness that arise from them.

Social Interaction: Informal communication that typically involves one-to-one or a small group discussing social issues of importance to them.

Usability: Interaction across the human-computer interface that is safe, effective, easy to learn and satisfying.

Chapter VII
Moments and Modes for Triggering Civic Participation at the Urban Level

Fiorella De Cindio
Università degli Studi di Milano, Italy

Ines Di Loreto
Università degli Studi di Milano, Italy

Cristian Peraboni
Università degli Studi di Milano, Italy

ABSTRACT

After more than a decade of e-participation initiatives at the urban level, what remains obscure is the alchemy—i.e., the "arcane" combination of elements—that triggers and keeps citizens' involvement in major decisions that affect the local community alive. The Community Informatics Lab's experience with the Milan Community Network since 1994 and its two more recent spin-off initiatives enable us to provide a tentative answer to this question. This chapter presents these experiments and looks at election campaigns and protests as triggers for (e-)participation. It also discusses these events as opportunities to engender more sustained participation aided by appropriate technology tools such as software that is deliberately conceived and designed to support participation and managed with the required expertise.

INTRODUCTION

After more than a decade of e-participation initiatives at the urban level, the alchemy—i.e., the "arcane" combination of elements—that spurs lasting citizen involvement in decisions important to the local community, is still obscure. Furthermore, as Venkatesh (2003) claims, participation cannot be assumed to start up in a social vacuum, nor in a technological vacuum. Based on our 15 years' experience, we believe that, rather than as a "continuum", participation should be seen as a discrete phenomenon, with peak moments when the local actors are more inclined to participate. This chapter discusses the phenomena observed during two of these recurring moments, the 2006 Milan municipal elections and a moment of protest at a Milan public school in June 2006. We also analyze the modes in which, using information and communication technology (ICT), the inclination to participate can be consolidated, beyond the hot moment, into a more ongoing practice of participation.

BACKGROUND

We take community networks (CNs) as the starting point of our analysis, because they can be considered pioneer experiences for supporting e-participation in local communities. Even though, over time, many community networks have declined or even disappeared (Luisi 2001; Schuler 2009, forthcoming), they remain landmarks, providing significant input for the design of socio-technical systems aimed at empowering active citizen participation. Community Networks as conceived in the 1990s (Schuler 1994; Bishop 1994; Silver 2004) were virtual (or online) communities, strongly rooted in a specific territory, whose shared focus of interest was 'public affairs'. They provided a framework for gathering *civic intelligence* (Schuler 2001), for supporting the development of people's projects (De Cindio, 2004), and for promoting public dialogue among citizens and between citizens and local institutions (e.g., Casapulla et al, 1998; Ranerup 2000; De Cindio et al., 2007). Kubiceck and Wagner's (2002) "ex post" analysis of CN development is helpful in understanding the evolution of CNs as generational succession, with one generation following upon another in a common line of tradition. They state that every generation of community networks is characterized by the advent of new technologies (which represent the formative "collective event" for each generation) and by changes in cultural context under the influences both of the preceding generation and of its own *Zeitgeist*. On the other hand, Selznick (1996) stresses that the emergence of community is based on *opportunity* for, and the *impulse* toward, comprehensive interaction, commitment, and responsibility. Extending these considerations, we claim that participation rises within a specific socio-technical context around particular opportunities and impulses.

In the following section, we therefore examine today's socio-technical context in order to show that all conditions for a new "generation" of participation are fulfilled. Against this backdrop, we then identify *moments* that can trigger participation and *modes* for transforming these opportunities into well-rooted participatory practice. We also provide examples of these moments and modes, analyzing them through some statistical indicators.

TRIGGERING CIVIC PARTICIPATION

In representative democracies, governing is done *for* rather than *by* the citizens. Political life is delegated to a separate sphere inhabited by a well-known class of professionals: government officials and politicians. In Western countries, this separation is recognized as one of the reasons of the increasing phenomenon of lower turnout (Ginsborg, 2006), which, in Europe, is manifest at all institutional levels—although, at least in our country (Italy), municipal elections still have good turnout. There are, however, several signs that citizens are eager to participate: the more traditional sign is the amount of time they volunteer with nonprofits, enabling these organizations to often act as stakeholder representatives. More recently, a new means of stake aggregation, that has found a place in people's everyday lives, reveals citizens' inclination toward involvement. We refer to the 'Web 2.0' phenomenon, a term that gained currency following the first *O'Reilly Media Web 2.0 conference* in 2004 (O'Reilly, 2005). As Musser and O'Reilly (2006) states, "Web 2.0 is much more than just pasting a new user interface onto an old application; it's a way of thinking, a new perspective" for living the web. Web 2.0 should be considered not a technical innovation but a new cultural step: transitioning from *isolated* personal websites to *interlinked* personal points of presence (Granieri, 2005) where individuals generate and distribute contents through their social networks (see e.g. Beer and Burrows, 2007). A large number of people (as well as politicians) have learned to grasp these new opportunities to communicate and participate. However, the way people have used Web 2.0 technologies so far, fails to fulfill Web 2.0's full potential. Young people—who have so far made up the bulk of users– have published their ideas, photos, videos, and the like, deeming Web 2.0 an environment for getting acquainted or making themselves known. On the other hand, politicians have started using the Web 2.0 environment with a showcase approach, only occasionally applying it to dialogs with voters/citizens. Despite these limits, the new socio-technical context offers a series of opportunities to spur participation and debate, especially at the local level.

Let us describe two moments suited to triggering public debate between citizens and politicians. We believe these moments can lay the foundation for longer-lasting participation with greater awareness. The first *moment* is the run-up to elections, when politicians are particularly willing to meet citizens and especially open to hearing out people's opinions so as to earn their vote. The second *moment* comes at no set time whenever citizens spring into action to protest some decision affecting their territory and lives. In both cases, appropriate *modes* and *tools* can engender a more stable and ongoing practice of participation.

Election Time

In representative democracies, people express their will at the moment of elections. Citizens play out their sovereignty by choosing representatives: depending on the specific electoral system, they may vote for a candidate, a party or a coalition, which ought to best represent their opinion and vision. The campaign running up to election day allows candidates to present their political vision and voters to make their choice. However, the electoral campaign is also the moment when politicians meet voters and listen to their problems, complaints, and suggestions. This is especially true in local elections, where citizens, individually or in groups, can envision problems, establish priorities among issues, propose solutions, and so on. In Italy, city-council elections—or district councils in the case of large cities—are usually lively moments. Voters may personally know one or more of the candidates, and people often feel that the choice of mayor, city council members, and district council members may influence their lives more directly than choices at the national level. We believe people's interest during election campaigns displays a participation potential that often stays latent in the context of representative democracy. Election campaigns can become a *moment* that brings this potential into effect, creating the practice of participation in local public life.

From Malaise to Protest

The rise of new media radically changed the way protest movements are organized: once, printed posters or leaflets, radio announcements, and word of mouth were the main tools for informing other citizens of an ongoing movement and related events. Now there are plenty of examples of net activism supporting protests (see, e.g., Meikle, 2002). Organizers of protests in Spain against the Aznar administration—which had attributed the March 11 train attack to the Basque nationalist separatist organization ETA—and of the 2006 student protests in France relied heavily on the web and cellphone texting. In Italy, too, in recent years, various protests against so-called 'large-infrastructure' public works projects have been organized mainly through the net. NOTAV (http://www.notav.info/, http://www.notavtorino.org/ and http://www.notav.eu/) is the de facto logo of the citizens' committees that want to halt construction of a high-speed TAV (for *Treno Alta Velocità*) railway in Piedmont. NOMOSE is the logo of the committee protesting against the planned construction of dams to protect Venice from rising water levels due to climate change (unfortunately, www.nomose.org/ recently expired). Several other examples exist in Italy, as well as elsewhere (for further examples see e.g. Rheingold, 2002 and Opp,Voss and Gern, 1995). These protest committees' websites are impressive: they contain documents, data, pictures, suggestions, opinions, and ideas; they demonstrate the *civic intelligence* that comes from active engagement (Schuler, 2001). These websites also show that the Internet is increasingly the glue that connects one experience to another (De Cindio and Schuler, 2007). Evidence that Web 2.0 technologies provide the framework for organizing protest that would be impossible without the Internet, arose September 8th, in Bologna and other Italian towns, when Beppe Grillo, an Italian comedian (and the most popular Italian blogger), used his blog (http://www.beppegrillo.it/english.php) to draw hundreds of thousands of citizens into the streets to demonstrate against the Italian political establishment (see Figure 1).

Both, election campaigns and protest moments produce a peak in citizens' active participation, which, however, does not easily consolidate into a sustained practice of participation. Politicians' openness to dialogs with the citizenry suddenly vanishes after the elections, and blogs started during the campaign lie abandoned. Citizens' protests either quickly succeed or lead to a stalemate in which dialogue peters out: little by little, the protest fades and the strongest interests often eventually win. We believe the *appropriate* use of *appropriate* participation technology can give rise to an ongoing practice of participation. To support this claim, let us present two cases from our experience. The first—ComunaliMilano2006—was deliberately promoted by the LIC (Civic Informatics Lab) during Milan's municipal elections in Spring 2006. The second case—partecipaMi—is the continuation of the first and provided the context for taking dialogue one step further when a protest in a school, thanks precisely to the partecipaMi website, managed to

Figure 1. Piazza Grande in Bologna on September 8, 2007

grow into a movement, open a dialogue with local government, and achieve positive results.

The beginning of an e-Participation Initiative: ComunaliMilano2006

The municipal elections held in Milan in the Spring of 2006, were seen as a significant political event. The role of Milan as capital of the wealthiest Italian region (one of the regions classified by the European Commission as Europe's growth engines), the nine-year administration of the same mayor, as well as other factors, attracted attention to the two major political coalitions' choice of mayoral candidates, to the issues they brought to public notice and to the lists of candidates they nominated for city council. In 2006 many candidates (not only the ten mayoral candidates, but also candidates for city council and for district councils) considered developing their own websites (and registering a domain name) an obvious part of their election campaign. However, these sites were scattered around the net. The www.ComunaliMilano2006.it website was designed to be a public square, organized in public forums, where citizens and candidates could meet to discuss issues of public interest. It provided a place where it was easy to find information on several candidates and links to their sites. ComunaliMilano2006 was promoted by the LIC (that also promoted and still manage the Milan Community Network— RCM) in order to renew its original challenge of creating an online space for public dialogue in the city of Milan. Even if the fruitful cooperation between the Milan Community Network and the municipality (Casapulla et al., 1998) ended in 2000, RCM is still a living online community where citizens share civic interests and discuss issues pertinent to urban life. To take advantage of the habit of discussing civic issues online, citizens registered on RCM can use their account to access the new portal. At the beginning of the new initiative, this provided an initial set of people with consolidated experience in online civic dialog.

ComunaliMilano2006 was designed as a public (virtual) square surrounded by the private spaces of the candidates (see Figure 2).

This metaphor inspired the design of the site's home page (Figure 3), where the main attention is on the public forums, with candidates' blogs following them.

Each candidate was given the chance to open a personal page divided into three sections (personal profile, reason for running, her/his platform), with the option of linking it to her/his website or to the on-site blog we invited each candidate to open. The thematic forums that make up the public square (e.g.: Mobility in Milan, Democracy

Figure 2. The metaphor behind ComunaliMilano2006

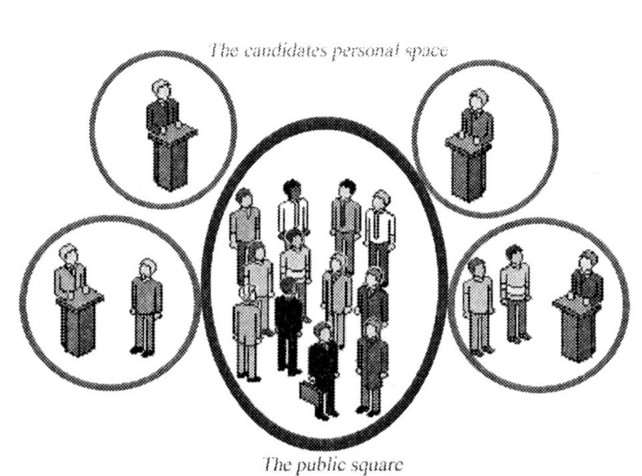

Figure 3. ComunaliMilano2006 home page

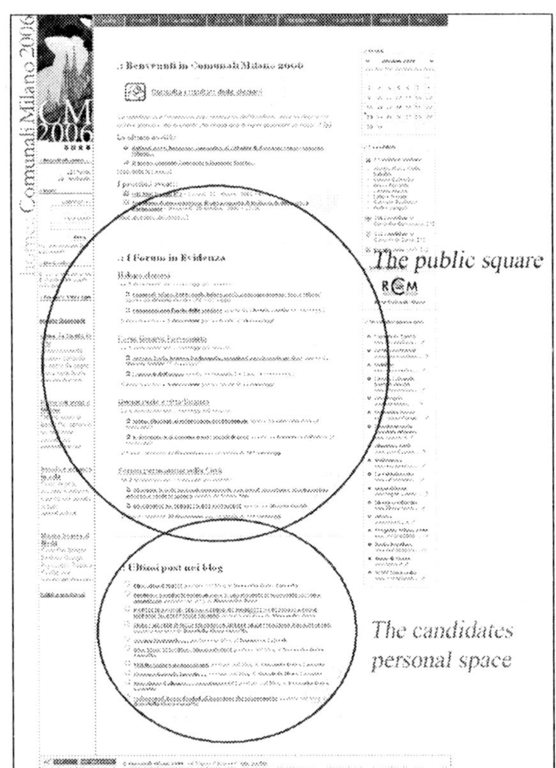

& Citizenship, Safety & the City) provide a few distinctive technical features to enhance *informed discussions*: they are organized in threads (see Figure 4) and message attachments are collected and organized all together in a documentation area (*information area*). Materials supporting discussion are thus easily available so discussion can be based on them. Visitors to the site can read the messages (and download materials), while registered citizens can either write messages or express their (level of) consensus by assigning a numerical value (from 1 to 5) both to messages and documents (see Figure 5). This is a kind of *weak participation* that prevents large sequences of messages just saying: "I agree/disagree with Tom" (as often happens in forums). The need to support such weak (or quick or implicit) participation was originally suggested to us by our previous experience with employees of the Province of Milan, who have long managed public forums hosted on the Milan Community Network (see De Cindio et al., 2007). Indeed, 41% of the citizens who participated in ComunaliMilano2006 did so by expressing consensus or disagreement about someone else's position expressed in a posting. Allowing people to express dis/agreement seems to be important for successful participation because it provides (technological) support for what Edward (2006) calls *different styles of citizenship*: one "stronger," more active, and another apparently "weaker". Supporting these different participation styles would guarantee a more inclusive and "democratic" environment for deliberative processes. It is worth mentioning that, based on the Milan Community Network experience (De Cindio et al., 2003), the active presence of a community manager assures that users respect appropriate netiquette (the RCM "Galateo" each registered user has to sign). The manager also acts as *facilitator* in discussions, linking related topics and copying news or documents found online, typically on the official City of Milan 'portal' and in the online version of the major local newspaper. Thanks to his work—he performs manually operations and activities proper of the Web 2.0 mood, such as linking external information sources and supporting social interactions between participants—ComunaliMilano2006 has become a portal to civic issues in Milan.

Another distinctive feature of the software developed to manage ComunaliMilano2006 is how forum topics were decided. Citizens could make proposals (e.g.: "let's discuss mobility in Milan" or "let's talk about the city seen from a woman's perspective") in a *brainstorming area*—that is to say a "simplified public forum" (it does not support replies) that gathers citizens' suggestions for new discussion topics. Other citizens could agree on a proposal, again by assigning a numerical value from one to five, or could suggest a different proposal. Of all the proposals, the highest-ranking ones were then "promoted" as topics for discussion in new dedicated forums. However, we must admit that this feature has been

Figure 4. Discussion viewed in a public forum

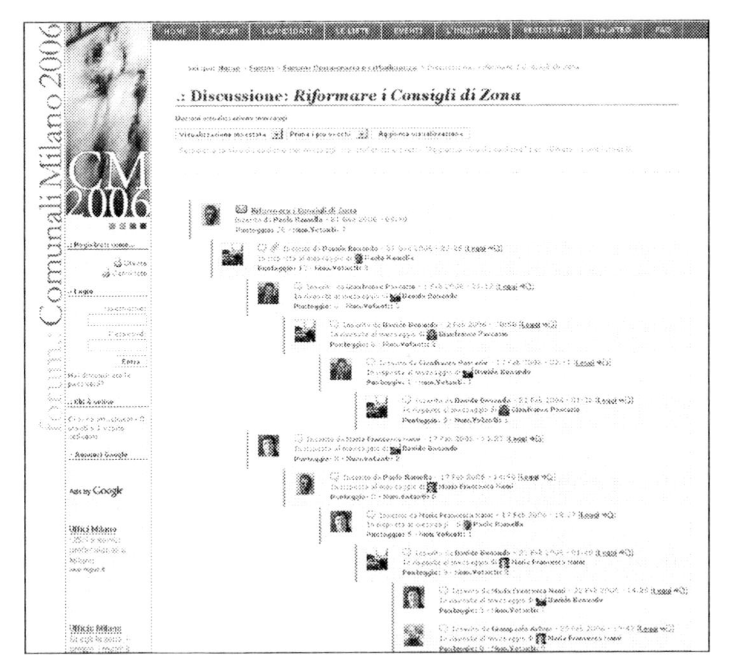

Figure 5. Rating choices (rating messages, documents, or proposals)

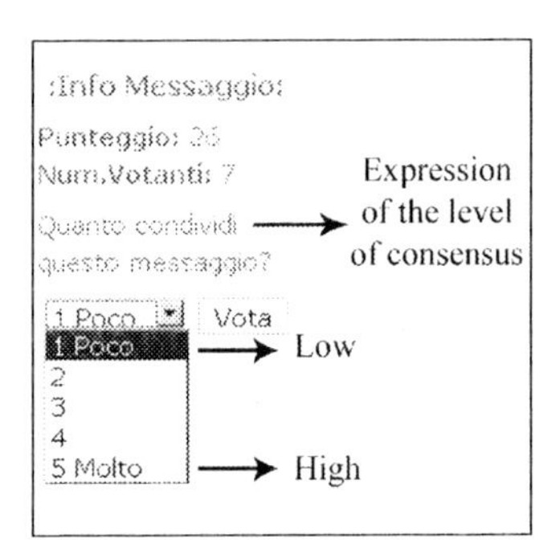

used less than we expected, for several, quite incidental, reasons.

On the other hand, the *event section* of the site was popular by users. Each registered citizen (candidate or not) could announce events by filling in a form (Figure 6).

Announcements were moderated and published when approved by the community manager, who also collected and published external event notices. Toward the end of the campaign, close to election day, the chance to announce events became increasingly attractive to candidates (Figure 7) and useful to citizens for finding out about meetings and debates that were not reported by print and broadcast media which can provide only a limited selection of such events. Overall, 398 "campaign events" were posted on the events calendar: 169 (~43%) were posted by candidates, 147 (~37%) by ComunaliMilano2006 staff, and 82 (~20%) by private citizens. We believe this simple but significant chance to be content providers (by collecting and sharing civic knowledge) inspires a citizens non-passive attitude that may encourage active participation.

The results of ComunaliMilano2006 were quite satisfactory: seven of the ten mayoral candidates (specifically all the major ones) published their pages, linking to their official sites. Of some 1600 candidates for city council, 239 (i.e. about 15%) prepared personal pages, 37 (15%) also requesting their own blog to engage in dialogue

Figure 6. Example of event announcement provided by a user

Figure 7. A sample of the events announced by candidates and citizens taking place few days before election day

with voters. Although the district councils have limited decision-making power in Milan, 311 candidates prepared personal pages and 63 of them also asked to open their own blogs. Altogether, 557 candidates subscribed to the site. As far as public forums are concerned, from November 15, 2005, when the site was launched, to July 2006, we opened ten forums with a total of 399 messages: 264 (66%) posted by citizens and 135 (34%) by candidates. Democracy & Citizenship (97 messages), Mobility in Milan (78 messages), and Research & Innovation (28 messages) had the largest number of messages. There were 734 posts and comments sent to the one hundred (37+63) blogs opened by candidates. Most of these postings were from candidates themselves—498 or nearly 68%—versus 236 (32%) from private citizens. We can therefore argue that people are more interested in discussing in the public square than in having conversations with individual candidates, whereas candidates prefer to communicate through their own blog rather than to get involved in peer-to-peer discussions. Finally, let us consider the site's hits, shown in Figure 8 in terms of page views. Obviously, as one might expect, we observe a peak in May 2006, the month immediately before the elections (held on May 29 and 30). June was still quite active because of the mayoral runoff and because people discussed election results. After dropping off in July and August, the traffic rose once again, which was by no means a foregone conclusion.

All told, these data show that setting up a site conceived and designed for facilitating public dialogue during the election campaign actually met a need felt both by candidates and by citizens. Although candidates remain more inclined to see the net mainly as a channel for informing citizens (overall, they wrote 225 forum messages versus 532 posts in their blogs), the site's structure somehow obliged them to also accept peer-to-peer public dialogue in the forums. This is to say that design choices influenced the practice of participation acquired during this experience. This was no fluke: in designing the site, we were well aware of politicians' reluctance to accept public dialogue with the citizenry (see e.g., De Cindio et al. 2004). However, to make it attractive to them, we included the personal blogs: this option proved especially appealing to 'minor' candidates, such

Figure 8. Monthly page views of the ComunaliMilano2006 website from its launch on November 15, 2007 to the end of January 2007

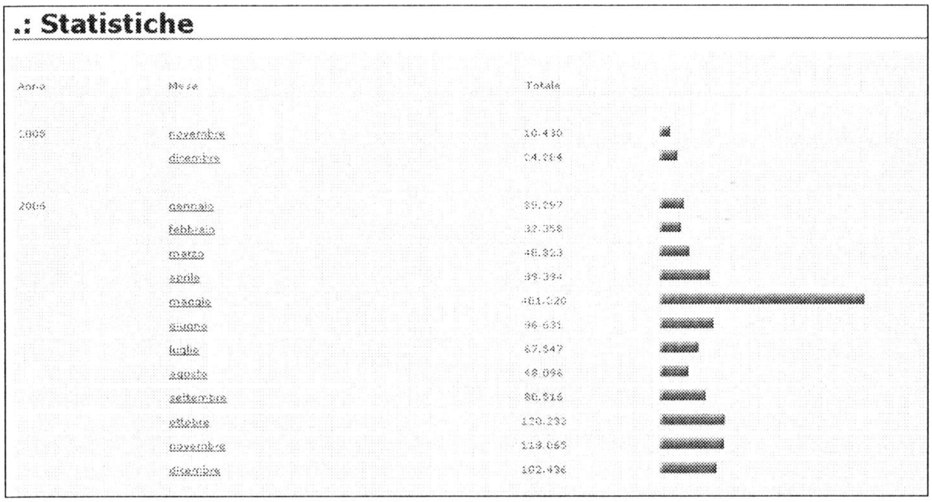

.: Statistiche

Anno	Mese	Totale
2005	novembre	10.430
	dicembre	24.284
2006	gennaio	35.297
	febbraio	32.358
	marzo	48.323
	aprile	89.394
	maggio	481.020
	giugno	96.631
	luglio	67.347
	agosto	48.096
	settembre	80.516
	ottobre	120.293
	novembre	118.065
	dicembre	102.436

as those who run for district councils. They often hesitate to invest large sums in the campaign, even the money needed for a website or blog. ComunaliMilano2006 gave them this chance for free, at the same time allowing them a means for online peer-to-peer dialogue with citizens. The candidates' participation in public square discussions was important to citizens too, giving them the feeling of being listened to by those who would govern the city. This feeling, and their desire to discuss issues dear to them, led citizens to actively participate in public debate, thus transforming the initial impulse (as shown in Figure 8) into a dialog-seeking habit. Discussions were actually largely dominated by individual citizens, which proved crucial after the elections, when elected candidates began to disappear.

Creating a Habit of Participation: PartecipaMi

Immediately after the elections, as ComunaliMilano2006 staff, we started pressing the newly elected candidates, especially city councilors, in order to consolidate the participation practice acquired during the ComunaliMilano2006 experience. Also, as Milan City portal (http://www.comune.milano.it) broadcasts information about City Council activities (agenda of the meetings, realtime streaming of the meetings, and the like), we proposed letting ComunaliMilano2006 integrates it with dialogue facilities. For instance, a topic on the city council's agenda might be discussed online in advance. This proposal was presented, after summer break, to the newly elected city council president, who had participated in ComunaliMilano2006, and to several district council presidents. Despite the interest shown by almost all of them, the proposal has not yet brought about an institutional relationship, primarily for reasons that depend on the political side. The fact that site page views remained high despite these problems attested to our success in creating a point of reference for those citizens and council members interested in discussing city issues. In particular, we noted that the site was becoming an online public space that fosters district-level debate, in which both those elected to the councils and some of those not elected have been participating.

At that point, we redesigned the site. While preserving most of ComunaliMilano2006's technical features, the new site took these new aspirations into account. On January 7, 2007, partecipaMi replaced ComunaliMilano2006.

The partecipaMi home page (Figure 9) identifies four main areas:

- The "Thematic Forums" area groups forums founded in ComunaliMilano2006. The three with the most recent messages are displayed on the home page, while the others are accessed through links.
- The "District Forums" area contains a map of the city's nine districts. Each sector links to a "Direct Line with the nth District Council," i.e., a topical public forum where citizens and district councilors discuss issues affecting the neighborhood.
- The "City Council Forum" area holds discussion related to major city-council business. Some council members occasionally chime in. To support discussion, the right frame of the home page contains a direct link to the upcoming city-council agenda.
- The "Permanent Forum on the City" replaces the brainstorming area as incubator for ideas and discussions that then find room elsewhere in the site, namely as a new thematic forum.

The design choices, especially graphic emphasis on Milan's nine districts on the home page, have proved to be appropriate. Most of the interest in partecipaMi occurs at the district level and has brought (as of January 2008) about 1740 registered citizens, some 50 district councilors, and six city council members. Two of the nine

Figure 9. The home page of partecipaMi

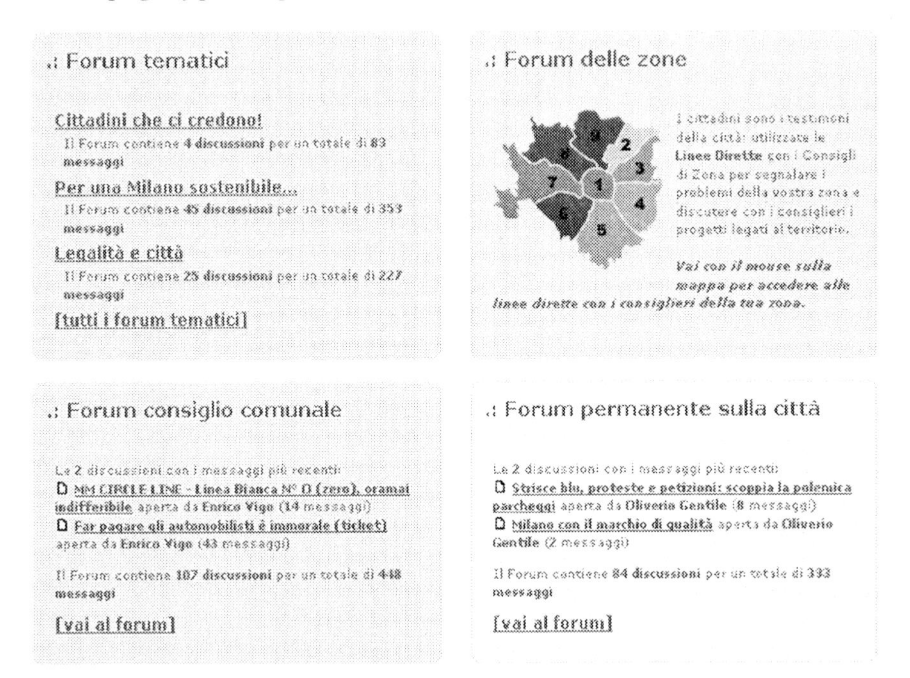

"District Direct Lines" are especially active (with 818 and 679 posts, respectively). Some thematic forums are also still alive, because they discuss hot topics in city life: "Democracy & Citizenship," for instance, contains 271 messages, while "Mobility in Milan" has 211. However, city and district councilors essentially avoid posting in these areas or in the "City Council Forum." It is worth noting that city councilors write in the district forums more often than in the "City Council Forum." Because of a substantial number of representatives' presence, the "District Direct Lines" increasingly become a channel for dialogue between citizens and their elected officials, one far more effective than traditional channels. As a result, a group of citizens decided to use the direct line with the Third District to make their protest heard.

A Protest Moment: The "Francis Bacon School"

The "Francis Bacon School" is a public elementary school in Milan, located in the Third District, near Francis Bacon square. As all public elementary schools in Milan, it is housed in a city-owned building. In late May 2007, a couple of weeks before the summer holidays, the city decided to allocate some unused rooms of the building to a branch office of the city's vital statistics bureau. Elementary-school parents grew anxious at the prospect of unknown adults being in the school building. Another critical issue was opening of the schoolyard gate with no specific surveillance. The parents decided to protest against the decision. Alongside traditional *modes* of protest (Figure 10), amplified by traditional media including local radio stations and newspapers, they decided to use partecipaMi, specifically the "Direct Line with the Third District" forum, as a further channel for asserting their position. On May 28, 2007, one of the parents started a new thread "Bacon School: the Bureau Office Smack in the Middle of School." Other parents backed her up either by replying or by expressing their agreement ('voting' on the message, as described above). Some were already registered on partecipaMi but most registered just to strengthen the protest. They also used the thread to let each other know when the district council was going to meet to discuss the issue, to analyze proceedings of the technical committee charged with studying the situation, and to discuss the inspections of the premises being carried out, as well as to organize their protest. When the partecipaMi community manager brought this thread to the attention of the alderman who had authorized the branch office, he posted a response. While taking the parents' concerns into consideration, he presented the reasons for his decision and claimed planned construction would obsolete their worries. However, protests did not stop and more parents registered on partecipaMi just to question his claims. At this point, someone suggested moving the branch office to the main district-council building, not far from the school. Several councilors, from the governing party as well as the opposition, supported this solution, both online and in official meetings. Finally, on June 4, the alderman announced online his decision to adopt the suggested solution and reported that he had ordered the branch office to move to the main district-council building. An event that could have been a defeat instead turned out to be a collective success. One parent, on behalf of the others, gave online kudos to the alderman "for having been amenable to reconsideration."

The parent's message ended: "And a HUGE THANK YOU to the Milan Community Network and www.partecipaMi.it, who proved instrumental in negotiating and working out a solution." In fact, this positive outcome had been facilitated by the online dialog.

Beyond our satisfaction at having helped solve a problem (a minor one to the majority party but real and felt problem to the families involved), partecipaMi's staff analyzed hit counters (Figure 11) and crucially noted that around the time of the protest there was a new increase in page views (though less pronounced than at election time). The trend grew steadily in the following two months, which witnessed hits outnumbering those of the months prior to the protest.

The fact that peak moments give rise to consolidation is confirmed by the trends of those registering on the site (Figure 12), of the number of blog postings (Figure 13) and of forum-message traffic (Figure 14). The two former trends, after a peak in the month before elections (May 2006), show clear linear increases, while forum traffic shows more-than-linear growth.

Figure 10. Parents protesting in front of 'Francis Bacon school'

Figure 11. Monthly page views from the beginning as ComunaliMilano2006 to now (January 2008)

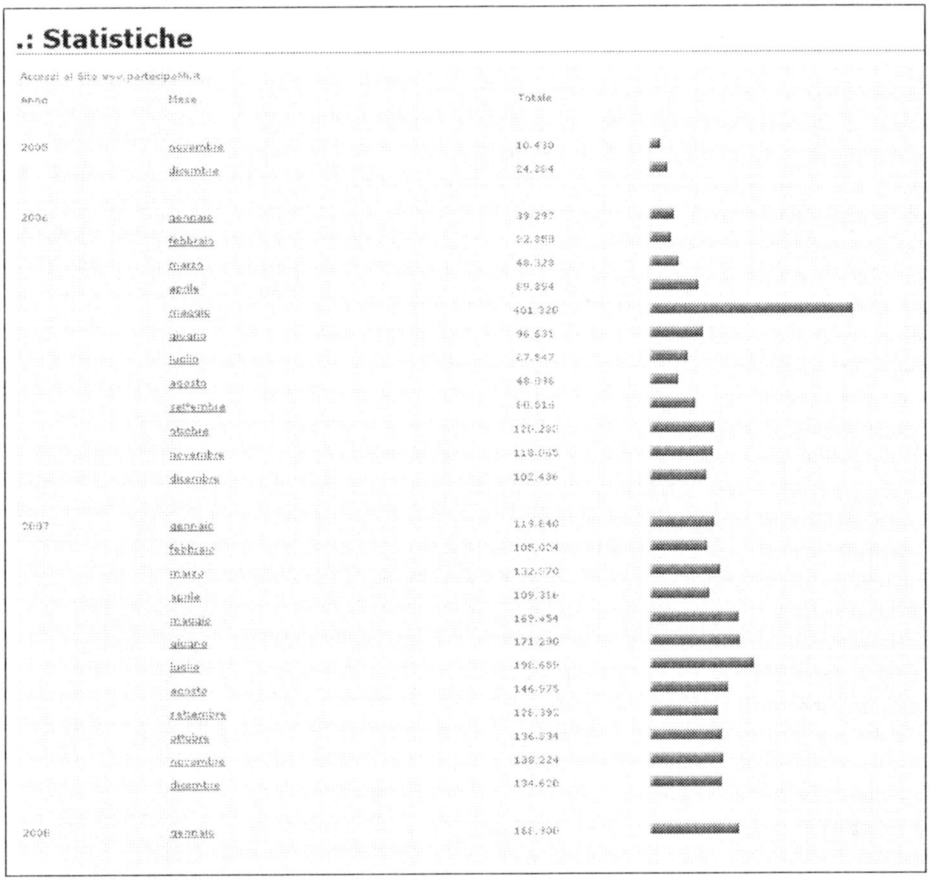

.: Statistiche

Accessi al Sito www.partecipaMi.it

Anno	Mese	Totale
2005	novembre	10.430
	dicembre	24.284
2006	gennaio	39.297
	febbraio	32.858
	marzo	48.328
	aprile	69.894
	maggio	401.320
	giugno	96.631
	luglio	67.947
	agosto	48.036
	settembre	60.916
	ottobre	120.280
	novembre	118.065
	dicembre	102.436
2007	gennaio	119.840
	febbraio	105.004
	marzo	132.070
	aprile	109.316
	maggio	169.454
	giugno	171.290
	luglio	198.689
	agosto	146.975
	settembre	128.392
	ottobre	136.834
	novembre	138.224
	dicembre	134.620
2008	gennaio	168.306

Figure 12. Trend of registrations on the site

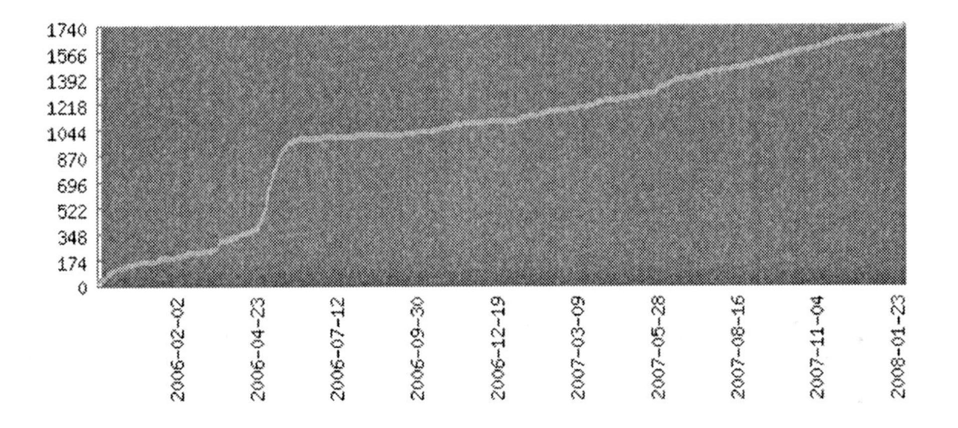

Figure 13. Trend of blog posts and comments

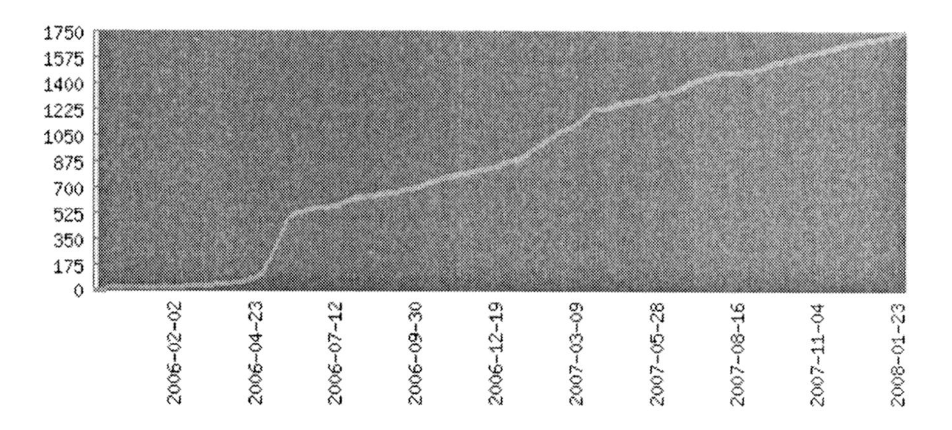

Figure 14. Trend of forum-message traffic

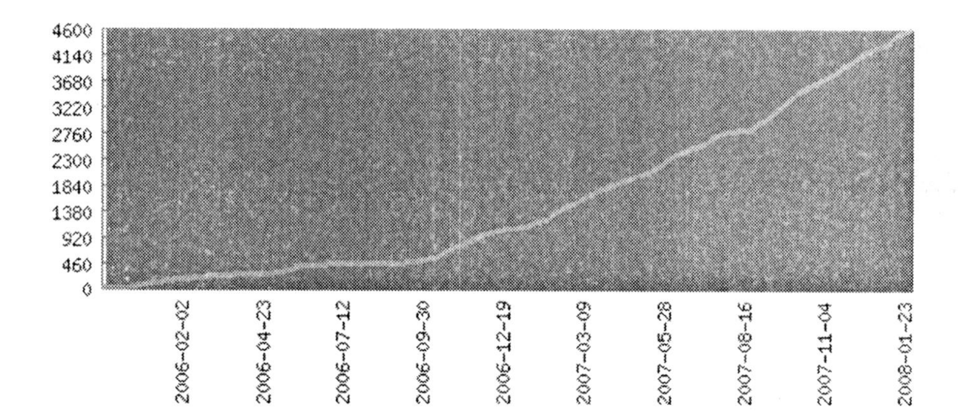

Consolidating peak participation into a more stable practice is not a matter of course. We believe this consolidation was possible due to certain choices made in designing the sites and the software they run on, such as the functionalities outlined above, the metaphor behind ComunaliMilano2006, and the like. If software conceived and designed for other purposes and for a different audience (as is most Web 2.0 software, whose main target is young, highly computer-literate people) is used, it is harder to establish participation processes like those outlined in this chapter. For example, while Beppe Grillo's blog (www.beppegrillo.it) mentioned above was effective in organizing the September 8 protest, it fails to provide the opportunities that enable protest to evolve into positive solutions to people's problems. Although this is likely due, in part, to Grillo's not intending to take the next step from protest to proposing solutions, we believe the limitation is intrinsic to the structure of a blog, where the blogger is owner, on a different level from those who post comments.

CONCLUSION

In this chapter, we have attempted to make the combination of elements that trigger and sustain civic participation less 'arcane.' Part of this alchemy derives from the fact that participation appears in peaks of social engagement rather than a continuum. Although technology is not the sole factor of success, the availability of dedicated software may help consolidate participation beyond the peak moments, towards a more sustained practice of participation.

We have so far identified certain *moments* and certain socio-technical *modes* that help reveal the alchemy of participation. The complex scenario we have reviewed yields two main points for future investigation. On the one hand, there may be other kinds of moments that trigger participation. For instance, one possibility we would like to test involves activating citizens in the online assessment of a decision. On the other hand, the development and testing of appropriate software tools must continue. For example, because e-petitioning is a mode that gives voice to citizens' problems, we are now improving the e-petitioning tools we have already developed, informed by the findings presented in this chapter.

However, we believe that a further step is required. Indeed, citizens' implicit demand appears to be to converge representative democracy with elements of direct or deliberative democracy. As John Stuart Mill long ago pointed out: "What can be done better by a body than by any individual is deliberation. When it is necessary or important to secure hearing and consideration to many conflicting opinions, a deliberative body is indispensable" (Mill, 1861/2004, p. 52). If this is true, what is needed are tools that support not only the expression of ideas, protests, and suggestions, but also support citizen involvement in the decisions relevant to their lives. The increasing attention to developing online deliberation technology is taking us in this direction (De Cindio and Schuler, 2007; Davies and Noveck, 2008; Schuler, 2008).

ACKNOWLEDGMENT

We wish to thank Oliverio Gentile, the community manager of ComunaliMilano2006 and partecipaMi, and Antonio De Marco, whose thesis project was ComunaliMilano2006. Without their relentless work, our projects would not have been possible, thus again demonstrating that the human element is essential to the functioning of such socio-technical systems. This work has been partly funded by the research project PRIN # 2006148797_002 "Soluzioni informatiche a supporto della cittadinanza digitale, della accountability democratica e dei processi deliberativi."

REFERENCES

Beer, D., & Burrows, R. (2007). Sociology and, of and in Web 2.0: Some Initial Considerations. *Sociological Research Online, 12*(5).

Bishop, A.P. (1994). *Emerging communities: Integrating networked information into library services.* University of Illinois at Urbana-Champaign.

Casapulla, G., De Cindio F., Gentile, O., & Sonnante, L. (1998). A Citizen-driven Civic Network as Stimulating Context for Designing On-line Public Services. In *Proceedings of the fifth Participatory Design Conference "Broadening Participation"* (pp. 65-74). Seattle.

Davies, T., Gangadharan, S.P., & Noveck, B.(2009, forthcoming). *Online deliberation: Design, research, and practice.* CSLI Publications/University of Chicago Press.

De Cindio, F., Gentile, O., Grew, P., & Redolfi, D. (2003). Community Networks: Rules of Behavior and Social Structure. *The Information Society.* Vol. 19(5) (pp. 395-406), November-December 2003.

De Cindio, F. (2004). The Role of Community Networks in Shaping the Network Society: Enabling People to Develop their Own Projects. In Schuler D. and Day P. (eds.): *Shaping the Network Society: The New Role of Civil Society in Cyberspace*, MIT Press.

De Cindio, F., De Pietro, L., & Freschi, A.C. (eds.) (2004). *E-democracy: modelli e strumenti delle forme di partecipazione emergenti nel panorama Italiano*, FORMEZ-Progetto CRC.

De Cindio, F., Peraboni, C., Ripamonti, L. A. (2007).Community Networks as lead users in online public services design, *Journal of Community Informatics, 3*(1).

De Cindio, F., & Schuler, D. (2007). Deliberation and Community Networks: A Strong Link Waiting to be Forged. In *Proceedings of CIRN Conference 2007 "Communities and Action"*, Prato 5-7 November 2007.

Edward, A. (2006). *Online Deliberative Policy Exercises and Styles of Citizenship: Issues of Democratic Design*. Position paper presented at the DEMOnet Workshop on eDeliberation Research, Leeds, UK, October 16th, 2006.

Kubicek, H., & Wagner, R.M. (2002). *Community networks in a generational perspective: The change of an electronic medium within three decades. Information*, Communication, and Society, Vol.5 (pp.291-319).

Ginsborg, P. (2006). La democrazia che non c'è. Einaudi, Turin (in Italian).

Granieri, G. (2005). Blog Generation. *LaTerza*. (in Italian)

Luisi, P. (2001). Tre buoni motivi per considerare finita la rete civica (così come l'abbiamo sempre conosciuta). In Luisi P. (ed.): Le reti civiche in Italia. Punto e a capo. (in Italian) Quaderni di Comunicazione Pubblica, CLUEB, Bologna.

Macintosh, A. (2006). Argument Maps to Support Deliberation. Position paper presented at the DEMOnet Workshop on eDeliberation Research, Leeds, UK, October 16th, 2006.

Mannheim, K. (1964). 'Das Problem der Generationen' , in K. H. Wolff (ed.)

Meikle, G. (2002). Future Active: Media Activism and the Internet. Routledge. London.

Mill, J. S. (1861) Considerations on Representative Government. London. (reprint: 2004: Kessinger Publishing).

Musser, J., & O'Reilly, T. (2006). Web 2.0—Principles and Best Practices. O'Reilly Radar, Tech. Rep., Nov. 2006.

Opp, K.-D., Voss, P., & Gern, C. (1995). *Origins of a Spontaneous Revolution: East Germany, 1989*. Ann Arbor: University of Michigan Press.

O'Reilly, T. (2005). Web 2.0: Compact Definition?, O'Reilly Radar blog, 1 October 2005 Retrieved on January 2007 from: http://radar.oreilly.com/archives/2005/10/web_20_compact_definition.html.

Rheingold, H. (2002). *Smart Mobs: The Next Social Revolution*. Cambridge, MA: Perseus.

Ranerup, A. (2000). On-line discussion forums in a Swedish local government context. In Gurstein M. (Ed.) *Community Informatics: enabling communities with information and communication technologies* (pp.359-379), Hershey,PA: Idea Group Publishing.

Schuler, D. (1994). *Community Networks: Building a New Participatory Medium*. Communications of the ACM, Vol.37(1) (pp.39-51).

Schuler, D. (2000). New communities and new-community networks. In Gurstein, M. (Ed.): *Community Informatics: Enabling Communities with Information and Communications Technologies*. Hershey,PA: Idea Group Publishing .

Schuler, D. (2001). *Cultivating society's civic intelligence: patterns for a new 'world brain'*. Journal of Information, Communication and Society, 4(2).

Schuler, D. (2009 forthcoming). Community Networks and the Evolution of Civic Intelligence. *AI and Society Journal.*

Schuler, D. (2009 forthcoming). Online deliberation in-the-small and in-the-large: Principles, assertions, experiments, and objectives. In Davies, T., Gangadharan, S.P., & Noveck, B. (Eds.) *Online Deliberation: Design, Research, and Practice*. CSLI Publications/University of Chicago Press.

Selznick, P. (1996). In search of community. In Jackson, W. and Vitek, W. (Eds.) Rooted in the land: *Essays on community and place* (pp.195-203). New Haven: Yale University Press.

Silver, D. (2004). The Soil of Cyberspace: Historical Archaeologies of the Blacksburg Electronic Village and the Seattle Community Network. In Schuler D. and Day P. (eds.): *Shaping the Network Society: The New Role of Civil Society in Cyberspace*, MIT Press.

Venkatesh, M. (2003). The Community Network Lifecycle: A Framework for Research and Action. *The Information Society.* Vol. 19(5) (pp.339-347).

KEY TERMS

Blog: A blog is a frequently updated site (often on daily basis) where posts have the form of journal or diary entries. The blogger's (the owner of a blog) posts often are opened for discussions, generating online conversations.

Community Network (CN): Community Networks (and related initiatives, such as "free nets" and "civic nets") are online enabling environments that promote citizens participation in community affaires (Schuler, 2000).

Community Manager: The community manager orchestrates the ongoing life and the structure of the online community. S/he also activates and coordinates the community's activities.

E-Participation: Participatory processes supported by ICT (information and communication technologies). E-participation is the use of ICT to broaden political participation by enabling citizens to connect with one another and with their elected representatives (Macintosh, 2006). Usually, e-participation is used as a macro-category that includes a variety of areas such as e-consultation, e-legislation, e-petition and e-deliberation.

Facilitator: An individual who enables groups and organizations to work more effectively, collaborate and achieve synergy. Within online discussions, a facilitator helps participants listen to each other and envisage solutions together.

Forum: Inspired from ancient *fora* (the public spaces in the middle of a Roman city which held public meeting or assembly for open discussion), online forums allows for messages to be posted and kept on a website for further (public) readings and discussions.

Moderator: In online communities, the moderator has the role of mediator among the actors in a discussion. S/he typically determines messages' approval according to the community rules. So her/his activity should not be seen as censorship, but as a necessity and a protective service to the community. (De Cindio et. al, 2003)

Web 2.0: The term "Web 2.0" was coined by O'Reilly Media at a conference in 2004 (O'Reilly, 2005) and it has become the label to refer to the next generation Web. The main characteristic of Web 2.0 is the central role that individuals play in creating social interactions, collaborating and sharing information online.

Section III

Engagement of Urban Communities

Chapter VIII
Fostering Communities in Urban Multi-Cultural Neighbourhoods:
Some Methodological Reflections

Michael Veith
University of Siegen, Germany

Kai Schubert
University of Siegen, Germany

Volker Wulf
University of Siegen, Germany

ABSTRACT

Societies face serious challenges when trying to integrate migrant communities. One-sided solutions do not pay tribute to the complexity of this subject and a single academic discipline provides no proper methodological approaches to the field. An inter-cultural computer club in an urban multi-cultural neighbourhood illustrates these phenomena: appropriate argumentations and models can only be found in a theoretical net of scientific disciplines. Categories in a complex socio-cultural field have to be uncovered. These categories can be explained with the help of the theoretical net. We develop a three-dimensional model combining empirical tools with the research strategy of participatory action research and grounded theory as a guide to theorizing the field. This model is introduced here as a means of socio-technical design and development.

INTRODUCTION

Migration is not a novel phenomenon in history. It has happened all the time and may have numerous reasons and causes. Each society as well as each generation is confronted with social and cultural changes which accompany migration. Those societies which lose people face different challenges than those which allow migrant groups to enter. Though migration movements *out of* a society are another interesting and important phenomenon to investigate from a socio-technical point of view, we are focusing here on the consequences following the integration of migrant groups *into* a society. With regard to the existing literature we can differentiate between four possible realisations of the analytic category of integration, i.e. assimilation, inclusion, exclusion, and segregation (cf. Berry 1992). Due to its various social circumstances, integration turns out to be (a) a very complex social phenomenon, (b) more an on-going process than a stable condition, and (c) subjected by continuous changes. Furthermore, it is bound to individuals and groups and is thus closely related to questions of socio-cultural identity.

Commonly recognized as necessities for on-going and successful integration is contact and communication among migrant communities and member communities of the reference society[1]. Especially schools appear to be a common place for these two prepositions, at least contact between all ethnic communities is an inescapable institutional factor (due to compulsory school attendance). Following this underlying condition—besides other motivational factors—we founded the intercultural computer club *come_IN* in cooperation with the elementary school *Marienschule* in Bonn, Germany. Here, parents are invited to accompany their children to the club and to work with them on hands-on projects which are relevant to all participants, i.e. these projects are negotiated in advance. All necessary work and all content is worked out and collected with the help of information and communication technology (ICT), mobile devices provide good support in documenting activities which take place outside the club room. By doing so, a commonly shared practice is motivated which leads to a heterogeneous community, spanning from multiple ethnicities, generations, and roles. A mixed cultural *spirit* is negotiated which mutually influences the socio-cultural nuance of participants' identity—integration is fostered.

In Stevens et al. (2004 and 2005) we tried to find evidence for a couple of these assumptions in a rather broad manner. These papers, however, make the attempt to set our undertaking on a methodological fundament. By doing so, we developed a three dimensional model of combining empirical tools with the research strategy of participatory action research and grounded theory as a guide to theorizing the field. This chapter is a mostly theoretical discussion about dealing with communities and technologies in practice. Its main contribution is to position the project come_IN in a theoretical as well as methodological framework. A rigorous participatory action research (PAR) approach established a firm basis for further research strategies within come_IN (for an overview on PAR, cf. Greenwood et al. 1993, or Kemmis and McTaggart 2004). As this further strategy we chose *grounded theory* (cf. Strauss and Corbin 1998). We will show how first steps of a more wide-spanned grounded theory (GT) within come_IN has been realized, how theory plays an important role within this undertaking, and how consequences are derived from the emerging complex net of concepts and categories. First results and empirical findings about come_IN can be found in Veith et al. (2007), where we focus on identity and role affiliation mediated by the club.

A cyclic action research program requires moments of theoretical reflection: is it still possible to return to the scientific discipline we started from (namely the fields of computer-supported cooperative work, computer-supported collaborative learning and human-computer interaction). Especially in a highly complex socio-technical field like come_IN, it is hardly possible to purely concentrate on questions of communities and technologies interaction. Thus, results which are relevant to multiple disciplines can be found and again the methodology must be regarded with intensive care. Again, GT is the most promising candidate for such an interdisciplinary scientific attempt that enabled us to carefully select insights from come_IN to demonstrate the interplay of research, action, design, and participation.

In the first chapter we introduce the field: the urban neighbourhood, the school, the club, its members, and the club-based projects. This is followed by theoretical preliminary consideration about integration, identity, and culture. We discuss some theoretical categories of social integration and our methodological approach. This is compared with empirical findings and experiences in the field. An outlook and overview about the future work regarding the project is given at the end.

THE INTERCULTURAL COMPUTER CLUB COME_IN

The Neighbourhood

Come_IN takes place in the so called "Bonner Altstadt", a old town neighbourhood within the city of Bonn. The Bonner Altstadt has a population of about 8,700 inhabitants[2]. The social and cultural structure of this district can be characterized as a colorful mixture of different communities. Statistical data characterizes this situation today: The quarter has a high rate of immigrants (~28% of the population, in comparison to 22% in Bonn as total) and a low education rate (35% have a Hauptschulabschluss[3] and 32% of those in employment are labourers). However, the German community in Bonn consists to a considerable part of academics, partly former students who stayed in the area after their graduation.

In the German context, elementary schools are important places where collocated but segregated communities meet. Since most kids attend the public elementary school in their local district, schools became one of the few places where people necessarily come into contact with different cultures. Therefore, schools in multicultural neighbourhoods face considerable challenges in dealing with a differentiated population of pupils.

The Elementary School

The computer club "come_IN" is conducted in cooperation with Marienschule, an elementary school in the Bonner Altstadt district. The focus lies on open and work oriented lessons, e.g. in small groups, workshops, projects, and so on. Each class room is equipped with 2-3 computers which can be used as resources in daily work. Pupils are taught in classes with mixed age-groups. Beyond the neighbourhood, the school has gained reputation for its innovative pedagogies and didactical practice. The club has 13 personal computers, one multi media computer, several digital video and photo cameras, a video projector, and several other digital devices, which are used for come_IN. There is a large, round table in the middle of the club room, which is used for discussions and other communicative events.

While innovative in its didactics, Marienschule experiences several serious problems in dealing with their highly differentiated pupils. Offering appropriate education to third generation children of Turkish origin turns out to be a serious challenge for a variety of reasons. A considerable part of these children starts school with little or even no German language skills. This is particularly surprising since they are the children of second generation Turks who often speak fluent German. Finally, the children of the Turkish origin seem to have less access to ICT than the children of German origin. This is not only a question of the availability of computer hardware and software in their homes, but also of the level of computer literacy within their families and networks of friends.

The Club come_IN

Due to these conditions the principal of Marienschule adopted our ideas of multicultural communities of practice (CoP) and became involved in the project. In the following, we will present the core concepts which were developed in an attempt to establish a multicultural CoP.

We decided to establish a shared practice across the ethnic communities by encouraging participants to jointly work on computer-supported projects. We assumed that dealing with computers and digital media would be attractive for many participants within the different ethnical communities. However, we believed that by just offering an infrastructure for a shared practice would not be enough to start the process (cf. Rohde 2004).

Influencing identities in the different ethnic communities would not only need to attract the students, but also the parents. Since the success of schooling is highly related to the social context of the children, schools and parents need to work together (Lanfranchi 2004). Thus, we introduced the rule that children may only come to the club if accompanied by at least one parent. By this rule, we used the attraction computers have for children to get parents involved in the process. In addition, it is hard for elementary school kids to manage complex projects by themselves. Conceptual support from their parents was needed to realize the envisioned project's outcomes. While establishing a project-related practice, we assumed that foreign and German parents would have more exposure to each other and therefore communicate with each other.

Members of come_IN

We are fortunate enough to have a wide range of societal dimensions within the club. These dimensions can be summarized with regard to age, ethnicity, and occupational background. Active participants like parents and 6- to 12-year-old children from different socio-cultural backgrounds (mostly from Turkish or German origin) work together with tutors form either educational or academic professions. To shed light on the socio-economic situation, the participants from the German community are mostly from white collar jobs (academia or managers) whereas members from the Turkish community are mostly observed to be pertaining to blue collar and more service oriented jobs, e.g. shop keepers or factory employees.

Projects Within come_IN

One of the first projects deals with a multimedia documentation of family histories. Right now, three generations of Turkish immigrants live in the neighbourhood. However, their cultural histories are only poorly, if at all, documented. The project's goal is to present these family histories together with German ones from the same neighbourhood. Such a collaborative history project may support the growth of a shared identity across the different communities.

Another project deals with sports, more particularly, soccer (or "football" as it is commonly called). Children, parents and teachers join forces in this project. In the beginning, children and their parents prepare a number of matches between different teams of pupils. During these matches, various ICT are used to record the project. Parents and children are improving their abilities in the use of ICT while creating stories, photo presentations and a film about the football matches. In addition, after the first matches, several Turkish parents arrange for another match against a Turkish football club (of which several come_IN members are also members) in Bonn. This will give the computer club more exposure to their urban neighbourhood, creating a new level of impact with an important potential for the future.

INTEGRATION, IDENTITY AND CULTURE

Integration, *identity*, and *migration* are three central terms in come_IN's terminology. As they are linked to several disciplines (as sociology, psychologies, social work, cultural studies etc.), they have to be examined from more than one scientific perspective. From this point of view, come_IN and all other attempts of active social integration are interdisciplinary projects. In the following, some basic aspects of the come_IN community are explained. Special attention is drawn to possible synergy effects which may occur beyond the borders of scientific discrimination. The main question at the end of these theoretical considerations must be the practical (which means design oriented) union of the topics discussed. All insights—in this case derived from three distinguishable disciplines—with no doubt have to provide implications for design matters. One question is: Can we design ICT in the field of come_IN? Of course, this question has several sub-questions which have to be answered in the first instance, but more about that later on. Let us now take a closer look at the term *integration* and how it can be understood.

Social Integration

The heterogeneous organization of social entities within a social system is one important and significant characterization of modern societies. Thereby, diversity and flexibility are dynamic factors which have great influence on the composition and functioning of social life. Members of society are continuously engaged in mechanisms of re-orientation and balancing. This permanent process of consistent alignment needs a social copula, an integrative motor. Otherwise, social cohesion is endangered. Émile Durkheim (1997) calls this *social putty* 'organic solidarity'. He related and observed this principle with regard to the division of labour, but for good reason it is worth to relate it also to other aspects of social life, as specification does not exclusively happen within working contexts. Here, we want to call Durkheim's organic solidarity *social integration*, or simplified integration.

Integration is, from this perspective, an on-going process, not a status. It links and inter-relates individuals and groups with other social entities on all levels of societal organization, i.e. micro-, meso-, and macro-level. The more complex and differentiating the developing system structure is, the more colourful and heterogeneous the social cohesion of a given social system, and vice versa. The individual and therefore the identity of the concerned subject is also confronted with this permanent interplay. Subjects' consciousnesses as well as actions are affected heavily by these balancing forces. Knowledge about and experiences with the social environment (which includes other individuals) stimulates this (re-)orientation. As a matter of fact, this is only possible because of the interaction of social entities, i.e. prominently but not exclusively *communication*. Communication, in turn, appears to be the means of negotiating social conflicts which naturally occur in heterogeneous environments. The reaching of consensus keeps the running of integration, whereas failing consensus as well as ignoring conflicts hinder the process of integration (for a deeper insight into conflict and consensus, please consider for instance Coser and Larsen 1976).

The argumentation above brings the following aspects of *integrative environments* (let us simply call them *communities*) to light:

1. diversity of members,
2. continuing contact of members,
3. knowledge and experiences among members,
4. communication as a means,
5. reaching of consensus to overcome conflicts. Implicitly mentioned is
6. consciousness of one's own identity, i.e. self-consciousness.

To sum up this paragraph, we find here some analytical categories which all arise from a classic sociological perspective. This point of view originally describes how the division of labor within a society is organized. With regard to learning we can find some similarities to the organization of communities of practice (CoP) (cf. Stevens et al. 2005 for the discussion of come_IN as a CoP). There, work is understood as one context of commonly shared practice, but not exclusively the only one (see Lave and Wenger 1991, and Wenger 1998). In come_IN social integration is further specified, namely by the social integration of migrant groups. Specification and heterogeneity here is broadened additionally. Geographical, linguistic and cultural differences tremendously increase the grade of diversity within the come_IN community (in comparison to mono-lingual and -cultural communities). In Veith et al. (2007) you can find, as mentioned above, first findings and results regarding this.

Social Identity: Why am I here?

Up to this point we looked at mechanisms which are strongly related to a large number of people, i.e. groups, communities, societies, etc. Integration, as we argued, is something that occurs when a community appears to be more than the sum of its members, i.e. it is a social phenomenon. Identity, as it is said, is something a person owns. For this purpose it is worthy to leave the sociological trace. For identity social-psychologists—besides others—provide a very interesting and fruitful explanation. It considers personal consciousness as well as the interplay of self-consciousness and social environment (see also Mead's (1934) concepts of 'me' and 'I'). As already mentioned in the previous paragraph, we presume that individuals live in an extremely heterogeneous and highly complex social world. While a person permanently interacts with her/his environment s/he "draws a picture" of her/hisself. This picture is affected by personal predispositions as well as by influential factors from the social world (cf. Tajfel 1982). For Henri Tajfel, identity is social identity, i.e. one's self-perception is stronger determined by a person's group membership than by personal predispositions. Group membership is an important driver of identity building. Within a group as well as between members of different groups, mechanisms of social comparison take place. Comparisons with the 'outside of a group' evolve possibilities of social differentiations, which in turn intensifies effects of intergroup attachments. We agree with Rohde (2004) who uncovers obvious parallels with the inner group concept of CoP, although these analogies are not explicitly mentioned in the underlying literature. Comparisons within a certain group, however, have two types of effect: First, they continuously manifest a person's self-perceived identity. And second, inner group comparisons stimulate the flexible balancing of a group's structure. Obviously, this shows similarities with attributes of social integration, which we mentioned in the previous section. Additionally, social categorisation and the building of social stereotypes accompany the process of integration (cf. Turner et al. 1987). Not only does an individual again reflect and manifest his/her own social identity. By allocating meaning to his- or herself and to the world (which means categorisation and stereotyping), a person is also able to deal with the complex social environment.

But where is the bridge between social identity (about which we were talking up to this point exclusively) and culture? And what is culture in our context? What actually is the content of the category 'group' (or 'community' as we may name it) which influences and determines a person's identity so strongly?

Migration and Culture

Following the cultural anthropologist Clifford Geertz, man is socially located in a net of meaning "he himself has spun" (Geertz 1973: 5). This net he calls culture. Obviously, culture in this sense cannot be distinguished from any social behaviour at all[4]. The concept of culture, for Geertz, is of semiotic nature. We assume that Tajfel as well as Lave and Wenger share this idea. The allocation of meaning (see above) as a social practice can clearly be identified as a semiotic element within Tajfel's theory. So, culture is flexible and has a plural form, i.e. different cultures exist. However, cultural differences are also common, as different social webs allocate different meaning to certain subjects, entities and so on. If, as it seems, culture is dynamic, flexible, irritated by on-going changes and therefore has to be continuously negotiated, it is subject within communities; culture becomes social reality in community practice.

Historically, Geertz examined "foreign" cultures in a way that he visits the countries in which he could get in contact with cultures different from his own. Nowadays, it is not that easy anymore. Migration interferes the bias character of culture. Whereas Geertz had "his" culture A in mind he could examine culture B locally separated in order to uncover parallels and differences in allocating meaning, migration mixes apparently clear cultural "borders". "Normal" social integration, Durkheim (1997) finds, is expanded by the mobility of migration. James Clifford (1988; 1999) describes a commonly experienced feeling which accompanies migration: the feeling of losing one's roots. Migration is a global phenomenon which allows global cultural contact. Apparently closed societies face serious challenges in dealing with the situation of integrating "rootless" communities (see Appadurai 1996). Homi Bhabha (1994) calls this "a productive provocation" against the "notion of pure, national identity". The social integration of ethnic minorities, thus, is a business for the migrant communities as well as for the mainstream society (represented by all communities within a common society), exactly the repeatedly mentioned interplays of homogeneity-heterogeneity, conflict-consensus, etc. (see above). Only to classify, name and differentiate new communities does not enrol the integrative process. Also assimilation cannot be the appropriate way to include socio-cultural diversity into society. Or with Appadurai's metaphor of "rootlessness": Roots have to grow again, they cannot simply be changed.

Summary of Theoretical Considerations

Coming back to our core field of interest, i.e. applied software development and communities and technologies, what implications can be drawn for come_IN? First of all, the theoretical considerations provide an overview about relevant criteria within the complex socio-cultural web of the intercultural computer club. Of course, concepts like CoP could be found from the early beginning (see again Stevens et al. 2005), but the whole importance of the project in a socio-cultural context came to light step by step.

If contact and identity are two elemental factors of social integration—at least with regard to the way we present it here—the following "parameters" of continuing and permanent integration can be proposed:

1. **Participation:** As one measure of contact, the gradual weight of public as well as institutional participation is of further interest.
2. **Communication:** Beyond contact, how frequently is the communication between all members of the community? What are the contents of the communication?
3. **Knowledge/Experience:** What is the result of contact and communication? What and how much do the participants know from and of each other? Do any common cultural basic allocations emerge?
4. **Acceptance:** In how far are these allocations commonly accepted?
5. **Conflicts/Consensus:** To what degree does the community allow diversity within the club? Does this vary? How do participants

deal with conflicts, which obviously must emerge when heterogeneous members interact with each other?

For several reasons it is not preferable to stick to purely theoretical considerations too much in hands-on projects like come_IN. Nevertheless, as basic orientations theoretical categories are a fair means to develop strategies to enter the field of application.

BEYOND DUCK AND COVER: APPROACHING THE FIELD

As software developers we are unknown in the field of social integration. On the other hand, ICT is dominating more and more aspects of every-day life. Though it has become naturally understood that working life is a subject for software developers (where they conduct work place studies), it is still a bit odd to find programmers in non-commercial fields of social life. As a result, our discipline does not provide a lot of theoretical or methodological armamentarium which could help to investigate novel fields nor does it teach how to design appropriate software to support social needs. Ironically, we are thus facing problems of integration, as well as migrant communities do. We are new in a rather stable system. We participate in a field which is characterized by numerous differing roles and sub-groups. "Researcher" is just one role, just one realisation of legitimate peripheral participation within the community of practice of come_IN. We find ourselves in the situation of being part of the community which we want to investigate, i.e. at least partly, we also explore ourselves.

As a result of this and other implications that arose from what was previously said, we find some practical circumstances, which play significant roles in the choice of an adequate methodology.

1. **Lack of theoretical grounding:** The subject of social integration mediated by ICT in the field of an intercultural computer club can hardly be associated clearly and definitely with one single scientific discipline. Furthermore, there are hardly any data from other projects available which would allow drawing parallels. An inductive, subject-oriented theory has to be developed initially. The theoretical considerations that we propose in this chapter (see section "Integration, Identity and Culture") can help us to narrow the range of relevant theories, but they can never fully explain how and why come_IN works how it does.
2. **Adjacency to common experience and context:** Never can there be only one theory covering the field of come_IN. Theories have to be established on the basis of the complex socio-cultural web of come_IN community. Thus, a method has to pay tribute to this contextualized net of significances, which produces comparable data and then allows us to make generalizations.
3. **Practical Relevance and Politics:** Come_IN, as a contribution to society, tries to surmount barriers like the digital divide, and some other fuzzy concepts like the 'cultural gap'. It tries to provide a means to increase equal opportunities for all participants. Additionally, we as applied computer scientists want to design software, we need 'practical results' in order to be 'productive'. Thus, we do not only analyze the field but do also influence it, with support, software and other forms of participation. All these points deny a clear-cut distance to the field, at least for those researchers who do the participatory job. Practical relevance is one of the main factors for our approach.

The methodological approach we want to draw from these three main conditions is called *grounded theory* (GT) (see for instance Strauss and Corbin 1998). It is a method of understanding a subject, of grounding a field. The field is theory and practice becomes theory. Practice does not exist because of or through theory. Theory derives from practice. From a developer's point of view, we want to argue that design also comes from practice, and should therefore arise from the interplay of context and user. By doing so, come_IN, as a subject, becomes a socio-technical community; its structure can be influenced on two levels, i.e. technical and organizational (cf. *integrated organization and technology development* (OTD): Wulf and Rohde 1995) in order to stimulate identity changes. Although interferences should be considered wisely and not haphazardly, they are the only appropriate means of participatory design in a come_IN community for the sake of equal opportunities for all participants.

Our methodological approach we applied to the field points out the strong interdependence between practice, academic point of view and interaction in the field. Basis for all undertakings within the project is practice, i.e. the come_IN community and its embeddedness into socio-cultural circumstances. Accruing from this fundament three scientific flanks come into focus, i.e. paedagogy (school), computer science (socio-technical support), and social work (neighbourhood)—all of which are representing one essential abutment for come_IN community. Furthermore, three dimensions of empirical investigation accompany the project, which mold—as a meaningful union of cyclic inter-subjective evaluation, the constructive motor of social change and development, i.e. organizational as well as technical change (see Table 1).

Now it is time to combine the theoretical considerations from the first sections of this chapter with the methodological reflections above. If we believe in the empirical power of our three-dimensional model—as we do—then only practice can prove us right or wrong.

SOCIAL INTEGRATION WITHIN COME_IN

As a matter of fact, the order of presentation we choose in this chapter does not represent the natural procedure of grounded theory. Here, it looks like we worked out theory first, tried to find an innovative way of negotiating meaning in the field, and then come back to theory by comparing practice with it. Actually, the act of in-depth theory building followed practical experience from come_IN, chronologically as well as with regard to content. Of course, we entered the field with some hypotheses (see Stevens et al. 2004 and 2005), but the main work on theory began after first empirical findings were gathered and we continued with each project cycle[5].

As an example, we chose social integration as a main topic for our investigations. By doing so, we follow the question of how ICT may foster social integration as a normative phenomenon. Please consider some theoretical categories from

Table 1. Three levels of methodological approach

Name	Means	Description
Empirical tools	Quantitative data, observations, interviews	Getting things done
Research strategy	(Participatory) action research, project orientation, narration	Understanding and forming practice
Theorizing concept	Grounded Theory / Design	Understanding and forming theory and design, design understood as the expression of socio-technical theory

Table 2. Theoretical categories of social integration within come_IN

Category	Main questions
Participation	How much participation can be made out?
Communication	How frequently is communication? What are the contents?
Knowledge/Experience	What is the result of contact and communication?
Acceptance	In how far are commonly negotiated meanings accepted?
Conflicts/Consensus	(How) do participants deal with conflicts?

section "Summary of Theoretical Considerations" again (Table 2):

Participation and Use of ICT

Participation turns out to be the main quantitative parameter within come_IN. Approximately 20-25 people attend the club each week. At a minimum about 14 members are always present, e.g. after holidays. At a maximum we welcome almost 40 people, e.g. during the football event. However, it is elemental to look at the participation of different members of the club. Some parts of the community attend the club frequently whereas other parts hesitate to participate regularly. The alleged fact that some would stay away from the club as they were not interested in the diversity within the club, cannot explain what actually is going on or not going on in come_IN. Many participants even say it is one of the great features of the club to get in contact with people from different socio-cultural backgrounds. More people from German origin join the club frequently than from Turkish origin. Social composition seems to be a model of explanation for the apparent disinterest of some (non-)members. Most of the Turkish members are workers whereas most Germans come from middle class. They describe their use of ICT quite differently. Most Germans are used to ICT from their work places, whereas Turkish men mostly know it from their leisure time. Interesting enough, some must learn how to work with the computer, some how to play with it. A third group even has to learn how to deal with ICT at all, which are many Turkish mothers which mainly gaze at the computers as unknown machines (see further below). Commonly shared practice within the club has to pay tribute to all three learning scenarios (working with, playing on, and respectively gazing at ICT). According to these different forms of use, the linking of computer work into the organization of projects is realized in come_IN. By using *theoretical sampling* it will be possible to further clarify the dependence between manner of use and participation. Observations, narrative interviews, and other means of empirical investigation will again be basis for this attempt. Especially the unclear assignments and the low participation in the phase of reflection need further explanation.

Communication and the Role of Common Language

German is the reference language within come_IN though it is, of course, allowed to speak other languages, i.e. mainly Turkish. Children turn out to be translators for Turkish parents, which seems to be a productive way to enable intercultural communication. However, kids as cultural brokers works out fine but is also recognized as being insufficient for deeper discussions (more further below). Topics that are communicated in come_IN mainly deal with commonly recognized interests, i.e. likes and dislikes. Discussions of how one thinks are rare, utterances of what is interesting are regular. Three main themes are in focus of come_IN activities, which are also supported by come_IN's socio-technical structure. (1) Kids; it is parenting and educating children which is interesting for all adults in the club. (2) Urban neighbourhood; it is the place where all members live what further fascinates the community. And (3) computers; it is ICT which also guarantees regular participation within the club. All members of come_IN consider these topics important for come_IN, but there are interesting correspondences between user group (see above) and favourite topics. Besides these three main topics of come_IN projects have been negotiated which may be akin to fields of computers, neighbourhood or kids, but set different foci. The come_IN community and the unofficial "steering committee"[6] also try to combine elements of all three main interests, e.g. the football project where ICT, neighbourhood and kids are involved.

Knowledge and Experience, or Breaking down walls?

The examination of knowledge and experiences within come_IN has to be recognized as a process. Most of come_IN members mentioned that knowledge about people from different backgrounds grew after time in come_IN. But it also appears that mainly German participants still now little about other members of the community, even if they are direct neighbours. First attempts have been made to overcome this lack, e.g. the football project, and a journey to Strasbourg, France, to visit the European Parliament. In the first case all participants could learn more about members' free time activities, favourite players, etc. whereas on the journey to Strasbourg people could gather experience from each other in direct interaction over two complete days. Still, knowledge and experiences within the club during normal club business is in deficit and not satisfying to the participants. With regard to further action in the field, the relations between communication, contact and experienced knowledge become obvious.

Acceptance: Rules and Diversity

The acceptance of rules and diversity happens on various levels of community interaction. The most dominating factor, nevertheless, is the school, i.e. the school principal and teachers within come_IN. As the old school principal has retired now (she is still member of the club), the new school principal has to find his way into come_IN first. Rules in come_IN (see above "The Club Come_IN") are mainly negotiated by the non-official steering committee. Its composition varies from time to time and it represents mostly those members that "stay a bit longer after club time is over", i.e. rules are made by those who are interested actively. The school, as host, decides who may come to the club and who not. Beyond this rudimental kind of rules, other *common practice* (much more likely than rules) is negotiated only seldom, which may have a couple of reasons (see later). However, diversity itself is not allocated negatively at all in come_IN. The main rule of interaction seems to be attention, i.e. as already mentioned, people sometimes tend to watch each other when they could talk.

Conflicts/Consensus: Avoid the Inevitable

As already recognizable from the findings presented in the sections above, conflicts are often avoided. Although conflicts are an integral parameter for social integration, there has no *culture* of arguing been established yet in come_IN, at least no intercultural one. Germans argue with Germans, as well as Turks argue with Turks, but other constellations have not been observed. There are several reasons which may explain this phenomenon.

1. **Shyness:** Come_IN members are cautious in dealing with people belonging to other ethnic groups than themselves. Of course, shyness itself cannot be an explanation. Much more likely, it depends on other problems, as:

2. **Lack of in-depth communication***:* As we have seen above, communication often ends where children cannot or shall not translate. This means that "adult topics" are discussed rarely. Interest in each other is there, but communication stops too early; it cannot become intimate.
3. **Rapid intervention:** The school's representatives tend to interrupt discussions quite early. Furthermore, plenum discussions are strictly guided by the principal.
4. **Passivity:** Due to the school principal's immense engagement within come_IN there seem to be some kind of passivity. The feeling of being served is not present but in a way existing latently.

As a result, consensus cannot be found as conflicts are avoided or ignored sometimes. As parameters of social integration—like communication, knowledge or participation—are highly dependent on the degree of conflicts in the community, attempts have to be made to encourage continuation and progress of the club's mission.

OUTLOOK AND CONCLUSION

In this chapter we tried to state and answer basic theoretical and methodological questions with regard to the specific and complex socio-technical practice within the intercultural computer club come_IN. Thereby, the focus was on the one hand on searching for adequate academic disciplines which may provide an appropriate theoretical ground for examining the field of social integration within come_IN. Social integration, on the other hand, as one instance for investigative focusing, served as a basic hypothesis to move into the field, i.e. ICT fosters social integration. As a further step, employing grounded theory shall lead to designing software, let us call it *grounded design*. We found basic categories with regard to social integration, in the field as well as in the literature. These categories and concepts are combined and inter-related to one another, spun to a net of significance. Software has to apply on specific points within the socio-technical structure of the come_IN community. Communication could be regarded as one important feature within come_IN which has to be fostered in the future. Language problems as well as problems with role affiliation have to be surmounted, technically and organizationally. It has to be mentioned that social integration is only one key category for come_IN. There are other categories which arose during our research so far, such as:

Sustainability: All activities in come_IN have to be examined with an explicit attention to continuity. Learning, also socio-cultural learning comes with repetition and sustainability. With regard to social integration, continuity becomes the essential driver for the process of integration. Integration happens when it continues. As the community changes over time, so does integration. These changes keep integration alive. This is evidence for the idea that our method never ends but may pass empirical milestones. Or to employ a technical metaphor (which shall be applied for the concept of grounded design), versions of empirical findings have to be made concrete from each theoretical sampling to the next.

Self-Organizing: As mentioned above, the head teacher of Marienschule did a great job *organizing* the community for over two years. Without her—that is what many members say—the club could not have survived. Of course, the same applies to many other participants but she really has been the most influential person in the club. Pessimistic thoughts emerged among come_IN members when she retired: Can we still go on? Who will lead us? Well, the club still exists. She still joins the club, but other participants now feel responsible for keeping the club running. A new cycle began with a new strategy: communication and contact (internal and external).

Come_IN was founded more than three years ago. Now it is time to initiate change. This is what has been and will be done:

1. A special group for women has been opened. In order to provide better opportunities for the *gazing user group* women shall be introduced to ICT according to their needs and competences. Additionally, linguistic barriers have to be surmounted, i.e. in helping some of the Turkish women finding partners to learn German, and to ground design applications for bilingual software.
2. More excursions are planned. Activities outside the club appeared to be a great motivator for participation. More mobile devices are needed in order to bring the journeys back into the club, e.g. notebooks, digital cameras, tape recorders, camcorders, etc.
3. Additional communication channels are needed. For over two years come_IN was busy to found its own identity, i.e. only a basic homepage has been created by a father of the club who works as a developer. Now, since come_IN even awarded a prize for its concept in German schools[7], the community concentrates on presenting the club to anyone who is interested.

To sum up, we as researchers accompany the club as equal participants (and are affected by dynamics of social integration, too). We do not dictate what software to use nor how members have to use it. What we do is reflecting, planning, doing, and observing project work, just like everyone in the club can. Of course, we do this with greater intensity and for different reasons than some other members. This will change, since there now is need for software which can provide support for those who cannot participate equally without extra aid. Our methodological approach makes it possible to generate concepts of transferability, i.e. we try to establish additional come_IN clubs elsewhere in Germany.

ACKNOWLEDGMENT

The research is funded by the German Ministry of Education and Research (BMBF) (Fkz: KB00905). We would like to thank all participants from the neighbourhood who helped us to implement our research approach. We are especially indebted to Ingrid and Klaus Kansy as well as Gunnar Stevens.

REFERENCES

Appadurai, A. (1996). *Modernity at Large—Cultural dimensions of globalization.* Minneapolis: Public Worlds.

Bhabha, H. (1994). *The Location of Culture.* London/New York: Routledge.

Berry, J. W. (1992). Acculturation and adaptation in a new society. *International Migration*, 30: 69-83.

Coser, L. A., & Larsen, O. N. (Ed.). (1976). *The Uses of Controversy in Sociology.* New York: Free Press.

Clifford, J. (1988). *The Predicament of Culture.* Cambridge: Harvard University Press.

Clifford, J. (1999). *Routes—travel and translation in the late twentieth century.* Cambridge: Harvard University Press.

Durkheim, É. (1997). *The division of labor in society.* With an Introduction by Lewis Coser. New York: Free Press.

Geertz, C. (1973). Towards an Interpretive Theory of Culture. In: *The interpretation of cultures. Basic Books*, 3-30.

Greenwood, D. J. & Whyte, W. F. & Harkavy, I. (1993). Participatory Action Research as a Process and as a Goal. *Human Relations*, 46 (2), 175-192.

Kemmis, S. & McTaggart, R. (2004). Participatory Action Research: Communicative action and the public sphere. In: Denzin, N. & Lincoln, Y. (Ed.). *Handbook of Qualitative Research.*, 3rd ed., Sage, Thousand Oaks CA, 559-604.

Lanfranchi, A. (2004). The success of immigrant children at school: Effects of early child care as transitional space. *International Journal of Early Childhood*, 36 (1), 72-73.

Lave, J. & Wenger, E. (1991). *Situated Learning: Legitimate Peripheral Participation.* Cambridge: University Press.

Mead, G. H. (1934). *Mind, self, and society. From the standpoint of a social behaviorist.* Chicago: University Press.

Portes, A. (2000). Social Capital: Its Origins and Applications in Modern Sociology. In: Lesser, El (ed.) *Knowledge and Social Capital: foundations and applications.* Butterworth-Heinemann, Boston et al., 69-88.

Rohde, M. (2004). Find what binds. Building social capital in an Iranian NGO community system. In: Huysman, M. & Wulf, V. (Ed.) *Social Capital and Information Technology.* Cambridge: MIT Press, 75-112.

Stevens, G. & Veith, M. & Wulf, V. (2004). Come_IN: Using computers to foster the integration of migrant communities. In: *SIGGROUP Bulletin: Special Issue on Community-Based Learning: Explorations into Theoretical Groundings, Empirical Findings and Computer Support,* 1: 70-76.

Stevens, G. & Veith, M. & Wulf, V. (2005). Bridging among ethnic communities by cross-cultural communities of practice. In: *Proceedings of the Second International Conference on Communities and Technologies (C&T 2005)*, Milan, Italy: Springer, 377-396.

Strauss, A. & Corbin, J. M. (1998). *Basics of Qualitative Research: Techniques and Procedures for Developing Grounded Theory*, 2nd ed. Sage, Thousand Oaks CA.

Tajfel, H. (1982). Social *identity and intergroup relations.* Cambridge: University Press.

Turner, J. C. & Hogg, M. A. & Oakes, P. J. & Reicher, S. D. & Wetherell, M. S. (Ed.) (1987). *Rediscovering the social group. A self-categorization theory.* Oxford: Blackwell.

Veith, M. (2006). *come_IN: Die Integration von türkisch-stämmigen Deutschen am Beispiel eines misch-kulturellen Computerclubs.* Unpublished master thesis, University of Siegen, Germany.

Veith, M. & Schubert, K. & Wulf, V. (2007). come_IN: Identity and Role Affiliation mediated by an Inter-Cultural Computer Club. In: *Proceedings of the IADIS International Conference e-Society 2007* (pp. 144-151). Lisbon: IADIS Press.

Wenger, E. (1998). *Communities of Practice. Learning, Meaning and Identity.* Cambridge: University Press.

Wulf, V. & Rohde, M. (1995). Towards an integrated organization and technology development. In: *Proceedings of the conference on Designing interactive systems: processes, practives, methods, & techniques* (pp. 55-64). New York: ACM Press.

KEY TERMS

Action Research: As action researchers we collaborate with practitioners to intervene in practice in order to solve concrete problems while expanding scientific knowledge. In the literature, action research represents an overarching class of research approaches, rather than a single monolithic research method. Action research is usually split into a reflective phase where problems and opportunities are analyzed and an intervention is planned, and a phase of action where the intervention is carried out.

Community of practice: A community of practice refers to a process of social learning that occurs, as well as shared sociocultural practices that emerge and evolve when people with similar or common goals interact as they strive towards those goals. One of our assumptions is, that an intercultural computer club like come_IN is a community of practice. In the club, one of our roles as 'researchers' is just one kind of legitimate peripheral participation within the community of practice of come_IN.

Learning: In our context we define learning as the sustainable process of acquiring abilities and knowledge regarding technical as well as social skills. It is always a sociocultural phenomenon which can be observed in different types of qualities, such as in Communities of Practice, Constructionism, Over the Shoulder Learning, or Learning by Doing. The ongoing gathering and interpretation of experience almost always accompanies the learning process.

Migration: In our context, migration is a key discriminant which divides our club members into two groups, that is, those with a migration background and those without. However, despite stereotyical belief, there is obviously no clear connection between ethnicity and migration. Only a few people may be considered pure locals, which raises the strong need for localization of support and the negotiation of cultural belonging.

Neighborhood: Urban Informatics for us apply on the local level of a neighborhood. Neighborhoods are fascinating as sociality in this context is not a matter of anonymity but of acquaintance. Place making as the socio-geographical process of local public interaction can be conducted by conflicts, which can only lead to a sustainable consensus if social integration is fostered on all levels within the socio-technical infrastructure.

Participation: As one measure of contact, the gradual weight of public as well as institutional participation is of further interest to us. Participation is the main quantitative parameter within come_IN. Commonly shared practice within the club has to pay tribute to all different kinds of participation and learning scenarios (e.g. working with, playing on, and respectively gazing at ICT). Using theoretical sampling helps to clarify the dependence between manner of use and participation.

Social Integration: We take a more holistic and even normative view on social integration in our study. For a society, it is crucial to follow the ideal of providing equal opportunities for as many of its members as possible. Therefore, social integration on the one hand means to overcome the technical and the social digital divide, and on the other hand, has to include all generations, all ethnic communities, and people from all socioeconomic backgrounds in this process.

ENDNOTES

1. With great emphasis on social capital, see Portes (2000); or for a classical source on social integration, see Durkheim (1997).
2. All official data are from the Statistical Office of the city of Bonn, January 2006
3. "Hauptschulabschluss" is the German certificate after completing 9 years of compulsory secondary schooling.
4. A presumption Geertz has in common with Max Weber. A fact Clifford Geertz explicitly points out (1973: 5).
5. For an overview on finding theoretical background for come_IN, please consider Veith (2006).
6. This nonofficial committee consists of teachers, researchers, and other volunteers as some encouraged parents or former pupils of the club who now go to secondary schools but became tutors of come_IN.
7. In October 2006 the project and the school awarded a prize of "Schulen ans Netz", a

German non-profit organization. "Schulen ans Netz" originally was founded 1996 by the German Ministry of Education and Research (BMBF) and the Deutsche Telekom (DTAG) to provide schools with Internet access.

Chapter IX
Beyond Safety Concerns:
On the Practical Applications of Urban Neighbourhood Video Cameras

Victor M. Gonzalez
University of Manchester, UK

Kenneth L. Kraemer
University of California, Irvine, USA

Luis A. Castro
University of Manchester, UK

ABSTRACT

The practical use of information technology devices in domestic and residential contexts often results in radical changes from their envisioned raison d'être. This study focuses on the context of household safety and security, and presents results from the analysis of the usage of video cameras in the public areas of an urban neighbourhood in Tecámac, Mexico. Moving beyond the original envisioned purpose of safety, residents of the community engaged in a process of technology appropriation, finding novel applications for the security cameras. These uses included supporting coordination among family members, providing enhanced communication with distant friends and family, looking after minors while playing outside, and showing the household to friends and colleagues. Our results illustrate that success in information technologies is a dynamic phenomenon and that technology appropriation has to be understood as a phenomenon that occurs at the level of the application of the device, rather than at the level of the device itself.

INTRODUCTION

The presence of personal computers and Internet access in the home is becoming increasingly more common in Mexico. A recent study by the Mexican Internet Association[2] reports a growth of 22.4% from the previous year in the number of computers connected to the Internet, totalling 4.78 million Mexican homes with Internet access as of May 2007 (Peña, 2007). This trend is in part due to lower prices of computer equipment and the availability of a number of credit schemes where Internet Service Providers (ISP) offer bundles in which computers are paid as part of the monthly service fee.

In addition, over the last few years, Mexico has experienced an unprecedented investment to build housing complexes that are affordable, particularly for the low and middle-income population. Through simplified mortgages schemes and minimum down payments, more people in Mexico have been able to afford their own house. In 2006, at the end of the previous administration, the Mexican federal government reported that 1.9 million new houses were built between 2002 and 2006, which represents 43% more than those built in the previous administration (Fox, 2006). This investment created a housing boom, in which fierce competition emerged between the main construction companies of the country, each trying to differentiate their products and add value in many different ways.

Among the many players in the Mexican housing development sector, one of them, Real Paraiso Residencial, pursued the idea of creating a novel house concept where Information Technology (IT) would not just play a central role, but would be part of the very definition of what constitutes a house. The vision included building a new generation of houses where computers and Internet access become a part of the basic infrastructure. Rather than following traditional paths leading to home automation, assisted living or smart homes, the aim was to support more 'down to earth' needs and concerns. One of those needs, and perhaps the most valuable in urban Mexican communities, is living in a safe and secure neighbourhood.

This chapter presents preliminary findings from our study conducted to understand the practicalities of implementing this particular vision of domestic computing in one of the housing developments of Real Paraiso Residencial: Real del Sol, in Tecámac, Mexico. In particular, we present findings regarding the efforts to create a safe and secure community using public video cameras that can be accessed by residents through a private intranet. Our analysis indicates that moving beyond the original purposes of the technology envisioned by developers, and in spite of an apparent failure to meet the original need residents of the community appropriated and domesticated the technology, finding novel and practical applications. We argue that our results are a good example of the importance of understanding the appropriation of information technologies as a phenomenon occurring at the level of the application of the device rather than at the level of the device itself as a whole, whether the device is a personal computer, a phone, or an e-mail tool (e.g. Microsoft Outlook) or, as in this particular case, the urban neighbourhood video camera.

CHARACTERISTICS OF THE STUDY AND METHODOLOGY

In the city of Tecámac, Mexico, Real Paraiso Residencial, a housing company in partnership with Conectha, an Internet Service Provider, built a residential complex (Real del Sol) consisting of around 2,000 houses equipped with a personal computer and broadband Internet access. By the end of February 2006, with the support of the aforementioned companies, we started a three-year study with families living in or about to move in to Real del Sol. The general purpose of our study is to analyze the way that this particular vision of home computing becomes materialized and socially constructed over time as a product of the interactions between neighbours, developers, and designers. Among other factors, our study explores the symbolic meaning of technologies in domestic settings, the role of technologies in supporting neighbourhood organization and management, and the integration of technological services into the daily practices of urban families.

A total of 34 residents were interviewed covering 27 households by the end of February 2006. Most of these families were interviewed again a year later (February 2007). All interviews were semi-structured and covered a set of topics including the factors motivating the purchase of the house, the experience of moving in, relationships with neighbours and the rest of the community, use of the technology to support domestic practices, as well as perspectives on how to make the technology more useful. The interviews were complemented with a number of observations of the community and informal interviews with staff from both Conectha and Real Paraiso Residencial, as well as with people from the city of Tecámac (not living at Real del Sol). These people included taxi drivers, shop owners and security staff. The data collected was then analyzed using a comparative approach aiming at identifying patterns among the responses and producing an integrated set of findings as suggested by the Grounded Theory approach (Strauss and Corbin, 1998; Charmaz, 2006).

REAL DEL SOL AND THE G7 HABITAT CONCEPT

Real del Sol housing development is located in the city of Tecámac. Although Tecámac is not part of the metropolitan area of Mexico City, it is close enough to allow for a reasonable commute to the city (approximately one hour driving, with good traffic conditions), since the majority of residents work there. The development offers three small parks, basketball courts, a secondary school, a primary school, a kindergarten, and other communal areas. To facilitate the estate's administration, the development is organized into *privadas*, groups of ten or twenty houses that are separated by gates. Residents in the *privadas* share some green areas and services (e.g., trash bins). Each house comes with standard utility services (e.g., water, electricity, gas), including Internet access thorough a wireless network that allows speeds up to 384 Kbps. Houses also include a personal computer for free or it is offered by Conectha at preferential prices. Figure 1 shows pictures of the development.

Real del Sol is a housing development guided by a particular idea of a tech-enabled home, called "Habitat of Seventh Generation" or G7 Habitat. The developers define a G7 Habitat as a household concept that provides seven basic elements: innovative design, financing support, post-sale link, connectivity, school link, shopping link, and security. Under this design concept, Information

Figure 1. Houses in Real del Sol (Privada Tarvos); a view of common areas

and Communications Technologies (ICTs) play not just a strategic role, but one where they are intrinsically linked and embedded into the basic idea of a household. Upfront, they envisioned houses where Internet access was one of the core utilities, as opposed to an optional service. They aimed at having all houses equipped with at least one personal computer connected to the Internet, with exclusive access to a set of information services tailored to serve the needs of the residents of Real del Sol. Conectha carried out this idea by designing and implementing a community intranet portal.

Using their password-protected personal accounts, residents of Real del Sol have access to a set of services on the intranet, including access to community information and announcements, educational content, on-line ordering of groceries from local shops, and access to public security video cameras placed around the complex. Beyond these local services, the intranet provides links to external sources such as job banks, adult education, and entertainment. Figure 2 shows a view of the intranet and the shopping portal.

The G7 Habitat concept was a very important differentiator for Real del Sol. It certainly served as a marketing tool and influenced the purchase decision; however it also determined the resident's expectations about their life at the development, the role played by information technologies, and the nature of their interaction with those technologies. Residents expected to live in a beautiful, modern, friendly, and, most importantly, secure neighbourhood. Technology was seen as an enabler to achieve that, particularly regarding the provision of services (e.g. online shopping and local information), the communication with other residents and the post-sale team, and the surveillance of the properties. Furthermore, residents were quite aware of the uniqueness of the initiative, and we perceived that early on they developed an exploratory approach while interacting with the technologies. They knew the services were new and evolving, as new opportunities and requirements were identified by developers. Consequently, most of the residents we met tended to engage in a process of discovery, exploring what was possible to achieve with what they had available. In this chapter, we present results of our analysis of how the residents of Real del Sol experienced the technology appropriation of public video cameras.

THE NEED FOR A SAFE AND SECURE HOME AND COMMUNITY

With a rampant increment in the number of kidnappings, burglaries and other forms of crime in the urban Mexican neighbourhoods, having safe and secure housing environments is a priority for most people and one of the main concerns

Figure 2. Community intranet in Real del Sol; shopping portal

for the inhabitants of housing developments in urban Mexico (Bergman, 2006; Ramos-Lira & Andrade-Palos, 1993). Many houses remain empty during the day because most adult members of the household work in the city, which makes the houses susceptible to burglaries. Furthermore, most housing areas lack police presence which exacerbates the problem. Burglaries even occur in cases where there is some sort of private security corps. Statistics from Mexican NGOs such as Mexico United Against Delinquency point out that in Mexico City, 20% of all robberies are committed in households and just one out of eight is reported to the authorities (Mexico Unido, 2006). Facing insecurity, many Mexican housing developments have opted to provide protection and a level of separation from the external world, with solutions ranging from alarm systems, CCTV (Closed Circuit television) and even "gated communities" in which the access to the community is restricted and physical barriers are erected to avoid intrusions. Similar trends are reported in other parts of Latin America and the rest of the world (Caldeira, 1996; Low, 2004).

Addressing the need for a secure home and community, the developers of Real del Sol installed video cameras all over the complex, covering each group of houses (*privada*) separately, as well as most public areas of the residential complex. Each camera is fixed at the top of a mast and protected with a cabinet for outdoor operation. The cameras are typical Internet-enabled CCTV cameras capturing low-resolution colour images in real time and connected to the community network via a wireless connection. The frame rate (12 fps) is not fast enough to recognize sudden movements by people, but it is still useful to monitor activity.

Each *privada* has one camera placed at the rear of the parking lot to maximize the visual field captured. The position of the cameras is static and there is no zoom functionality. Designers foresaw residents using the cameras to monitor activity outside their properties, as well as the four communal areas. Using a user ID and password, residents have access to the cameras through the community intranet. The intranet itself is accessible from any networked computer, making it possible to access the cameras from workplaces or from any other location. Cameras operate 24/7 and are not monitored by professional security staff. Figure 3 shows images of the cameras placed in *privada* "Antares II" and one of the public parks.

Given the significance of the topic and because it was one of the elements supported by the G7 Habitat concept, part of the interviews with the residents focused on the issue of household security and the role cameras played to serve that need. The responses from the residents confirmed the relevance of security as a major concern. We found that living in a secure community was among the

Figure 3. Images from public video cameras in Real del Sol

top three reasons for purchasing a house in Real de Sol. Many of the residents commented that while living in their previous neighbourhood, they were victims of robbery, burglary or even kidnapping. Because those memories were still fresh, security was a top criterion when looking for a new home. Consequently, from this perspective, having access to security cameras in Real del Sol made complete sense and influenced the purchase decision.

Security remained an important issue even after residents moved in to Real del Sol. During the first weeks after moving, residents installed protection for doors and windows and many of them added an extra lock to the entrance door. These were considered standard security measures and even highly recommended by the Real de Sol post-sale team. The fact that the residents moved into a new community increased the need to take extra precautions, which were later reduced as they met their neighbours and were able to identify who lived there, thus being able to spot any strangers. As time passed by, we found that the residents of the community (Real del Sol) developed a comfortable sense of security, but still remained concerned about being an easy target for burglars or vandals coming from the neighbouring areas. Therefore, the need for video surveillance remained a valuable and important application for the residents and significantly motivated their initial access to the intranet.

APPROPRIATION AND PRACTICAL USE OF PUBLIC VIDEO CAMERAS

The residents of Real del Sol expressed mixed opinions when talking about their experiences accessing and using the public video cameras provided by Conectha. Clearly, any positive or negative opinion regarding the application of the public video cameras depended on the particular context of the family being interviewed. A family with small children, where one of the adults is a full time house maker, pointed to the value of the cameras that was very different from that expressed by a single professional male who travels regularly. We comment on these differences later on in this chapter. However, beyond the characteristics of the household residents, we found that a more important factor in determining whether the cameras were perceived as useful or not was the extent to which people allowed themselves to explore and play with the technology, appropriating it and finding an application (or applications) that suited their needs. Following Dourish, we understand appropriation as the way in which technologies are adopted, adapted and incorporated into working use (Dourish, 2003). In other words, technology appropriation can be described as the effort of users to make sense of the technology within their own contexts; in this case, the domestic one. It can be argued that this exploratory nature of the technology appropriation occurring with the public video cameras of Real del Sol emerged as a direct result of the novelty of the G7 Habitat concept. Residents knew that their houses were special and technology was supposed to play a major role as opposed to 'conventional' houses. Consequently, this scenario produced a more open and discovery-oriented interaction with the services provided. This was particularly evident in the case of the public video cameras and perhaps even more latent as the technology itself was previously unknown by most residents.

Use of Video Cameras and Anticipation Before Moving In

Before moving in, most residents did not have access to similar kinds of surveillance technology. Hence, it was natural to be curious about it and this was, indeed, one of the main reasons for residents to take the first steps to explore the intranet. In fact, all interviewees said they had accessed the video camera service at least once. In contrast, many have never checked the e-mail account provided by Conectha or browsed some of the sections of the intranet.

We found that the interaction of residents with the video cameras started before moving into their properties and as soon as they had access to the community intranet. Access was granted at the moment people received their property. Although many could not move in immediately, they started using the intranet to obtain more information about their new house and to communicate with the post-sale team. Consequently, motivated by curiosity, many interviewees reported to have accessed the cameras from their previous homes or workplaces. They mentioned that they wanted to see how things were in Tecámac, and although they were not able to witness that much, observing simple things such as the weather was interesting enough to access the cameras from time to time. The following comments from one of the interviewees show this behaviour:

When we were in the process of acquiring the house, I used to access the portal a lot to look at the public cameras and say 'Oh, look, they are doing this or adding this new thing' or 'Oh look, it is raining, or the weather is like this' so, it really grabbed my attention, see?" [Female resident, Married, 34 years old].

In two particular cases, residents had access to the cameras immediately after they signed the contract and were able to observe their house as it was erected. That reduced the need for regular visits to the property and seemed to increase their sense of ownership and control over the construction. One of the residents reported that after not noticing any changes in their property for some days, they called the construction company to inquire about it.

Once people moved in, the novelty of using the cameras as a "window to Real del Sol" did not make as much sense as before, and this type of use eventually vanished.

Use and Disappointment After Moving In

We found that at some point after moving in, residents started using the cameras for the original purpose envisioned: monitoring their house for security purposes. In four cases, residents reported using the video cameras while being away during a business trip. This was the case of one of the residents who works as a tour guide and has to travel constantly:

I use the camera to have a look at my house because this year I have been away a lot, and nobody stays here. So, despite that this is a relatively safe area, you always have to be checking what is happening in your house, to check that everything is normal, everything is fine, and to check that the people that you see in the privada are the ones living here [Male resident, Single, 32 years old]

In clear contrast with residents who find the cameras useful for surveillance purposes, the majority of residents in our study expressed clear disappointment when they referred to this particular application. Many factors contributed to this disappointment. Firstly, residents realized that security personnel were not monitoring the cameras. Residents erroneously assumed that, since there were some security guards at the entrances and walking around the complex, they would have access to the cameras and monitor any suspicious activity. Such an idea was untrue, and no personnel were assigned that particular duty. From the beginning, Conectha assumed that it would be neighbours themselves monitoring the cameras, but this assumption was not clearly divulged among the neighbours. The feeling of disappointment increased as soon as the residents realized that even if they decided to monitor the cameras themselves, the level of detail and quality of the image was not optimal. Furthermore, many felt frustrated because the position of the camera in their *privada* made it very difficult to see their houses. Lacking a fish-eye view, houses too close or too far from the camera are really difficult to see. However, perhaps the most important reason why many residents did not perceive the value of the cameras for security purposes was the fact that none of the activity was recorded. Due to this, the possibility of video analysis in case of robberies did not exist. In sum, as a result of misunderstanding and limited functionality of the video cameras, most residents did not find them very valuable and felt that what they had been offered did not entirely match what they received.

Unexpected Findings

Further discussion with residents led us to unexpected findings. In spite of the disappointment with the cameras, residents did not entirely reject the video cameras but rejected its application for security purposes. Most of our interviews reached a point where discussion about the cameras ended with residents expressing frustration over them. At that point we suggested removing the cameras and, surprisingly, most residents objected to this suggestion. The residents quickly expressed a good number of reasons to keep the cameras, which pointed to a number of useful applications. Those applications, interestingly, were clearly far beyond safety concerns. From the analysis of data, we could identify four main applications of the cameras as expressed by the residents: (1) supporting coordination among family members, (2) providing enhanced communication with distant friends and family, (3) looking after minors while playing outside, and (4) showing the household to friends and colleagues. Each one of the applications is discussed separately below.

Supporting Coordination Among Family Members

With the cameras pointing at the parking lot, it was possible to see when people or cars arrived or left the *privada*. Some residents mentioned that the cameras facilitated the coordination among family members, as they could access the cameras from their workplace and easily know who is home just by looking at the car parked outside the house or a bicycle leaning on the door. Indeed, this extra information was useful to decide when to contact people or to check their availability. For instance, this type of application was articulated by one of the residents, a financial adviser living with his wife and two children in Real de Sol:

I travel constantly and I use it by connecting to the Internet and by checking if my wife's car is there I can tell if she's, it's there. If the car is not there then why would I call her if I can see she's not there, right? I also can tell my wife whether the children are playing or not, when they walk out of the privada I call her 'Hey, do you know Pepetoño is outside?' And she doesn't realize that, right? Then it's a little bit to help her, right? [Male resident, married, late 40's].

Enhanced Communication with Distant Friends and Family

We also found that the public video cameras were used to supplement conventional means of communication (e.g., telephone). This was particularly relevant when residents communicated with other members of the family or friends living far away in Mexico or abroad. Residents mentioned that sometimes they use the cameras to wave at people they are talking to on the phone so they can be seen. One of the interviewees, a mother of two teenagers in her late thirties, showed great emotion when she mentioned this particular use of the camera as she has a boyfriend who lives far away in a Southern Mexican state. She said that they communicate regularly by phone, e-mail and instant messenger. However, the video camera serves to complement the interaction. She mentioned that the public video camera is often used when she is on the phone with her boyfriend. She usually stands outside the house at an observable spot so he can see her. In other cases, the cameras provide a sense of presence that allows for sharing of quotidian aspects of the life of the family with others. For instance, other resident, a young housewife in her early thirties with two small daughters, referred to the usefulness of the cameras for her parents living in southern California, USA. The parents had not met their 3 year-old youngest granddaughter, and frequently called to ask the girls to go and play at one of the parks nearby so they can see them. This form of communication has become an everyday thing for them to the extent that the grandmother is aware of the routines of the family as if they were neighbours. As it is explained by her daughter:

My mother calls me all the time, even one day she called me on my cell phone and she said: 'Where are you?', and I said 'I am just getting into the house', and she responded 'Yes, I just saw you, you are wearing this and that, and the girls this and that, right?' It was fun. So, in spite that they are far away, we have constant communication. My mom knows about what time I get back from picking the girls up from school or things like that, and she can use the cameras to see us and call us when we are here. [Female, married, early 30s]

Looking After Minors While Playing Outside

Residents reported that one of the benefits of the cameras located in the common areas was monitoring their children while they played. One mother of two boys mentioned that she keeps the camera's stream of the public park near her house running most of the time. Before she grants her sons permission to go outside and play, she makes sure that other kids are around by using the camera, as she prefers to have her kids accompanied in case something happens or a stranger is around. She always keeps an eye on the camera while the kids play there. Many other residents expressed similar usages for the cameras and some of them even monitor their children from their workplaces while they play in the parks, as was expressed in the quote presented before. Thus, the video cameras provide eyes on the street similar to those provided by senior citizens in older housing areas in urban neighbourhoods (Jacobs, 1961).

Showing the House to Friends and Colleagues

After moving in, some of residents were using the cameras not only to 'keep an eye' on their children or their home, but also for showing their houses to their relatives and friends. Many residents mentioned that while visiting family in other cities, they were asked to access the video cameras to show their new houses. The following quote is illustrative of this type of application:

"When I was in the office working I used to tell them 'Do you want to see my house?' So I went on to the web page and we saw, for instance, the cameras. My sister lives in Queretaro and we sometimes used that method" [Female, married, 28 years old].

This indicated that for some residents at the early stages after their purchase the slogan was not "Tu casa conectada al Mundo" (your house connected to the world), which was used by Real Paraiso Residencial to advertise the housing complex, but instead "Tu casa a la vista del Mundo" (your house to be seen by the world). Residents said that they felt proud of their properties and the fact that they could show others their property in real time increased this feeling.

DISCUSSION

The way that residents in Real del Sol interacted and used the public video cameras has a number of implications for understanding the role of information and communication technologies in housing communities, in general, and the appropriation of technology in domestic settings, in particular. We will discuss them in this section.

The G7 Habitat concept represents a particular approach for home and community computing that has been proven successful in certain areas. As a marketing tool the success has been clear, but beyond that, we can perceive that overall, residents are happy with the service in terms of connectivity and customer support. Residents are able to contact Conectha directly by going to their office located at one of the shopping areas of Real del Sol, and their requests are attended to promptly.

On the other hand, the information services provided by the Intranet have been less successful. This is an area where Conectha has been working with a very flexible scheme of operation, and the services have been improved and changed over time. For instance, the tools provided to support communication among neighbours, which at the beginning consisted of a simple page where residents could post messages, has evolved more recently into a complete system for the administration and organization of each *privada*. Using the system, the residents can record and check the payment of maintenance fees, post information and messages, and coordinate projects with their neighbours such as painting the parking spaces, or planting new trees. Similarly, the post-sale link that previously included just contact information, forms to report problems, and documents with general guidelines, now includes a number of short instructional videos with themes such as property maintenance, domestic finances, health, home decoration, baby care, and others.

These changes were in part motivated by Conectha's desire to provide better applications, but also by the fact that they could afford experimenting with ideas and applications. As the project moved on, it was clear to us that developers themselves were discovering what was possible to achieve with the technology and adjusted their plans to any opportunity they foresaw. In contrast with other projects where a scientific motivation directs interventions (e.g. Carroll and Rosson, 1996), the companies behind the G7 Habitat concept aimed at identifying business opportunities, but at the same time they were not constrained by wanting to achieve short term benefits, giving enough room for innovation and development. Consequently, we believe that we have not yet seen a full appropriation of most intranet's applications, and it will be in the next stages of our research where we will be able to assess their usage and effects.

We found that the technology appropriation experienced by residents with the public video cameras can be characterized as a creative discovery exercise. This was in part influenced by the aura of novelty around the G7 Habitat concept. Our findings show that this technological appropriation can be better understood at the level of the application of the device, rather than looking at the device as a whole. Following Degele, we can see that because information technologies are flexible, malleable, multipurpose, and open, their appropriation can be a creative act, where new roles are invented, beyond the existing margins originally established by the designers (Degele, 1997). In the case of the public video cameras in Real del Sol, a creativity act occurred not necessarily as a result of the lack of success of the technology as a security tool, but due to its affordances and those of the context in which they were implemented. Cameras were publicly available from any computer in the world, and therefore residents could easily foresee the possibility of giving access to others and communicating with them through this channel. Similarly, the cameras show views of the community beyond the limits of the *privada*, opening the possibilities for residents to look at their new homes before moving, or showing it to friends and families.

We can say then that by creating these new roles for the video cameras, residents changed points of reference, and took a new stance on the application of technology, changing its *raison d'entre* and defining a new way to assess their value (Degele, 1997). Those who found new uses for the video cameras learned to disassociate its value with a single application and saw the cameras as multipurpose devices. They were able to discover that despite the poor security function, the cameras were good for other functions. Consequently, in spite of the fact that cameras were introduced with a clear (and unique purpose), people who found new usages gradually redefined their relationships with the cameras, accepting their limitations for certain applications, and their advantages for others.

Similar technology appropriations have been reported to occur with other technologies such as the telephone, which moved from being considered a tool to support brief communication among professionals, to become a tool to support extended and casual social interactions among friends and families (Fischer, 1992). The e-mail client is another example, which now is considered by many people as an essential tool for supporting task-management, going beyond its original application as a communication tool (Ducheneaut & Bellotti, 2001). As a result of the creative interaction of people with technology, phones, e-mail clients, and in our case, public video cameras, have found new application niches, and it is at this level that we can understand their success or failure.

Not surprisingly, we found that the actual identification of a good application for the cameras was necessarily delimited by the family circumstances and their context. For instance, most people who found the video cameras useful as a security tool were those living alone, travelling constantly, and spending limited time socializing with neighbours due to work commitments. Thus, it seems that less social interaction within the community resulted in people becoming more concerned about their household while being away. Similarly, those couples with no children showed little or no interest in the cameras at the public parks, in contrast with those who used them for looking after the kids playing outside while in their home or from their workplace. Clearly, applications such as enhancing communication with distant family or friends were useful for just a few people with those particular circumstances.

Our results indicate that experimenting and discovering new applications for the public video cameras has been a very individualistic experience thus far. Residents interviewed do not report having conversations with other residents about their use of the public video cameras, nor do they learn from others the applications they use. Consequently, it is likely that the more people share this information, the more video cameras will be used for those purposes.

Beyond the particular applications, our findings confirm the relevance of understanding how social presence, in particular, video social presence, is experienced within the context of domestic and residential settings. Previous research efforts on video-based presence have been mainly centred on the workplace, supporting the awareness for remote collaborators (e.g. Dourish et al., 1996). Other research efforts have focused on supporting communication between members of the family or emotional partners, but they do not emphasize video-based communication (Keller et al., 2004; Markopoulos et al., 2004). Our findings indicate the value of video to enhance other forms of communication (e.g., phone), operating as a window through which one can observe loved ones, or just a particular place and its conditions (raining, sunny, and so forth). Using video in this way goes beyond surveillance and becomes a tool through which people can feel closer to the place, their families and friends. Interestingly, our informants expressed no concerns regarding privacy. They grant access to people they trust and consequently, they are not worried about their boyfriends, sisters or parents having access to the video cameras and watching them. We believe that this situation would be different if the cameras were broadcasting images from inside their homes. However, because this is not the case, what others can see is what they could see if they were neighbours living in the same *privada*.

Finally, our findings echo what other researchers have found when studying housing developments designed with security in mind: a safe community is not the one that is actually protected, but the one that *seems* to be protected. When studying gated communities in New York and Texas, Setha Low found that the sole provision of gates, guards or low walls is some times enough to provide some sense of privacy and security (Low, 2004). In our case, many residents interviewed felt that having these cameras does not contribute towards creating a safer community, while other residents saw some value, given that their mere presence might be an effective way to scare away potential burglars. Whether crime is reduced by having this sort of scarecrow is still uncertain, gated access to the *privadas* and some other visible protections can contribute to this effect.

CONCLUSION

Peace of mind, as stated by Real del Sol residents, is a priority, and one of the main reasons for moving into the complex. Given the concerns with regards to security, it is likely that this need will be attended to either through cooperation among neighbours or as a service from Conectha. Some neighbours said that they will be contracting more and better security cameras from Conectha in the near future, whereas others are planning to install alarms or other security devices in their houses. Conectha staff has also commented on their plans to provide a new type of service where video recordings will be available on demand.

Looking forward, we believe that the concerns with security are likely to play a major role in incentivising the communication and organization among neighbours. Currently, as a result of some recent robberies and vandalism, neighbours have become more active by organizing meetings to implement projects that aim at creating a more secure community. Solutions range from hiring private security guards, to erecting physical walls that will limit access to the complex. However, coordination to achieve those plans is not always easy as people are not always available for meetings, which suggest that some alternatives must be implemented to allow for participation. During the interviews, residents mentioned that they have thought about creating an e-mail distribution list for those purposes. Our future investigation in the community will reveal the evolution on this and many other aspects of the interactions of residents with the technology and its domestication.

REFERENCES

Bergman, M. (2006). Crime and Citizen Security in Latin America: The Challenges for New Scholarship. *Latin American Research Review*, 41(2), 213-227

Caldeira, T. (1996). Fortified enclaves: the new urban segregation. *Public Culture*, 8(2), pp. 303–328

Carroll, J. & Rosson, M. (1996). Developing the Blacksburg Electronic Village. *Communications of the ACM*, 39(12), 69-74

Charmaz, C. (2006). *Constructing Grounded Theory: Practical Guide Through Qualitative Analysis*. London: Sage.

Degele, N. (1997). Appropriation of Technology as a Creative Process. *Creativity and Innovation Management Journal*, 6(2), 89-93

Ducheneaut, N. & Bellotti, V. (2001). Email as habitat: An exploration of embedded personal information management. *ACM Interactions*. September-October; 30-38.

Dourish, P. (2003). The Appropriation of Interactive Technologies: Some Lessons from Placeless

Documents. *Computer Supported Cooperative Work (CSCW),* 12(4), 465-490.

Dourish, P., Adler, A., Belloti, V., and Henderson, A. (1996), Your Place or Mine? Learning from Long-Term Use of Audio-Video Communication. *Computer-Supported Cooperative Work (CSCW),* 5(1), 33-62

Fischer, C. (1992). *America Calling: A Social History of the Telephone to 1940.* Berkeley: University of California Press

Fox, V. (2006). Sexto Informe de Gobierno.*Gobierno de Mexico: Presidencia de la Republica.* September 2006

Jacobs, J. (1961). The Death and Life of Great American Cities. New York: Random House.

Keller, I., W. van der Hoog, & P.J. Stappers. (2004). Gust of me: reconnecting mother and son. IEEE *Pervasive Computing,* 3(1), 22-27.

Low, S. (2004). *Behind the Gates: Life, Security, and the Pursuit of Happiness in Fortress America.* New York: Routledge

Markopoulos, P., N. Romero, J. van Baren, W. Ijsselsteijn, B. de Ruyter, and B. Farshchian. (2004). Keeping in touch with the family: home and away with the ASTRA awareness system. *Proceedings of the Conference on Human Factors in Computing Systems (CHI 2004). Vienna, Austria:* ACM Press New York, NY, USA.

Mexico Unido por la Democracia (NGO), Retrieved in March 2006 from http://www.mexicounido.org/

Peña, A. (2007). Usuarios de Internet en Mexico 2007. *Asociación Mexicana de Internet AMIPI, AC.* May 2007.

Ramos-Lira, L. & Andrade-Palos, P. (1993). Fear of Victimization in Mexico. *Journal of Community and Applied Social Psychology,* 3(1), 41-51

Strauss, A. & Corbin J. (1998). *Basics of Qualitative Research: Grounded Theory Procedures and Techniques.* Newbury Park, CA: Sage Publications.

KEY TERMS

CCTV: Closed Circuit Television

G7 Concept: Concept created by a Mexican housing company which comprises seven basic housing elements: innovative design, financing, post-sale link, connectivity, school link, shopping link, and security.

Gated Community: These communities have restricted access to residents and visitors only. They comprise a community surrounded by physical barriers (~3m. fences) which are erected to avoid unauthorized intrusions.

Intranet: A restricted-access network that works like the Web. Usually owned and managed by a company. The intranet enables a company to share its resources with the neighbours without confidential information being made available to everyone with Internet access.

Privada: A Cluster of 10 or 20 houses (similar to a cul-de-sac) separated from the rest of the community by gates. These privadas share common areas such as trash bin area or green areas.

Technology Appropriation: The effort of users to make sense of the technology within their own contexts.

ENDNOTES

1. An earlier version of this paper was presented at the Family and Communication Technologies (FACT) Workshop, Northumbria University, United Kingdom, May 24, 2007.
2. A private group analyzing Internet trends and consumer habits in Mexico

Chapter X
The Figmentum Project: Appropriating Information and Communication Technologies to Animate Our Urban Fabric

Colleen Morgan
Australasian Cooperative Research Centre for Interaction Design, Australia

Debra Polson
Australasian Cooperative Research Centre for Interaction Design, Australia

ABSTRACT

This chapter explores how we may design located information and communication technologies (ICTs) to foster community sentiment. It focuses explicitly on possibilities for ICTs to create new modalities of place through exploring key factors such as shared experiences, shared knowledge and shared authorship. To contextualise this discussion in a real world setting, this chapter presents FIGMENTUM, a situated generative art application that was developed for and installed in a new urban development. FIGMENTUM is a non-service based application that aims to trigger emotional and representational place-based communities. Out of this practice-led research comes a theory and a process for designing creative place-based ICTs to animate our urban communities.

INTRODUCTION

This chapter explores how we may design site specific information and communication technologies (ICTs) to foster community sentiment. It focuses explicitly on possibilities for ICTs to create new modalities of place through exploring key factors such as shared experiences, shared knowledge and shared authorship. Modalities of place refer to the process by which numerous social, cultural, functional and emotional operators shape individual comprehension of place (Sandin, 2003). To con-

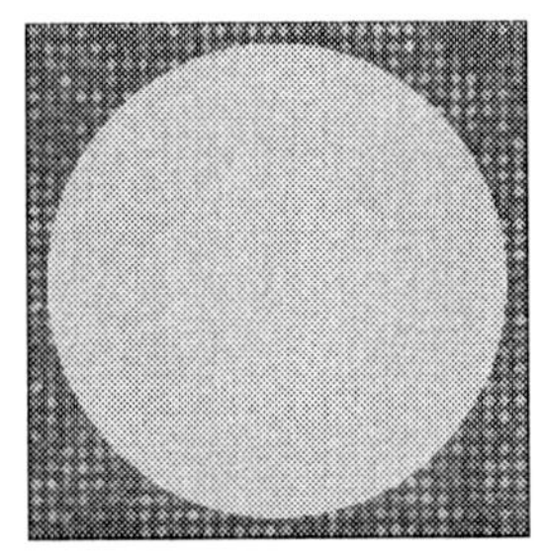

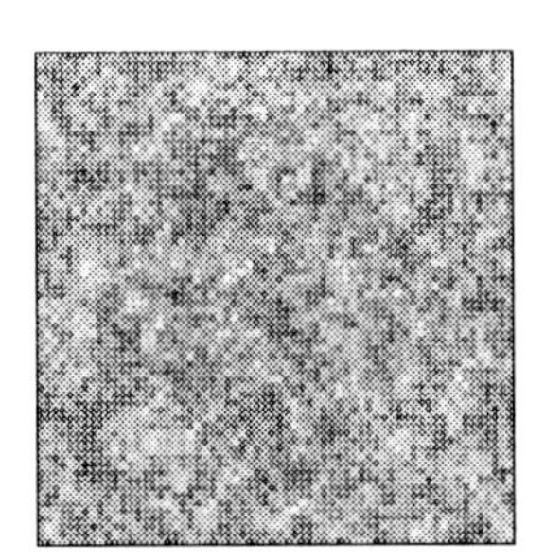
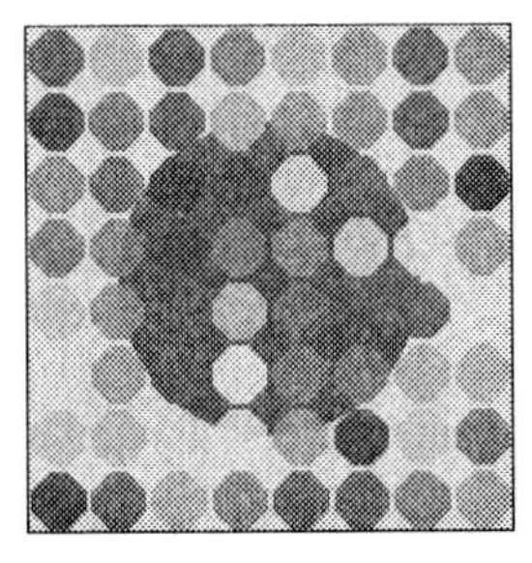

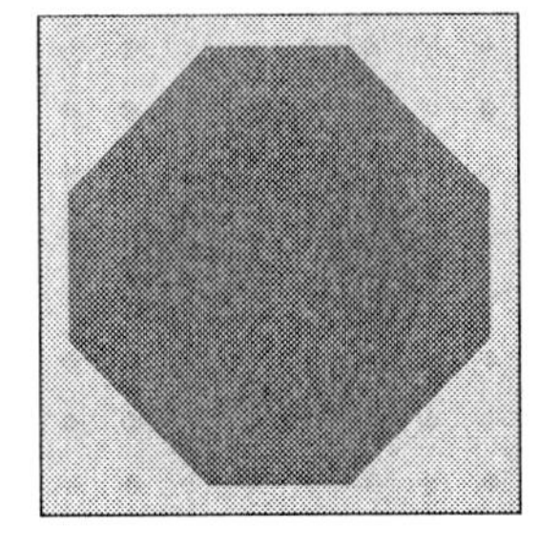

textualise this discussion in a real world setting, this chapter presents *FIGMENTUM,* a situated generative art application that was developed for and installed in a new urban development. *FIGMENTUM* aims to trigger emotional and representational imagined communities. Unlike many current community based urban informatics projects that take the form of online community notice boards, user profiles and the like, *FIGMENTUM* is a non-service based application. Instead of service-based ICT applications, our research focuses on the emotional and highly intangible cognitive processes that contribute to imaginings of community solidarity. It also emphasises the need for community based ICTs to provide motivating factors for social interaction within communities. An evaluation of *FIGMENTUM* reveals the critical need for place-based ICT applications to be custom designed to suit the social, cultural, spatial, technical and temporal characteristics of individual sites to successfully augment experiences of place. Although this field is in its infancy, it is clear that new ICT applications have the potential to be valuable tools for animating our urban fabric. This chapter works towards a theory for designing creative place-based applications that may provide enriched experiences within a community.

To design applications that may foster urban communities we must first work from within a framework of understanding peoples' relationships to place. To construct this framework, this chapter explores the notions of place attachment, imagined communities and social capital. This chapter then examines how social, cultural, functional and emotional operators contribute to experiences or 'modalities' of place. Understanding how modalities of place are constructed with these operators allows us to embed the knowledge developed from our framework into the process of designing and implementing community based ICTs. As such the chapter moves on to discuss the process of combining the core values derived from this theoretical framework with design principles drawn from existing community based ICT projects to design *FIGMENTUM.* In conclusion, a summary of how *FIGMENTUM* was received by the community and an analysis of the research outcomes is presented, followed by recommendations for future research.

BACKGROUND

This study helps to gauge the capacity for ICTs to foster community sentiment and social capital.

It approaches the field with the perspective that significant research is still required to explore and define the potentials of ICTs and to establish processes for designing and implementing interventions to satisfy these potentials. The Australian Department of Communications, Information Technology and the Arts supports this standpoint noting, "There is vast potential to use ICT to build social capital and contribute to community development and formation. However, [...] it is largely untapped and unrecognised in many areas. For ICT use to move beyond bonding – to harness its power for bridging and linking to resources that enhance economic and social development – it needs more attention to the type of social capital being developed" (DCITA, 2005, p. 9).

In a key point of departure, this research focuses upon how we may design ICTs to encourage the emotional and highly intangible cognitive processes that contribute to imaginings of community solidarity. We seek to identify foundational characteristics that generate social capital and how ICTs may support these characteristics. Currently, a considerable number of community ICT research projects are concerned with evaluating how ICTs in the form of social networking systems can be effective in strengthening community sentiment (Foth, 2006, Arnold et. al, 2003). We suggest that before community ICT research explores service based applications such as social networking systems, we must turn our attention to how ICTs may contribute to the key elements of social capital such as trust, shared experiences, shared knowledge and shared authorship. As has been highlighted in reference to master planned communities, it is not simply a case of build it and they will come (Foth, 2004; Gilchrist, 2004; Pinkett, 2003). We argue that such is also the case in relation the social networking systems that are beginning to be plugged in to these communities. Without a foundational incentive for social interaction, the value of social networking systems is undermined. It is precisely this lack of a foundation for social interaction that we feel designers of community ICT interventions should seek to overcome. Once we understand how ICTs may build a foundation for social interaction we will be in a position to more fully realise the value of social networking systems.

This research aligns itself with the perspective that technologies are socially constructed. Graham and Marvin (1996) identify four key paradigms in studying the relationships between people and technology. The first is *technological determinism* which views technological evolution to shape social evolution. Second is a blend of *futurism and utopianism*. This perspective emphasises the positive effects of technologies and possibilities for social, economic, and political liberation. A *dystopian* perspective does the opposite by stressing the possibilities for social, economic, and political oppression. The forth perspective is what Graham and Marvin title, the *social construction of technology*. This approach views technological evolution as a result of human agency. From this perspective, technology is driven by a social process by which a collection of individual human decisions shape technologies and therefore their potential impact on society. Of the four perspectives, this research aligns with the fourth perspective. As designers, we must engage in a practice whereby we acknowledge societal and contextual factors and actively design for positive social outcomes.

New urbanism is another theoretical domain that underpins this research. New urbanism is an urban planning philosophy that has emerged in response to current modes of urban sprawl and suburban expansion (De Villiers, 1997). The new urbanist ideal represents "a rediscovery of planning and architectural traditions...where life centers around a courthouse, common plaza, train station or main street (Bressi, 1994, p. xxv). Of the key aspects of the new urbanist philosophy, this research particularly adopts the notion that communities should be activated by central public spaces. While new urbanists have been criticised for being overly nostalgic and aspiring towards an idealistic historical urban condition that may have never existed (Ross, 2000), the importance of public spaces is an aspect of new urbanism supported across urban design theory. Public spaces for social interaction are essential for building community ties (Jacobs, 1961; Oldenburg, 2001) . As such the role public art may play in igniting and agitating community interaction becomes significant. Our research in place-based ICTs aims to explore how new technologies can leverage public spaces for greater social animation.

PLACE ATTACHMENT, IMAGINED COMMUNTIES AND SOCIAL CAPITAL

The concept of *place attachment* is a valuable tool for developing a framework for understanding people's relationships to place as it refers to the emotional, functional, and social ties people develop *within* a community and *towards* a particular place (Hummon, 1992). There are a number of key aspects of peoples' attachment to places. For this study we have identified them as *community satisfaction, community attachment,* and *imagined community identity* (Hummon, 1992, p. 254). *Community satisfaction* refers to how members are able to consciously view their community. In large part community satisfaction can be seen as functional attachment to place and is categorised in reference to objective community characteristics such as housing quality, recreational facilities, size, and population density (Hummon, 1992: 255). *Community attachment* refers to "the study of emotional investment in place" (Hummon, 1992: 256). Often referred to as emotional place attachment, community attachment represents a significantly ontological facet of place attachment and is determined by subjective characteristics of individual experience. This form of place attachment is chiefly determined by length of residency, social involvement, organisational membership, and proximity to friends and family.

The third aspect, *imagined community identity* refers to a highly intangible perspective of place attachment. It "explores the ways locales are imbued with personal and social meanings, and how such symbolic locals can serve in turn as an important sign or locus for the self" (Hummon, 1992: 258). The formation of this aspect of place attachment is largely an unconscious process for community members and groups and therefore is the least understood or clear aspect of place attachment (Hummon, 1992). While this aspect is the most difficult to define and understand, it is the aspect that deserves most attention when designing for community engagement. It is argued here that the development of an imagined community identity provides a motivating foundation for social interaction and as such it is essential that community ICTs foster imagined community identities.

Benedict Anderson's notion of an *imagined community* is a concept largely overlooked in conceptualisations of place attachment although it provides valuable insights into the highly intangible aspects of peoples' attachment to place. Anderson writes, "all communities larger than primordial villages of face-to-face contact (and perhaps even these) are imagined...it is imagined because the members of even the smallest nation will never know most of their fellow-members, meet them, or even hear of them, yet in the minds of each lives the image of their communion" (1983: 15). We suggest that it is this communion that underpins peoples' motivation to become emotionally invested and socially active in their community. Theorisations of imagined communities identify several key contributors to this strong image of communion such as shared experience, shared knowledge and community boundaries. Sharing knowledge and experience is key to the formation of imagined communities as it equips community members with shared ideas, information and a unified field of exchange that provides the basis for everyday social communication and interaction (Anderson, 1983, p. 44; Walmsley,

2000, p. 10). To further ensure community solidarity, this communication must be implicated in the boundaries of a community. Anderson notes that while a nation is largely conceptualised through imagined means, it is imagined as "inherently limited and sovereign...The nation is imagined as limited because even the largest of them encompassing perhaps a billion living human beings, has finite, if elastic, boundaries, beyond which other nations [or communities] lie" (1983: 15). In order to foster community imaginings and motivate social interaction, ICTs must be designed to afford each of the key contributors; shared experience, shared knowledge and community boundaries.

The notion of social capital complements the notion of imagined communities and helps us identify the value in building community solidarity. Robert Putnam notes that social capital is defined by the "features of social organization such as networks, norms, and social trust that facilitate coordination and cooperation for mutual benefit" (1995, p. 67). Pierre Bourdieu provides a definition of social capital that places a greater emphasis upon networks of relationships. He defines social capital as "the aggregate of the actual or potential resources which are linked to possession of a durable network of more or less institutionalised relationships of mutual acquaintance and recognition" (1986, p. 243). James Coleman further contributes to a definition of social capital noting that it "is not lodged either in the actors themselves or in physical implements of production" (2003, p. 162); rather it sits in the communication and relationships between and among actors. Research that contributes to an understanding of the relationships between social capital and technologies will contribute to greater social interaction and inclusion, and healthier communities (Adkins & Foth, 2005). While social capital is an extremely 'slippery' term; trust, social norms and community closure have each been identified as critical components contributing to social capital (Dekker & Uslaner, 2001; Coleman, 2003). These components coexist in complex relationships to one another. "The causal arrows among civic involvement, reciprocity, honesty, and social trust are as tangled as well-tossed spaghetti" (Putnam, 2000, p. 137). Reciprocation of social norms is a significant factor in the creation of social trust and furthermore community closure is an important device for cross-checking and reinforcing social norms (Coleman, 2003, p. 166). In order to foster social capital, a community-based ICT should aim to foster social trust by reinforcing community boundaries and social norms.

The above analysis of place attachment, imagined communities and social capital provides us with a theoretical framework from which we can draw core values for creating new modalities of place that motivate social interaction. New modalities of place are created through a process by which social, cultural, functional and emotional operators are used or acted upon to shape individual comprehension of place (Sandin, 2003). We understand that community based ICTs should aim to reinforce community boundaries and social norms, and promote shared experiences and shared knowledge. In order to apply these core values, they must be combined with a set of design principles that inform how they can be practically applied in order to create positive modes of place.

TECHNOLOGY, COMMUNITY, AND THE DESIGN PROCESS

To develop a set of design principles that would allow us to apply the core values derived from our theoretical exploration, we had to consider the community being designed for and what form of ICT would be most applicable. We also had to observe and learn from existing community based ICT projects.

We chose to design *FIGMENTUM* for The Kelvin Grove Urban Village (KGUV), a new urban development in inner Brisbane, Australia. The village was designed as "a master-planned community, bringing together residential, educational, retail, health, recreational and business opportunities into a vibrant new precinct" (http://www.kgurbanvillage.com.au/, accessed January 21, 2008). Our research project focused specifically on the Queensland University of Technology, Creative Industries Precinct that exists within the KGUV. The Queensland University of Technology describes the precinct as "Australia's first site dedicated to creative experimentation and commercial development in the creative industries. It provides a unique opportunity for designers, artists, researchers, educators and entrepreneurs to easily connect and collaborate with others to create new work, develop new ideas and grow the creative industries sector in Queensland" (http://www.ciprecinct.qut.edu.au/idea/, accessed January 21, 2008). Being members of the precinct community ourselves, we were aware that the Creative Industries Precinct had been pitched as a creative hub not only for the university community but also for the wider village. We found however this promise of unique opportunities for connections and collaboration to be rather idealistic. In our experience we found that access to sharing ideas and creative works was largely limited to 'polished' exhibitions by professional artists and graduating students, hindering opportunities for informal collaboration throughout the creative process. Furthermore, invitation to these exhibitions was largely faculty based limiting opportunities for exposure to the surrounding urban village or even cross-disciplinary collaboration. These exhibitions were mostly housed in 'The Block', a large, purpose build exhibition space. While these exhibitions are valuable we felt that other spaces, designed specifically for sharing artworks on a more informal, everyday basis were not being utilised. The Precinct features several foyer spaces with mounted plasma screens and/or wall projections that are ideal for presenting experimental works or works in progress.

Additionally the newness of the Creative Industries Precinct and the wider KGUV meant that the community lacked existing social ties and a cultural history, and therefore lacked a sense of community sentiment. As members of the new community, we saw our research project as an opportunity to utilise the resources and spaces of the Creative Industries Precinct and to begin building collective memories for the community in order to provide a foundation for social interaction and build upon the social capital embedded in the community.

The next step in our process was to select a form of ICT that would allow us to apply our research in practice, to apply and test our theoretical framework. In combining theory and practical application, this research project is a practice based action research project. It applies the methodological approach of action research. Action research, as described by Reason and Bradbury, "seeks to bring together action and reflection, theory and practice, in participation with others, in the pursuit of practical solutions to issues of pressing concern to people, and more generally the flourishing of individual persons and their communities" (2001, p. 1). While we acknowledge that many forms of ICT applications may have suited this research project we chose to focus our research upon the potential of generative art systems due to their dynamic and flexible nature and their aesthetic appeal. As defined by Ernest Edmonds, a generative art system is "an art system that evolves in response to the interpretation of participant interaction with...a software agent" (2003, p. 23). The decision to focus upon generative arts came out of an explicit decision to explore the emotional and intangible processes that contribute to imaginings of community solidarity. Generative art systems allow us to trigger emotional and representational community imaginings through a visual display that can be encoded with information contributed by community members. They are governed by rule sets that determine how information is encoded in the visual display and as such can be programmed to foster specific aspects of community solidarity. Generative art works can be co-authored and viewed by many simultaneously. They are able to offer a community *shared experiences*, *shared ownership* and *shared knowledge*, three factors that are essential to the creation of a shared community imaginings. The Community Informatics Research and Applications Unit (CIRA) acknowledge the potential for arts to engage people with new technologies and make note of the sense of empowerment experienced when creating and being creative (Keeble & Loader, 2001). Co-authored creative arts create this sense of empowerment by heightening perceived access to cultural production and providing a sense of belonging. For all of these benefits, we chose to utilise the generative art system, *Active Pixels* developed by Andrew Brown and Daniel Mafe (2006) for our research project. *Active Pixels* is a generative art system that allows artists to create dynamic screen-based artworks. The works are created through a process by which the artist alters system variables including but not limited to; the shape, shape size, colour, transparency, and frame rate of the work. The artist can also create layered works where they can alter each of these variables for separate layers. At the beginning of our research project *Active Pixels* was being used by the developers for their own artistic practices. We saw an opportunity to use the system as a tool to create community authored works. We understood that we could develop a rule set that would determine how information was read by the system and subsequently which variables of the work would be altered and in what manner.

To inform this rule set and the design how people would interface with work we conducted an audit of existing community based ICT projects. These works were selected using a criteria set to ensure their relevance. This consisted of four key questions derived from our theoretical framework:

1. Does the work employ an ICT?
2. Does the work significantly contribute to the imagined connectedness of the community members through providing shared information and shared experiences?
3. Does the work allow community members to contribute to the work?
4. Is the work implicated in creating a closed community or supporting community boundaries that already exist?

With the assistance of this criteria set, four key works were selected for the audit, namely, *Urballoon (*http://www.urballoon.com), *Urban Tapestries (*http://urbantapestries.net/), *Tejp (http://civ.idc.cs.chalmers.se/projects/pps/tejp/)*, and *The D-Tower (*http://www.d-toren.nl/site/). The observations made from the works audit were then refined into a set of key design principles to inform the development of *The Figmentum Project.* These principles were that the work must:

- Communicate consistent information that is relevant to all on a community-wide basis.
- Be sustained for an appropriate length of time as a provision for widespread-shared experiences.
- Allow users to contribute information to the work and ideally compile these contributions to generate co-authorship.
- Allow users to 'read' the work so as to decode the information it represents.
- Be easily accessible and able to be accessed at a variety of times.
- Demand only little time and prior knowledge from participants.

Using these design principles and the core values derived from our theoretical framework, we were able to determine the internal logic or rule set that would govern user interaction with the generated artwork. A process of development and user testing followed before the final product *FIGMENTUM* was installed and open to community contribution.

Figure 1. FIGMENTUM; The Creative Industries Precinct

FIGMENTUM

FIGMENTUM is a 'living wall' that employs a generative art system to visually represent the mood of the community in which it is embedded. Community members interact with the work by responding to a set of questions via the project website, http://figmentum.acid.net.au. These questions pertain to their current mood and their relationship to site of the work. The responses of the community members are read and according to the system rule set, encoded in the live artwork display. This interaction was designed to allow community members to share in their everyday experiences, and to be co-authors of an artwork.

In detail, each of the five questions posed by the work correlate with a variable within the artwork:

The questions are posed so as to represent these relationships. For example when the question 'How hungry are you?' is posed, the respondent registers their response by moving a marker along a slider of the colour spectrum. The respondent may choose to register their mood as red, green, pink and so on. Every two minutes the system aggregates the responses, determines the average response to each of the questions and encodes the artwork with the updated data. If the average response to 'How hungry are you?' was the colour red, the background colour of the artwork would turn red. A conscious decision was made to make the responses interpretive rather than prescriptive. Rather than pre-determine that the colour pink would represent a happy mood or that a circle represented less hunger than a hexagon, how the respondents represented their responses and interpreted the artwork was left open to interpretation, reflection and discussion.

Furthermore, the rule set allows the artwork to represent the numbers of people contributing to the work and to track responses over a period of time. The number of people contributing to the artwork is represented by the number of shapes appearing on the screen with more of shapes indicating more responses. Each of the layers of the work represents a two-minute period of time, the foremost layer displaying the responses from

Table 1. FIGMENTUM Response/System Variable Relationships

Question relating to:	Correlating Artwork Variable
Mood	The colours present in the artwork
Hunger Level	The transparency of the artwork
Energy Level	The frame rate of the artwork
Frequency of visits to the site	Movement within the artwork
Feeling towards the site	Types of shapes present in the artwork

the last two minutes, the second layer representing the two minutes prior and so forth through the five layers.

By viewing the artwork, community members can measure their everyday experiences against those of their community. For example they can compare their energy or hunger levels against the rest of the community. In asking participants to respond to questions that are specific to their community, the work also aims to demarcate and reinforce the boundaries of the Creative Industries Precinct Community. The work allows people to become active participants, collectively creating a cultural artefact for their community. It reflects the community back on to itself in ways that are normally not possible, intensifying opportunities for shared imaginings and increased social ties.

To practically apply the theoretical framework and design principles we had developed through our research, *FIGMENTUM* was installed for a period of one working week in building Z2 of the Creative Industries Precinct. The artwork was displayed live on several plasma screens and one wall-sized projection across two foyer spaces on level three and four of the building. Members of the Creative Industries Precinct community were able to contribute to the artwork via the project's website. As such specific data concerning who the respondents were was not able to be collected. However, it is reasonable to assume that high proportion of the respondents were students and staff of Creative Industries subjects that are held in Building Z2. These subjects range across schools and include subjects from fashion, communication design, media and communications, film and television, dance, animation, and music. Throughout the course of the week, 444 responses were registered by the system. An exact number of individual respondents cannot be determined as people could contribute to the work numerous times. Furthermore, although the IP address of the computers used when responding was collected, a large number of computers used were those housed in the computer labs of Z2 which are communal computers that an be used by any of QUT's staff or students. Throughout the week a significant amount of time (three to five hours a day) was taken to observe and conduct informal interviews with people in the site, both those who contributed to the work and those who simply viewed it.

FIGMENTUM was designed specifically to provide a foundation for social interaction. While the work attempted to embody foundations for fostering shared community imaginings, limitations in the design of the interaction and of the site in which the work was embedded restricted engagement of the community. In practically applying our theoretical framework, we found that it is not enough to simply embed theory into a design. Rather a complex understanding of relationships between the work and the contextual operators of a locale must also contribute to the design of a social intervention. An evaluation of *FIGMENTUM* underscores the need for a process of design for place-based ICTs that acknowledges and emphasises the contextual operators of a site. In order to generate new modalities of place, ICTs must utilise and act upon existing social, cultural, technical, temporal, and spatial relationships that shape individual comprehension of place (Sandin, 2003).

Our observational data revealed how the existing relationships of the site impacted upon how people engaged with *FIGMENTUM.* Most notably the functionality and temporality of the site were seen to limit people's engagement with the work. The work was displayed in foyer spaces to lecture and tutorial rooms. As such the behavioural expectation of those in the site was of quiet, sensible behaviour. Due to their functionality, the spaces are conducive to internal cognitions such as reading, writing, listening to music or daydreaming, thus limiting outward engagement with others or the site itself. The spaces are also highly transitory in nature with students coming and going from classes. Several people expressed concern over the time it would take to contribute to the work as they had impending classes or had to leave for other appointments. Furthermore the diverse range of subjects housed in the lecture and tutorial spaces meant that the students who were occupying the space were from varied schools and as such a high number of the students appeared not to know one another. While those using the site often did not seem to directly know one another, they were members of the same peer group, young university students. The peer driven nature of the site was seen to contribute to self-conscious behaviour. While we observed that by and large it was individuals occupying the space, it was groups of two or more that were more likely to engage in and contribute to the work. Individuals were seen to avoid the work for fear of getting the interaction wrong in front of their peers. It was not until others, generally in groups, had approached the work that we saw an increase in engagement from individuals. Each of the contextual factors discussed here impacted upon how people interacted with the work, decreasing the time and effort people would invest in engaging with the work.

As the results of our observational data reveal, the complex contextual operators of a site play a large role in shaping how people engage with their surroundings and hence the success of a social intervention. Designers of community ICTs not only have to identify core values and design principles to inform the design process, they must also conduct an analysis of the complex relationships at play within the intended site. This analysis allows designers to first select sites that are conducive to community engagement and secondly design for maximum engagement considering the specifics of the site. In the case of *FIGMENTUM*, the level of engagement required for interaction was not suited to the site in which it was embedded. To be more successful in the site, the interaction of *FIGMENTUM* would have had to be more intuitive and demand a lower level of engagement, allowing the community to contribute to the work simply by going about their everyday action. At present the interaction of *FIGMENTUM* is much more suited to a site such as a café, food hall or courtyard. A site that is a social hub and one in which people are not primarily in transit and are able to invest time and critically engage with the site. Along with embedding core values and design principles into the design of an ICT, it is a complex process to design applications that demand new forms of engagement that what is already experienced at a site.

In addition to providing possibilities for community building, this research has revealed that ICTs can act as tools for investigating the limitations and opportunities of particular sites. Through demanding particular engagement, social interventions can highlight the complex social, cultural, spatial, and temporal relationships at play in a site and how these may encourage or deter particular activities. In the case of the Creative Industries Precinct, spaces in Building Z2 are intended to facilitate unique opportunities for collaboration and the generation of new ideas. In questioning the reality of this promise and intentionally causing trouble for the site, *FIGMENTUM* revealed a number of limitations. The opportunities for casual, yet critical engagement around creative works promised by these spaces are limited by the forced dual functionality of the spaces. As discussed above, the social, cultural, spatial and temporal relationships that are constructed by the educational functionality of the spaces are vastly different to those required for casual, critical engagement in creative works. Through helping us to identify the opportunities and limitations of particular sites, ICT based urban intervention can help us in the future to demand more of our community spaces.

While our research enabled us to clearly identify some of the social, cultural, spatial and temporal relationships at play in the site and hence recognise flaws in the intended dual use of the spaces, we were also able to identify weaknesses in the implementation of the work. *FIGMENTUM* aimed to provide a casual and informal means through which people could contribute to a community owned work. This was an explicit attempt to move away from formal and moderated works towards informal works that engage people on an everyday basis whilst empowering them as creators of cultural artefacts. Through observing how people interacted with the work we were able to identify limitations of the interaction design. The most evident was a lack of *onsite* opportunities for contributing to the work. The interaction was designed so that people could contribute to the work via the project website. Initially we had not provided an interface that would allow people to contribute to the work while being physically collocated. We soon realised this led to a fragmented experience of the work that distanced the contributors from the artwork. In response we provided a laptop at the site of the work so as to facilitate onsite contribution. While the introduction of a laptop increased the number of contributors to the work, the interfaces of the website were designed for remote interaction, leading to occasional confusion.

Further compounding to this confusion was the complex way in which the artwork was encoded with information from the contributors. From our observations it became clear that the interfaces did not effectively communicate the relationships between what the user contributed and the appearance of the artwork. To our knowledge there had not been any previous works that required contributions from an audience. This lack of historic interaction meant that clear and simple interaction design was essential. In actual fact the interaction design led to fragmented and sometimes confused experiences of the work. To improve on this interaction, future iterations of *FIGMENTUM* must include improved interfaces for contribution that clearly describe how the artwork is encoded with results of user input. While the visual representation of *FIGMENTUM* was well resolved into an aesthetically pleasing display, it is a difficult challenge to resolve a user interface that facilitates a process of contribution that supports a desired user experience.

CONCLUSIONS AND FUTURE RESEARCH

FIGMENTUM was successful in visually representing the community back onto itself in ways that have the potential to foster community sentiment. As a new contribution to the field, this research has explored opportunities for fostering highly intangible and emotional experiences of community. It has presented core values and key design principles to be considered when designing community based ICTs. Furthermore it has revealed ICTs as useful tools for identifying opportunities and limitations of spaces. Above all this research underscores the need to understand how people experience particular places in order to design ICTs that are able to generate new modalities of place and motivate social interaction.

In a continuing cycle of iterative design, we will continue to refine and implement *FIGMENTUM*. In response to our current findings, future research will explore several key areas. These will include selecting a new site for the work that is more conducive to the desired engagement and revising the interaction design of the work so as to consider the contextual operators of the site and hence maximise possibilities for engagement. This revision of the interaction design will place a strong emphasise on improving interfaces for contribution and simplifying how information is encoded in the work. Furthermore, future iterations will see the work installed for a prolonged period of time, allowing time for people to learn how to read the work and for the work to become part of the everyday experience of the community. Outside our own research, we feel that further research is required in order to realise the potential for ICTs to contribute to communities that are rich in social capital. As our research has attempted to, this research must contribute to an understanding of the process required to design community based ICTs. By creating and installing *FIGMENTUM* we have begun to identify the processes of design required for producing place-based ICTs that may animate our urban fabric. As researchers and designers we must continue to actively seek the desired social outcomes that technologies can help us to achieve.

ACKNOWLEDGMENT

We would like to acknowledge the wonderful support and advise of Barbara Adkins and other members of the *FIGMENTUM* development team; Craig Gibbons, Andrew Brown and Dan Mafe. This research was supported by Queensland University of Technology and ACID, the Australasian CRC for Interaction Design, established and supported under the

Cooperative Research Centres Programme through the Australian Government Department of Innovation, Industry, Science and Research.

REFERENCES

Adkins, B. & Foth, M. (2005). A Research Design to Build Effective Partnerships Between City Planners, Developers, Government and Urban Neighbourhood Communities. *The Journal of Community Informatics*, 2(2), 116 – 133.

Anderson, B. (1983). *Imagined Communities: Reflections on the Origin and Spread of Nationalism*. London: Verso.

Arnold, M., Gibbs, M. and Wright, P. (2003) Intranets and the Creation of Local Community: 'Yes, an intranet is all very well, but do we still get free beer and a barbeque?'. In M. Huysman, E. Wenger, V. WulfKluwer (Eds.) *Communities and Technologies*. Dordrecht: Kluwer Academic Publishers.

Bourdieu, P. (1983). Forms of capital. In J. G. Richardson (Ed.) *Handbook of theory and Research for the Sociology of Education* (pp. 241-258). New York: Greenwood Press.

Bressi, T. (1994). Planning the American Dream. In P. Katz (Ed.), *The New Urbanism, Toward an Architecture of Community* (pp. xxv-xlii). New York: McGraw-Hill.

Dekker, P. & Uslander, E. (2001). The 'Social' in Social Capital. In P. Dekker & Uslander (Eds.) *Social Capital and Participation in Everyday Life* (pp. 176-187). London: Routledge.

DCITA. (2005) *The Role of ICT in Building Communities and Social Capital: A Discussion Paper.* Retrieved Sept 24, 2007, from http://www.dcita.gov.au/communications_for_consumers/funding_programs__and__support/community_connectivity/the_role_of_ict_in_building_communities_and_social_capital_a_discussion_paper

Edmonds, E. (2003). Logics for Constructing Generative Art Systems. *Digital Creativity*, 14 (1): 23-28. http://gateway.library.qut.edu.au/login?url=http:// search.ebscohost.com/login.aspx?direct=true&db=inh&AN=7647583&site=ehost-live (accessed April 23, 2006, from EBSCOhost: Inspec database).

Foth, M. (2004). Designing Networks for Sustainable Neighbourhoods: A Case of a Student Apartment Complex. In G. Johanson & L. Stillman (Eds.), *Community Informatics Research Network (CIRN) 2004 colloquium and conference proceedings. 29 Sep – 1 Oct 2004* (Vol. 1, pp. 161-172). Prato, Italy.

Foth, Marcus (2006) Facilitating Social Networking in Inner-City Neighborhoods. *Computer* 39(9), 44-50.

Gilchrist, A. (2004). *The Well-Connected Community: A Networking Approach to Community Development.* The Policy Press.

Graham, S. & Marvin, S. (1996) *Telecommunications and the City: Electronic Spaces, Urban Places.* London: Routledge.

Hummon, D. (1992). Community Attachment: Local Sentiment and Sense of Place. In I. Altman & S. M. Low (Eds.), *Place Attachment* (pp. 253-278). New York: Plenum Press.

Jacobs, J. (1961). *The Death and Life of Great American Cities.* New York: Vintage Books.

Keeble, L. & Loader, B. D. (2001). *Community Informatics: Shaping Computer-Mediated Social Relations.* New York: Routledge.

Oldenburg, R. (2001). *Celebrating the Third Place.* New York: Marlowe & Co.

Pinkett, R. D. (2003). Community Technology and Community Building: Early Results from the Creating Community Connections Project. *The Information Society,* 19(5), 365 – 379.

Putnam, R. D. (1995). Bowling alone: America's declining social capital. *Journal of Democracy,* 6(1), 65-78.

Queensland University of Technology (2005). Creative Industries Precinct: The Idea. Retrieved June 21, 2007, from http://www.ciprecinct.qut.edu.au

Reason, P & Bradbury, H. (2001). *Handbook of Action Research: Participative Inquiry and Practice.* London: Sage Publications.

Ross, A. (2000) *The Celebration Chronicles: Life, Liberty and the Pursuit of Property Values in Disney's New Town.* London: Verso.

Sandin, G. (2003). *Modalities of Place: On Polarisation and Exclusion in Concepts of Place and in Site-Specific Art.* Lund, Sweden: Lund University.

State of Queensland (Department of Housing) and Queensland University of Technology. (2007). Kelvin Grove Urban Village: Be Part of a Clever Community. Retrieved June 21, 2007, from http://www.kgurbanvillage.com.au

Walmsley, D. J. (2000). Community, Place and Cyperspace. *Australian Geographer,* 31(1), 5 – 19.

WORKS CITED

Active Pixels, Developed by A. Brown and D. Mafé. Brisbane: Australasian CRC for Interaction Design [Java Application].

Tejp, Developed by Re:form and Play Studios, Viktoria Institute, Interactive Institute and Future Applications Lab, http://civ.idc.cs.chalmers.se/projects/pps/tejp/ (accessed March 25, 2006).

The D-Tower, Developed by V2_Lab and Institute for the Unstable Media http://www.d-toren.nl/site/read.htm, http://lab.v2.nl/projects/dtower.html (accessed March 25, 2006).

Urballoon, Developed by Eyebeam, www.urballoon.com (accessed March 25, 2006).

Urban Tapestries, Developed by HP Labs and Proboscis http://urbantapestries.net/ (accessed March 25, 2006).

KEY TERMS

Generative Art Systems: These consist of an artwork; usually displayed on a digital screen, the appearance of which is governed by a rule set developed by the curator that determines how the system interprets the information it receives. The curator can either predetermine this information, or it can be created through user interaction. Ernest Edmonds defines a generative art system as "an art system that evolves in response to the interpretation of participant interaction with the work by a software agent" (2003, p. 23).

Imagined Communities: The ontological community solidarity. The notion is best captured in Andersons quote; "all communities larger than primordial villages of face-to-face contact (and perhaps even these) are imagined…it is imagined because the members of even the smallest nation will never know most of their fellow-members, meet them, or even hear of them, yet in the minds of each lives the image of their communion" (1983, p.15).

New Urbanism: This is an urban design ideal that returns to planning and architectural traditions for the design of small communities. The ideal focuses on the liveability of suburban areas, emphasising the need for central community hubs and a layout that provides walkable access to everyday needs and services.

Modalities of Place: The process by which numerous social, cultural, functional and emotional operators shape individual comprehension of place (Sandin, 2003). The capacity for technologies to create new experiences of place is dependent on the way the operators of a location are used or acted upon in order to determine the mode of the place.

Place Attachment: The emotional, functional, and social ties people develop *within* a community and *towards* a particular place (Hummon, 1992). Place attachment is a complex and interdisciplinary notion and has been theorised from diverse perspectives ranging from architecture and urban planning to psychology and sociology. In the context of this chapter place attachment consists of three key aspects; *community satisfaction, community attachment,* and *imagined community identity* (Hummon, 1992, p. 254).

Social Capital: The capital, actual and potential, that is embedded in social relationships and networks. Pierre Bourdieu defines social capital as "the aggregate of the actual or potential resources which are linked to possession of a durable network of more or less institutionalised relationships of mutual acquaintance and recognition" (1986, p. 243).

The Social Construction of Technology: This is a perspective on the evolution of technology that emphasises human agency. Identified by Graham and Marvin (1996), this perspective views the evolution of technology as a social process by which a collection of individual human decisions shape how technologies will impact upon society.

Chapter XI
Voices from Beyond:
Ephemeral Histories, Locative Media and the Volatile Interface

Barbara Crow
York University, Canada

Michael Longford
York University, Canada

Kim Sawchuk
Concordia University, Canada

Andrea Zeffiro
Concordia University, Canada

ABSTRACT

The Mobile Media Lab (MML) is a Canadian interdisciplinary research team exploring wireless communications, mobile technologies and locative media practices. By developing interactive mobile experiences, we observe and reflect on the dynamics inherent in wireless immersive environments connected to a growing tendency towards ubiquitous computing or pervasive media. Our projects, whilst rooted in digital ephemera, treat physical territory as an active and volatile interface creating networked situations to connect the physical to the virtual. Our intention is to use media to quietly augment everyday life and to initiate novel ways of telling stories of the past by harnessing digitally rendered images, text and sounds. In our chapter, we will focus on two projects, Urban Archaeology: Sampling the Park and The Haunting, which were part of our work done under the rubric of the Mobile Digital Commons Network (MDCN, 2004-2007). We will use the phrase voices from beyond as a trope in our reflections upon the deployment of mobile media technologies and use of locative media practice to intentionally blur past and present moments. As we argue, archival fragments and ghostly images can be presented via handheld devices to use the power, potential and public intimacy of media dependent upon the presence of electro-

magnetic spectrum. In addition to key texts on locative media, we draw on Benjamin's understanding of history as a sensibility whereby the past and present co-mingle in the minds and embodied memories of human subjects, Darin Barney's notion of the "vanishing table" as an alternative means for engagement in technologically mediated zones of interaction, and writing on communications theory that deals with the spectral qualities of new media (Sconce; Durham Peters; Ronell).

INTRODUCTION

Sound of phone dialing—ringing—answered. Houdini speaks:

It's Harry Houdini. Be careful who you trust.

Things are not always as they seem, my friend.

Never trust the dead.

[sound file in the digital version]

In *Speaking into the Air: A History of the Idea of Communication,* John Durham Peters (2000) draws upon the literary insights of Franz Kafka to offer these thoughts on the ghostly qualities of communication:

As Kafka notes in an epigraph to this book, those who build new media to eliminate the spectral element between people only create more ample breeding ground for the ghosts. A cheerful sense at the weirdness of all attempts at communication offers a far saner way to think and live. (pp. 29-30)

It is with attentiveness to fostering a new media practice that is haunted by the past, imbued with a sense of place, and rooted in an exploration of the weirdness of communication that we offer our reflections on two locative media projects created in Montréal, *Urban Archeology: Sampling the Park* and *The Haunting. Urban Archeology* was produced for Parc Émilie-Gamelin, a highly contentious and socially charged small urban square in a densely populated downtown Montréal neighborhood known as *Centre-Sud. The Haunting* was devised as a game of ghost capture for the large urban greenspace of Mount Royal Park that is at the heart of the city. These were conceived within and created as a part of the work done under the rubric of the Mobile Digital Commons Network (MDCN, 2004-2007).

In both projects our desire as designers and researchers was to use hand-held media devices networked to global positioning systems to initiate novel ways of telling stories of the past. While the past was brought into the present in both *Urban Archeology* and *The Haunting*, an important narratological difference distinguishes them. *Urban Archeology*, with no discernable teleological goal, acted as a location-based experience akin to a site-specific installation. *The Haunting*, a location-based ghost capture game, was predicated on a clearly specified set of objectives for users, conceived of as players, and was scripted within the parameters of game design (Salen & Zimmerman, 2004). By creating the possibility for users to access digitally rendered images, text and sound our intention was to quietly disrupt and supplement the habitual uses of these spaces.

In the sections that follow, the phrase *voices from beyond* will as act as a trope (Burke, 1969, p. 503) to structure our self-reflections on these two projects for it captures three elements essential to our design approach: the use of historical documents to animate the present, our adoption of user-testing and participatory design methods in our locative media practices, and our ethics of place. These places, which were 'seeded' with content at various global positioning coordinates disrupted the users' sense of being in a present moment. Images, texts and sounds revivified the past, if only in a glimmering instant that soon faded away. In this sense, we take a page from the book of Walter Benjamin (1968) who suggested that: "[t]o articulate the past historically does not mean to recognize it 'the way it really was'"(Ranke). It means to seize hold of a memory as it flashes up in a moment of danger" (p. 255).

METHODOLOGICAL RISKS AND THEORETICAL COMMITMENTS: INTERPRETATIVE, REFLEXIVE METHODOLOGIES

As mentioned both of these projects took place under the umbrella of a larger organization known as The Mobile Digital Commons Network (MDCN). The MDCN sponsored a number of multi-disciplinary collaborative projects that involved researchers with expertise from design, engineering, the social sciences, and the humanities. Michael and Barbara were part of the initial research and design team who produced *Urban Archeology*. Kim and Andrea joined the Montréal MDCN team after the *Urban Archeology* project was completed. All four of us participated in the research and design of *The Haunting*. This direct involvement in *Urban Archeology* and *The Haunting* underlies our decision to focus on these two case studies. Our direct involvement in these case studies provides us with an experiential and intimate insider knowledge. We are fully aware that our connection to these projects also may limit our critical reading of their shortfalls, however, in offering our description and reflections upon our production practices our approach is influenced by those traditions in qualitative methods promoting the self-reflexivity of the researcher (Tillmann-Healy, 2003; Richardson, 2000), feminist writings that foreground the positive dimensions of active participation in a community or a practice (Reinharz, 1992), and methodologies that grapple with one's embodied and affective involvement in a research practice (de Garis, 1999). Theoretically and methodologically, we are situated within humanities-based approaches to new media and technology (Friedberg, 2006; Hansen, 2004) and cultural studies (Slack and Wise, 2005). Finally, our approach to the use of technology in the environment takes a page from actor-network theory's assertion that in any networked situation neither the human subject nor the technology is the only force to consider. As such, theorists such as Haraway, drawing from Latour, use the term actant (Haraway, 1996; Latour, 2005).

To be even more specific, and to return to questions of methods and theory, we draw upon sociologist Laurel Richardson's assertion that "writing is a method of inquiry" (2000, p. 923). This position advocates experimentation in presenting the results of one's research pointing to the history of the generic conventions that underlie the supposedly objective, descriptive prose style promoted within the social sciences (pp. 923-925). As Richardson pithily argues, such a style bears the epistemological traces of an allegiance to its positivist history, which sees the natural sciences as the natural model for the social sciences to emulate. In developing what she terms a "Creative Analytic Practice" (p. 929) for conducting research Richardson promotes the use of metaphors to explore and "crystallize" one's ideas. Following Richardson, and the work of George Lakoff and Mark Johnson (1980) on metaphors, we see metaphor as intrinsic to the process of thinking and critical reflection: it is not a lyrical add-on.

As mentioned, the trope "voices from beyond" is key. This figure of speech is meant to invoke the mythic history and apocryphal tales (Sconce, 2000) of ghostly hauntings that we gleefully have employed. In this we tip our hats to the growing body of work on the new media that questions its newness by invoking the relationship of present practices and discourses to its precursors (Gitelman, 2006; Marvin, 1988). These important histories unsettle the claim that the new media is as new as it might like to proclaim. This trope also signals our commitment to disclosing how, by using mobile wireless technologies, the present can be imbued with the past.

In designing these two projects, we have used historical materials pertinent to these places to furnish images, texts, and sounds as it is our desire to make users aware of the history of the place that they may walk through everyday, a history that in some cases is literally buried beneath their feet. Critical to our conceptualization of these historically based locative-media projects is a commitment to a historical sensibility whereby the past and present co-mingle in the minds and embodied memories of human subjects. As Walter Benjamin's (1968) ruminates: "History is the subject of a structure whose site is not homogeneous empty time, but time filled by the presence of the now" (p. 261). Benjamin's poetical and eclectic thinking on history dovetails with David Lowenthal's (1985) discussion of the past as "a foreign country." It is a place we desire to visit, but to which we never have full access. Our perspective explores how these technologies may be used to bring these stories to the surface, to provide a tentative connection to the past, and act as a means to connect users to each other as they partake in one of our location-based projects.

Locating Locative Media

Our design work with mobile, wireless technologies is situated within an emergent new media practice, locative media,[1] which attaches location specific data garnered from global positioning satellite (GPS) coordinates and mobile communication technologies. Locative media is an artistic sub-branch of ubiquitous computing research that was initiated at a workshop held at RIXC in Riga, Latvia entitled, "Mapping the Zone." According to the event's organizing group, the idea behind the workshop was twofold. First, it was "an explicit acknowledgment of Virilio's idea that 'one cannot understand the development of information technology without understanding the evolution of military strategy'" (Locative Media, p. 4). Second the intention was to resituate these artistic events outside of the global market and military contexts from which these technologies emerged (Tarkka, in press, p. 3; Galloway & Ward, 2006). The specific term locative media was coined by Karlis Kalnins, who argued that "locative is a case not a place" (Kalnins, 2004; Tuters & Varnelis, 2006, p. 1). The locative case, in Finnish, corresponds roughly in English to the *preposition* 'in', 'at', or 'by', indicating the types of proximity or relationality that we have to a given territory.

Since this initial meeting, a number of artist projects have taken up these ideas and technologies including:

- The *Tactical Sound Garden* a platform for playing sounds in city streets (http://www.tacticalsoundgarden.net/)
- *Blogmapper*, a location-tagged weblog (http://www.blogmapper.com/);
- *Urban Tapestries*, a project using cell phones to tag and access GPS co-ordinates with community based content http://urbantapestries.net/;
    - locative games played on mobile phones such as *Bot Fighters,* an example of one numerous location based games played on mobile phones (http://www.gamespot.com/mobile/action/botfighters/) and;
- Research centered projects such as *Mobile Bristol* (http://www.mobilebristol.com/flash.html) and the *Mobile Digital Commons Network (*http://www.mdcn.ca).

Most of these projects annotate a space with particular content and many aspire to instigate public authoring using GPS coordinates to tag a location with user-generated content. The use of cellular telephones, Bluetooth frequencies, infrared sensors, GPS satellites ties locative media practitioners to ubiquitous computing, or ubi-comp.

Ubi-comp highlights a desire or a concern with the pervasiveness and invisibility of media technology in the built or the natural environment. As microprocessors and micro-chips decrease in size and increase in power and capability they may be inserted into an array of objects embedded with sensors—from singing birthday cards, to wireless MP3 players, to digital thermometers, to traffic lights. This very pervasiveness fosters great concerns about their deployment for surveillance and information gathering purposes. For example, the passage of information between technologies may be unbeknownst to those who traverse these invisible data networks (Kaplan, 1997; Zeffiro, 2006a). And, as the Basel Action Networks' documentation of micro-technology dumpsites indicates, when the ecological impacts of these technologies are ignored, this pervasiveness can be toxic (http://www.ban.org/) .

Malcolm McCullough (2004) advises that with, "the rise of pervasive computing, more applications must enhance, and not undermine, our perceptions of grounding place" (p. 174). With *Urban Archeology* and *The Haunting*, our intent was precisely to enhance, via a locative experience, the subtle nuances of being *of* and being *with* a distinctive place. While locative media *harkens* on the term 'location' both in theory and practice, locative media is not simply location-based media. Although our projects implement location-based technologies, our practices are aligned with an active engagement with a particular place. It also invokes the tradition of artist inquiry already mentioned. What became apparent through our research is that one of the key issues in locative-media is where and how a space or place may matter to a game or project that uses mobile wireless technologies, such as cellular telephones or global positioning systems.

The Place of Mobile Media and the Volatile Interface

It is often argued that ubiquitous technologies erase our sense of place and create what Mark Augé called, a *non-lieux* (1995). Rowan Wilken (2005), for example, suggests that mobile phones may operate 'independent of place' and "where they are not exactly independent of place, they appear immune to place, serving to insulate their users from the geographical place they are actually in" (p. 4). It is our assertion that the use of technology with respect to place matters, deeply, and we have tried to build this into our design and use of these technologies in both projects. Indeed, our locative media practice reveals the spirited aliveness of the environments that we work in. Such ever-shifting conditions of production and reception for locative media projects that are outdoors provide the context for unforeseen interactions. This is particularly crucial in a context like Canada where temperatures can range from minus forty degrees Celsius in the winter to plus thirty in the summer (Sawchuk, 2007).

In this sense, rather than erasing place, we think of the GPS coordinates we demarcate and the technologies that connect users to these invisible zones for interaction as creating a focal-point for communication in a specific place, what Canadian communications theorist Darin Barney (2004) has called a "Vanishing Table". Just as the table at a dinner party creates a shared occasion for communication, the locative media experiences we create temporary linkages between a variety of protagonists to forge connections.

We term our understanding of place as a territory for interaction, the volatile interface. This phrase references feminist theorist Elizabeth Grosz' (1994) understanding of the unruly, materiality of the body in *Volatile Bodies* (1994) and the idea of new media interfaces as a "responsive environment" (Krueger, 1997). Our projects, whilst rooted in digital ephemera, treat physical territory, not as the inert ground for our design machinations, but as an active inter-actant in the assemblage of networked situations that come together when one stages a locative media event. As we worked on these projects, it became apparent that the connection of story to place, and our understanding of place as an active, and living interface needed to absolutely integrate with our practice of designing for a locative media experience: this entailed doing historical research on location, talking to community groups involved in these places and spending significant time in these environments, rather than designing from the comforts of a desk or an office.

Communication as Ritual: Calling All Ghosts

In addition to these theoretical debates on locative media, ubiquitous computing, environment as a volatile interface, and the importance of place, our thinking as designers, researchers and writers draws on those histories of communications that detail the ineffable, ghostly qualities of communications and the media. The idea of 'haunted media' has been used by scholars such as John Durham Peters (2000) and Avital Ronell (1989) to discuss the legacy of spiritualism in the history of technology, often suppressed in official accounts or forgotten in the desire to promote communications as a rational endeavour of ever-greater scientific and engineering refinements. We are inspired by these approaches for many reasons, not least of all because they see communications, to paraphrase the words of James Carey (1989), as a ritual activity that can create a sense of belonging to a place or a community and not merely as an act of transmitting information.

Jeffrey Sconce (2000), for one, is critical of the recurrence of the tropes like the "ghost in the machines" for new media studies. We agree that our use and investment in this term may carry the danger of too uncritical a position that may, in the words of one reviewer of our paper, "make it more difficult to understand the often very mundane and historically continuous ways in which emerging communication technologies 'actually' work." This, we hope, is mitigated by our description of the technologies and design practices we used to create our project. We also deliberately court this danger. We take the paranoid fantasies that people have about the media seriously, and we would argue that wanting to experience them may come from a place of pleasure, which is not always diametrically opposed to learning. Rather than rejecting terms which evoke "the weirdness of communication" to quote Durham Peters' paraphrasing of Martin Heidegger, we decided to use this possibility in our design to give a glimpse and glimmer into the past rather than didactively recounting a complete story of these places.

In the pages that follow we describe these two projects, one fully completed (*Urban Archeology*) the other in existence as a working prototype (*The Haunting*). Given our emphasis on place, we provide "thick descriptions" (Geertz, 1973) of these spaces to reflect upon our main theme: paying attention to the voices from beyond. In offering this detailed description and comparative discussion we detail our intentions as producers and we use our experience to critically reflect upon our design intentions.

1. Urban Archeology: Sampling the Park: A Layering of Voices

Urban Archeology: Sampling the Park, which was designed for Parc Émilie-Gamelin in 2005, explores the social history of this Montréal city square by providing glimpses into the past. Parc Émilie-Gamelin is nested in a prominent intersection in one of the densest areas of Montréal. Sometimes known as Square Berri, it is somewhere between a park and something much more intensely urban, as a nexus for a variety of mobilities. The park is bounded by busy streets crowded with vehicles and pedestrian traffic. The southwest corner of the park provides access to the Berri UQAM metro station, a primary point of intersection for Montréal subways. The corner also houses a small police post. Directly to the north is the Greyhound Bus Terminal, a major transportation artery providing access to and from the city.

It is a site layered with activity that brings together diverse populations from all parts of the city. For example, to the west is the Université du Québec à Montréal, Québec's largest university providing a temporary home to over 40,000 francophone students while to the east is Montréal's

Figure. 1. Parc Émilie-Gamelin, Montréal - Aerial View

Figure 2. Urban Archeology-3D Map of hotspots tagged with GPS coordinates (Illustration: A. Morris)

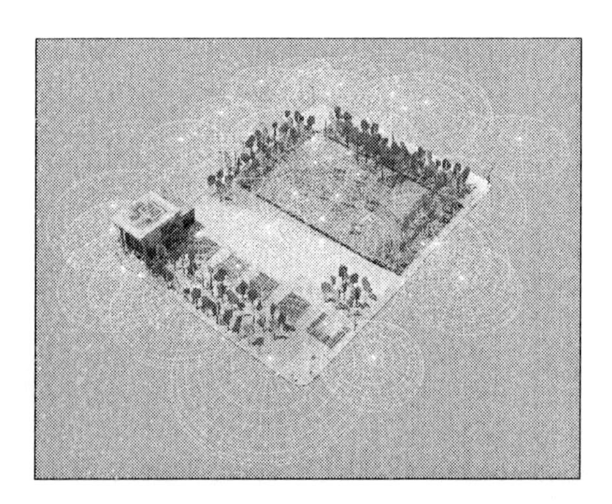

gay village. The park is bordered on all sides by a diverse number of businesses: including fast-food restaurants, cafés, boutiques, a mall complex and a landmark building, the Archambault music store. It is decidedly located in the east part of the city, which has traditionally been a space of the French working class and the urban underclass. Ste. Catherine is a street that has had notoriety in the history of Montréal as one of the centres for the sex trade industry (Straw, 2007).

The park changes seasonally, hosting many of the festivals that make Montréal a major tourist attraction in the summer. To the south end is a large concrete area where festival activities take place, all of which require city permits. During the winter, this space becomes a skating rink. To the north end is a grassy slope, which is traversed in the middle with a terraced stream of water that flows from large aluminum sculptures by Melvin Charney at the top of the slope, celebrating urban architecture and the built environment. The installation of Charney's sculptures and redesign of the park was the result of public spending in the 1980s initiated by the movement to 'revitalize' the city.

The park is emblematic of a number of competing interests between the city, capital, small businesses, the nearby residents and institutions, and the different populations inhabiting the park itself. The space has been and continues to resist the attempts of various municipal governments to gentrify and sanitize the area of what are perceived by some as urban blight. For example, there recently has been an attempt to rid the park of dogs in order to clear out the homeless population, many of whom own dogs for protection and company. Located nearby are a number of local community groups like the Cactus Needle Exchange, Dans La Rue, STELLA, and Action Sero-Zero.[2] The history and the locale of the park augment the various tensions involving social class that persist in Canada. The challenge for *Urban Archeology* was to represent these contradictions to evoke this history of contestation.

Our process of research involved gathering material from the city archives and talking to various individuals about the past and present circumstances of the park. As our archival research revealed, the square has witnessed a series of dramatic changes. For example, for 120 years leading into the early sixties, the park was the site of a Catholic mission founded by Sister Émilie Gamelin, for whom the park was renamed in 1988. The design team met with Sister Therese, the head archivist in the order of the Sisters of Providence, who in turn introduced us to Sister Yvette who had worked at the mission for over 20 years. Both Sister Therese and Yvette toured the park with the design team and recounted stories of the soup kitchen, pharmacy, hospital for the poor, and home for impoverished elder women that had functioned at the site. Microfilm articles from old newspapers, reported how the mission was destroyed by fire and purchased in 1962 to make way for an envisioned 'Plaza of the East'. This project eventually fell through leaving the site to serve as a parking lot for almost 20 years. The work of Émilie-Gamelin is memorialized with a representational bronze statue located inside of the metro station.

Figure 3. Archival research: Photos of Sister Émilie-Gamelin. (Photo: M. Longford)

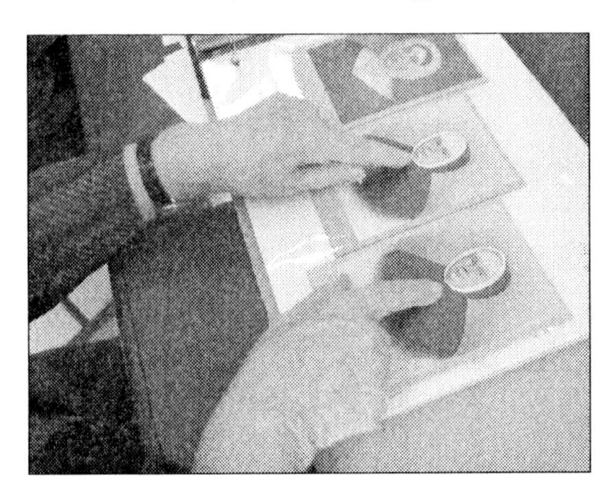

Figure 4. The mission after it was destroyed by fire in 1962. (Photo: Archive of The Sisters of Providence)

This archival material was sampled and integrated with more recent content gathered from the park which were mixed and shaped into a variety of sound and image collages. An authoring environment developed by Mobile Bristol (MB) [3] a research group at the University of Bristol and HP Labs in the UK, was used as the platform of dissemination. The Mobile Bristol Toolkit provides a 'drag & drop' Graphic User Interface (GUI) for attaching media files such as sound, text and image to GPS coordinates. The authored experience can then be downloaded to a hand-held personal digital device produced by HP, known as the iPAQ, and played back in real-time and space using headphones and a GPS receiver.

Using the programming layer of the software, the sound experience can be augmented and choreographed in a number of ways. For example, using 'source points,' we can create graduated sound experiences, by changing volume from the periphery of a sound region to its central source point. You can also attach more than one sound to a location and have it playback in a specific sequence depending on one's location in the park. These layered sound experiences added a degree of dimensionality creating the impression of walking through an event unfolding in space. In addition to the sound, image files can also be attached and played back on the iPAQ in a browser window or using Macromedia Flash.

Figures 5 & 6. Users in the park with GPS and iPAQ. (Photos: M. Longford)

In all, Parc Émilie-Gamelin was seeded with 40 hotspots treated as 'content wells' that could be accessed by the user. We did not conceive of the user walking through a linear historical narrative, rather we opted to give them snippets and fragments in order to rupture present time with moments from the past. As Walter Benjamin (1968) would say, "The true picture of the past flits by. The past can be seized only as an image which flashes up at the instant when it can be recognized and is never seen again" (p. 225). A collage of voices and sounds that were comprised of oral narratives, recent events, hidden histories, and the diegetic sounds of the park itself, were presented abstract at times and at other moments almost in a literal photo-journalistic fashion. For example, at one seeded spot, the user would cross a sound tunnel conforming to the underground subway cutting below the park, sampled and mixed as a kind of aural collage, with accompanying images of its mid-sixties construction.

In contrast to a project like Mobile Bristol's 'Riot' (http://www.mobilebristol.co.uk/QueenSq.html) which used a similar tool to bring people into the history of a particular event and provided the overarching narrative, we had no singular narrative, but were exploring multiple historical moments within a given space. They were bound by an event at a time and a place: we were bringing into focus, in a sense, the layering of events through time. The point is not just to render the 'invisible' visible but to invoke the feeling of sedimented layers of time. Like the interface we speak of, our bid at historicizing is also volatile: we are less concerned with seamlessness and more interested in "the vertigo of the radically multiple (not subjective) inside viewpoint" (Kwinter, 2001, p. 41) *within* a particular place. Mediated through the iPAQ, historical fragments are the slippery bedrock on which the present stands.

2. Other Voices: Lessons from User-Integrated Testing

There were several public demonstrations of the *Urban Archeology* project in 2004 and 2005. We decided to get more formal feedback on the project and set up a pilot demonstration of the project in October 2006 with a group of first year university students recruited from a class in research methodologies. While not a representative sample by any means, the group members were varied in terms of gender, language and cultural backgrounds. Although limited in scope, this user-test provided invaluable feedback that challenged several theoretical and design assumptions we were working with, and which pointed to the value of iterative and participatory user-tests as a valuable design tool in future projects.

Figures 7 & 8. Users in the park with GPS and iPAQ. (Photos: M. Longford)

As we only had 8 iPAQs we necessarily had to divide the groups into two. The first group was sent into the park with little or no instruction, other than technical information on how to use the iPAQ and turn on the GPS receiver they were wearing on their arms. The second group, who had to wait for the others to return, was treated to a brief introduction to the project, including a presentation of the map of the park with the hotspots programmed on it. Both groups were followed by focus group interviews in which we wanted to know what they had experienced and how they would describe it.

Immediately it became evident that there was a significant difference between the interest the two groups had in *Urban Archeology*. The first group sent into the park with only technical information, returned their iPAQs and headsets to the park in less than 10 minutes. The second group spent an average of 40 minutes in the park, roaming through it to search out our lovingly constructed, richly researched content. What we learned immediately was that the kind of unfolding of the place and the curiosity to see how much content was not there naturally. Students with no directions, backstory or information regarding the 40 hotspots simply gave up: they had no cognitive coordinates to keep them moving. Consequently for these students, the space provided little intrigue. They were not *with* the space; they did not indulge in its nuances and in turn, they disallowed engagement. The second group, which had been kept waiting, however, had enough information about the project. They sought out and explored the different hotspots: for some it even became a game to find all 40. Further probing revealed other important design assumptions that we had not considered whilst producing *Urban Archeology*. We discovered that this new media was, in McLuhanesque fashion, haunted by the vestiges of old media. In both groups, these users compared this technology to their other media experiences, most notably the museum audio guide. As such, some expected crisp, clean information that would reveal some hidden dimension of history in a linear manner.

When we asked the students at the end if they would describe the experience as one of augmented reality, which is the language from new media we were using, several were dubious about this terminology. This was for several reasons. Some queried whether having a large sound insulated headset on did not cut one off from the park and most importantly the people in it. Many expressed discomfort at wandering a park, with so many of Montréal's homeless with expensive electronics equipment. In other words, the experience called forth the class contradictions embedded in the technology, beyond the mere representations we offered.

Revealed in questioning users about their experiences was a perspective very different from our own, one worth considering in terms of location-based media design intended for a heterogeneous public. In analyzing the frustrations voiced by participants, specifically the discontinuity between their physical and perceptual experience, participants were also cued into the social tensions of the park.

In other words, while it was not our initial intention, projects such as *Urban Archeology* can initiate a process of acknowledging the existence of "the digital divide" experienced as ownership and possession of technology as well as access to the know-how. Participants became aware of their status as 'outsider' to the space. This unease is not equivalent to failure. On the contrary, it brings forth the political potential for location-based experiences to actively employ Brechtian "distanciation" in order that, as Tomás Gutiérrez Alea (1988) explains, "the spectator arrives at—a state of astonishment or surprise in the face of daily reality" (p. 44).

These lessons from *Urban Archeology* including the need for on-going user-testing on location, designing for the specificity of a space, creating narrative and visual cues to propel the user through a space, and the need to entrain users in the use of the technology would gradually be brought into our next project, *The Haunting.*

3. The Haunting: Ghosts in the Machine and on the Mountain

The Haunting was a prototype for a location based cell phone game with a historical twist in on-going development for Parc Mont Royal in Montréal. Situated in the centre of Montréal, Mont Royal is a large hill, roughly 500 acres, with three distinct peaks: Colline de la Croix, Colline d'Outremont, and Westmount Mount. The highest point, Colline de la Croix, is 233 meters above sea level and is visible from afar, creating a very distinct visual topography. 'The Mountain' as it is known to Montréalers, serves for many as a geographical point of reference for situating oneself in the city. A large cross sits on Colline de la Croix and is visible from the south, east and north of the city.

One's position to the mountain may also be descriptive of social status, which is tied to the history of neighbourhoods. Nestled on its western base is Westmount, historically the Anglophone centre of Montréal, and a highly affluent community. Directly to the east, towards Parc Émilie-Gamelin, is Plateau Mont Royal, still largely francophone and currently experiencing rapid gentrification. One of the most prominent features of Mount Royal are three cemeteries, one Catholic, one Protestant, and one Jewish which lie on the north-facing side. The presence of these cemeteries, the spiritual significance of the Mountain to Montréalers, and the visual presence of two suggestively symbolic landmarks: an illuminated cross and a giant red radio tower that resembled a giant pitchfork against the sky—those things invited a ghost story. Writing a ghost story allowed us to address one of the environmental challenges of working with small screens outdoors: eliminating the glare of the sun which makes seeing images on a phone or iPAQ difficult.

Inaugurated in 1876, the park was designed by Fredrick Law Olmstead, who is also responsible for Central Park in New York City. Similar to Central Park, Parc Mount Royal is the largest green space in the city and is host to numerous cultural and athletic activities year round. Locals regularly flock to the Mountain for leisure, exercise, and to partake of numerous activities organized by the City and community groups. One such group, *Friends of the Mountain*, are responsible for protecting natural ecosystems that coexist with city inhabitants who use the park. They also lead the charge to keep the Mountain a public space against recurrent attempts by speculators and entrepreneurs to develop it. Finally, the Mountain is a popular destination for tourists, who often venture to the park for photographic opportunities at The Belvedere, a panoramic lookout over the centre of the city and the St. Lawrence River.

Mount Royal Park has figured prominently in the social and political history of Montréal. Wealthy Montréalers fled to the park during the cholera epidemic at the turn of the century. The annual St. Jean Baptiste festival, adopted as symbol of the Quebec independence movement was celebrated for years in the park. Today, events such as the regular Sunday Tam Tams, organized unofficially by drumming enthusiasts from all of the cultural and immigrant communities that comprise Montréal, challenge local bureaucratic structures and city by-laws. Parc Mount Royal was chosen as a site for *The Haunting* because of this rich history, cultural roots and spiritual tie to Montréalers.

In designing for *The Haunting*, we imagined the park as a giant Ouija board and the cell phone as the latest technology to act as a medium for communicating with the dead. Media debris, flickering screens, unearthly vibrations, and screaming cellulars inhabit the 'Summit of Spirits' surrounding the cross at the top of the mountain overlooking the city. Interaction scenarios, alternative mapping techniques, spontaneous public performance, and location based play structures rooted in non-linear narrative are explored in mobile experience design.

A glowing ring of cell phones, a contemporary version of swapping ghost stories around a virtual fire, inspired the atmospheric backstory to *The Haunting*. Based on the feedback from *Urban Archeology*, we also decided to provide our game with a narrative voice.

Figure 9. The Haunting: Using the metaphor of the Ouija Board as an interface for Mount Royal Park, Montreal. (Illustration: M. Landry)

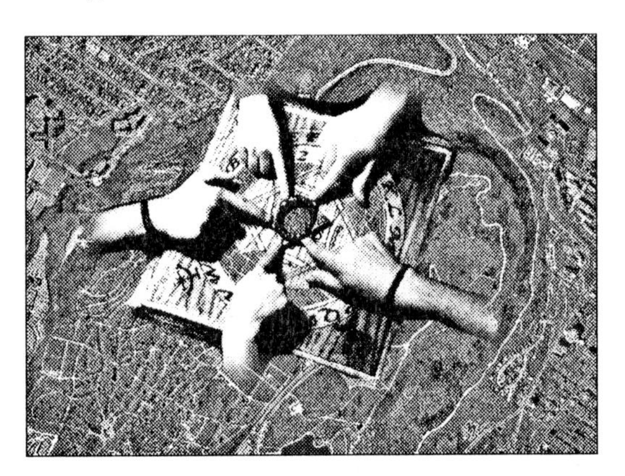

Figure 10. Players in Mount Royal Park. (Illustration: L. Palmer)

A working prototype for a game, *The Haunting* is intended for groups of two players per phone, with six groups in the filed at a time. It can be played by any age group, although as it is intended to be played after sunset. The whole game lasts about an hour, depending on the skill of the players. It may be played as a competitive game against another person or team. At the onset of game play, players receive a phone call from Alma, the operator and voice of VFB (Voices from Beyond) Mobility. Alma asks players to assist VFB engineers in neutralizing a series of disturbances that are threatening cellular communication networks throughout the city. Alma instructs players on how to neutralize these disturbances by capturing ghosts using a cell phone and continues, throughout the experience, to act as a meta-narrator giving players instruction on the phone itself. In addition to assistance from Alma, a map is provided on the cell phone so that players may navigate the game space in real time. Using GPS to play a game of virtual 'hot and cold,' the map displays the location of disturbances in relation to a player's position. The game is finished once all the ghosts are captured and expelled from the phone.

The game space, is divided into three sections or layers in a forested area at the top of the mountain bounded by the Olmstead Trail which approximately 3.8 km in length. In this, we literally remapped the Mountain as a potential space for interactive play. Level one, or 'Summoning' referred to as, 'The Forest of Shadows,' introduces players to the story and navigating the footpaths that through low level encounters with ghostly disturbances utilizing sound and vibration. Level two or 'Infestation' occurs around the 'Devil's Pitchfork,' named for the largest radio tower in the city located at the north west edge of the game space that resembles a tuning fork bathed in the red glow of navigation beacons mounted on the antennae. Interactions are more complex as disturbances become increasingly agitated and the ghosts inhabit the phone with sound and im-

Figure 11. The Haunting: Map showing game levels and location of the hotspots seeded with ghosts. (Illustration: R. Fenwick, Co&Co Design)

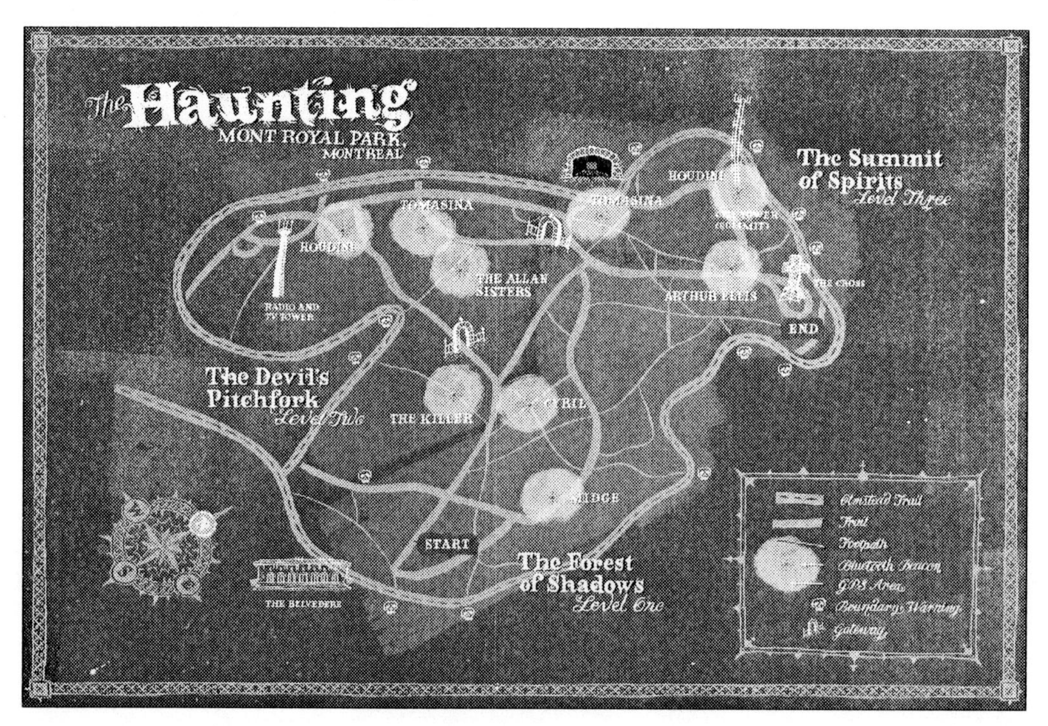

ages. Finally, in level three or 'Possession,' players learn that, "things are not always as they seem" as they realize their phones are now possessed and neutralizing disturbances becomes increasingly difficult as ghosts resist capture and attempt to escape the phone. Players discover they are being stalked by a malicious spirit and are forced to run to the cross at the 'Summit of the Spirits' in order to escape the forest and perform a kind of exorcism to release and liberate the spirits from their phones. Throughout the second and third layers of the game, communication with Alma and VFB Mobility is increasingly intermittent as the technology is rendered unreliable. Thus, the goal of *The Haunting* is to locate the disturbances and find the ghosts buried on the mountain, capture them in the phone, and later liberate them at the summit of the mountain, so that they may rest in peace and save the city from certain cellular chaos.

As they roam the mountain, players have nine encounters with spirits. Five of the spirits are based on real historical characters with a connection to Montréal, two of whom are buried in the cemeteries on the mountain. The game is structured around the gruesome tale of the last execution performed in 1935 by Arthur English, Canada's first official hangman, which went horribly wrong. As a result of miscalculations by English when measuring the length of the rope for "the drop," Thomasina Sarao, who was convicted of killing her husband in an insurance scam, was decapitated.[4] The paranormal disturbances on the mountain are caused by the eternal struggle as Thomasina seeks to unite her body to her head and escape the mountain and Arthur English who continually stalks her wanting another chance to hang her in an effort to 'get it right.' Players are duped into helping Thomasina and find themselves threatened by Arthur's noose.

The third character to complete the triad is Harry Houdini, the magician, escape artist, and debunker of spiritualists. Three years before Thomasina's execution, Houdini invited a student at McGill University, located at the base of the mountain, to punch him in the stomach as demonstration of his physical strength. Unfortunately, he was not ready to receive the blow and it is widely believed it ruptured his appendix and he died three days later in Chicago (Kalush & Sloman, 2006). Houdini, acts as a guiding spirit warning players to be careful of the malevolent ghosts that inhabit the mountain and assisting them with their escape. Other characters include, the Allan Sisters, who were lost at sea the Lusitania in 1915, one is buried on the mountain and the other, whose body was never recovered continues to roam the mountain in search of her sister. Three fictional characters based on epidemics and other deaths connected to the history of the city are a part of the game including the 'Killer', a surly monosyllabic old ghost named Cyril, and a kindly grandmother named Midge.

What distinguishes *The Haunting* technically from *Urban Archeology* is the use of cellular phones, Bluetooth beacons, and GPS for game play. For *The Haunting*, we wrote our own authoring software the Mobile Experience Engine (MEE).[5] The MEE uses XML to describe the layers of interaction to make up the game. The XML file is then fed into a code generator written in C++ that generates the game application for downloading onto the phone. Players encounter each ghost seeded with a Bluetooth beacon located at the heart of a series of concentric circles representing a paranormal disturbance. The beacon, made up of Bluetooth radio, a microcontroller, a power supply, and an LED housed in a silicone shell is placed in a tree. The topography of the terrain, the length and difficulty of the trail, and the time needed to deliver content determined the diameter of the circles ranging from 15 to 50 metres.

As players 'cross over' the outer ring of the concentric circle, tagged with a GPS coordinate, triggers content on the phone 'summoning' a ghost. The second stage of infestation indicated as ghosts begin to inhabit the phone with sounds and images triggered on inner rings by adjusting

Figure 12. The Haunting: Players in Mount Royal Park. (Illustration: L. Palmer)

Figure 13. VFB Gear: Coffins contain cell phone, GPS receiver, & flashlight. (Coffin: R. Prenovault Photo: J. Delisle)

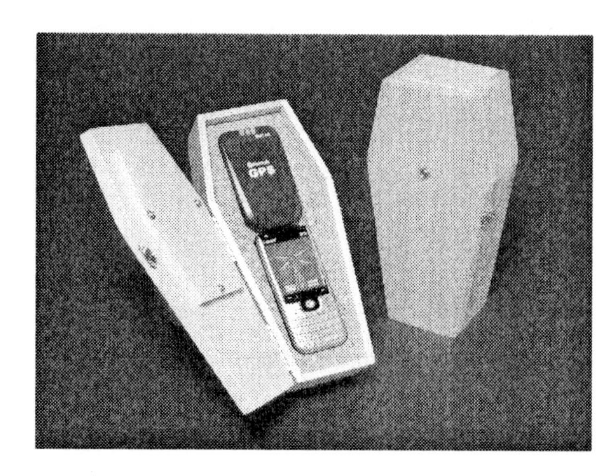

the transmission range on the Bluetooth radio in the beacon. The ghost is fully present as players approach the epi-centre of the concentric circle signaled by an apparition in the form of a blue light. The phone sniffs the presence of a spirit at the outer edge of the Bluetooth radio range and an LED makes its presence visible. The world of spirits and radio waves collide as players are bathed in a pulsating blue glow emanating from the surrounding trees. Simultaneously, the third stage, possession, is triggered by a timer event on the phone, launching the capture sequence by asking the player to press the number six on the keypad. The ghosts are sucked into the phone and captured amidst vibrations and a collage of swirling sounds and images. The light, no longer pulsating, but glowing brightly from the beacon cuts out and the player is returned to the solitude of the nighttime forest. Using GPS and Bluetooth beacons in a networked environment, this project treats the territory of the mountain as a potent and lively interface by encouraging players to stop for a moment. Users are asked to pay attention to the connections, disjunctures and happenstance moments created when a small screen, held in the hand, is suddenly alight and alive within the overwhelming setting of the park at night. At least this is what should happen when conditions are perfect. They never are.

In our encounters with the technologies on the mountain, the weather, the place, and the users revealed how these technologies are not seamless and transparent, as they often promised and propounded, but extremely sensitive, at least at this moment environmental factors brought a volatility to the technology: cold weather sucked our batteries of power, small keypads were difficult to push with precision while wearing the thick mittens that were keeping our hands warm, and we regularly encountered satellite drift as GPS coordinates drift and change depending on the rotation of the earth (Sawchuk, 2007). The 'Devil's Pitchfork', a necessity for communications in the city, created incredible interference on the Mountain. The interference on the cell phone, a manifestation of Jeffrey Sconce's (2000) notion of haunted media, produced "faint, wispy doubles of the 'real' figures on the screen, specters who mimic their living counterparts, not so much as shadows, but as disembodied echoes seemingly from another plane or dimension" (p. 124). Thus, in providing a "living link to distant vistas," we playfully confronted popular conceptions of liveliness and the relationship between the 'there' and 'not-there' in relation to cell phone usage.

We attempted to choreograph this volatility into our planning, including our on-going field tests of the technology and design where we walked the Mountain through the fall, winter and summer filming, photographing, and plotting distances between each level of the interactions we desired to create. The instability and force of this environment remained present in at all levels of our production process which required we work, film and coordinate our software and hardware to specific trees and pathways on the Mountain. These same paths disappeared under the snows of winter, only to re-emerge in the spring.

This environmental instability was exacerbated by our decision to move the programming of interactive experiences to cell phones by supporting the creation of an new open source software, The Mobile Experience Engine (MEE). The attempt presented us with a number of difficulties and challenges, which we continue to live with: our software system remains unstable. The Mobile Experience Engine was developed in part, as a response to the complicated state of the mobile platform world. Mobile platforms, such as cell phones and PDAs, run on as many as eight different operating systems. This means that mobile devices with similar attributes have widely differing capabilities and the capacity to streamline applications is complicated as the tools must be continuously re-designed. As a software tool for generating code, the MEE creates a layer of code that can be compiled for different platforms. Nevertheless, there proved to be continual negotiations between what we, as designers, hoped to create and what was feasible from the point of view of the engineers within the time frame of the grant. This required a constant negotiation on the parts of designers and engineers and on the production process as a whole.

Conclusion: Voices from Beyond at the Vanishing Table

In conclusion, these outdoor surroundings provide a rich but unstable setting for new media design. Every place has its own challenges, personality, demands and rules that as designers and users we are asked to dialogue with, listen to, and respect. Any place is home to a variety of interactants (Latour, 2005): humans, animals, flora, atmospheric conditions and technologies. In our case we encountered cyclists, skiers and hikers; squirrels, dogs and raccoons; trees, heavy underbrush, roots and stumps; sun, rain, snow, shifts in topography whilst juggling handheld phones looking for beacons and invisible GPS hotspots along pathways and in the trees. The presence of many voices from beyond includes the physical terrain we encountered: an active, ongoing trickster usurping the designs and interactions that we may create from the safety and warmth of our studios. In this way, the invisible hotspots demarcating potential zones of interaction act as a focal point for a communicative occurrence and are designed to interact *with* a particular place and the beings who inhabit it, including of course the people we ran across or who found us. In what became the most poignant moments in *Urban Archeology*, one participant talked of his cancerous tumours as he wandered through the park where he now lived after arriving in Montréal, by bus, for treatment at the nearby St. Luc Hospital.

The narratives that we encountered and which unfolded were not arbitrary anecdotes. These historical descriptions are deeply tied to very specific places and cannot be transported anywhere. Rather than erasing place, one can invoke it to challenge the often wide-spread and contagious urban amnesia that tends to bury the past with bulldozers. Historicizing place in the present offers a potential to resymbolize place in conjunction with its past, thereby politicizing the present (Zeffiro, 2006b). In every sense, imbuing a specific place *with* historical fragments and mediated via locative media substantiates the volatile interface. The fragmented aural histories call attention to the volatility of the environmental interface.

These locative experiences, as we have suggested, act as a kind of table for gathering or a meeting ground for a number of different spirits. Here we draw upon Darin Barney's discussion of technology as a kind of "vanishing table" a phrase with deliciously ghostly resonance. These hotspots, thought of as a vanishing table act as the *raison d'être* for a variety of interactions that can create temporary moments of community: between players; between players and the technologies held in their hand; between players and the very ground upon which they walk or the chilly air they pass through, which is alive with ethereal clouds of electromagnetic spectrum. As Barney writes, "It is certain that community is impossible without communication; it may also be the case that communication is meaningless without a world. To comprehend the relationship between digital technologies and community we must hold these two prepositions together" (p. 63). Barney's thinking on this subject captures the paradox of the locative media experience. A table both separates the diners from each other and provides the means to be connected at a common ritual, that of eating. Likewise the information and images transmitted via these technological assemblages both interrupt the habitual movements of potential users through the space, but also provide a reason for them to gather in order to see anew what may have become routine. A hotspot must be loud enough to arrest the movements of users temporarily, yet quiet enough to encourage a connection to that particular location that has been seeded with image, text or sound.

To return to the words of Durham Peters for a moment, we do not want to eliminate the spectral elements between people, the moments of unease or discomfort that may be produced because of the social context of class, laid bare in *Urban Archeology*. We are not concerned with the lack of precision of our hotspots. Nor are we daunted by the variability of the volatile interfaces that are the ever-shifting environments in which we locate our projects, particularly in spaces like the Mountain. We choose to design for locative media because the very challenge of the volatile interface puts our faith in technology's ability to seamlessly deliver information via these invisible networks into abeyance. These environments avail us of the possibility for chance, indeterminacy, and random happenings to take place that may at times usurp our carefully pre-programmed events. This is fully commensurate with our desire to create an ample breeding ground for the ghosts of history to the newness often associated with locative media.

ACKNOWLEDGMENT

We would like to thank the various individuals and institutions that worked and funded this project.

Urban Archeology Team Credits: Amitava Biswas, Barbara Crow, Kajin Goh, Ile Sans Fil, Jennifer Gabrys, Michael Longford, Bita Mahdaviani, Antoine Morris, and William Straw.

The Haunting Team Credits: Co-Principle Investigator: Michael Longford; Creative Leads & Project Direction: Michael Longford, Kim Sawchuk, David McIntosh, Barbara Crow; Project Managers: Andrea Zeffiro, Patricio Davila; Concept, Script Development & Interaction Scenarios; David Gauthier, Patricio Davila, Anna Friz; Sound Designer: Anna Friz ; Video Production: Thibaut Duverneix; Interface Research & Design: Marie-Claude Landry, Leanna Palmer, Anton Nazarko, Marit-Saskia Wahrendorf, Nevena Niagolova; MEE Engineers: David Gauthier, Tom Donaldson, Ken Leung; Bluetooth Beacon (Hardware & Software): David Gauthier, Geoffrey Jones, André Arnold; Map & Website: Co&Co Design Collective; Prop Design & Fabrication: Robert Prenovault; Costume Design: Andrea Zeffiro; Character Voices & Images: Marjorie Beveridge, Jan N. Desrosiers, Christine Duncan, Anna Friz, David Gauthier, Nevena Niagolova, Christine Plaza, Robert Prenovault, Paul Ternes, Andrea Zeffiro, Jennie Ziemianin; Historical Research: Jeff Bolingbroke, Wai Kok; Production Interns:

Fiona Chung, Jan Drewniak, Christine Plaza, Philip Sportel, Jennifer Ziemianin; and Charrette Participants: Neil Barratt, Lucy Belanger, Jérome Delapierre, Patrick Saad, Adam Brandejs, John Pavacic, Mark Poon, and Janice Leung.

Funders: Heritage Canada, Hexagram, Banff New Media Institute, Concordia University, York University, Ontario College of Art & Design, and Fonds de recherché sur la socetie et la culture.

REFERENCES

Alea, T. G. (1988). Identification and distancing: Aristotle and Brecht. *The Viewer's Dialectic*. Trans. Julia Lesage. (pp. 42-51). Havana: José Marti Publishing House.

Augé, M. (1995). *Non-places: Introduction to an anthropology of supermodernity*. Trans. J. Howe. London & New York: Verso.

Barney, D. (2004). The vanishing table, or community in a world that is no world. *TOPIA: Canadian Journal of Cultural Studies*. 11(Spring), 49-66.

Benjamin,W. (1968). Theses on the philosophy of history. In H. Arendt (Ed.), *Illuminations: Essays and reflections*. Trans. Henry Zohn. (pp. 253-264). New York: Schoken Books.

Brain, M., & Harris, T. (2006). *How GPS receivers work: How stuff works*, http://electronics.howstuffworks.com/gps.htm.

Burke, K. (1969). *A grammar of motives*. Berkeley, California: University of California Press.

Carey, J. (1989). *Communication as culture*. Boston: Allen Unwin.

Fortier, A. (1996). Troubles in the field: The use of personal experiences as sources of knowledge. *Critique of Anthropology*. 16(3), 303-323.

Freidberg, A. (2006). *The virtual window: From Albertit to Microsoft*. Cambridge: MIT Press.

Galloway, A., & Ward, M. (2004). Locative media as socialising and spatializing practice: Learning from Archeology. *Leonardo*. 14(3). Retrieved June 30, 2007 from http://leoalmanac.org/journal/Vol_14/lea_v14_n03-04/gallowayward.asp.

Gitelman, L. (2006). *Always already new: Media, History and the data of culture*. Cambridge: MIT Press.

Geertz, C. (1973). *The interpretation of cultures*. New York: Basic Books.

Grosz, E. (1994). *Volatile bodies: Toward a corporeal feminism*. Bloomington: Indiana University Press.

Hansen, M. (2004). *New philosophy for new media*. Cambridge: MIT Press.

Haraway, D. (1991). *Simians, cyborgs, women: The reinvention of nature*. New York: Routledge, Chapman and Hall, Inc.

Kalnins, K. (2004). "[Locative] locative is a case not a place." Retrieved March 13, 2006, from http://db.x-i.net/ locative/2004/000385.html.

Kalush, W. & Sloman, L. (2006) *The secret life of Houdini: The making of America's first superhero*. New York, NY: Simon & Schuster, Inc.

Kaplan, C. (2006). Dead reckoning: Aerial perception and the social construction of targets. *Vectors: Journal of Culture and Technology in a Dynamic Environment*. Retrieved November 10, 2006, from http://www.vectorsjournal.org/index.php?page=12&viewIssue=4.

Krueger, M. W. (2003). Responsive environments. In N. Wardip-Fruin and N. Montfort (Eds.), *The new media reader*. (pp. 379-389). Cambridge: MIT Press.

Kwinter, S. (2001). *Architectures of time: Toward a theory of the even in modernist culture*. Cambridge: MIT Press.

Lakoff, G. & M. Johnson. (1980). *Metaphors we live by.* Chicago: Chicago University Press.

Latour, B. (2005). *Reassembling the social: An introduction to actor-network-theory.* Toronto, ON: Oxford University Press.

Locative Media (and Ad-Hoc Social Networks). Online posting. 17 Feb. 2006. On-line:http://locative.x-i.net/intro.html

Longford, M. (2006). Territory as Interface: Design for Mobile Experiences. *Wi: journal of the mobile digital commons network*, 2(1), http://wi.hexagram.ca/.

Lowenthal, D. (1985). *The past is a foreign country.* New York: Cambridge University Press.

Marvin, C. (1988). *When old technologies were new: Thinking about electric communication in the late nineteenth century.* New York: Oxford University Press.

McCullough, M. (2005). *Digital ground: Architecture, pervasive computing, and environmental knowing.* Cambridge & London: MIT Press.

Monmonier, M. (2002). *Spying with maps: Surveillance technologies and the future of privacy.* Chicago & London: The University of Chicago Press.

Peters, J. D. (2000). *Speaking into the air: A history of the idea of communication.* Chicago: University of Chicago Press.

Richardson, L. (2000). Writing: A method of inquiry. In N. K. Denzin and Y. S. Lincoln, (Eds.), *Handbook of qualitative research, second edition.* (pp. 923-948). Thousand Oaks: SAGE Publications.

Reinharz, S. (1992). *Feminist methods in social research.* New York: Oxford University Press.

Salen, K., & Zimmerman, E. (2004). *Rules of play: Game design fundamentals.* Cambridge: MIT Press.

Sawchuk, K. (2007, March). Out in the cold: Endothermic embodiment and locative media. Paper presented at the Mobile Nation Conference, Toronto, ON.

Sconce, J. *Haunted media: Electronic presence from telegraphy to television.* Durham: Duke University Press, 2000.

Slack, J. D & J. MacGregor. 2005. *Wise culture and technology: A primer.* New York: Peter Lang.

Special Issue. (2006). Locative media. *Leonardo Electronic Almanac.* 14(3).

Straw, W. (2007). Embedded memories. In C. Acland (Ed.), *Residual Media.* (pp. 3-15). Minneapolis: University of Minnesota Press.

Tarkka, M. (in press). "Labours of location: Acting in the pervasive media space." In B. Crow, M. Longford and K. Sawchuk (Eds.), *Sampling the spectrum: The politics, practices, and poetics of mobile communication,* Toronto, ON: University of Toronto Press.

Tillman-Healy, L. (2003). Friendship as method, *Qualitative Inquiry.* 9(5),729-749.

Tuters, M., & Varnelis, K. (2006). Beyond locative media. Retrieved February 27, 2007 from http://netpublics.annenberg.edu/locative_media/beyond_locative_media.

Weiser, M. (193, July). Some computer science problems in ubiquitous computing. Communications of the ACM.

Wilken, R. (2005). From stabilitas loci to mobilitas loci: Networked mobility and the transformation of place. *Fibreculture Journal* (6).

Young, B. S., & James, G. (2003). *Respectable burial: Montréal's Mount Royal cemetery.* Montréal: McGill-Queens University Press.

Zeffiro, A. (2006a). The persistence of surveillance: The panoptic potential of locative media. *Wi: Journal of the Mobile Digital Commons Network*, http://www.wi-not.ca.

Zeffiro, A. (2006b, Ontario). The locative body: Mediation and urban embodiment. Paper presented at the national meeting of the Humanities and Technology Association, New York, New York.

KEY TERMS

Bluetooth Beacon: Bluetooth technology is a short-range radio technology that allows the wireless networking of computational devices, data can be exchanged between mobile technologies (i.e. Mobile phones, PDAs, laptops) which can be linked at a distance up to 10 meters. Traditionally, a beacon serves as an aid to navigation. For *The Haunting* project, a Bluetooth radio, coupled with a microcontroller, a power supply, and an LED, was placed "in-situ" in the park. The beacons were used to help users identify content hotpots and navigate the park at night. As users approached a beacon with a cell phone, the phone would discover the beacon via Bluetooth, which in turn would trigger content on the phone and switch on an LED for the duration of the interaction.

Electro-magnetic Spectrum: Spectrum refers to the transmission and regulation of airwaves into frequencies. The electromagnetic spectrum is organized by frequencies according to the length of the waves carrying long or short communication signals. These frequencies are allocated in bands referring to services on an exclusive or shared basis (ITU 2004). There are extensive regulations regarding service category (fixed service, mobile service), and service type (this describes types of transmissions and emissions).

GPS: The global positioning system (GPS) is a worldwide satellite-based radio-navigation positioning system that was developed by the United States Department of Defense and is conveniently operated by the Air Force (Monmonier, 2002, p. 12). This worldwide MEO (medium or middle, earth orbit) satellite navigational system consists of a constellation of 24 satellites, which orbit the earth twice every 24 hours (Monmonier, 2002, p. 13-14; Brain & Harris, 2006, p.1). A GPS receiver acquires positionality using two pieces of information: 1) the location of at least three satellites; and 2) the distance between its position on the ground and each of those satellites (Brain & Holmes, 2006, p. 2). This operation is based on the three-dimensional triangulation of intersecting circles (Monmonier 2002, p. 12, 174, 181), and each circle expresses a range of locations equidistant from one of the satellites. It is the point of intersection shared by the circles that situates the location of a receiver.

iPAQ: The iPAQ (International Physical Activity Questionnaire) is a hand-held portable digital computing device with wireless capabilities such as Bluetooth and GPS. It is also referred to as a Personal Digital Assistant (PDA). It was created by Compaq in 2000 and later bought and further developed by Hewlett Packard.

Location Based Media: This refers to fixed media artifacts that take into account both the specific geographic, historic and cultural significance of a place and our cognitive interaction with that place. In other words, our understanding place is informed as much by "how" we experience a place as "what" we experience in that place.

Locative Media: Locative media attaches digital media to global positioning satellite (GPS) coordinates accessed by mobile communication technologies. Locative media is an artistic sub-branch of ubiquitous computing research initiated at a "Mapping the Zone" workshop held at RIXC in Riga, Latvia, 2003. Its early practitioners, such as Marc Tuters and Karlis Kalnins, were interested in the relationship between the military developments of this technology and the types of proximity or relationality it can reveal about spaces and places. To date, it has been taken up by a number of artists, researchers, and activists interested in extending the shared potential of the technology by creating applications allowing users share local histories and community based information.

Site-specific Installation: A site-specific installation is a term derived from contemporary art, which refers to a work created for a specific space or place in time, which is often 3-dimensional and includes performance elements. The work takes into account the viewer's entire sensory experience coupled with the specific geographic, historical, and cultural significance of place. There is an ephemeral aspect to site-installations in that they are often temporary and once removed from the site in which it was installed can only be accessed through documentation.

Ubi-comp: Mark Weiser coined the term "ubiquitous computing," while working at Xerox Palo Alto Research Center (PARC). He argued that in the third wave of computing, technology would recede into the background of our lives. By deploying a network of small task specific sensor based computational devices embedded in the environment we can create ambient aware or smart environments. Using mobile devices we can interact with ubicomp environments what is sometimes referred to in physical computing as the "Internet of things."

ENDNOTES

1. For a recent overview of locative media, see special issue by *Leonardo*, http://www.leoalmanac.org/journal/vol_14/lea_v14_n03-04/essays.asp
2. Dans la Rue, is an organization offering outreach programmes for youths, http://www.danslarue.com/; Stella, is a non-profit organization for sex trade workers in Montreal, http://www.chezstella.org/; Séro Zero is an HIV-prevention group for bisexual and gay men.
3. http://www.bbc.co.uk/bristol/content/madeinbristol/2003/09/15/wireless.shtml
4. http://www.virtualmuseum.capm.phpid=story_line&lg=English&fl=&ex=00000224&sl=4198&pos=1
5. The MEE is an open source software development kit for creating advanced applications and media-rich experiences on mobile communication devices. http://www.open-mee.org.

Chapter XII
Embedding an Ecology Notion in the Social Production of Urban Space

Helen Klaebe
Queensland University of Technology, Australia

Barbara Adkins
Queensland University of Technology, Australia

Marcus Foth
Queensland University of Technology, Australia

Greg Hearn
Queensland University of Technology, Australia

ABSTRACT

This chapter defines, explores and Illustrates research at the intersection of people, place and technology in cities. First, we theorise the notion of ecology in the social production of space to continue our response to the quest of making sense of an environment characterised by different stakeholders and actors as well as technical, social and discursive elements that operate across dynamic time and space constraints. Second, we describe and rationalise our research approach, which is designed to illuminate the processes at play in the social production of space from three different perspectives. We illustrate the application of our model in a discussion of a case study of community networking and community engagement in an Australian urban renewal site. Three specific interventions that are loosely positioned at the exchange of each perspective are then discussed in detail, namely: Sharing Stories; Social Patchwork and History Lines; and City Flocks.

INTRODUCTION

The intersection of urban and new media studies is a dynamic field of practice and research. There are a number of reasons why this is so. Technically these are both highly innovative domains, and the rate of change is significant and challenging. Urban life and media platforms are both in the midst of paradigm shifts. Theoretically, both fields can be understood as sites of signification and structuration of the social field—and because they both evidence such change they are potent laboratories for advancing understanding. The pragmatic corollary is that policy makers and corporate investors are also highly engaged in the intersection.

Apart from the complexity of maneuvering through the often differing agendas of researchers and practitioners and of private and public sector agencies that operate at this intersection, the objective of advancing understanding is also challenged by a plethora of different and sometimes differing theories. Yet, universally useful contributions to knowledge can be achieved if urban cultural studies, urban sociology, urban technology and human-computer interaction, urban architecture and planning, etc., overcome language and conceptual barriers. A cross-disciplinary approach requires effort to create models which help to overcome phenomenologically isolated attempts at explaining the city. Such models would ideally be cross-fertilised by the findings and insights of each party in order to recognise and play tribute to the interdependencies of people, place and technology in urban environments. We propose the notion of ecology (Hearn & Foth, 2007) as a foundation to develop a model depicting the processes that occur at the intersection of the city and new media.

In the context of the field of urban planning and development, the promise of digital content and new media has been seen as potentially serving new urbanist visions of developing and supporting social relationships that contribute to the sustainability of communities. As Carroll et al. (2007) have argued, recent critiques of assumptions underpinning this vision have pointed to the following outcomes as 'most in demand', and simultaneously most difficult to deliver:

- Community (Anderson, 2006; DeFilippis, Fisher, & Shragge, 2006; Delanty, 2000; Gleeson, 2004; Willson, 2006);
- Diversity (Talen, 2006; Wood & Landry, 2007);
- Participation (Hanzl, 2007; Sanoff, 2005; Stern & Dillman, 2006);
- Sustainability (Gleeson, Darbas, & Lawson, 2004; Van den Dobbelsteen & de Wilde, 2004);
- Identity (Al-Hathloul & Aslam Mughal, 1999; Oktay, 2002; Teo & Huang, 1996);
- Culture and History (Antrop, 2004; Burgess, Foth, & Klaebe, 2006; Klaebe, Foth, Burgess, & Bilandzic, 2007).

It is critical that the emergence of urban informatics as a multidisciplinary research cluster is founded on a theoretical and methodological framework capable of interrogating all these relationships and the assumptions that currently underpin them. As Sterne has warned in relation to research pertaining to the field of technology more generally,

the force of the 'preconstructed'—as Pierre Bourdieu has called it—weighs heavily upon anyone who chooses to study technology, since the choice of a technological object of study is already itself shaped by a socially organized field of choices. There are many forces in place that encourage us to ask certain questions of technologies, to define technology in certain ways to the exclusion of others, and to accept the terms of public debate as the basis for our research programs. (Sterne, 2003, p. 368)

In this respect, if we are to promote an analytical focus on the capacities and possibilities of digital content and new media to meet the challenges of community, participation, sustainability, identity and so on, it is important to employ frameworks that permit systematic study of these relationships. We agree with Grabher (2004) who points to analytical advantages in resisting assumptions around passive adaptation to environments, and permitting a focus on networks, intricate interdependencies, temporary and permanent relationships and diverse loyalties and logics. We propose that these qualities are best integrated in an ecological model which we will develop in this chapter.

This chapter defines, explores and exemplifies our work at the intersection of people, place and technology in cities. First we introduce the notion of ecology in the social production of space to continue our response to the challenge of making sense of an environment characterised by different stakeholders and actors as well as technical, social and discursive elements that operate across dynamic time and space constraints (Foth & Adkins, 2006). Second we describe and rationalise our research approach which is designed to illuminate the processes at play from three different perspectives on the social production of space. We illustrate the application of our model in a discussion of a case study of community networking and community engagement in an Australian urban renewal site—the Kelvin Grove Urban Village (KGUV). Three specific interventions that are loosely positioned at the exchange of each perspective are then discussed in detail, namely: *Sharing Stories*; *Social Patchwork* and *History Lines*; and *City Flocks*.

THEORETICAL FRAMEWORK

Our theorisation is conscious of the extent to which technical, social and discursive elements of urban interaction work across (1) online and offline communication modalities; (2) local and global contexts; and (3) collective and networked interaction paradigms (Foth & Hearn, 2007; Hearn & Foth, 2007). The distinction between online and offline modes of communication—and thus, online and offline communities—is blurring. Social networks generated and maintained with the help of ICTs move seamlessly between online and offline modes (Foth & Hearn, 2007; Mesch & Levanon, 2003). Additionally, studies of Internet use and everyday life have found that the modes of communication afforded by Internet applications are being integrated into a mix of online and offline communication strategies used to maintain social networks (Wellman & Haythornthwaite, 2002).

Urban communicative ecologies operate within a global context increasingly dominated by Web 2.0 services (eg., search engines, instant messenger networks, auction sites and social networking systems). The notion of 'glocalization'—introduced by Robertson (1995) and later re-applied by Wellman (2001; 2002)—is useful here because it emphasises the need to develop locally meaningful ways of using this global service infrastructure rather than trying to compete with existing global sites and content. Studies have highlighted a range of opportunities for the development of local (and location-aware) services as well as locally produced and consumed content (Boase, Horrigan, Wellman, & Rainie, 2006; Burgess et al., 2006).

The similarly increasing ease with which people move between collective and networked community interaction paradigms hints at opportunities for new media services that can accommodate both kinds of interaction and afford the user a smooth transition between the two. Collective interactions ('community activism') relate to discussions about place; for example, community events, street rejuvenation initiatives and body corporate affairs. Networked community interactions ('social networking') relate to place-based sociability and features that for example, seek to raise awareness of who lives in the neighbourhood, provide opportunities for residents to find out about each other, and initiate contact.

In the introduction we pointed to the complexities of the relationships that need to be captured in understanding the role of information and communication technologies in the vision underpinning contemporary urban villages. Specifically it responds to recent calls for a more nuanced understanding of the patterns of relationships underpinning their uses in urban contexts. As Crang et al. (2006) observe, research in this area requires a more specific focus on the everyday uses of ICTs and the interactions of multiple technologies in everyday practices. In this respect, we argue that an ecological model enables a conceptualisation that opens up the possibility of diverse adaptations to a specific environment, and the different logics, practices and interdependencies involved. However, Crang is equally emphatic about the salience of social difference in the use of ICTs in urban contexts, pointing to research that underlines the episodic and instrumental use of them in contexts of deprivation and pervasive use in wealthier households (Crang et al., 2006). In this respect, we must provide for the possibility that different ecological configurations are related to the positioning of social agents in the field of urban life.

What is required, then, is a model that can capture ecological relationships in the context of the production of social difference in urban contexts. We extend previous conceptual work on ecological models (Adkins, Foth, Summerville, & Higgs, 2007; Dvir & Pasher, 2004; Foth & Adkins, 2006; Foth & Hearn, 2007; Hearn & Foth, 2007) by exploring Lefebvre's (1991) model in *The Production of Space*. It provides some conceptual tools to locate these ecologies as occurring in the context of different levels of space production. A central differentiation in his model is between 'representations of space' and 'representational spaces'. Representations of space refers to, "conceptualised space: the space of scientists, planners, urbanists, technocratic subdividers and social engineers [...] all of whom identify what is lived and what is perceived with what is conceived". Representational spaces on the other hand refers to, "space as directly lived through its associated images and symbols, and hence the space of the 'inhabitants' and 'users'" (Lefebvre, 1991, p. 38). In terms of this conceptual distinction the object of knowledge in the study of urban space,

> *is precisely the fragmented and uncertain connection between elaborated representations of space on the one hand and representational spaces [...] on the other; and this object implies (and explains) a subject—that subject in whom lived, perceived and conceived (known) come together in a spatial practice.* (Lefebvre, 1991, p. 230)

The model thus provides a context in which urban experience can be understood in terms of differential levels and kinds of power and constraint in the production of space and the relationships between them. Inequities can then be understood as involving different configurations of these relationships. Ecologies of ICT use in this framework are then understood in terms of the adaptations and interdependencies occurring in specific contexts underpinned by relationships of spatial use.

It is now possible to locate the vision of contemporary master-planning projects pertaining to, for example, 'community', 'identity' or 'culture and history', as produced at the level of the way space is conceived through planning and development practices, at the level of the way space is perceived through marketing material and spatial practice, and at the level of everyday experiences of residents as well as in the interrelationships between them (see Figure 1). This provides a framework in which the meaning of these visions can vary across different kinds of residents and can conflict with the conceived meanings of planners and developers. Studying ecologies of the use of new media and ICT from this perspective promises to illuminate the conceptual connections between these technologies and the field of urban life.

In the following part of the chapter, we discuss three initiatives that combine research and community development goals. They are positioned to facilitate a conceptual exchange between the three ecological domains of conceived, perceived and lived space, that is, *Sharing Stories*, *History*

Figure 1. Embedding an ecology notion in Lefebvre's triad

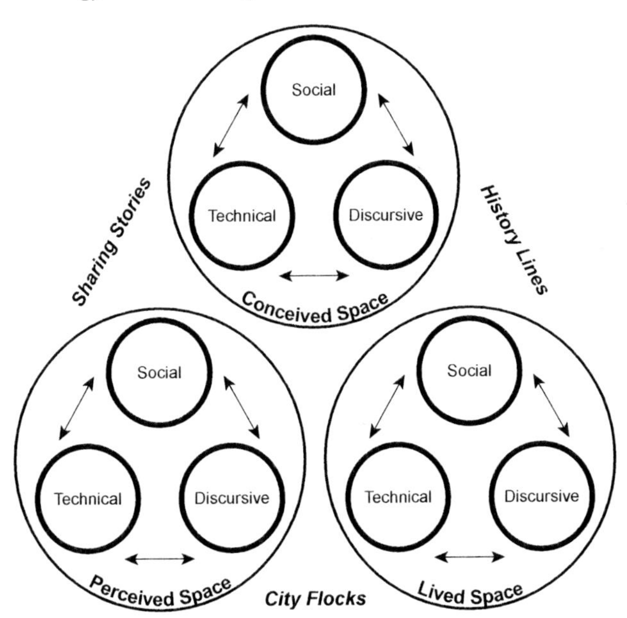

Lines and *City Flocks*. However, before we start, we introduce the emerging space of our study.

THE URBAN VILLAGE

The Kelvin Grove Urban Village (KGUV) is the Queensland Government's flagship urban renewal project. Through its Department of Housing, and in partnership with Queensland University of Technology, this 16 hectare master-planned community (see Figure 2) seeks to demonstrate best practice in sustainable, mixed-use urban development. By 'linking learning with enterprise and creative industry with community', the KGUV is designed to evolve as a diverse city fringe neighbourhood. Situated 2 km from Brisbane's CBD, it is based on a traditional village design, with a town centre and shops on the main streets. Since planning for the Village started in 2000 and construction started in 2002, AUD 1 billion had already been committed to deliver a heterogeneous design that brings together infrastructure with educational, cultural, residential, health, retail, recreational and business facilities within one precinct.

The following numbers and statistics illustrate the progress and development trajectory of the KGUV:

- When completed, there will be over 8,000 sqm (GFA) of retail space and in excess of 82,000 sqm (GFA) of commercial space located throughout KGUV.
- In 2007, there were 375 residential units (including 7 townhouses and 155 affordable housing units) in the KGUV. This is anticipated to exceed 1,000 two-bedroom equivalent units once the Village is complete (including student and senior accommodation).
- In 2007, there were 10,800 students and 1,800 staff based at the Kelvin Grove campus of QUT, and a total of 1,663 students and approx. 150 staff at Kelvin Grove State College.
- In 2006, 22,000 people attended exhibitions and performances at QUT's Creative Industries Precinct. Additionally, the KGUV-based theatre company has presented 137 performances to 40,000 patrons, plus there were various other events and productions in 2007.

Figure 2. Aerial view courtesy of the Kelvin Grove Urban Village development group

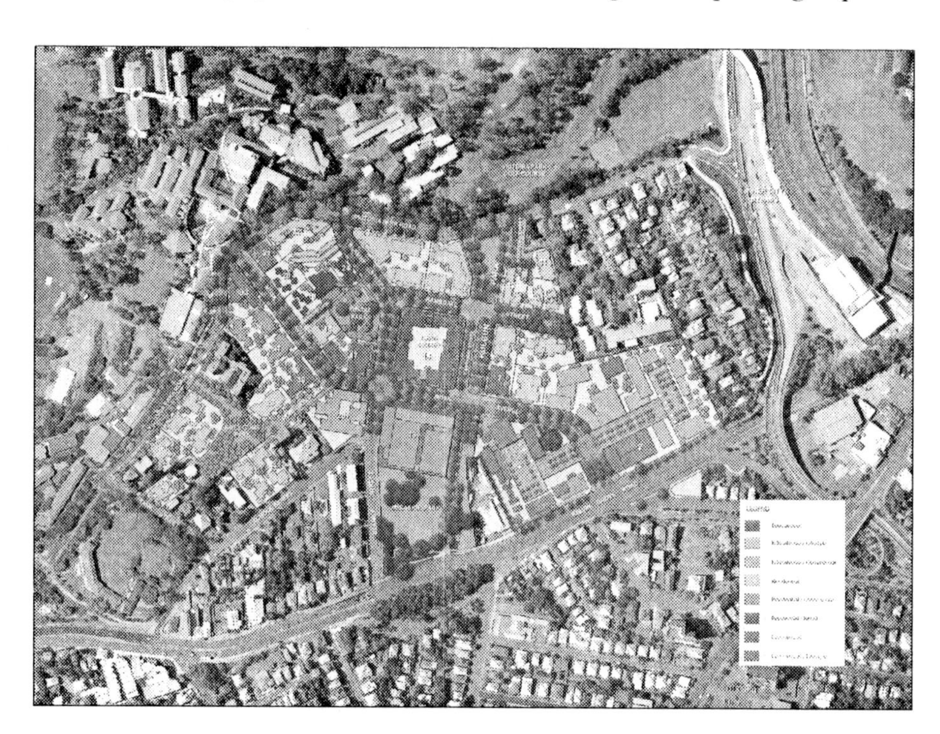

Technical connectivity is established by a 'triple play' (ie., phone, TV, data) fibre-to-the-home and fibre-to-the-node network operated by a carrier within the KGUV. The services can include low or nil cost large bandwidths (for example, Internet Protocol at 100 Mbits/s) within and between points in the KGUV, fibre or wireless network access and quality of service management for multimedia over Internet Protocol. Internet and world wide web access are at commercial broadband speeds and prices. Wireless hotspots allow users to access the Internet in parks, cafés and other locations around the Village. The implementation of the AUD 700,000 infrastructure investment started in 2005. These pipes, wires, ducts and antennas provide the technical connectivity, yet the majority of the infrastructure and certainly the social effect is invisible or unnoticeable. The communication strategies and policies in the KGUV master plan call for ideas and strategies to enable, foster and showcase the social benefits of this infrastructure 'beyond access' (Foth & Podkalicka, 2007).

Our diverse research interests are positioned under the collective umbrella of *New Media in the Urban Village*. The Department of Housing acknowledges that the strategic design of the built environment and access to the ICT infrastructure are necessary but not sufficient neither to ensure 'effective use' (Gurstein, 2003) nor social sustainability. Therefore the master plan calls for the research and development of appropriate interventions, measures and systems which can provide mechanisms to help link the people and businesses that 'live, learn, work and play' at the KGUV, including residents of the KGUV and nearby areas (including affordable housing residents, seniors and students); university staff and students living or studying in the KGUV and nearby areas; businesses and their customers; and visitors. Our suite of research projects are aimed at responding to this call. We now introduce some of these projects, how they respond to the objectives of the KGUV master plan and how they are guided by and feed back into our theoretical underpinnings.

NEW MEDIA IN THE URBAN VILLAGE

The three initiatives discussed hereafter form part of a larger research project that examines the role of new media and ICT in place making efforts to ensure social sustainability of a master-planned urban renewal site. Apart from the projects presented in the following section, a number of affiliated studies are also underway, for example, an international exchange of experiences studying urban social networks (Foth, Gonzalez, & Taylor, 2006), an exploration of health communication to understand the link between the design of the built environment and residents' well-being (Carroll et al., 2007), and a design intervention to display visual evidence of connectivity (Young, Foth, & Matthes, 2007). New research trajectories are about to start examining ways to use new media to digitally augment social networks of urban residents (Foth, 2006) and the role of narrative and digital storytelling to inform urban planning (Foth, Hearn, & Klaebe, 2007).

Sharing Stories

What about just the ordinariness of everyday life not being thought of as important? People move on and then no one is there to take ownership for it. So there are no custodians of the evidence. How can we change that? (Gibson, 2005)

Sharing Stories (Klaebe & Foth, 2007) exemplifies how traditional and new media can work effectively together and in fact compliment each other to broaden community inclusion. This multi-layered public history became a research vehicle used to expand the social interaction far beyond that of a community-based history project alone, so as to include the possibilities of global networks using new media applications. Leveraging opportunity, while negotiating and embracing a multidisciplinary new media approach proved rewarding for participants and local residents.

Kelvin Grove has always been a gathering point. While never densely populated, the land was once a meeting place for Indigenous clans, and in the last century, significant military and educational institutions were located there. In 1998 with the closure of the military barracks, the land was purchased by the Queensland Department of Housing. Together with QUT, planning began to transform the site into a creative urban village, lucratively located only two kilometres from the CBD that would include a mix of commercial, retail, university and residential land use. A triple bottom line approach was embraced as core to the master plan—incorporating economic, environmental and social sustainability.

Genuine creativity involves the capacity to think problems afresh or from first principles; to discover common threads amidst the seemingly chaotic and disparate; to experiment; to dare to be original; the capacity to rewrite rules; to visualise future scenarios; and perhaps most importantly, to work at the edge of one competences rather than the centre of them. (Landry, 2001)

The *Sharing Stories* history project was designed as a social engagement strategy that would become a longitudinal component of the development site from the outset. Its purpose was to collate the history of the site itself, to give a reference point and a context to future residents, to capture 'history in the making' as the ethos of the development was cutting edge. Furthermore, it was also essential to embrace the communities adjacent to the site, so they too were taken on the journey as their physical surroundings were to change so dramatically.

New technologies in communication are altering both the form and the content possible for historical discourse, with the processes of transmission arguably becoming less conventionally text-based, and instead offering visual and progressively more individuated options. Increasingly, visual life-story alternatives are being explored. *Sharing Stories* is predominantly an exercise in augmenting a participatory public history. This type of project and online visual display of the content created is still a relatively new approach to local community history and even more groundbreaking as a social engagement strategy to be incorporated in urban development. *Sharing Stories* engaged with the community using 'on the ground' interaction, seeking to be as inclusive as possible with its approach to participatory involvement. By accessing existing local social networks (schools, clubs, alumni groups, etc.) the profile of the project was raised and then promoted through public broadcasting coverage (local television and newspaper stories throughout the three year period)—all of which contributed to a ripple effect that continually drew more interest and contributions to the project as a whole.

Digital storytelling (Burgess, 2006; Klaebe et al., 2007; Lambert, 2002) was an innovative, alternative way of using new media to engage community in the *Sharing Stories* project. In the context of this project, digital storytelling is a combination of a personally narrated piece of writing (audio track), photographic images and sometimes music (or other such aural ambience) unified to produce a 2-3 minute autobiographical micro-documentary film. Traditionally, digital stories are produced in intensive workshops and this was the strategy employed for this project in 2004. Commonly, the thinking behind advocating this type of approach to create digital stories is to allow participants without access to new media the opportunity of using the technology in a hands-on way, so as to become part of the production team that produces their aesthetically coherent, broadcast quality story to a wider, public audience.

A one-to-many strategy included staging regular public events, so as to take the community on the journey of change in their locale, as the urban development process commenced. The primary aim of the *Sharing Stories* project was to capture the history and so a website was created to house content, as an evolving 'living archive' that would be accessible to interested parties locally and beyond. Fragments of the oral history and photographic collection, together with short story narratives produced from the historical research that was concurrently underway was seen as a strategic approach to keeping the community interested, informed and involved over the subsequent years of the development. Public events could be promoted online and afterwards, portions could be archived on the website. Public events held included: professional and local school visual art exhibitions, digital storytelling screenings and photographic exhibitions.

Throughout the *Sharing Stories* project it was noted that both 'real' and 'virtual' contact and exposure were critical in producing an effective participatory public history. The website stimulated interest in the public participating in the project and attendance at public exhibitions and similarly, the locally produced 'on the ground' activities created content and interest in the online representation. The project is thus an example of an initiative which combines opportunities for research data collection and analysis with community development outcomes. It was funded at the planning and development stages as a vehicle to inform the representations of space whilst preserving the history and heritage of KGUV. The local and historical knowledge that *Sharing Stories* produced has informed the marketing and public relations material and technical documentation of KGUV, but has also found its way into tangible representations of space in the form of for example, plaques embedded in the foot paths and other signage with historic anecdotes and citations.

History Lines

With the Web, you can find out what other people mean. You can find out where they are coming from. The Web can help people understand each other. (Berners-Lee, 2000)

An urban development does not occur overnight. A locale undergoing reformation rapidly changes shape. While the *Sharing Stories* phase

Figure 3. History lines

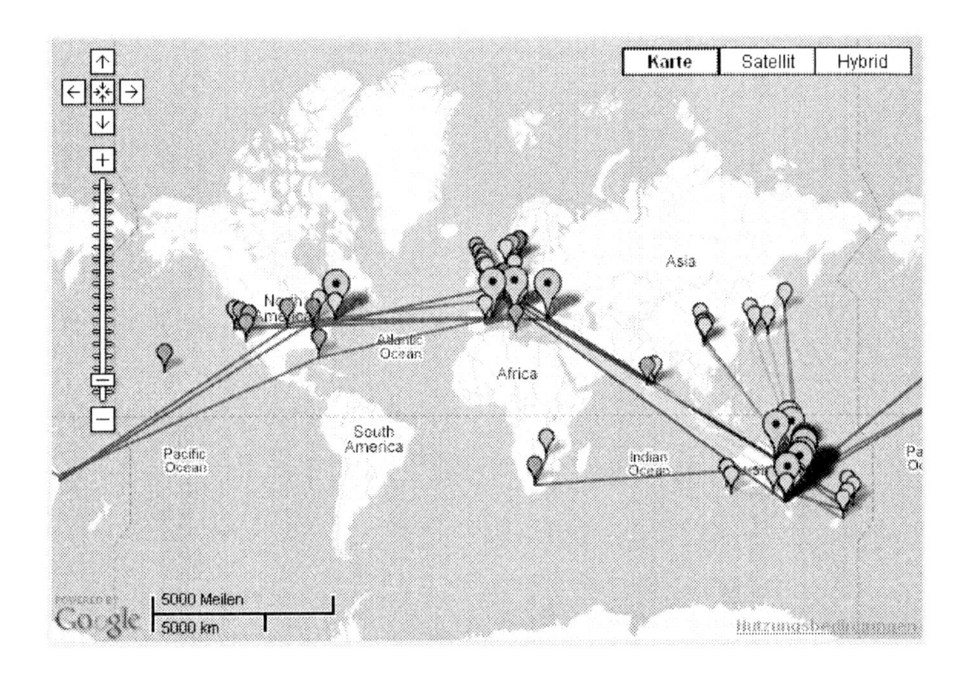

concluded in December 2006, a new phase of rejuvenation is already beginning to occur. It was only in the latter months of 2006 for instance that new residents began to reside in the Kelvin Grove Urban Village. Thousands more are expected over the next few years, all of which will be living in apartment style accommodation. Research around social sustainability in urban developments turned its attention to capturing the migrational churn of the people coming into the neighbourhood. What are their stories? Where are they from and why are they moving to an inner-city location? And more importantly for some researchers, how can new media applications be used to engage the incoming population, so that the locale becomes *their* story and *their* emerging history?

As the new population grows, the Kelvin Grove Urban Village attracts researchers who are grappling with these issues. Their backgrounds diversely combine to possess sociology, anthropology, history, education, health, IT, media/communication, and cultural studies. These researchers work together informally to share ideas and data, as well as to conduct joint workshops and focus groups, reducing 'research fatigue' of new residents. Applications developed by some of these researchers (Klaebe et al., 2007) in the 'Social Patchwork' project, include *History Lines* and *City Flocks* as two examples which we discuss here.

The *Social Patchwork* project involves translating narratives into formats that are Web 2.0 amenable. Researchers aim to develop applications that are useful in the urban development, as well as build historically orientated prototypes that encourage socially sustainable community engagement, to build and strengthen local communities and identities within community. Narrative based applications for example that can support intercreativity, as opposed to conventional interactivity (Meikle, 2002). New media approaches, guided by interpretive narratives, can utilise shared networks, shared interests and can be linked to measure the migrational churn of the suburb.

History Lines brings a cross section of new residents together in an activity using narratives, digital maps and location markers pinpointing where they have lived during the course of their life (see Figure 3). Participants can map their life journey thus far by recording personal narratives of place, while also narrating their relationship with past communities. When stories and locations are collated together, overlapping and common lines of location emerge. The material can be used anonymously as content for an exhibition, as well as a link on a neighbourhood portal to stimulate interest in community networking.

Weyea & Geith, in chapter IV of this book, call for research to 'identify effective information tools to enable citizens to shape what their communities look like' and 'use community data, locative media and social software to enable effective local action'. Our objective with *History Lines* aligns with this research agenda in that it can be an aid in measuring the migrational churn of participating residents and become a tool for urban planners to give them a better understanding of why and where people have lived throughout their lives. Our experimental design is positioned to feed experiences of the lived space into the planning and development process of the conceived space. We think it can deliver a better understanding of the role of narrative-based new media innovations in support of a more participatory urban planning process (Foth, Hearn et al., 2007; Foth, Odendaal, & Hearn, 2007).

City Flocks

City Flocks—developed by Mark Bilandzic (Bilandzic, Foth, & De Luca, 2008)—is a mobile information service for public urban places. The system is managed by local urban residents and is designed to tap into their tacit social knowledge. *City Flocks* (see Figure 4) allows participants to operate their mobile phones or computers to leave and access virtual recommendations about community facilities, making them easily accessible for other people employing user-generated tags. Residents can also voice-link to other residents, as participants can choose to nominate the mode of contact and expertise they are willing to share. *City Flocks* also allows users to plot their life's journey, but is primarily a rich resource for travellers and/or new residents to an urban location.

Figure 4. City flocks screen

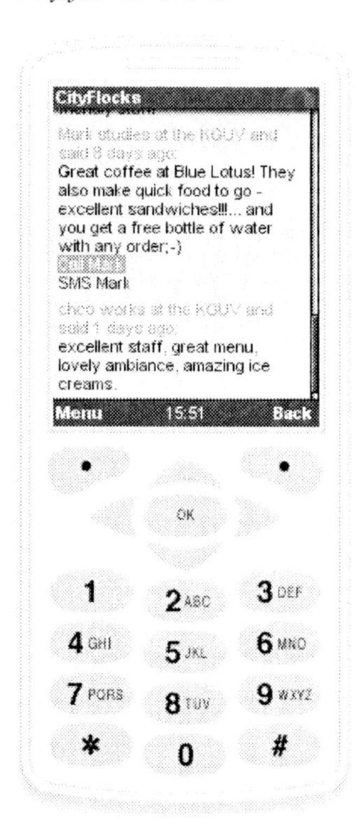

Both *History Lines* and *City Flocks* share networks and these shared interests can be linked using folksonomy tagging as opposed to traditional taxonomy directories (Beer & Burrows, 2007) so that they can be utilised by participants both globally and locally. Each application can also be used to encourage connections in the 'real' world. For instance *City Flocks* encourages users to contact fellow residents who have local knowledge of their neighbourhood and are happy to be contacted by newcomers either by email, SMS or by telephoning in order to gain first-hand tacit advice 'in person'. Groups that share 'history', for instance participants who have lived in Sydney and now live at Kelvin

Grove, can find and contact each other to meet socially at a local café, thus using virtual connections creatively to meet in the 'real' world. This application is similar to many other social networking applications including *peuplade.fr, placebase.com, communitywalk.com, theorganiccity.com, urbancurators.com* and *nearbie.com*. However, *City Flock*'s focus is to broaden the scope of the interface to include mobile devices and encourage users to interact directly with each other, rather than mediated via a website. Within the conceptual framework of the social production of space, we see *City Flocks* as a tool to link and balance the spatial practices in the perceived space of KGUV with the actual experiences of the lived space.

CONCLUSION

While both *History Lines* and *City Flocks* are in early stages of development, feedback from focus groups, urban developers and social planners has been encouraging. Whether this kind of online narrative-based testimony will be historically significant in the future is unclear; what is clear however is the fresh inter-creative way in which participants can freely leave their virtual 'mark' on a geographical location. *Sharing Stories* represents an exercise in interactivity, both on and offline. The content created offline could be later used and reflected upon in an online environment. *City Flocks* and *History Lines* comparatively demonstrate online and offline intercreativity, but in these examples the offline connection experience can augment more local, personal or face-to-face interaction. Communicative ecologies are both the enabler and outcome of the capacity to share interests, to 'find each other' and continue a virtual connection in reality. The use of these applications represents new configurations of online and offline relationships enabled by ICT that enhance the capacity of residents to appropriate represented space through their own representational practices. Together, however, they raise the question of the extent to which different kinds of users are able to assert their role in representational space and assert a level of agency over the represented space of the village. There is a key role for communicative ecologies to investigate the extent to which these services are amenable to use by people from different kinds of backgrounds—pervasive or episodic users, for example (Crang et al., 2006)—drawing attention to the configuration of relationships and interdependencies that enable people to assert a position in representational space.

We employed the notion of ecology to establish a holistic framework which allows us to differentiate interdependencies between forms of social interaction, technologies used by urban residents, as well as contextual and discursive aspects whilst at the same time keeping the bigger picture in mind. The design goal of the Urban Village to achieve and maintain a steady-state equilibrium of social sustainability requires further analytical and empirical work. Are there some features that are necessary and sufficient to maintain a socially sustainable equilibrium in a communicative ecology? Are there other factors which are detrimental and cause a gradual withering away of social relations and connective tissue? The interpretation of the various 'ingredients' which make up the Urban Village invokes a coming together of people, place and technology in an urban environment which is an inbetween, that is, not solely random, serendipitous, accidental, yet not solely master-planned and socially engineered development site either. The ecology notion points at an organic process which Gilchrist (2000) describes as 'human horticulture'.

In the introduction we chalked out a field of 'difficult to deliver' desires surrounding community, diversity, participation, sustainability, identity, culture and history. Reflecting back on these challenges, the ecology notion and the associated interventions which we trialled in this study prompt a critical and ongoing rethinking of the concept 'community' and its relation and contribution to the desired image of an 'Urban Village'. It is imperative to unpack the facets of the term 'community' and their individual meaning for different stakeholders and purposes at different times and places. One of the principles which guided the design of our interventions was the consideration of the diversity of the urban environment, both in socio-cultural as well as built environment terms. Rather than attempting an umbrella approach which would have regarded the residents of the Urban Village as a collective group united by their collocation, we tried to draw on their diversity, history and individual ability to express themselves creatively. Our preliminary observations and experiences allow us to argue that nurturing individual identities in this manner does support a local culture of participation, interaction and engagement conducive to engendering a sense of an urban village atmosphere. This culture of shared experiences combines some traditional place making efforts (Walljasper, 2007) with novel ideas employing new media, digital storytelling and social networking fosters the emergence of a socially sustainable urban development.

ACKNOWLEDGMENT

This research is supported under the Australian Research Council's Discovery Projects funding scheme (project number DP0663854) and Dr Marcus Foth is the recipient of an Australian Postdoctoral Fellowship. Further support has been received from the Queensland Government's Department of Housing. The authors would like to thank Mark Bilandzic, Jaz Choi, Aneta Podkalicka, Angela Button, Julie-Anne Carroll, Nicole Garcia, Peter Browning for supporting this research project and the anonymous reviewers for valuable comments on earlier versions of this chapter.

REFERENCES

Adkins, B., Foth, M., Summerville, J., & Higgs, P. (2007). Ecologies of Innovation: Symbolic Aspects of Cross-Organizational Linkages in the Design Sector in an Australian Inner-City Area. *American Behavioral Scientist, 50*(7), 922-934.

Al-Hathloul, S., & Aslam Mughal, M. (1999). Creating identity in new communities: case studies from Saudi Arabia. *Landscape and Urban Planning, 44*(4), 199-218.

Anderson, B. (2006). *Imagined Communities: Reflections on the Origin and Spread of Nationalism* (Rev. ed.). London: Verso.

Antrop, M. (2004). Landscape change and the urbanization process in Europe. *Landscape and Urban Planning, 67*(1-4), 9-26.

Beer, D., & Burrows, R. (2007). Sociology and, of and in Web 2.0: Some Initial Considerations. *Sociological Research Online, 12*(5).

Berners-Lee, T. (2000). *Weaving the Web: The past, present and future of the World Wide Web by its inventor.* London: Texere.

Bilandzic, M., Foth, M., & De Luca, A. (2008, Feb 25-27). *CityFlocks: Designing Social Navigation for Urban Mobile Information Systems.* Paper presented at ACM SIGCHI Designing Interactive Systems (DIS), Cape Town, South Africa.

Boase, J., Horrigan, J. B., Wellman, B., & Rainie, L. (2006). *The Strength of Internet Ties.* Washington, DC: Pew Internet & American Life Project.

Burgess, J. (2006). Hearing Ordinary Voices: Cultural Studies, Vernacular Creativity and Digital Storytelling. *Continuum: Journal of Media & Cultural Studies, 20*(2), 201-214.

Burgess, J., Foth, M., & Klaebe, H. (2006, Sep 25-26). *Everyday Creativity as Civic Engagement: A Cultural Citizenship View of New Media.* Paper presented at the Communications Policy & Research Forum, Sydney, NSW.

Carroll, J.-A., Adkins, B., Foth, M., & Parker, E. (2007, Sep 6-8). *The Kelvin Grove Urban Village: What aspects of design are important for connecting people, place, and health?* Paper presented at the International Urban Design Conference, Gold Coast, QLD.

Crang, M., Crosbie, T., & Graham, S. (2006). Variable Geometries of Connection: Urban Digital Divides and the Uses of Information Technology. *Urban Studies, 43*(13), 2551-2570.

DeFilippis, J., Fisher, R., & Shragge, E. (2006). Neither Romance Nor Regulation: Re-evaluating Community. *International Journal of Urban and Regional Research, 30*(3), 673-689.

Delanty, G. (2000). Postmodernism and the Possibility of Community. In *Modernity and Postmodernity: Knowledge, Power and the Self* (pp. 114-130). London: Sage.

Dvir, R., & Pasher, E. (2004). Innovation engines for knowledge cities: an innovation ecology perspective. *Journal of Knowledge Management, 8*(5), 16-27.

Foth, M. (2006). Research to Inform the Design of Social Technology for Master-Planned Communities. In J. Ljungberg & M. Andersson (Eds.), *Proceedings 14th European Conference on Information Systems (ECIS), June 12-14*. Göteborg, Sweden.

Foth, M., & Adkins, B. (2006). A Research Design to Build Effective Partnerships between City Planners, Developers, Government and Urban Neighbourhood Communities. *Journal of Community Informatics, 2*(2), 116-133.

Foth, M., Gonzalez, V. M., & Taylor, W. (2006, Nov 22-24). *Designing for Place-Based Social Interaction of Urban Residents in México, South Africa and Australia.* Paper presented at the OZCHI Conference 2006, Sydney, NSW.

Foth, M., & Hearn, G. (2007). Networked Individualism of Urban Residents: Discovering the communicative ecology in inner-city apartment buildings. *Information, Communication & Society, 10*(5), 749-772.

Foth, M., Hearn, G., & Klaebe, H. (2007, Sep 9-12). *Embedding Digital Narratives and New Media in Urban Planning.* Paper presented at the Digital Resources for the Humanities and Arts (DRHA) Conference, Dartington, Totnes, UK.

Foth, M., Odendaal, N., & Hearn, G. (2007, Oct 15-16). *The View from Everywhere: Towards an Epistemology for Urbanites.* Paper presented at the 4th International Conference on Intellectual Capital, Knowledge Management and Organisational Learning (ICICKM), Cape Town, South Africa.

Foth, M., & Podkalicka, A. (2007). Communication Policies for Urban Village Connections: Beyond Access? In F. Papandrea & M. Armstrong (Eds.), *Proceedings Communications Policy & Research Forum (CPRF)* (pp. 356-369). Sydney, NSW.

Gibson, R. (2005, May 26). *Imagination and the Historical Impulse in Response to a Past Full of Disappearance.* Paper presented at the Centre for Public Culture and Ideas, Brisbane.

Gilchrist, A. (2000). The well-connected community: networking to the 'edge of chaos'. *Community Development Journal, 35*(3), 264-275.

Gleeson, B. (2004). Deprogramming Planning: Collaboration and Inclusion in New Urban Development. *Urban Policy and Research, 22*(3), 315-322.

Gleeson, B., Darbas, T., & Lawson, S. (2004). Governance, Sustainability and Recent Australian Metropolitan Strategies: A Socio-theoretic Analysis. *Urban Policy and Research, 22*(4), 345-366.

Grabher, G. (2004). Learning in projects, remembering in networks? Communality, sociality, and connectivity in project ecologies. *European Urban and Regional Studies, 11*(2), 99-119.

Gurstein, M. (2003). Effective use: A community informatics strategy beyond the digital divide. *First Monday, 8*(12).

Hanzl, M. (2007). Information technology as a tool for public participation in urban planning: a review of experiments and potentials. *Design Studies, 28*(3), 289-307.

Hearn, G., & Foth, M. (Eds.). (2007). *Communicative Ecologies. Special issue of the Electronic Journal of Communication, 17(1-2).* New York: Communication Institute for Online Scholarship.

Klaebe, H., & Foth, M. (2007). Connecting Communities Using New Media: The Sharing Stories Project. In L. Stillman & G. Johanson (Eds.), *Constructing and Sharing Memory: Community Informatics, Identity and Empowerment* (pp. 143-153). Newcastle, UK: Cambridge Scholars Publishing.

Klaebe, H., Foth, M., Burgess, J., & Bilandzic, M. (2007, Sep 23-26). *Digital Storytelling and History Lines: Community Engagement in a Master-Planned Development.* Paper presented at the 13th International Conference on Virtual Systems and Multimedia (VSMM'07), Brisbane, QLD.

Lambert, J. (2002). *Digital Storytelling: Capturing Lives, Creating Community.* Berkeley, CA: Digital Diner Press.

Landry, C. (2001, Sep 5-7). *Tapping the Potential of Neighbourhoods: The Power of Culture and Creativity.* Paper presented at the International conference on Revitalizing Urban Neighbourhoods, Copenhagen.

Lefebvre, H. (1991). *The Production of Space* (D. Nicholson-Smith, Trans.). Oxford: Blackwell.

Meikle, G. (2002). *Future active: Media activism and the Internet.* New York: Routledge.

Mesch, G. S., & Levanon, Y. (2003). Community Networking and Locally-Based Social Ties in Two Suburban Localities. *City and Community, 2*(4), 335-351.

Oktay, D. (2002). The quest for urban identity in the changing context of the city: Northern. *Cities, 19*(4), 261-271.

Polanyi, M. (1966). *The Tacit Dimension.* Gloucester, MA: Peter Smith.

Robertson, R. (1995). Glocalization: Time-Space and Homogeneity-Heterogeneity. In M. Featherstone, S. Lash & R. Robertson (Eds.), *Global Modernities* (pp. 25-44). London: Sage.

Sanoff, H. (2005). Community participation in riverfront development. *CoDesign, 1*(1), 61-78.

Stern, M. J., & Dillman, D. A. (2006). Community Participation, Social Ties, and Use of the Internet. *City & Community, 5*(4), 409-424.

Sterne, J. (2003). Bourdieu, Technique and Technology. *Cultural Studies, 17*(3/4), 367-389.

Talen, E. (2006). Design for Diversity: Evaluating the Context of Socially Mixed Neighbourhoods. *Journal of Urban Design, 11*(1), 1-32.

Teo, P., & Huang, S. (1996). A sense of place in public housing: A case study of Pasir Ris, Singapore. *Habitat International, 20*(2), 307-325.

Van den Dobbelsteen, A., & de Wilde, S. (2004). Space use optimisation and sustainability: Environmental assessment of space use concepts. *Journal of Environmental Management, 73*(2), 81-88.

Walljasper, J. (2007). *The Great Neighborhood Book: A Do-it-Yourself Guide to Placemaking.* Gabriola Island, BC, Canada: New Society.

Wellman, B. (2001). Physical Place and Cyberplace: The Rise of Personalized Networking. *International Journal of Urban and Regional Research, 25*(2), 227-252.

Wellman, B. (2002). Little Boxes, Glocalization, and Networked Individualism. In M. Tanabe, P. van den Besselaar & T. Ishida (Eds.), *Digital Cities II: Second Kyoto Workshop on Digital Cities* (LNCS 2362, pp. 10-25). Heidelberg, Germany: Springer.

Wellman, B., & Haythornthwaite, C. A. (Eds.). (2002). *The Internet in Everyday Life*. Oxford, UK: Blackwell.

Willson, M. A. (2006). *Technically Together: Rethinking Community within Techno-Society*. New York: Peter Lang.

Wood, P., & Landry, C. (2007). *The Intercultural City: Planning for Diversity Advantage*. London: Earthscan.

Young, G. T., Foth, M., & Matthes, N. Y. (2007). Virtual Fish: Visual Evidence of Connectivity in a Master-Planned Urban Community. In B. Thomas & M. Billinghurst (Eds.), *Proceedings of OZCHI 2007* (pp. 219-222). Adelaide, SA: University of South Australia.

KEY TERMS

Communicative Ecology, as defined by Hearn & Foth (2007), comprises a technological layer which consists of the devices and connecting media that enable communication and interaction. A social layer which consists of people and social modes of organising those people—which might include, for example, everything from friendship groups to more formal community organisations, as well as companies or legal entities. And a discursive layer which is the content of communication—that is, the ideas or themes that constitute the known social universe that the ecology operates in.

Collective Interaction is characterised by a shared goal or common purpose, a focus on the community rather than the individual. The interaction is more public and formal than private and informal, and resembles many-to-many broadcasts. The mode of interaction is often asynchronous, permanent and hierarchically structured. Technology that supports collective interaction includes online discussion boards and mailing list.

Digital Storytelling refers to a specific tradition based around the production of digital stories in intensive collaborative workshops. The outcome is a short autobiographical narrative recorded as a voiceover, combined with photographic images (often sourced from the participants' own photo albums) and sometimes music (or other sonic ambience). These textual elements are combined to produce a 2-3 minute video. This form of digital storytelling originated in the late 1990s at the University of California at Berkeley's Center for Digital Storytelling (www.storycenter.org), headed by Dana Atchley and Joe Lambert.

Local Knowledge: Knowledge, or even knowing, is the justified belief that something is true. Knowledge is thus different from opinion. Local knowledge refers to facts and information acquired by a person which are relevant to a specific locale or have been elicited from a place-based context. It can also include specific skills or experiences made in a particular location. In this regard, local knowledge can be tacitly held, that is, knowledge we draw upon to perform and act but we may not be able to easily and explicitly articulate it: "We can know things, and important things, that we cannot tell" (Polanyi, 1966).

Master-Planned Communities are urban developments guided by a central planning document that outline strategic design principles and specifications pertaining to road infrastructure, building design, zoning, technology and social and community facilities. They are usually built on vacant land and thus in contrast with the type of ad-hoc organic growth of existing city settlements.

Networked Interaction is characterised by an interest in personal social networking and a focus on individual relationships. The interaction is more private and informal than public and formal, and resembles a peer-to-peer switchboard. The mode of interaction if often synchronous, transitory and appears chaotic from the outside. Technology that supports networked interaction includes instant messengers, email and SMS.

Triple Play infrastructure combines broadband Internet access, television reception and telephone communication over a single broadband connection, usually a fibre optic network. It is a marketing term which refers to a business model offering a bundle package of all three services accessible over the same network infrastructure.

Section IV
Location, Navigation and Space

Chapter XIII
Cityware:
Urban Computing to Bridge Online and Real-World Social Networks

Vassilis Kostakos
University of Bath, UK

Eamonn O'Neill
University of Bath, UK

ABSTRACT

In this paper, we describe a platform that enables us to systematically study online social networks alongside their real-world counterparts. Our system, entitled Cityware, merges users' online social data, made available through Facebook, with mobility traces captured via Bluetooth scanning. Furthermore, our system enables users to contribute their own mobility traces, thus allowing users to form and participate in a community. In addition to describing Cityware's architecture, we discuss the type of data we are collecting, and the analyses our platform enables, as well as users' reactions and thoughts.

INTRODUCTION

The formalised study of network graphs is considered to have begun by Euler's famous solution to the Seven Bridges of Königsberg problem in 1736 (Biggs et al., 1986). In his solution, Euler represented the four landmasses and seven bridges of Königsberg, now Kaliningrad, as four nodes and seven links respectively. Thus, he was able to prove that no route crosses each bridge only once. Graph theory has greatly advanced every since, mostly focusing on mathematical proofs and theorems on graph topology, trees and cycles.

While graphs have been used to explore relationships between social entities for over a century, it was not until the 1950's that this became a systematic, and ultimately scientific process. Some of the first studies to engage in social network analysis are the kinship studies of Elizabeth Bott (Bott, 1957) and the urbanisation studies pioneered by Max Gluckman in Zambia (Gluckman & Aronoff, 1976). Similarly, Granovetter's work (1973) lay the foundations for the small world hypothesis, suggesting that everyone is within six degrees of separation, while Wellman's work gave some evidence of how large-scale social changes have affected the nature of personal communities and the support they provide (1979). Since then, social network analysis has moved from being a suggestive metaphor to becoming an analytic approach, with its own theories and research methods. In the 1970's, Freeman developed a multitude of metrics for analysing social and communication networks (e.g. 2004), thus boosting commercial interest in the area due to companies aiming to optimise their procedures and operations. In the last decade, the identification of mathematical principles such as the small-world and scaling phenomena (Barabasi & Albert, 1999; Watts & Strogatz, 1998), underpinning many natural and man-made systems, have sparked further interest in the study of networks.

The systems design community has also been interested in the study of social networks as well as online social networks. Typical research topics in the area include the effect of social engagement on behaviour (e.g. Millen & Patterson, 2002), the issue of identity and projected identity (Lee & Nass, 2003), as well as the design of socio-technical systems (Herrmann et al., 2004). The recent proliferation of online social networking system such as Facebook, Dodgeball and MySpace, has provided researchers with platforms for carrying out research into online social behaviour, and a journal devoted to this topic (http://www.elsevier.com/locate/socnet). In the Urban Computing domain, such studies have looked at the effect of social incentives and contextual information on the use of public transportation (Booher et al., 2007), the relationship between users' online profiles and their online behaviour (Lampe et al., 2007), the various trust issues that emerge from using such systems (Riegelsberber & Vasalou, 2007), how such systems can help strengthen neighbourhoods (Foth, 2006), and the development of systematic grounds to base our designs (Kostakos et al., 2006).

To make inferences from online behaviour datasets, researchers still have to collect data from the real world and relate it to the online data. Thus, while social networking websites make it easy to capture large amounts of data, researchers still need to employ interviews, focus groups, questionnaires, or any other method that enables them to relate online with real world data.

In this paper we describe the development of the Cityware platform, which aims to bridge the gap between online and physical social networks. It allows users and researchers to explore an amalgamation of online and physical social networks. The key strength of our platform is that it allows the collection of vast amounts of quantitative data, both from the online and real worlds, which is immediately linked, synchronised, and available for further analysis. Furthermore, our platform enables both end users and researchers to gain a better understanding of the relationship between online and urban social networks. Here we describe the architecture of our platform, the types of data it makes available to users and researchers, the typical user-oriented scenarios that are beginning to emerge, and our planned research-oriented scenarios.

CITYWARE

Our platform is a massively distributed system, spanning both the online and physical worlds. Its architecture uniquely allows it to expand and contract in real time, while also enabling live data analysis. The main components of the platform are: people's Bluetooth-enabled devices, Cityware nodes, Cityware servers, Facebook servers, Facebook application. An overview of this architecture is shown in Figure 1.

Infrastructure

In many ways the most vital element of our platform is people's Bluetooth enabled mobile devices,

Figure 1. Overview of the cityware platform

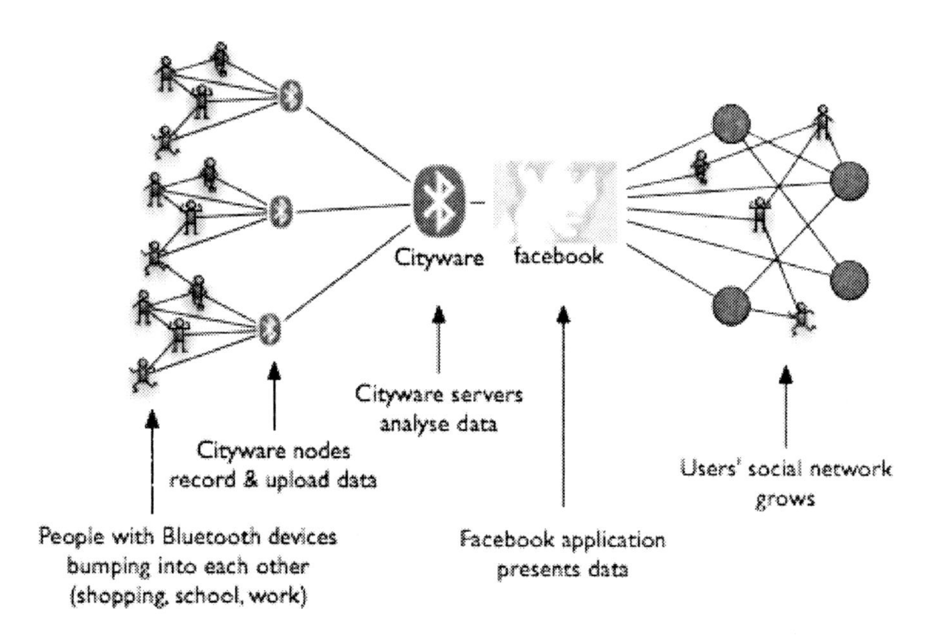

such as mobile phones, PDAs or laptops. For any data to be collected, users must have switched on their Bluetooth devices, and set them to "discoverable" mode. From empirical observations, we know that, at least in certain cities in the UK, about 7.5% of observed pedestrians had Bluetooth switched on and set to discoverable (O'Neil et al., 2006). More crucially, however, Bluetooth matches very closely to people's movement, as it typically has a short range (10 or 100 meters).

The presence of discoverable Bluetooth devices is captured via the deployment of Cityware nodes. These nodes are computers that carry out constant scanning for the unique identifier of Bluetooth devices, thus recording details about the devices in the immediate vicinity. The advantage of this approach is that the users' mobile devices do not need to run any special software; simply enabling Bluetooth is adequate.

Initially, we deployed a small number of nodes as part of a pilot study. However, we also released open-source software that allows users to turn their Windows, Linux, and Mac OS X computers into nodes. Additionally, we modified the open-source application WirelessRope (Nicolai et al., 2005) to make it compatible with our platform, thus enabling mobile phones themselves to become Cityware nodes. So far, our platform has attracted nearly a thousand individuals for Europe, America, Asia and Australia who have set up their own nodes and are uploading data to our servers.

Analysis

The method we use to scan for Bluetooth devices generates discrete data about the presence of devices in the environment. A visualisation of our data, which we have termed timeline, can be seen in Figure 2. Here, the graph represent a specific scanning site, and each dot represents a discovery event, i.e. a point in time (x-axis) when the Bluetooth scanner picked up a specific device in the environment. By applying filters, we can

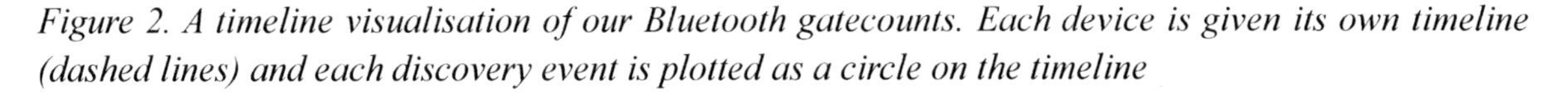

Figure 2. A timeline visualisation of our Bluetooth gatecounts. Each device is given its own timeline (dashed lines) and each discovery event is plotted as a circle on the timeline

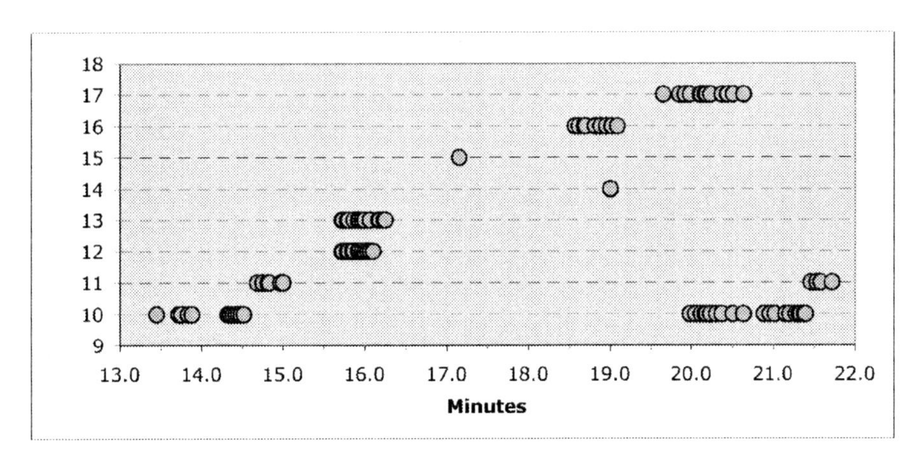

see that, for example, device 16 was present in the environment between approximately 18.5 minutes and 19.5 minutes.

To study the patterns social interaction in our data, we first need to identify instances where two or more devices were present at the same place and the same time. For example, in Figure 2 we see that devices 12 and 13 encountered each other. Such encounters are effectively opportunities for networking, both social and wireless. We developed filters that analyse our data and give us instances of devices encountering each other at each Cityware node location. These results take the form of records:

device1_id, device2_id, date, time, duration, Cityware node location

At this stage in our analysis we have a long list of such records, describing which devices encountered each other and by which scanner. For example, in Figure 2 we see that devices 12 and 13 encountered each other at 15.5 minutes and were together for approximately 1 minute. This list of encounters is a textual representation of the patterns of encounters across the scanning locations. To further study the patterns and structure hidden within this list, we transform it into a social network graph as follows: assuming that each device from our dataset becomes a node in the social graph, then the list of encounters indicates which nodes are connected. Proceeding in this manner, we are able to generate a social graph per Cityware node, as well as a social graph containing data from all Cityware nodes.

For illustration purposes, in Figure 3 we show the graph generated from one node's scan data, located inside a pub. In this graph, each observed Bluetooth device is represented as a node in the graph, and connected nodes indicate devices that encountered each other at some point. We note that the scanner (i.e. Cityware node) is not represented in this graph. By visual inspection we can verify that most devices are linked to the main core, whilst some devices are islands. The latter indicates cases where a device has never encountered any other device. Additionally, the size of nodes represents the total amount of time that a device has spend in this location, while the colour of the nodes (blue to red) indicates the betweenness of a node (from 0 to 1 respectively).

One of our initial observations is that due to the sheer number of nodes in the graphs, the visualisations themselves help little in analysing our data because of the visual clutter. However, by transforming our data into graph form, we are

Figure 3. A graph visualisation of the encounters that we recorded by one Cityware node

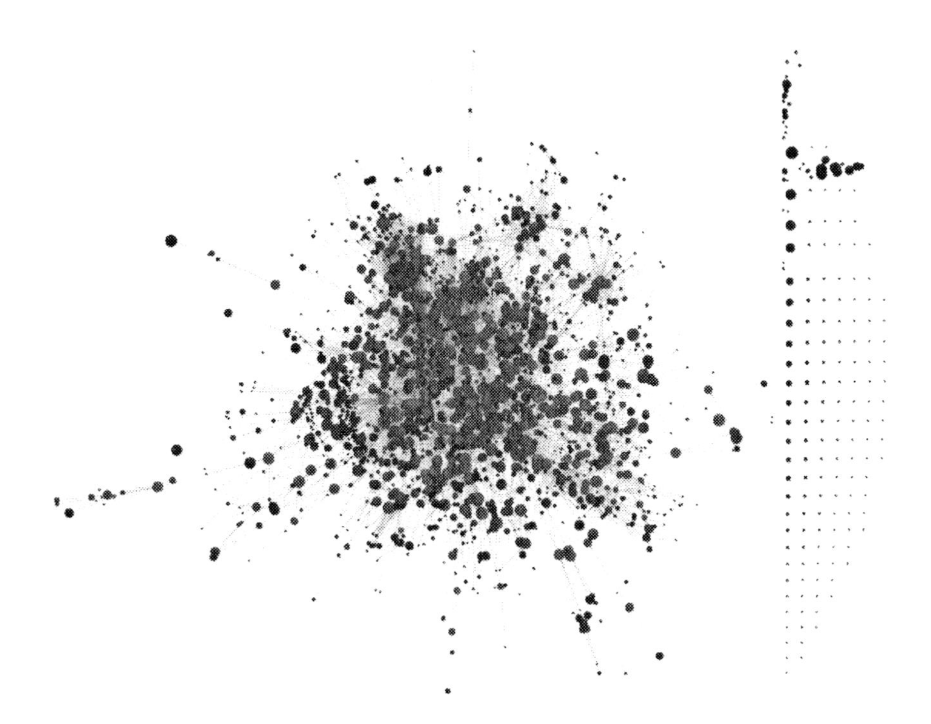

able to run a number of well-established analysis algorithms (e.g. centrality measures, community detection, etc.) using existing software such as Pajek, Ucinet, and iGraph. It is the results of these analyses that we present users through our Facebook application.

User Interface

Our platform relies on the Facebook system (http://www.facebok.com) to present data to users. Facebook is an online social networking website, where people upload profiles and explicitly link themselves to others via annotated links (e.g. flatmate, work-mate, went to school together, etc.) Our user interface has been deeply integrated with the Facebook system itself, matching its look and feel and using a number of Facebook's capabilities, such as the ability to display a list of the user's established friends, and the ability to send notifications using the Facebook mechanisms. A screen-shot of our user interface is shown in Figure 4.

To use our system, users must have a Facebook account, and additionally they must decide to add the Cityware application to their Facebook profile. This can only be done by logging on to Facebook and ticking the Cityware application from the list of available applications. The next step in using our application is for users to register their devices. This involves typing into our system the Bluetooth identifier of their device. Users may associate more than one Bluetooth device with their Facebook profile.

Once this link has been established between Bluetooth data and a users' Facebook profile, our system is able to display the user's encounters, sorted either by recency, duration, or frequency. These encounters represent an ego-centric textual representation of the social network captured by the scanners (e.g. Figure 3). Given that each user represents a node in this graph, our Facebook application enables users to see their graph "neigh-

Figure 4. Screen-shot of the Cityware user interface

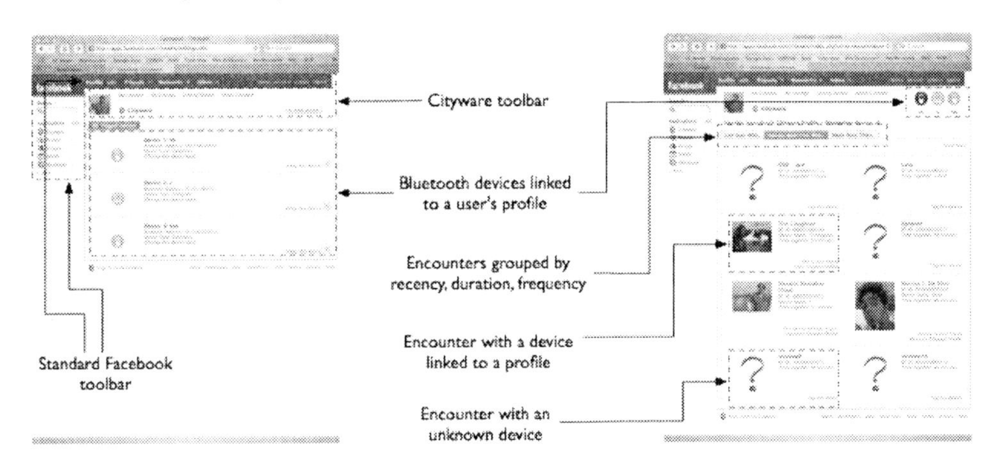

bours", and their "neighbours neighbours" in three distinct ways: who they met most recently, who they spent most time with, and who they meet most frequently.

For each encounter, our system displays the Bluetooth name of the device (as recorded by the Cityware nodes). If a user recognises a device as belonging to someone they know, they are able to "tag" that device, thus linking it to a Facebook account and to that account's owner. If this happens, the owner of the newly tagged device is notified via Facebook's own built-in mechanisms.

The end result is that users are presented with a list of encounters that have taken place in the real world, with some of those encountered devices being linked to Facebook profiles. For such devices our system can display the owner's picture as a well as link to that person's profile.

Our platform's distinctive characteristic is that it provides information that both end-users and researchers can use. This is because end-users see and explore data that is directly related to them (i.e. who they meet, and related statistics), while researchers have access to the "big picture", thus being able to explore and understand aggregated behaviour. Additionally, the self-registration and tagging mechanisms provide the crucial links between online and real-world networks. Effectively, our system enables users to annotate our dataset, thus enriching it with all the information that users make available via their Facebook profiles.

USER FEEDBACK

Facebook has its own built-in discussion board mechanisms to facilitate public and private conversations. These are accessible only by logging on to Facebook. These mechanisms have proven to be invaluable for collecting and categorising user feedback. A community has begun to form around our platform, with members using the discussion board to help other users with technical difficulties, suggesting design ideas, and holding debates. The size of this community is about 4000 people, including those who are running Cityware nodes and those who have added the Cityware application.

Prominent amongst the discussion topics is troubleshooting. Many users have posted questions in relation to the node software installation, making sure the software runs constantly, how bluetooth works, as well as how Cityware works. Fellow users have responded to these queries, suggesting that a peer-support community is being formed around Cityware.

While Cityware was officially released in late July 2007, it was not until mid-August that it became widely popular, mostly due to a web article published by the BBC (Waters, 2007). Since then, we have observed an interesting phenomenon amongst users of Cityware. As if feeling somehow "connected" or part of the same social group, our users are eager in establishing new nodes all over the world. A big part of the online discussion evolves around users proudly stating that they have established "yet another node", thus making their town or city part of Cityware. Additionally, users are eagerly posting messages requesting to know if there are any nodes near where they live. This enthusiasm is not different from what has been observed in other recent social phenomena (c.f. Rheingold, 2002).

A further interesting aspect of the feedback we have collected has to do with the context in which users are setting up nodes. While some users have reported establishing nodes in their homes, others have done so in their workplaces. Furthermore, some users of our application own shops and establishments (such as nighclubs) in which they have installed Cityware nodes. A feature that was heavily requested by users was the use of a map to visually locate Cityware nodes. Since we had not developed such functionality, we instructed users to mark their nodes on the public website http://www.wikimapia.com. This enables users all over the world to locate, as well as mark, Cityware nodes, post comments about them, or even attach pictures.

Privacy is a much-debated topic amongst users of Cityware. While some users are being critical of Cityware's privacy implications, many are supportive. We should note that the discussion board is not public, but rather only for self-selected users of Cityware, and as such may not be representative of the general public. Certain users have expressed concern about people being tracked about a city, and having their preferences and routines being inferred by a malicious party. In response, other users commented that anyone can at any time opt-out of Cityware by switching Bluetooth to "invisible". Additionally, it was highlighted that authorities can track people who simply own a mobile phone, regardless of Cityware. Furthermore, users commented that location is not being made available by our system, but nevertheless could be inferred. Another user noted that people are already disclosing information about themselves via their Facebook profile, and that Cityware can expose only that information. A good synopsis was offered by a user who wrote: "There are two groups of people here – one group that willingly submits to this, and the other group, that are totally opposed to any tracking/recording." This comment very well reflects our understanding of user's reactions towards our system. We feel that the reactions are mixed, with some feeling very positive and others very negative towards our systems, but no consensus having been reached at the moment.

RESEARCH POTENTIAL

While end users of Cityware are enjoying the functionality of our system, we are quite interested in the research possibilities that our platform has enabled. To quickly summarise some properties of our system as of late 2008: 3000 people have added Cityware to their Facebook profile, 450 nodes have been registered, while roughly 100,000 unique Bluetooth devices have been recorded by all Cityware nodes over a period of 4 months.

The dataset being collected by Cityware nodes is extremely rich as it describes people's visiting and encounter patterns across space and time. While comparable datasets, such as the Crawdad project (Crawdad, 2007), are available to the scientific community, it is only when such quantitative data can be linked to qualitative data that interesting research possibilities open up. While Cityware collects large amounts of quantitative data on people's movement and encounters, it also has access to the extremely rich qualitative data that people make available through their Facebook profiles.

Typically, Facebook users provide a wealth of information on their profile, including their demographics and preferences. More crucially, however, users annotate their relationships with people they know. Friends can be marked, for example, as colleagues, house-mates, or relatives. Additionally, a relationship can be annotated with dates, locations or organisations that may be relevant.

By combining the wealth of user-supplied qualitative data with the large amounts of quantitative data collected by Cityware nodes, we can begin to explore new research approaches to social metrics, system design, security, and even epidemiology. The logical next step for our research would be to compare people's movement and encounters with the qualitative data provided by users. For example, we can begin to empirically understand how people spend their time: with friends, family, or colleagues? Do these patterns change over time, seasons, or countries? Additionally, we want to explore if "friendship", "house-mate", or any other type of relationship systematically manifests the same Bluetooth patterns. This would lead the way for developing context-aware systems that can automatically classify a user's social network into friends, colleagues, etc.

Furthermore, such systems can make use of increased amounts of implicit, rather than explicit user input, which can enable them to adapt their behaviour appropriately and in certain cases understand and predict user needs. Hence, another area we are exploring related to making use of such data to make predictions about the users' behaviour, and accordingly adapt any software they may be using. At the moment we have distributed node software that runs on mobile phones. This software could act upon predictions about user behaviour and adapt any of the phone's functionality. Crucially, user feedback about the validity of predictions can easily be related back to our servers for further analysis.

A further research strategy is to explore the usefulness of our system for enhancing the security and privacy of users. We can conceptualise our dataset as a world map of relationships between users, annotated by users. This map may be used to inform users of security-related decisions they face (such as making a wireless payment) when entering a new context, such as a restaurant in a city they are visiting for the first time. Our servers can identify user comments about such a place, but more importantly assign weight to such comments based on the user's "social proximity" to the authors of these comment.

Finally, the data collected by Cityware is an invaluable source for understanding how mobility and encounter patterns can help in the diffusion of ideas, innovations and viruses (Kostakos et al., 2007). This could be achieved by exploring aggregate diffusion patterns over time, and exploring how different types of information (e.g 1Kb vs 1Mb) or viruses (biological / digital) would spread through the network of encounters and people. We note that this data is a result of public observation, hence we argue that the data can be readily used by the observers. Thus, local or national governments can use such data to develop and evaluate immunisation strategies to combat biological viruses. Similarly, telecoms operators and handset manufacturers can assess the effectiveness of their infrastructure against digital viruses that can spread via the Internet, GPRS, SMS/MMS, and Bluetooth.

CONCLUSION AND ONGOING WORK

In this paper we have described the Cityware platform, how users have reacted to it, and the potential for research strategies that it has enabled. As part of our ongoing work we are developing visualisations that both end users and researchers can utilise for better understanding the various patterns and properties of our dataset. We are also considering the development of software that will allow users to automatically geo-tag their data if they have a compatible GPS receiver. Furthermore, we are in the process of correlating aggregate encounter patterns with user-specified properties of those encounters. Finally we are examining the potential viral spread through users' encounters, and relating viral spread to user-specified qualitative data.

REFERENCES

Barabási, A.L., Albert, R. (1999). Emergence of scaling in random networks. *Science,* 286, 509-512.

Booher, J. M., Chennupati, B., Onesti, N. S., and Royer, D. P. (2007). Facebook ride connect. *CHI 2007 Extended Abstracts*, ACM Press, New York, NY, 2043-2048.

Bott E. (1957). *Family and Social Network. Roles, Norms and External Relationships in Ordinary Urban Families*. London, Tavistock Publishers.

Biggs, N., Lloyd, E. and Wilson, R. (1986). *Graph Theory 1736-1936*. Oxford University Press.

Crawdad project. http://crawdad.cs.dartmouth.edu. Last access 22 August 2007.

Foth, M. (2006). Facilitating Social Networking in Inner-City Neighborhoods. Computer 39, 9 (Sep. 2006), 44-50.

Freeman, L. (2004). *The Development of Social Network Analysis: A Study in the Sociology of Science*. Vancouver, BC, Canada: Empirical Press.

Gluckman, M. & Aronoff, M. J. (1976). *Freedom and constraint : a memorial tribute to Max Gluckman*. Assen: Van Gorcum.

Granovetter, M. (1973). The strength of weak ties. *American Journal of Sociology*, 78(6), 1360-1380.

Herrmann, T., Kunau, G., Loser, K., and Menold, N. 2004. Socio-technical walkthrough: designing technology along work processes. *Proc. Conference on Participatory Design (PDC)*, ACM, New York, NY, 132-141.

Kostakos, V. and O'Neill, E. (2007). Quantifying the effects of space on encounter. *Proc. Space Syntax Symposium 2007*, Istanbul, pp. 9701-9709.

Kostakos, V., O'Neill, E., and Penn, A. (2006). Designing Urban Pervasive Systems. Computer 39, 9 (Sep. 2006), 52-59.

Kostakos, V., O'Neill, E., Penn, A. (2007). *Brief encounter networks*. arXiv:0709.0223

Lampe, C. A., Ellison, N., and Steinfield, C. (2007). A familiar face(book): profile elements as signals in an online social network. *Proc. SIGCHI Conference on Human Factors in Computing Systems (CHI)*, ACM Press, New York, NY, 435-444.

Lee, K. M. and Nass, C. 2003. Designing social presence of social actors in human computer interaction. *Proc. SIGCHI Conference on Human Factors in Computing Systems (CHI)*, ACM, New York, NY, 289-296.

Millen, D. R. and Patterson, J. F. (2002). Stimulating social engagement in a community network. *Proc. Conference on Computer Supported Cooperative Work (CSCW)*, ACM, New York, NY, 306-313.

Nicolai T., Yoneki E., Behrens N., Kenn H., (2005). Exploring Social Context with the Wireless Rope. *LNCS,* 4277:874-883.

O'Neill, E., Kostakos, V., Kindberg, T., Fatah gen. Schiek, A., Penn, A., Stanton Fraser, D. and Jones, T. (2006). Instrumenting the city: developing methods for observing and understanding the digital cityscape. *Proc. Ubicomp 2006*, 315-332.

Rheingold, H. (2002). *Smart Mobs: The Next Social Revolution*. Cambridge, MA: Perseus.

Riegelsberger, J. and Vasalou, A. (2007). Trust 2.1: advancing the trust debate. *Proc. SIGCHI Conference on Human Factors in Computing Systems (CHI), Extended Abstracts*, ACM Press, New York, NY, 2137-2140.

Waters, D. (2007). Bluetooth helps Facebook friends. BBC, http://news.bbc.co.uk/2/hi/technology/6949473.stm (Last access 22 Auust 2007).

Watts, D.J., Strogatz, S.H. (1998) Collective dynamics of small-world networks. *Nature*, 393, 440.

KEY TERMS

Aggregate Patterns of [Behaviour/Encounter/Diffusion]: On an individual level each person behaves in distinct and unique ways, having specific objectives in mind. Yet, when analysed at an aggregate level, communities and cities exhibit non-random patterns that emerge from the combination of each distinct person's activities. Such patterns are known as aggregate patterns, and can describe how people encounter each other, or how information is diffused and spread through the community.

Bluetooth Identifier: A unique 12-digit hexadecimal number used by Bluetooth components for identification.

Massively Distributed System: A real-time computer system with large numbers of physical and logical components spanning great geographic distances.

Social Network: A structure that represents social relationships. The strutter typically consists of nodes and links between the nodes, and the nodes represent people while the links represent a specific type of relationship such as friendship, marriage, or financial relationship.

Urban Computing: A research field focused on the development of computer systems that are to be used in urban space. Typically, such systems entail fixed, mobile and embedded components.

Chapter XIV
Information Places:
Navigating Interfaces between Physical and Digital Space

Katharine S. Willis
Bauhaus-University Weimar, Germany

Jens Geelhaar
Bauhaus-University Weimar, Germany

ABSTRACT

In our everyday lives, we are surrounded by information which weaves itself silently into the very fabric of our existence. Much of the time we act in the world based on recognising qualities of information which are relevant to us in the particular situation we are in. These qualities are very often spatial in nature, and in addition to information in the environment itself, we also access representations of space, such as maps and guides. Increasingly, such forms of spatial information are delivered on mobile devices, which enable a different relationship with our spatial world. We will discuss an empirical study which attempts to understand how people acquire and act on digital spatial information. In conclusion, we will draw on the outcomes of the study to discuss how we might better embed and integrate information in place so that it enables a more relational and shared experience in the interaction between people and their spatial setting.

INTRODUCTION

Telephone calls worldwide on both landlines and mobile phones contained 17.3 exabytes of new information if stored in digital form; this represents 98% of the total of all information transmitted in electronic information flows, most of it person to person.

(Lyman et al 2003)

Information surrounds us. We use information, we create information and information allows us to communicate across time and space. But to make information tangible we need to classify, to categorise, to contextualise and to define it. In organizing data we add the knowledge of the receiver which enables the exchange of meaning. This we call interaction with information. In this way information does not exist as isolated, distinct data, but as a form of communication which is constantly affected by the setting in which it is created, gathered, manipulated and retrieved. When we interact with information we do not act in a vacuum, but based on a background of experience, using memories and qualities of the real world to guide us. Interaction is a continually negotiated two way process, that has been described by Pask in this example "A painting does not move. But our interaction with it is dynamic for we scan it with our eyes, we attend to it selectively and our perceptual processes build up images of parts of it. Of course a painting does not respond to us either [...] but our internal representation of the picture, our active perception of it, does respond and does engage in an internal conversation with the part of our mind responsible for immediate awareness." (Pask, 1971, p. 78). We act so as to simplify cognitive tasks by leaning on the structures in our environment. We rely on the external scaffolding of categorised information formats; such as maps and models, diagrams and traffic signs. We learn to use the world around us to assist us, so that not all thinking is done inside the head and much of it instead takes place within the context of real world situations. To understand how we interact with information we need to include the wider scenario of a person as they act in a real-world environment, but also taking account of the fact that this is a social and spatial environment which includes records and traces of prior actions in the form of communication systems (languages), storage systems (libraries), transport systems (roads), and spatial systems (the built world) (Morville, 2005). In fact the last category is often underrated; that of the spatial quality of information. Information is often understood both in terms of where it is located and consequently how it can be retrieved or found. We navigate through information, both metaphorically and in actuality, constantly deciding on what is useful to us and what we can ignore. In this sense there is always much more information available to us in the environment than we pay attention to. Just imagine a typical street, with a person walking along traveling from one shop to another. A multitude of information is present in the environment, much of it dynamic; passing cars, flashing shop signs, visual landmarks and of course other people. We learn to read subtle messages to enable us to make decisions, whether these are generated from conscious or sub-conscious choices. Spatial categorisation is ubiquitous in our language and how we organise our understanding of the world.

The quote at the beginning of this section introduces a new factor in how we interact with information. So far, interaction with computers has focused mainly on the communication between humans and devices. Going back as far as the introduction of telephone, communication devices have created new ways of creating and interacting with information. A critical quality of this communication is that its relation to space is changed. Information is no longer intimately tied to the place in which it is created or accessed, and is instead often experienced at a distance. A number of theorists from many different backgrounds have discussed some of the impacts that this has on how we interact with information in space. The sociologist Castells introduced the concept of the 'space of flows' where the "structure and meaning of the space of flows are not related to any place, but to the relationships constructed in and around the network processing the specific flows of communication" (Castells et al 2007, p 172). He further highlights how mobile communication changes the location reference, so that the space of interaction is defined entirely within the flows of communication. McCullough extends this discussion into how ubiquitous technologies do not obviate the need for meaningful descriptions of place but require new ways of grounding digital information, so that they do not undermine ways of acting in the physical world. He proposes that in embedding computing in the physical world, opportunities arise for providing "better filtered information about where one is without the cost of carrying information on where one is not" (McCullough, p. 190). Höök, Benyon and Munro (Hook et al 2003, p. 3) interpreted this in the context of human's interaction with computers, and highlighted the many subtle ways in which we use cues from our environment and from the behaviour of others as a source of information to guide us. Extending an original concept introduced by Dourish and Chalmers, the authors termed this concept 'social navigation', and outlined how this sees people as inhabiting and moving through information space based on social background. All of these authors introduce a counterpoint to the often dominant cliche of the digital world; that networks enable people to interact with anyone, anywhere, at any time and in any place. As Batty has pointed out this portrays an idea of information which has no relationship to time and space, and as such illustrates our crude vision of the emerging digital world (Batty, 2000). Instead, the focus of this chapter is on the relationship between information and place and focuses on investigating how humans interact, adapting to access the right amount and the right information in the right time and the right place. In order to investigate how information and place are interconnected we will first outline the nature and complexities of this relationship. An empirical study is then described which looks at how the delivery of information in place in digital formats affects how people perceive and act in space. This study looks in particular at the nature of representations of space and how spatial learning is affected depending on whether a traditional cartographic map or a map delivering dynamic spatial information on a mobile device. The results of this study are discussed in the context of how we might find better ways of supporting interaction with place-based information, such that it creates more possibilities for it to be interwoven and connected with real world experience and situations.

INFORMATION IN PLACE

Everything is related to everything else, but near things are more related than distant things

(Tobler, 1970 p. 236)

Information is attached to place in many ways; it may be literally inscribed in text as a sign or it may exist more subtly in the form of cues or physical structure of the environment. In fact our built environment is structured in ways that means that people can navigate through it without getting lost. These structural qualities can either be embedded in the urban form by the differentiated pattern of development over time, or put in place more deliberately. For example, in London, sightlines to St Paul's Cathedral from vantage points around the city are protected by planning law RGPA3. This helps the landmark to act as point of distant visual orientation in the often confusing layout of the city (see Figure 1.).

In this sense our environment can be considered as having 'legible' qualities, a term coined by the Urban Planner Kevin Lynch to describe how the form of built space is interpreted much like a language. He highlighted how "an ordered environment [...] may serve as a broad frame of reference, an organiser of activity, belief or knowledge [...] like any good framework, such a structure gives the individual a possibility of choice and a starting point for the acquisition of further information" (Lynch 1960, p. 4). In fact as far as Greek times, we have used the legible properties of the environment to help us to structure information. The Greeks invented a method called the mnemonic, which uses the structure of space to enhance human memory. In this method

Figure 1. Strategic view of St Paul's Cathedral from Richmond, 10 miles away

the first step is to memorise a series of loci or places. Images from a speech or other written work which are to be remembered are then placed in the imagination in these places. As the orator makes a speech he or she mentally enacts moving through the space in their imagination, recalling at the memorised places the images where they were placed in the sequence (Yates, 1996 p 18). Once saved to memory, the position of the loci may be used again and again for a whole range of material. The mnemonic illustrates not just the power of spatial relations to support knowledge, but also the essentially visual nature of the way in which space is experienced. Although we experience spaces through many senses, we tend to associate visual characteristics with places. For this reason we find salient or legible elements in the environment easier to remember and to orient ourselves in our daily lives. In this way information exists in places, but certain types of predominantly visual form and structure are more easily identified and acted upon. They make it easier to form associations with other aspects of place; and weave more smoothly together with the social meanings in space. Thus a dominant landmark in a city becomes an obvious meeting point for friends and so on.

Information also exists as representations of space, readily coded means of conveying certain aspects of space. We are also used to referring to such abstract representations of space. These include maps, guides and spatial descriptions which provide spatial information displayed in a categorised manner. Spatial relations are generally preserved in maps, but only certain specific types of information are selectively displayed. However because of the way information is selected in a map it can mean that it is interpreted incorrectly. For instance, when a person refers to a map whilst they are in an environment one of the first things they will try to do is to locate their position on the map. This involves translating the information available to them in the environment such as street names or prominent landmarks and matching them with the information displayed on the map. Often this is a simple task, but sometimes a person may make an error in matching the two sets of information or may simply not be able to find features in the environment that are presented on the map. This condition highlights 'the problem of mapping', when information in two different reference systems do not match or correlate resulting in the person either making an error or being unable to make a decision. This occurs in cartographic maps, but it is also prevalent in spatial information displayed on mobile devices. Yet digital information is increasingly delivered in spatial contexts where it is tailored to information specific to a location: a type of computing referred to as Location-Based Services (LBS). Typically such technologies use positional information generated from a GPS chip, and enable devices such as satellite navigation devices. In Edwin Hutchin's book 'Cognition in the Wild' (Hutchins, 1995) he introduces the metaphor of navigation as computation. He claims that navigation and computation can be correlated because they are both essentially about "operating within an information processing system" (p. 50). The consequence of this is that navigators often make mistakes because they imagine that the reality presented to them by the information matches the reality in the physical and social world. Accuracy within a system doesn't necessarily deliver an account of reality that can be acted upon in a meaningful way. Despite the perceived accuracy of such systems to register and identify a location, mapping problems occur with such devices in terms of the way information is interpreted, such as that described in the newspaper report in *The Evening Standard*, 22nd March 2007 below:

The school outing to Hampton Court Palace ought to have been a fairly simple journey. The driver chose to rely exclusively on his satellite navigation system, and 60 children spent the whole day being driven round in circles after it directed him to a narrow street in the north of the capital. A 63-mile journey that should have taken 90 minutes took eight hours. The incident brought a warning from the AA that drivers had begun following satnav directions 'like robots' and needed to have at least some idea of where they were going before setting out.

The practical problem described this report above suggests that sometimes when people use mobile maps to assist them in navigating through the city they can make large, even dangerous errors and they may also not be able to recognize these errors. It appears that the driver in this example was not able to match his knowledge and strategies for navigating the urban environment with that delivered by navigation assistance in a successful manner. It might also be assumed that as a consequence he did not actively learn about the city through which he was traveling, instead becoming merely a passive receiver of information. This is not an isolated incident; there are numerous subjective reports of people making basic navigational errors when using mobile maps; cars driving off cliffs, lorries driving down inaccessible country tracks etc. It has developed to such an extent that in the UK the Department of Transport has resorted to installing 'ignore your satnav' signs (Daily Mail, 2007) at recognized danger points. Although these examples relate to car-based navigation systems, recent empirical studies have shown that the implications also hold true for pedestrian navigation with mobile maps (Muenzer et al 20067, Aslan et al 2006).

So why does this happen? Let us first relate this back to how we learn about the spatial environment. This occurs through a series of psychological transformations by which an individual acquires, codes, recalls and decodes information about relative locations and attributes in a spatial environment (Downs & Stea 1974). It is a dynamic process and occurs following successive experiences of sequential routes, where knowledge about the environment is integrated into configurational survey representations (Siegel & White 1975). At a global level we use overview type knowledge that distinguishes between here and there and regions around and between them. We use this information to determine a plan for traveling between one place and another. Once we set out on a path, we need local representations of our surroundings to help us make decisions at key points. All of these types of knowledge are used to try to prevent us from losing our way and arriving at our destination. However we also use spatial assistance of many kinds to augment the knowledge in our heads, these include maps and also more recently developed mobile maps of navigation assistance supported by GPS. Mobile maps seem to provide an ideal solution to the problem of getting local information on where to go whilst completing a wayfinding task; this information is incremental since it is delivered in stages, rather than a paper map source which provides all the information in a stable format and is usually studied primarily in the planning stage of a task. In order to better understand the nature of spatial knowledge acquisition with mobile maps an empirical experiment was undertaken which evaluated spatial knowledge acquisition in a large-scale environmental setting by comparing participant's who had learned the environment from a map and those who had learned it using a mobile map.

INTERACTING WITH PLACE-BASED INFORMATION: LEARNING FROM EMPIRICAL STUDIES

From a map people acquire survey knowledge encoding global spatial relations
(Thorndyke & Hayes Roth, 1982)

Often we use not just clues from the physical and social space immediately available to us to guide us in our everyday lives, but also visual representations of space. These sources of spatial information, often in the form of maps, provide a third information source about the environment, aside from that residing in the person and the environment itself, and are a symbolic depiction highlighting relationships between elements of that space. The critical feature of a map is that it provides knowledge about space that cannot be gathered by the person in a short period of time. Map-like knowledge is usually presented from an allocentric perspective or a bird's eye view. Research has shown that this type of knowledge is only acquired following long experience of an environment (Downs & Stea, 1974, Siegel & White, 1975). Thus a map essentially provides detailed information to a person unfamiliar to a place that normally would take months or years to acquire. A further feature of maps is that they are typically used to plan a trip and then referred to for checking during a journey often only when the person becomes lost or disorientated. They help an individual develop an image of the route they are going to take in their mind, and then try to follow this route by matching expected features experienced in the environment with the features of the memorized image of the map. This is a different to the model of interaction with LBS-type information which is typically not referred to pre-journey, but is instead received incrementally during the trip with the device itself matching features of the map with the environmental position calculated by GPS technology. Consequently, although visually the interface of a cartographic map and that of an LBS map may be essentially similar the way the person acts and responds to the information is fundamentally different. In order to understand how individuals acquire spatial knowledge about the environment when they use mobile map devices, we undertook a field study which compared how the environment is learned depending on the delivery of different types of information. The study recreated critical features of the seminal study by the cognitive psychologists Thorndyke and Hayes Roth (Thorndyke & Hayes Roth 1983), which evaluated the differences between how people learn about space depending on whether they derived their knowledge from two typical sources of information about external space; either from navigation (i.e. walking around over a period of time), or from a cartographic paper map. The original study found that people who had learned a space from a map had access to a bird's eye view of the environment which enabled good recall of straight line distances between destinations in the space, but poor recall when trying to estimate distances along streets or paths. Conversely people who had learned from direct experience had better orientation abilities and could make good estimates of the distance between destinations along paths. In the study where we compared how people learned through maps and mobile maps differences were also found. We will not describe full details of the study methods and procedure (see Willis et al 2007 for full description), but will instead outline the experimental methods and discuss the results and the implications for learning with spatial information delivered on mobile devices. The setting for the main part of the experiment was a small-scale urban setting in Bremen, Germany, and is approximately 400m x 300m in size (see

Figure 2. The map of the environment used in the map condition

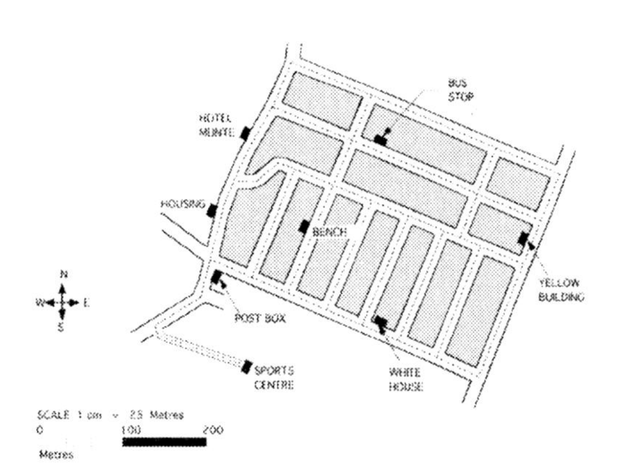

Figure 3. Mobile device with map interface

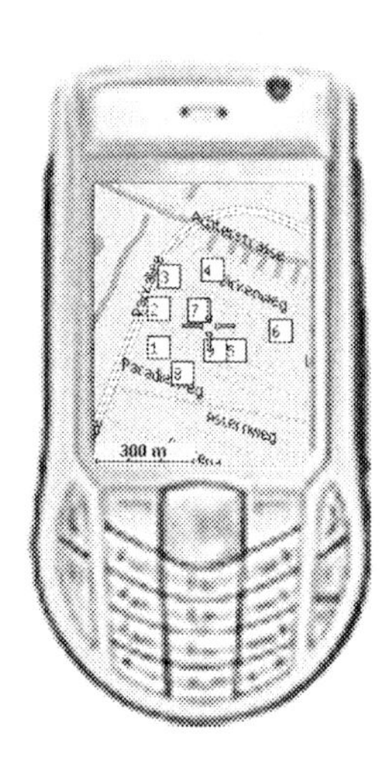

Figure 2.). The setting comprises a rectilinear grid path structure with several prominent landmarks and numerous repetitive allotment plots with small houses laid out in rows. There were no visual clues or landmarks that enabled global orientation, and most landmarks were not visible from one location to another.

The experiment procedure involved map participants learning a map of the environment in which they were to be tested whilst sitting in a closed office. Each map-learning participant was told that she was to learn the map of the environmental setting including the layout of the paths, and the names and locations of the landmarks. The map participants learning was tested using a study-recall procedure. Having referred to the map for two minutes it was taken away and participants were asked to redraw the features of the map, and this was repeated until the critical features were reproduced reasonable accurately by the participant. In contrast the mobile map participants navigated the environment using the mobile map in order to learn it (see Fig ure 3.) although they completed the same testing procedure as the map participants.

Following the learning phase participants completed a series of estimation tasks in the environment to assess their spatial knowledge. When the participants were tested for their knowledge of both orientation and distance between eight destinations in the environment the first outcome was that both sets of participants had good memory for the orientation. The accuracy of metric knowledge between the two groups was different, with mobile map users making more pronounced errors. At first these results seems strange; how can an individual acquire knowledge about spatial relations, but have such poor metric knowledge? Map learners acquired a bird's eye view of the environment which enables them, despite a very short period of learning, to demonstrate good spatial knowledge acquisition. The map participants in a sense mentally envisaged a layer of their internal map laid or stretched over the environment, and used this to frame their estimations. In order to try to understand the different type of knowledge for the two participant groups, the maps drawn in the learning phase were studied for schematic differences. There were qualitative differences between the features and quality of the maps depending on whether the participant had learned the environment using the map or the mobile map. Consequently, it appears that the basic quality and features of knowledge acquired by the two groups in the learning phase of the experiment is different.

Figure 4. Example of 1st map drawn by mobile map participant in learning task

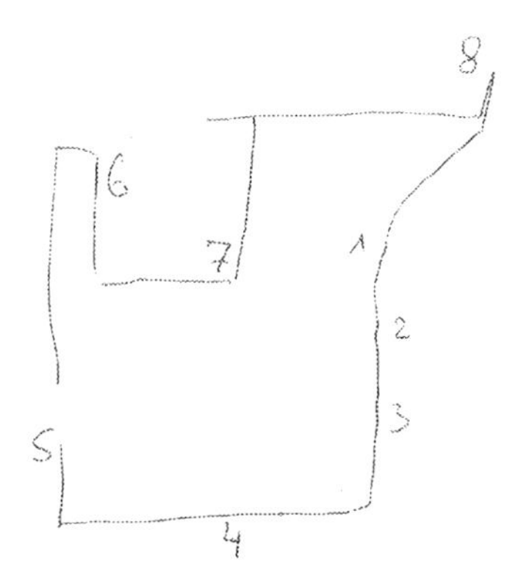

Figure 5. Example of 1st map drawn by map participant in learning task

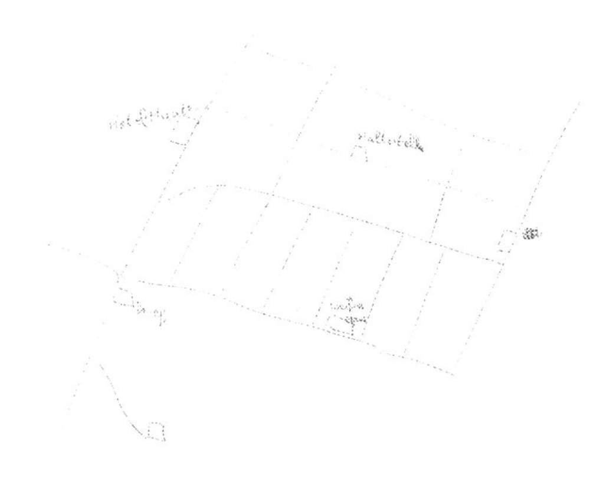

Mobile map participant (for example see Figure 4.) typically indicated a route, with a series of sequential landmarks. In this case the configuration of the environment is not shown, and no features were indicated which were not on the route. None of the mobile map participants redrew the map with the layout facing true 'north-up' despite the fact that they had accessed a map representation which was orientated 'north up' throughout the learning task. In addition all the participants rotated the map so that the paths were orientated parallel the x,y axis whereas the paths on the map are at orientated at an angle of approximately twenty degrees to the y axis. In contrast the map participants (for example see Figure 5.) typically redrew the map to depict a clear survey structure onto which landmarks and features are placed. All of the map participants redrew the map 'north-up'. This suggests that the map participants very quickly assimilated a strong and stable image of the environment in the learning task.

Interestingly, it was not the graphic representation of the mobile map that distinguished it from a cartographic map in the effect on performance, but rather the delivery and attention it required of the individual. Learning is an effortful task that requires conscious attention. Mobile maps require attention from the individual whilst they are in the environment because the information is both automatically changing and updating, and also because the individual can interact with the information (e.g. by changing the scale of the map interface). For example a study by Lindberg and Gaerling found that when participants were engaged in a secondary task whilst navigating a route were less able to keep track of where the learned locations than a group who did not have a concurrent task (Lindberg & Gaerling, 1983). With a mobile map the fact that it constantly updates the users position on the map itself, thus effectively offering a dynamic, constantly changing set of information seems to create a very sequential form of spatial knowledge. This meant that their attention was being divided between the device and the environmental setting, which affected their memory. In addition since the user had the ability to change the scale of the map, meaning that they did not consistently view the same map representation throughout the task, this created a type of focused 'keyhole' information acquisition. Rather than the mobile map interface disappearing into the background it seems to have the opposite affect; creating a conflict, with attention to the features of the real environment being divided and shared with attention to the interface. Since both offered different forms of information; the mobile map a survey type representation, and the environment an egocentric perspective, the participants were in a constant process of trying to resolve the information from the map and the information from their view environment, and this was cognitively demanding. The map participants also had to translate the two perspectives to make their judgments, but paradoxically the learned map representation did not demand their attention because since it offered a single static clear representation it enabled more cognitive offloading where the individual could focus on matching cues in the environment to their internal map.

Similarly an older study by Held and Hein (Held et al, 1963) that learning in a spatial environment is hindered when the subject experiences it in a passive mode i.e. being led through the environment without being able to make self-motivated choices. When a subject is actively engaged in the environment then they are stimulated or motivated to learn and gain knowledge about the space in which they are moving. A mobile map essentially enables and even encourages someone using it to switch off, and to become the passive receiver of information, and as such does not support learning. As described in the newspaper report earlier in this section there are many examples of people making apparently stupid navigational mistakes whilst using a mobile map because they essentially disengage with spatial clues in the real world and allow the device to make decisions. In this way not only do they fail to actively participate mentally in the spatial task at hand, they are also not motivated to learn about the environment which they are in which has implications for long term actions. If people continue to use digital spatial information delivered incrementally in the environment then there is a real concern that people will become disengaged from the spatial world.

BINDING TOGETHER PLACE-BASED INFORMATION AND DIGITAL INFORMATION

"It's not down in any map; true places never are"

(Melville, 1851, p. 54)

The empirical study described above outlines some of the limitations of interacting and learning with information from mobile maps. In fact in our everyday lives we are overloaded with digital information for helping us guide our way through spatial environments: digital route maps, tourist guides, web site recommenders, as well as geo-tagged information in the space itself. If we are to design better interfaces between information and place it is necessary to reflect on why mobile maps might not be the best solution. On a broader level the key quality of such maps is that they provide global, homogenous and unambiguous forms of information, which has lost its relational quality with both the person and the environment. Since this information is often provided through standardised platforms it is by necessity not personalised, but based on matching requirements with large databases of existing information. This arises because ontologies and tags are essentially ways of classifying and ordering. But "a classified and hierarchically ordered set of pluralities, of variants, has none of the sting of the miscellaneous and uncoordinated plurals of our actual world" (Dewey, 1925, p. 49). We are immersed in need to categorise things, people and places. Space is one of the prime examples of our need to structure experience, as it is subject to homogenous global definitions and rigid infrastructures such as latitude and longitude, the cardinal directions, addresses, post-codes and territorial boundaries. Maps are a fundamental example of how spatial information presented in an abstract manner permits only a certain way of acting in the world (Brown, 2004). This is grounded in the fact that space is traditionally perceived and understood as metric or Cartesian in nature; quantifiable and bounded. However the experience of space as mediated by mobile and wireless technologies transcends these concepts of space. Space enacted through such technologies conforms to a different concept of bounded-ness(Willis, 2008). Instead of some form of definable extent, space is instead experienced more in terms of regions that are not only defined by spatial extents but also by patterns of informational or social access. Consequently, collectively defining boundaries becomes part of the pattern of communication; for example the common practice of asking for and reporting location at the beginning of a mobile phone call. Boundaries are still an omni-present characteristic of space, but moving in and out of bounded zones can occur much like the flicking of a switch, rather than involving some form of graduated change. Similarly displacement in layered media spaces is not confined to the physical properties of 'real world' objects, but also extends to include the specific ranges of technologies. For instance Bluetooth enables interaction within a radius of approximately ten metres, whereas WiFi nodes offer access within a range of up to one hundred meters. As such the definition of interaction in a space of communication flows is structured around spatial nodes of opportunity. But perhaps the key characteristic of such spaces is that the concept of linkage is intensified, and in many ways more subtle and differentiated levels of connectivity frame interaction. Mobile and wireless technologies typically operate in infrastructures where connections are characterised by a whole array of weak and strong links, rather than being defined by the nature of their direct physical proximity. As these networked infrastructures start to dominate over physical spaces, so the relational quality of people and space becomes less a case of physical presence and more a degree of how people considered themselves related to places and spaces; their attachment, or lack of, to the situation they are in. In this sense space is enacted (Varela, 1991), which acknowledges the co-creation of observer and observed through the construction of their relation. It does not exist a priori in some form of contained and constrained structure, and is instead constructed dependent on a whole range of changeable circumstances, most of which are local in relation and effect.

In fact our daily experience of space is not global, but highly personalized and tightly interwoven with features of our social lives and dynamic factors such as time. Space is experienced and enacted not as an abstract quality but as practiced experience. In everyday life this manifests itself in a number of ways. For instance when we travel to a new or semi-familiar place often what we really want to learn about a place is local knowledge - the short cut, the best restaurant or the history of the neighbourhood (Bilandzic et al. 2008). This information resides in people, as memories and knowledge, not in abstract sources of information. In order to tap into this resource of local knowledge about place it is critical to understand where such knowledge about a place resides. Often the best way to find your way in an unfamiliar place is to ask someone. In asking a local about the place you are in, it creates a sense of shared experience. It bridges the gap between the image and lived experience. It creates a situation where knowledge which they have about the place can be shared and valued. In order to realize the value and deliverability of local information what is required is a way of gathering or stimulating the authoring of such local information in a publicly available format. Public distributed authoring is already a common way for people to share information, through formats such as blogs and photo sharing websites. But what is needed is to extend this approach to capturing and sharing local place information, which will require people to adjust their traditional views of spatial information as being created and delivered by experts. Spaces will be seen as useful, not based on functional qualities, but instead based on social recommendations whether these people are friends or strangers. Obviously out of these changed attitude to authoring information, issues of privacy will come into play, as well as blurred distinctions between concepts of what constitutes public and private spaces. But the primary structural change in terms of how space is viewed is that it needs to be perceived not as static, metric and contained but instead narrative in form and constructed out of everyday practice.

CONCLUSION

The information embedded in the spaces of our everyday lives guides us in almost everything we do. However we also refer to abstract information about space such as paper maps and also more recently spatial information delivered on digital devices. We discussed an empirical study which investigated and compared how people acquire knowledge when using maps and mobile maps. This found that people who used a mobile map performed comparatively poorly, suggesting that they had failed to learn about a space in a meaningful way. This was described as being due to the fact that when information is delivered incrementally in the environment this affects the attention and also encourages the person to become passive in the interaction. We concluded by outlining the wider issues associated with interacting with spatial information that is essentially global and abstract and suggested that to create opportunities for enacting space in a more relational and local manner. In this sense, the challenge is less a case of putting information back in its place, but of putting place back into information.

Indexicality: In this chapter we have discussed how the nature of interaction with information is changing. Mobile and wireless technologies enable a whole range of ways to interact with local and appropriate information, that weaves more meaningful into people's everyday action in space. The key question is how to define the protocols of interaction between the mobile devices which detect change and the static located sensors or emitters which deliver information. In fact the push-pull of information in space is not clearly delineated between things we carry with us, and sensors in the environment, but the two qualities of interaction to exist. Yet the conditions created for the push and pull of information related to specific spaces will need to become far more subtly defined and explored. In order for this to happen we need better methods of characterizing the distinction between quality of places in a way that enables recognition way place is experienced in both digital and physical terms.

REFERENCES

Aslan, I., Schwalm, M., Baus, J., Krüger, A. & Schwartz, T. (2006). Acquisition of Spatial Knowledge in Location Aware Mobile Pedestrian Navigation Systems. *Proceedings of Mobile HCI 2006*, (pp. 105-108). New York: ACM Press.

Batty M. (1990). Editorial: Invisible Cities. *Environment and Planning B: Planning and Design*. Vol. 17, (pp. 127-130). London: Pion.

Bilandzic, M., Foth, M., & De Luca, A. (2008, Feb 25-27). *CityFlocks: Designing Social Navigation for Urban Mobile Information Systems*. Paper to be presented at the ACM SIGCHI Designing Interactive Systems (DIS) Conference, Cape Town, South Africa.

Brown, B. & Laurier, E., (2004). Maps and Journeys: An Ethnomethodologival Investigation. *Cartographica*, 4 (3), 17-33.

Castells, M., Fernandez-Ardevol, M., Linchuan Qiu, J. & Sey, A. (2007). *Mobile Communication and Society: A Global Perspective*. Cambridge, MA: MIT Press.

Dewey, J. (1925). *Experience and Nature*. Chicago: Open Court.

Downs, M. & Stea, D. (Eds) (1974). *Image and Environment, Cognitive Mapping and Spatial Behavior.* Chicago: Aldine Publishing.

Held, R., & Hein, A. (1963). Movement Produced Stimulation in the Development of Visually Guided Behavior. *Journal of Comparative and Physiological Psychology*, 56, 872-876.

Höök, K, Benyon, D & Munro A.J. (Eds.) (2003). *Designing Information Spaces: The Social Navigation Approach.* UK: Springer-Verlag

How Satnav Took our School Trip Up a Back Alley. *The Evening Standard, 22nd March 2007.* Retrieved 22nd March 2007 from http://www.dailymail.co.uk/pages/live/articles/news/news.html?in_article_id=443993&in_page_id=1770.

Hutchins, E. (1995). *Cognition in the Wild.* Cambridge, MA: MIT Press.

First 'ignore your sat nav' roadsigns go up. *The Mail on Sunday, 10th February 2007.* Retrieved 10th February 2007 from http://www.dailymail.co.uk/pages/live/articles/news/news.html?in_article_id=436983&in_page_id=1770

Lindberg, F. & Gaerling, T. (1983). Acquisition of different types of Locational Information in Cognitive Maps: Automatic or Effortful Processing. *Psychological Research*, No. 45, 19-38.

Lyman, P. & Varian, H. (2003). How Much Information 2003. *School of Information Management and Systems, UC, Berkeley.* Retrieved November 2007 from http://www.sims.berkeley.edu/how-much-info-2003.

Lynch, K (1960). *The Image of the City.* Cambridge, MA: MIT Press.

McCullough, M. (2005). *Digital Ground.* Cambridge, MA: MIT Press.

Melville, H. (1851). *The Whale.* London: Richard Bentley.

Morville, P (2005). *Ambient Findability: What We Find Changes Who We Become.* Sebastopol, CA: O'Reilly.

Münzer, S. Zimmera, H., Schwalma, M., Bausb, J. & Aslanb, I (2006). Computer-assisted navigation and the acquisition of route and survey knowledge. *Journal of Environmental Psychology.* 26 (4), 300-308.

Siegel, A. & White, S. (1975). The Development of Spatial representations of Large-Scale Environments. In H. Reese (Ed.), *Advances in Child Development and Behavior*, 10, (pp. 10-55). New York: Academic Press.

Thorndyke, P. & Hayes-Roth, B. (1982). Differences in Spatial Knowledge Acquired from Maps and Navigation. *Cognitive Psychology*, 14, 560-589.

Pask, G. (1971). A Comment, a case history, a plan, in Jasia Reichardt,. *Cybernetics, Art and Ideas* (pp. 76–99). London: Studio Vista.

Tobler, W. R. (1970). A Computer Movie Simulating Urban Growth in the Detroit Region. Economic. *Geography,* 46, 234–40.

Varela, F., Thompson, E. & Rosch, E. (1991). *The Embodied Mind: Cognitive Science and Human Experience.* Cambridge, MA: MIT Press.

Willis, K., Holescher, C., & Wilbertz, G. (2007). Understanding Mobile Spatial Interaction in Urban Environments. In *Proceedings of 3rd IET International Conference on Intelligent Environments (IE07)* (pp 61-68). Ulm, Germany: IEE Press

Willis, K (2008 to appear). Spaces, Settings and Connections in Aurigi, A., De Cindio, F. *Augmented Urban Spaces: Articulating the Physical and Electronic City.* UK: Ashgate Press.

Yates, F. (1966). *The Art of Memory.* London: Pimlico.

KEY TERMS

Bluetooth: A form of digital transmission which enables many devices to be easily interconnected using a short-range wireless connection.

GPS: Global Positioning System a constellation of twenty-four satellites that make it possible for people with ground receivers (satnav) to pinpoint their geographic location.

Landmark: A geographic feature or built structure that is easily recognizable, and is often used to assist orientation.

LBS: Location based services are wireless 'mobile content' services which are to provide location-specific information to mobile users moving from location to location.

Mobile map: A term for a map application supported by GPS running on a mobile device.

Satnav: A satellite navigation system capable of receiving and displaying GPS data.

WiFi: Also known as Wireless fidelity refers to certain kinds of wireless local area networks, or WLAN (as opposed to LAN, or computers that are networked together with wires).

Chapter XV
A Visual Approach to Locative Urban Information

Viktor Bedö
University Pécs, Hungary

ABSTRACT

This chapter contributes to the ongoing effort to understand the nature of locative urban information by proposing that locative urban information is a kind of problem that necessitates the use of visual instruments, such as maps integrated into spatial annotation systems. The thesis is that the dynamics of the movement and behavior of messages appearing, disappearing, and spreading on the urban maps provide clues as to what extent a specific type of information is dependent on urban space for context, i.e., its level of location-sensitivity. A parallel is drawn between the interpretation of dynamic patterns appearing on urban maps and scientific discovery supported by the use of visual instruments. In order to illustrate how the question of locativity arises when developing technologies for urban life, a short examination of BlueSpot, a locative media project in Budapest, is provided.

INTRODUCTION

This paper contributes to the ongoing effort to understand the nature of locative information. Messages such as "Get out at the next stop!" or "Text me when you get there!" (Hemment, 2007), can be interpreted only in the context of a concrete place at a concrete moment in time. In contrast, for example, an abstract scientific formula such as "$e=mc^2$", is put into context by other formulae and by scientific method which tell you how to use these formulae. These examples illustrate these two opposite poles; messages for which location is needed as context and messages that can be understood with no reference to location. Between these poles there is a range of situation/location-sensitivity of messages, referred to as the level of locativity. Unfortunately, however, we do not yet possess sound instruments for measuring the level of location-sensitivity, therefore we do not know where to place specific types of information or messages in this range. In other words, it is not clear according what algorithm a location-sensitive search engine ought to rank web results when taking the exact time and location of the web search into account.

This paper proposes using interactive urban maps (such as the maps integrated into projects such as Plazes[1], denCity[2] or Urban Tapestries[3]) as research instruments in order to learn more about which situations and topics are location-sensitive. The thesis is that the dynamics of the movement and behavior of messages appearing, disappearing, and spreading on the urban maps provide clues as to what extent a specific type of information is dependent on urban space for context, i.e., its level of location-sensitivity. Firstly, a short explanation will be presented on how urban space can be seen as providing a context for mobile messages through the use of interactive urban maps. Secondly, the conditions under which the distribution patterns of locative messages emerge on the maps of spatial annotation systems will be discussed, in support of the thesis. Thirdly, a parallel is drawn between the interpretation of dynamic patterns appearing on urban maps and of scientific discovery supported by the use of visual instruments. In conclusion the paper argues that research methods involving human vision are necessary in order to answer questions concerning locative urban information. In order to illustrate at which point the level of locativity plays a crucial role, BlueSpot, a Budapest based locative media project is examined.

BACKGROUND

The question, "If most people are only tourists for about two weeks of the year, what location-sensitive services are being devised for the other fifty weeks?" (Giles, & Thelwall, 2006, p. 9) bridges at least two of the larger research fields: *Mobile Studies*[4] and *Locative Media*. As the research series in Mobile Studies suggests, while networked ICT devices liberate community formation and the flow of knowledge from geographical space (Meyrowitz, 2005), information of a primarily practical nature transmitted by mobile communication remains situation- and location-sensitive (Nyíri, 2003). *Locative media* projects originating from new media arts, address a very similar concern as:

Locative media may be understood to mean media in which context is crucial, in that the media pertains to specific location and time, the point of spatio-temporal 'capture', dissemination or some point in between. The term "locative media" initially appeared (...in 2003...) as a tentative category for new media art that sought to explore the intersection of the virtual space of the Internet with (...) physical space. (...) The term locative media has (...) been associated with mobility, collaborative mapping, and emergent forms of social networking. (Hemment, Evans, Humphries, & Raento, 2006)

Mapping issues in locative media refer on the one hand to tracing people, information and objects (OpenStreetMap[5]) and on the other hand, to spatial annotation and geotagging (Plazes, denCity, Urban Tapestries)(cf., Tuters, & Varnelis, 2006).

Mobile Studies already suggests that information about concrete places is best conveyed by representations with a spatial logic, such as images (Nyíri, 2005; Tversky, 2003). In recent years the research series examined the link between mobile communication and visual communication from the perspective of MMS messages, a format which can include both images and words. Concerning MMS use, it was found that while sharing practical and mobile knowledge is often far more effective when using images rather than words (Nyíri, 2003), a rapid spreading of MMS is hindered by several factors. The inertia of our cultural heritage, which is rooted in literacy, is one of these factors (Kondor, 2005). Another reason is that images sent by MMS are very often so personal that only the closest friends can interpret them (Scifo, 2005); only those who know the very situation-specific context the mobile photos were taken in or in which they should be viewed (Koskinen, 2005). Implicitly, and in some cases explicitly, these findings raise the question of context as a framework for interpretation; not only the circle of close friends, but also the possibility of localization is context-creating.

As information about places is best conveyed by images, the interactive urban map (as a specific kind of image) is a method of creating context for mobile messages (Bedö, 2007). In this way, the question of how the image appears in mobile communication is augmented by the question of how the mobile message appears on a specific kind of image; the map.

As the Institute for the Future report on context-awareness frames it, "in the simplest terms, context awareness just means having information about the immediate situation—the people, roles, activities, times, places, devices, and software that define the situation. But context awareness is also about meaning and meaning-making..." (Vian, Liebhold, & Townsend, 2006, p. 1). In the case of words, it goes without saying that the meaning of every word depends on its context: while a solitary word is hard to interpret, it gains a more definite meaning when it is embedded in the context of a sentence. In terms of pictures, Nyíri (2003) points out that visual animations are analogous to sentences: an animation (a series of pictures ordered along a linear time sequence) is more capable of representing a logical structure and thus allowing more definite interpretation and providing richer meaning than a static picture. It must be noted that the map as an image can provide the context for both text and image information. Dual Coding Theory suggests that during mental processing, blending and mutual activation can occur both between a *verbal* and an *imagery* system (Sadowski, & Paivio, 2001). It follows that a map can constitute a context for text as well as for images, or for any format. Since a word carries a richer meaning in the context of a sentence, a message like "Text me if you get there" can carry richer meaning if read in the context of urban space that is visualized with the help of interactive urban maps.

This paper argues that the interactive urban map—linking the fields of mobile studies and locative media—can not only be conceptualized as a context-creating media for communicating locative information, it can also be used as research instrument to reveal more about the nature of locative information and thus lead to an answer on how to gauge the level of location-sensitivity of mobile messages. Before looking at how the pattern of messages on maps can do this, a short examination of BlueSpot, a locative media project in Budapest, is provided in order to illustrate how the question of locativity arises when developing technologies for urban life.

BLUESPOT CASE STUDY

The Emergent City Action Group[7] (ECA) was created to deal with the question of how the emergence of urban self-organizing processes can be enhanced using information and communication technologies. The group was formed by experts and researchers from the social sciences, natural sciences, humanities, economics and design. The aim was to develop principles and models that could be integrated into future urban planning strategies. The group was founded on the following premises: (1) A city is a spatio-temporal self-organizing system. A crucial feature of self-organization is the employment of feedback mechanisms, including feedback between different levels/hierarchies of organization. (2) Cities contain information which is dependent on the context of concrete place and time, e.g. locative information.

This second premise was examined in the *Background* section, so let us briefly look at the first one. The term self-organization derives from the sciences. The analogy of biological self-organization applied to urban structures became widely recognized through Jane Jabobs' *The Death and Life of Great American Cities* in the 1960s (Jacobs, 1960). Self-organization means that the entities of a system interact with each other in such a way that order emerges without central control (Obornyi, 2006; cf. Townsend, 2000). Neighborhood segregation (e.g. relatively homogeneous areas in urban residential zones according, for example, to age, number of children, wealth, education, and ethnicity) is an example of spatial patterns produced by self-organizing urban processes. Also, forms of temporal self-organization occur with the use of mobile communication technologies, as the coordination of social activity often relies on ad hoc mobile negotiation instead of keeping to predefined schedules (Nyíri, 2006; Townsend, 2007). Instead of arranging an appointment at 7pm, for example, friends can iterate the meeting time by using relative time measures such as "in 30 minutes", and "after you have finished". So, temporal self-organization manifests itself in personalized time windows which open up in the overall framework of fixed public time (Nyíri, 2006).

In the course of the preliminary research phase, the ECA collected and reviewed domestic and international projects and initiatives dealing with information technologies and social software in urban space. In a series of brainstorming sessions, project members and chosen experts collected ideas about the potential applications of communication technology (both high-tech and low-tech) for enhancing urban self-organization. The collected projects and ideas mainly followed the following pattern: mapping and tracking people, objects, processes, sharing (shopping), forums (for house, neighborhood, and city), spare time socializing, and finding friends in bars. Unlike purely forum-based solutions, the group of locative community projects take into account the spatial dimension of social and urban processes (denCity, Plazes, Urban Tapestries), and are apt to introduce new principles, thus setting up new rules for urban self-organization.

More specifically, the following question crystallized: What kind of information is most sensitive to the spatial dimension of urban processes? As explained above, information of a primarily practical nature is certainly more sensitive to urban context than abstract scientific knowledge, but are there any means to gauge the level of this context-sensitivity, or in other words, locativity?

The preliminary research of the ECA revealed a lack of a principle or tool for measuring the level of location-sensitivity. As a consequence, the ECA designed a new locative communication platform named BlueSpot which introduced a simple new principle and let serendipity make the highest possible impact. BlueSpot is a communication platform that enables people to address places instead of people. In order to maximize inclusiveness, a technology was chosen that allows everybody with a Bluetooth or java enabled phone to participate for free (no GPS or mobile Internet data contract was needed). In order to maximize serendipity e.g., the emergence of unexpected uses, BlueSpot was made topic-neutral. Except for the feature that short text messages can be sent to places, there is no predetermined or expected pattern of use.

The BlueSpot infrastructure consists of around 50 hotspots located in Budapest which are connected to a central server by the Internet. The hotspots are indicated by physical signs at the particular locations.

When a user is within the range of a hotspot

Figure 1. BlueSpot locations in Budapest

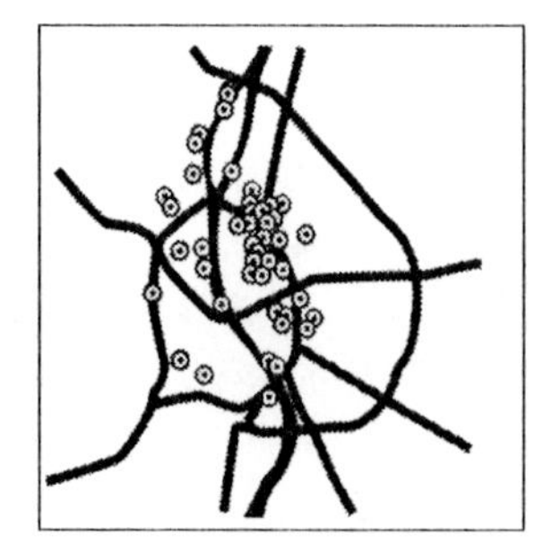

Figure 2. BlueSpot hotspot and mobile client

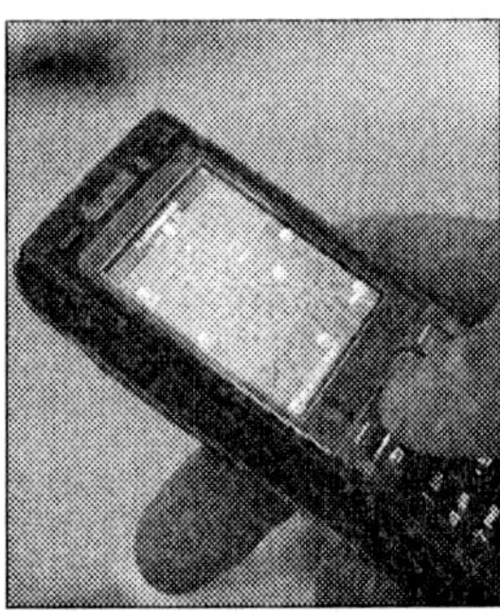

and has Bluetooth activated on his or her mobile device, the system will send the user the client software. After authorization, the program will install itself on the mobile, and the user can start sending messages right away, or access the list of messages on a chosen location. The user might send a message to any distant location, or the very location where he or she is currently at. The messages on every location are accessible from every other location.[6]

BlueSpot was a toolset for citizens on the one hand and an experimental research instrument for ECA on the other. The system ran from July 9th to December 18th 2007. The total number of message authors was 465, who altogether sent 985 messages. Every message sent was stored in a database that will be published in future (taking privacy issues into account). Although the number of messages sent through the system over the duration of the experiment was smaller than optimum, a visualization of the data (currently being planned) will reveal directions in which data was sent and the spatial distribution of topics.

MAPPING HYBRID URBAN SPACE

In the chapter "The kind of problem a city is" in the book mentioned above, Jacobs metaphorically refers to the microscope as a paradigmatic instrument for examining urban processes: in order to gain more understanding about urban processes one should "look for 'unaverage' clues involving very small quantities, which reveal the way larger and more 'average' quantities are operating." (Jacobs, 1962, p. 454). Nowadays urban maps, such as the maps of spatial annotation systems, provide us with a bigger picture: we can literally look directly at distribution patterns of messages or movements of people to reveal urban processes. It is necessary to give a short explanation of the underlying processes and mechanisms which produce the patterns on the map we will be confronted with.

The use of mobile communication technologies merge the borders between physical and digital space and create a hybrid space:

Because many mobile devices are constantly connected to the Internet, (...) users do not perceive physical and digital spaces as separate entities and do not have the feeling of "entering" the Internet, or being immersed in digital spaces, as was generally the case when one needed to sit down in front of a computer screen and dial a connection. (De Souza e Silva, 2006, p. 263)

The physical city consisting of streets, antennae, rooftops, trees, buildings, sensors, masts (Galloway & Ward, 2006) but also radio waves (Sant, 2006), seamless and/or *seamful* Wifi and GPS coverage (Sharpe, 2006) becomes a "substrate upon which layers of information and data can be referenced" (Townsend, 2007, p. 312). Formerly ephemeral aspects of the city such as personal messages and shared impressions become persistent (Giles, & Thelwall, 2006), and gain recognizable structure through information and data embedded into or augmenting urban space, through geotagging and spatial annotation. In addition to housing blocks and streets, social phenomena are also gaining 'viscosity' (Shirvanee, 2006). They guide the direction of, and have an effect on, the speed of moving physical objects as well as the spreading of information and encounters of people. Mobile services enable 'social navigation' i.e., the navigation of the city based on other people's recommendations or past decisions (Bilandzic, Foth, & De Luca, 2008). Even the simple fact that someone is talking to someone else on a mobile phone, visibly alters their route and speed when crossing a plaza (Höflich, 2005).

As Kitchin (2007) points out, hybrid space does not fit into the classical cartographic paradigm, according to which a map is an ontologically fixed representation of a terrain or process. Nowadays, interactive urban maps are capable of depicting geographical space as well as dynamic real-time information from both physical and virtual space. We have to note that as a consequence, the issues of visualizing abstract information (cf., Ware, 2000; cf., Card, 1999) merge with cartographic problems (cf., Dodge, & Kitchin, 2001; cf., Kitchin, 2007). The intended function of the map determines the projection, simplifications, and distortions applied (Tversky, 2003) which also affects the choice between the topographical or topological layout. Traditional street maps follow the topographical layout of a city. Network- or cyberspace-visualizations are topological maps, meaning that distance is measured in the number of steps necessary to get from one node to another node in the network (Dodge, 2001). The classic example of a map between these extremes is the London tube map: topographical relations are distorted in order to emphasize topological characteristics. In this paper we have to concentrate on maps that represent geographical/topological layout, as we want to use the map a tool for answering a question about urban places/locations.

The interaction between the urban map and the user is constant as the user shapes the map through his or her movement, action and search strategies. These changes can immediately be displayed on the map, thus creating real-time feedback. As Kitchin suggests, in terms of the new cartographic paradigm: mapping is spacing. The result is that the map is not merely a representation (Galloway, & Ward, 2006) or model, but to a certain extent constitutes urban space itself.

THE KIND OF PROBLEM A MAP CAN SOLVE

Instead of analyzing aspects of navigation in this paper, the emphasis is on the use of maps as research instruments to gain a better understanding of the nature of locative urban information. The maps used as examples are maps of spatial annotation systems, and within this, the focus is on systems that are used with mobile communication devices such as handheld computers or, more commonly, mobile phones. The spatio-temporal dynamics of the distribution, spreading, disappearing and movement of messages on the spatial annotation systems' map ought to reveal to what extent a specific kind of information or message is coupled to the ever-changing context of urban space. The term 'mobile communications' contains multiple meanings: it means that information contained in the message moves through space without the help of a delivery person or a transport system like a rail; it also means that people move, carrying with them information in, for example, the form of books; more recently it has also come to mean that people are moving and are at the same time using wired and wireless information and communication technologies (Poster, 2005). Both people and messages can be tracked and mapped. While the movement of the message and information changes the urban context, the movement of people also permanently changes the urban context through new encounters. People need to access different information in different contexts, so messages will travel and move with the changing context. These dynamics appear on the interactive urban maps such as maps of spatial annotation systems.

What we see on these maps are bubbles and flags containing or representing text, image or sound messages, and people. Messages appear, spread or disappear according to the built-in parameters of the map and/or the personal settings of the user. Preset or personally manageable parameters often include filtering according to

Figure 3. Screenshots of the Plazes map at different times of a day

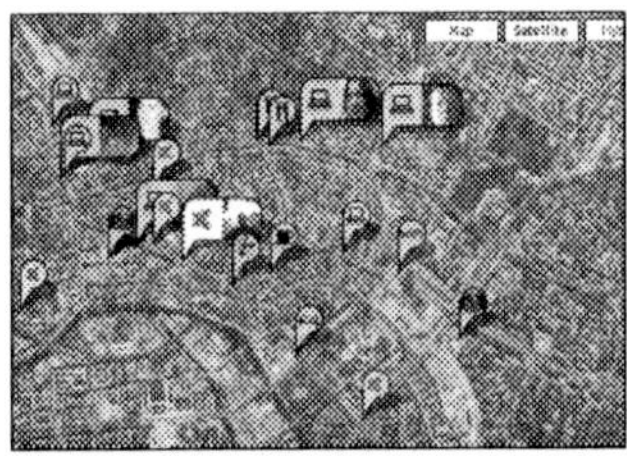

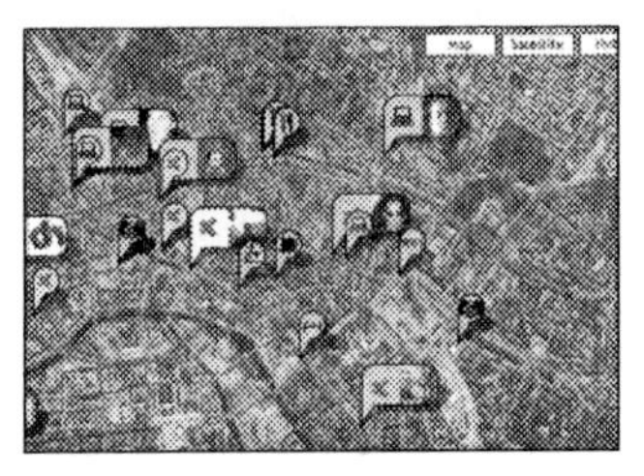

Figure 4. Screenshot of one possible setting of a denCity map

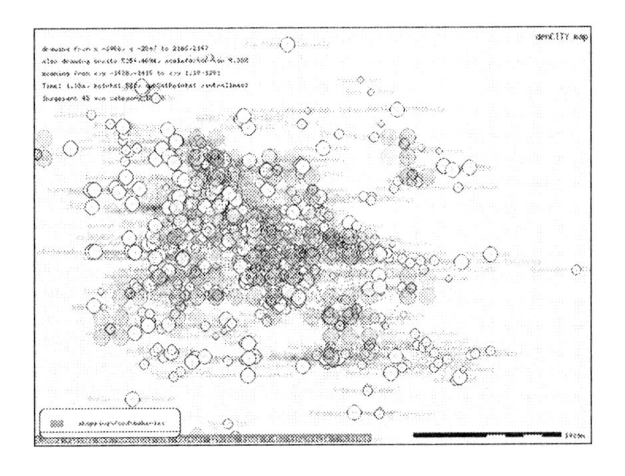

topic (denCity), geographical radius of relevance (denCity) or the choice between public and private (Plazes). Patterns drawn by this movement of messages on the map let us gauge which kind of information topics are most sensitive in the permanently changing context of urban space: messages that are independent of the ever-changing constellations of moving people, of self-organizing activities through exchanging messages, will not follow these processes of spatio-temporal dynamics. Patterns on the map which are static or random might be location-bound, but not situation-sensitive or locative. On the other hand, the kind of messages that are situation-sensitive, e.g., locative, will draw traces and patterns of spatio-temporal self-organization onto the map. What we are looking for are unaverage clues—to follow Jacobs' precept –, but also recurring spatial and temporal patterns.

How do we recognize unaverage clues or patterns? When medics learn the visual skills to interpret spots on an X-ray or neuroscientists learn to interpret FMRI brain scans they do it by watching an expert. Only experts are "experienced in looking at a wide range of such images and [understand] the parameters of the equipment..." (Kemp, 2006, p. 320), and so are able to detect and identify the patterns. However, as discussed by this paper, interactive urban maps are new instruments, there is no expert who can show us what patterns we have to look for (nor can we be sure which parameters and filters to use for the map, e.g., the settings of the instrument). As in the case of scientific discovery employing visual instruments, separating recurring patterns from random dynamics on the map is specifically a visual task, where there is no substitute for human vision. Detecting symmetry (Hargittai, & Hargittai, 2000), the discovery of the double helix structure of DNA on the basis of images of X-ray diffraction (Hargittai, & Hargittai, 2000; Kemp, 2006), and the identification of the mysterious patches on the moon's surface seen through the telescope (Kemp, 2006) are all instances of specifically visual acts in scientific discovery. The former natural scientist, now art historian, Martin Kemp introduces structural intuition, as the principle by which artists and scientists extract order from the chaos of visual phenomena. Structural intuition is, on the one hand, shaped by the gravitational pull of physiological and cognitive structures (reaching back to pre- or subverbal deep cognitive structures) and on the other by the cultural habits and conventions of human vision. Attempts, however, to entirely formalize this act of 'seeing' have failed in the course of the history of vision, as Kemp (2006) asserts.

FUTURE TRENDS

A better understanding of the nature of locative urban information fosters the development of numerous innovations such as location-sensitive web search algorithms or the location-sensitive filtering of information. McCullough, addresses this issue from the interaction-design point of view and calls attention to the need of an 'urban markup' that incorporates a conceptual semantic structure for urban contexts and that can be shared by locative services (McCullough, 2006). Referring to public authoring in the case of the World Wide Web, McCullough advocates a bottom-up way to characterize sites, operations of context-based tasks, and other specific-activity domains, such as the conference room, stadium, hotel lobby, and express lane. Individual and social activities, traces of mobile communication collectively drawn on the map of spatial annotation systems, reveal urban contexts and situations through recurring and identifiable spatio-temporal patterns. The visual act of identifying these patterns—possibly combined with other methods such as linguistic analysis—can lead to the identification of urban situations and contexts at street, neighbourhood, or city level, which can be integrated into ontologies as suggested by McCullough.

CONCLUSION

The interpretation, relevance and sharing of locative urban information is dependent on urban space as context. In order to understand more about the nature of this kind of information we need to employ methods and instruments which deal with the spatial dimension. Scientific research often makes use of visual instruments, above all in fields that have to deal with complex spatio-temporal patterns. As we have seen in this paper, processes in hybrid urban space are complex spatio-temporal processes, thus visual instruments are required in order to deal with them. Maps integrated into spatial annotation systems can represent these instruments for researchers of mobile communication and locative media. They can also provide clues as to which kind of information is most sensitive to the spatial dimension of urban processes. Since the use of maps as research instruments is based on the identification of unexpected clues and specific spatio-temporal patterns, the role of human vision can not be substituted by other methods.

ACKNOWLEDGMENT

Completion of this work was supported by the PhD Scholarship of the University Pécs and a Research Scholarship of the DAAD - German Academic Exchange Service. I would like to thank the editor Marcus Foth and three anonymous reviewers for their constructive comments on earlier drafts.

REFERENCES

Bedö, V. (2007). Maps as tools of thinking. In K. Nyíri (Ed.), *Mobile Studies: Paradigms and perspectives* (pp. 123-133). Vienna: Passagen.

Bilandzic, M., Foth, M., & De Luca, A. (2008). CityFlocks: Designing social navigation for urban mobile information systems. In J. van der Schijff, G. Marsden. & P. Kotze (Eds.). *Proceedings ACM Designing Interactive Systems (DIS).* Cape Town, South Africa. Retrieved February 2, 2008, from http://eprints.qut.edu.au/archive/00010871/.

Card, S. K., Mackinlay, J., & Shneiderman, B. (Eds.). (1999.). *Readings in information visualization: Using vision to think.* San Diego, CA: Academic Press.

De Souza e Silva, A. (2006). From cyber to hybrid: Mobile technologies as interfaces of hybrid spaces. *Space and Culture* 9(3), 261-278. Retrieved February 2, 2008, from http://sac.sagepub.com/cgi/content/abstract/9/3/261

Dodge, M., & Kitchin, R. (2001). *Atlas of Cyberspace.* Harlow, UK: Addison-Wesley.

Galloway, A. & Ward, M. (2006). Locative media as socialising and spatializing practice: Learning from archaeology. *Leonardo Electronic Almanac.* 14(3). Retrieved February 2, 2008, from http://leoalmanac.org/journal/Vol_14/lea_v14_n03-04/gallowayward.asp

Giles, L., & Thelwall, S. (2006). *Urban Tapestries: Public authoring, place and mobility* (Research Rep. 2nd ed.). London: Proboscis. Retrieved February 2, 2008, from http://socialtapestries.net/outcomes/reports/UT_Report_2006.pdf

Hargittai, I., & Hargittai M., (2000). *In our own image: Personal symmetry in discovery.* New York: Kluwer.

Hemment, D., Evans, J., Humphries, T., & Raento M. (2006). Locative media and pervasive surveillance: The Loca project. In Gibbons, J., & Winwood, K. (Eds.), *Hothaus Papers: Perspectives and paradigms in media arts.* Birmingham, UK: VIVID/Article Press. Retrieved February 2, 2008, from http://www.drewhemment.com/2006/locative_media_and_pervasive_surveillance_the_loca_projectby_drew_hemment_john_evans_theo_humphries_mika_raento.html

Hemment, D. (2007, September). *Locative Arts and Locative Activism.* Paper presented at the Locative Media Summer Conference, Siegen, Germany.

Höflich, J. (2005). In K. Nyíri (Ed.). *A sense of place: The global and the local in mobile communication* (pp. 159-168). Vienna: Passagen.

Jacobs, J. (1962). *The death and life of great American cities.* London: Jonathan Cape.

Kemp, M. (2006). *Seen/unseen: Art, science, and intuition from Leonardo to the Hubble telescope,* Oxford, UK: Oxford University Press.

Kitchin, R. (2007, February). *Mapping Code/Spaces.* Paper presented at the conference Mapping Anthropotechnical Spaces, Berlin, Germany.

Kondor, Zs. (2005). The iconic turn in metaphysics. In K. Nyíri (Ed.), *A sense of place: The global and the local in mobile communication* (pp. 395-404). Vienna: Passagen.

McCullough, M. (2006). On urban markup: Frames of reference in location models for participatory urbanism. *Leonardo Electronic Almanac.* 14(3). Retrieved February 2, 2008, from http://leoalmanac.org/journal/Vol_14/lea_v14_n03-04/mmccullough.asp

Koskinen, I. (2005). Seeing with mobile images: Towards perpetual visual contact. In K. Nyíri (Ed.), *A sense of place: The global and the local in mobile communication* (pp. 339-347). Vienna: Passagen.

Meyrowtiz, J. (2005). The rise of glocality: New senses of place and identity in the global village. In K. Nyíri (Ed.), *A sense of place: The global and the local in mobile communication* (pp. 21-30). Vienna: Passagen.

Nyíri, K. (2003). Pictorial meaning and mobile communication. In K. Nyíri (Ed.), *Mobile communication: Essays on cognition and community* (pp. 157-184). Vienna: Passagen.

Nyíri, K. (2005). Images of Home. In K. Nyíri (Ed.), *A sense of place: The global and the local in mobile communication* (pp. 375-381). Vienna: Passagen.

Nyíri, K. (2006). Time and communication. In F. Stadler, & M. Stöltzner (Eds.), *Time and history. Zeit und geschichte* (pp. 301-316). Frankfurt: Ontos Verlag.

Obornyi, B. (2006). A város térbeli fejlődése [Spatial growth of cities]. Unpublished manuscript, Emergent City Action Group. Retrieved February 2, 2008, from http://emergentbudapest.org/en4_2_1.html

Poster, M. (2005). Digitally local: Communications technologies and space. In K. Nyíri (Ed.), *A sense of place: The global and the local in mobile communication* (pp. 31-41). Vienna: Passagen.

Sadowski, M., & Paivio, A. (2001). *Imagery and text: A dual coding theory of reading and writing.* Mahwah, NJ: Lawrence Erlbaum Associates.

Sant, A. (2006). Trace: Mapping the emerging urban landscape. *Leonardo Electronic Almanac.* 14(3). Retrieved February 2, 2008, from http://leoalmanac.org/journal/Vol_14/lea_v14_n03-04/asant.asp

Scifo, B. (2005). The domestication of cameraphone and MMS communication: The early experiences of young Italians. In K. Nyíri (Ed.), *A sense of place: The global and the local in mobile communication* (pp. 363-373). Vienna: Passagen.

Sharpe, L. (2006). Swimming in the grey zones: Locating the other spaces in mobile art. *Leonardo Electronic Almanac.* 14(3). Retrieved February 2, 2008, from http://leoalmanac.org/journal/vol_14/lea_v14_n03-04/lsharpe.asp

Shirvanee, L. (2006). Locative viscosity: Traces of social histories in public space. *Leonardo Electronic Almanac.* 14(3). Retrieved February 2, 2008, from http://leoalmanac.org/journal/vol_14/lea_v14_n03-04/lshirvanee.asp

Townsend, A. (2000). Life in the real-time city: Mobile telephones and urban metabolism. *Journal of Urban Technology.* 2(7), 85-104. Retrieved February 2, 2008, from http://urban.blogs.com/research/JUT-LifeRealTime.pdf

Townsend, A. (2007, September). *Thinking in telepathic cities.* Paper presented at the conference Towards a Philosophy of Telecommunication Convergence, Budapest, Hungary. Retrieved February 2, 2008, from http://www.socialscience.t-mobile.hu/2007/prepro2007_szin.pdf

Tuters, M. & Varnelis, K. (2006). Beyond locative media. *Networked Publics.* Retrieved February 2, 2008, from http://networkedpublics.org/locative_media/beyond_locative_media

Tversky, B. (2003). Some ways graphics communicate. In Nyíri, K. (Ed.), *Mobile communication: Essays on cognition and community* (pp. 143-155). Vienna: Passagen.

Vian, K., Liebhold, M., & Townsend, A. (2006). *The many faces of context awareness: A spectrum of technologies, applications, and impacts* (Research Rep. SR-1014). Palo Alto, CA: Institute for the Future, Technology Horizon Program.

Ware, C. (2000). *Information visualization: Perception for design.* San Francisco: Kaufman.

KEY TERMS

BlueSpot is a Budapest-based locative media experiment initiated by the interdisciplinary research team, the Emergent City Action Group. The Bluetooth hotspot-based platform allows users to address messages to places in the city using their Bluetooth and Java enabled mobile phones. URL: http://bluespot.hu/.

Context Awareness, besides recognizing and taking into account factors defining a situation such as people, roles, activities, times, places, devices, and software, also refers to imbuing meaning, based on these factors (Vian, 2006). Location in urban space, for example, can provide a context that facilitates the interpretation of locative information.

Hybrid Urban Space refers to the condition of urban space where the use of mobile communication technologies merge the borders between physical and digital space and create a hybrid space (De Souza e Silva, 2006). The hybrid urban space also results in a new cartographic paradigm (Kitchin, 2007).

Locative Information is a term derived from locative media. As locative media pertains to a specific location and time (Hemment, 2006), locative information can be interpreted in the context of a concrete place at a concrete moment in time, or its relevance is a function of the concrete situation.

Mobile Studies is a term coined in 2006 by the Hungarian research team Communications in the 21st Century, who conduct humanities and social science research into the impact of mobile communication on human thinking and social life (Nyíri, 2007, p.11). With the involvement of international academics, research has been pursued since 2001 in Hungary within the framework of the project, conducted jointly by T-Mobile Hungary and the Hungarian Academy of Sciences. URL: http://socialscience.t-mobile.hu/

Spatial Annotation Systems refer to systems which allow users to associate messages or any information with places in geographical space via computers or mobile communication devices such as PDAs or mobiles. Depending on the system used, the location to which a message will be attached can be either selected manually on a map or supported by locating services, for example, based on GPS or mobile cell information.

Urban Self-Organization is a term derived from the natural sciences. The analogy of biological self-organization applied to urban structures became widely recognized through Jane Jabobs' *The Death and Life of Great American Cities* in the 1960s (Jacobs, 1960). Self-organization means that the entities of a system interact with each other in such a way that order emerges without central control.

ENDNOTES

1. http://plazes.com/
2. http://dencity.konzeptrezept.de/
3. http://urbantapestries.net/
4. http://www.socialscience.t-mobile.hu/
5. http://www.openstreetmap.org/
6. Founded in 2005, supported by grants of the Hungarian National Office for Research and Technology, and the Media Research Center at Budapest Technical University (http://mokk.bme.hu/)
7. For more technological details, see http://emergentbudapest.org/en3_1_1.html

Chapter XVI
Navigation Becomes Travel Scouting:
The Augmented Spaces of Car Navigation Systems

Tristan Thielmann
University of Siegen, Germany

ABSTRACT

Car navigation systems, based on "augmented reality," no longer direct the driver through traffic by simply using arrows, but represent the environment true to reality. The constitutional moment of this medium is the constant oscillation between environmental space and two-dimensional projection space. Using the words of Walter Benjamin, one could also speak of a transparent translation of the world that should not obscure the original. In contrast to the prior generation of navigation systems, the orientation points of the "augmented map" are also fully linked with databases of other available information suppliers. Temporal information, in addition to spatial information, is becoming increasingly important with features such as real time gridlock reports aided by highway sensors and guidance to the nearest event. Does the future lie in the fusion of travel guides and navigation systems? This paper argues that future developments in urban informatics resulting from the convergence of cartographic, media and communication technologies can be inferred based on the increasing phenomenon of mobile augmented reality applications.

INTRODUCTION

Wherever we go in modern urbanized spaces, we are directed by software. Urban regions do not only shape and configure global new media infrastructure investment and global Internet and mobile phone traffic. "The modern city exists as a haze of software instructions. Nearly every urban practice is becoming mediated by code" (Amin & Thrift, 2002, p. 125). Urban informatics research therefore needs to engage much more powerfully with the complex intra-urban and inter-urban geographies that so starkly define the production, consumption and use of artefacts, technologies and practices. In other words, we are increasingly seeing new landscapes of code that are now beginning to make their own emergent ways.

Software is part of the paraphernalia of everyday urban life, "a kind of extended phenotype in which the environment we have made speaks back to us" (Thrift & French, 2002, p. 329). So, maybe instead of understanding software as a kind of urban infrastructure, we can see it as an extension of human spaces, as an intermediary passing information from one place to another so efficiently that the journey appears as a movement without friction (Latour, 2005, pp. 37-42). And here I point to software's ability to act as a means of providing a new and complex form of augmented spatiality. Car navigation systems are a good example for the overlaying of informational environments onto the landscape, as they are one of the key pervasive computing applications that deeply influence the '**remediation**' of urban life (Graham, 2004).

Although I recognise the history of car navigation systems (French, 2006; Thielmann, 2008) and the range of ethnomethodological work on how people navigate in cars (eg. Brown & O'Hara, 2003; Laurier, 2004), this is not my primary concern. Instead of investigating the integrated navigation support for various modes of transportation (Arikawa, Konomi & Ohnishi, 2007) or the automatic production of driving spaces by actively shaping road environments and driver behaviour through GPS-tracking (Dodge & Kitchen, 2006), I want to focus on the augmented spaces of car navigation systems.

This paper suggests that we need to unpack the '**augmented space**' term, introduced by Lev Manovich (2006), as a general aesthetic paradigm of the urban informational landscape. It explores three different spatial augmentations created by car navigation systems: augmented maps, augmented subjectivity and augmented navigation.

AUGMENTED MAPS: A SHORT REVIEW OF CAR NAVIGATION SYSTEMS

*Navigation technology has made travel routine. We will make it a unique experience using the same technology! [...] Each trip is an individual story that should not be told, but experienced. The virtual travel guide is aimed at sustainably increasing the strength and diversity of these experiences. It opens up new spaces for experiences that far exceed the possibilities of representation offered by the usual tourist guides. With this, they smooth the path for a modern way of travel—**Travel Scouting**.* (iPublish, 2006, p. 6f., own translation)

The advertisement for the "first virtual travel guide to the world" (iPublish, 2006, own translation), the *Merian scout Navigator*—which was launched in 2007—is aimed at combating the loss of individual travel created by vehicle navigation systems and the classic travel guides that led to a standardization of the "tourist gaze" (Urry, 1990).

The approach sketched out here for a subjective, strategic 'undermining' of a cartographic trope (Crang, 1998, p. 62ff.) should not, however, obscure the fact that this use of maps, the social practice of spatial reading, is no more capable of undermining the ideological-discursive function of the map. On the contrary, the *Merian scout Navigator* combines the contents of the classic travel guide (restaurant and hotel recommendations, places to go, information on culture, country and people etc.) with mobile navigation. If one drives past a place of interest, the system offers, for example, a "drive by-audio guide", that gives a short audio presentation on the topic. In other words, the map's discursive function is completed by such a multimedia system.

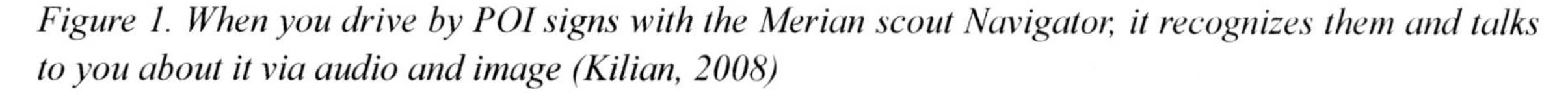

Figure 1. When you drive by POI signs with the Merian scout Navigator, it recognizes them and talks to you about it via audio and image (Kilian, 2008)

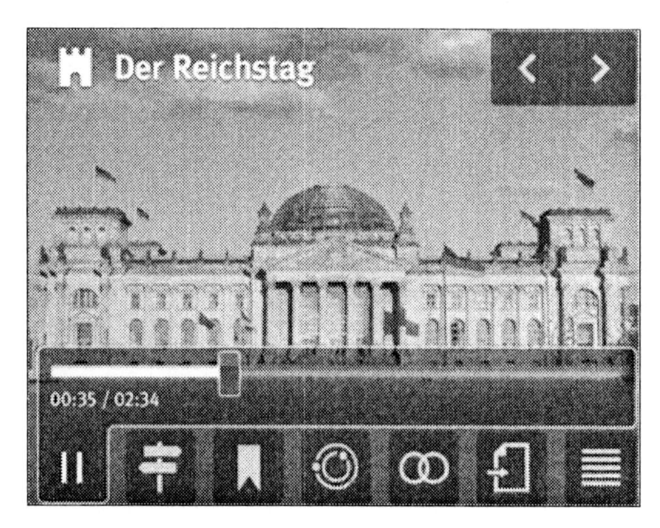

The system provider *iPublish*, a daughter company of the German *Ganske Publishing Group*, can thereby not only fall back on the interactive contents of the travel magazine *Merian*, but also the gourmet magazine *Der Feinschmecker*, the city lifestyle magazine *Prinz* or the audio books published by the *Hoffmann & Campe* publishing house. For example, while driving through the Uckermark, the home of the German Chancellor Angela Merkel, it is possible to listen to her biography *Mein Weg*, or to simply let oneself be navigated to the closest party.. This is where the convergence of cartography and media applications, which is one of the main foci in urban informatics research, is particularly evident.

A similar kind of convergence had already been developed through the "first hypermedia program" (Manovich, 2001, p. 259), the ***Aspen Movie Map***, that made available panoramic camera experiments, thousands of still frames and audio in addition to basic travel footage (Naimark, 1997). Furthermore, "the Movie Map is a comprehensive alternative to both conventional maps and travelogues. [...] The experience is one of direct immersion in vicarious travel: a visit without being there" (Bolt, 1984, p. 72f.).

"The 1978 project *Aspen Movie Map*, designed at the MIT Architecture Machine Group, headed by Nicholas Negroponte (which later expanded into MIT Media Laboratory) is acknowledged as the first publicly shown interactive virtual navigable space" (Manovich, 2001, p. 259) and, furthermore, is probably the first computer-controlled car navigation system.[1] This program used videodisks to allow the user to 'drive' on the streets of Aspen, Colorado and to choose new directions at each crossroad touching left and right arrows graphically overlaid on a touch-sensitive screen. It was also possible to stop, view houses, meet people, or even to change the season (Brand, 1988). The ***Aspen Movie Map*** offered a virtual travel through pre-recorded spaces, very much like what Google offers now with its *Streetview* service.

Realization of the *Movie Map* was accomplished by driving through the town of Aspen in the US state of Colorado and taking one frame per every 10 feet of travel by a set of truck-mounted 16mm stop-frame cameras (Naimark, 1997). This footage was then stored on videodisk. Based on the information provided by a playback system of several laserdisc players, a computer, and a touch-screen display, the corresponding photo series appeared on a color TV

Figure 2. Aspen Movie Map 1978-83 (Bolt, 1984)

monitor. A second horizontally-mounted monitor indicated the current position and direction taken with an arrow on an electronic topographical map, while the track of the route changed from red to green as you travelled.

This corresponds in essence with the type of representation we are used to in current navigation systems. The tiny translucent insert map at the top-middle of the video screen is also very similar to the head-up display that superimposes navigational information on a driver's field of view—for example, presently offered for *BWM 5* and *6 series* (see figure 3 and 4).[2] What is also remarkable about the ***Aspen Movie Map*** is that the user could zoom in on any street intersection of Aspen by switching between an aerial photo, a street map or a detailed map view, very much like the current functionality of *Google Maps* and *Google Earth.*

Thus the idea of developing a virtual space on a large scale based on real space and with the aid of photos and video recordings, was initially not further pursued in those navigation systems sold since 1985. For example, the *ETAK Navigator*, the first commercially available car navigation system—initially only configured for the road network in California—used representation procedures that had been tried and tested in cartography (French, 2006; Thielmann, 2008).

The ***Aspen Movie Map*** combined the overview representation of a map with the self-guided experience of a city that simultaneously promises an immersion experience. Such simultaneous views, called 'split-screen mode', are now offered only by the top-class car navigation systems. For example, the *BMW* driving assistant system *ConnectedDrive* (see figure 5) shows real-life images on weather conditions, hotels, restaurants, parking garages etc. on one side and cartographic information on the other side.

The future of mobile navigation lies in the '**augmented map**', the hybrid representation of cartographic, real and virtual image information—whether on a head-up or a conventional display. Becker's Traffic Assist Z 201, for instance,

Figure 3 and 4. BMW navigation system with head-up display, 2003

composes the digital image from three maps at one time (see Figure 6). The user can switch between a conventional topographic map, a bird's-eye-view and a three-dimensional representation, including '3-D terrain view' and '3-D city view', with an image quality that helps the driver visualise gradients, slopes or approaching curves in the road.

Continental (formerly Siemens VDO) is also developing a three-dimensional map representation for Europe, reportedly "with a picture quality never before achieved" (Siemens-VDO 2005), to further facilitate (dis-)orientation for the driver. This is based on a so-called 'digital terrain model', in which digitized satellite or aerial photographs are laid over a grid of the earth's surface in order to create a photo-realistic representation of urban scenarios (see Figure 7). In contrast to the prior generation of navigation systems, the orientation points on the *Continental* are not isolated on the map, but are fully linked with the entire system and databases of other available information suppliers. It is therefore possible to obtain tourist information on important buildings, find out opening times and address data, make telephone connections or determine a linkage to the destination.

Temporal information in addition to spatial information is becoming increasingly relevant in new navigation systems (Telematics Research Group 2006), with features such as real time gridlock reports aided by highway sensors (*TMC Pro*) or guidance to the nearest open drugstore—for example, possible with *BMW ConnectedDrive*. Also, Volkswagen and Google Earth together developed an open navigation system (see figure 8) that integrates the Internet with real-time data, such as weather reports or current fuel prices (Mays, 2006). They permit "spaces [to] become places as they become 'time thickened'", as Mike Crang (1998, p. 103) describes. Such **connected navigation** systems have more in common with travel photography than with the original objective of the **Global Positioning System** (GPS), that is, the accurate targeting of ballistic missiles. It is therefore no surprise that the future seems to lie in the fusion of travel guides and navigation systems.

AUGMENTED SUBJECTIVITY

According to Jody Berland (1996, p. 125), the dream of traveling or the creation of travel records meet up with the passion for precise measurements in satellite imaging (and mapping). **GPS** not only registers the positional coordinates, but also records the high points, notable characteristics and special happenings on a trip—those personal experiences that are not entered onto conventional maps. For example, the three-dimensional navigation systems actually developed by Becker and Continental (see Figures 6 and 7) are less geared towards realistic perceptions of the surroundings, but more aimed at visualization of landscape and

Figure 5. BMW ConnectedDrive: The mobility service "Park Info," 2004

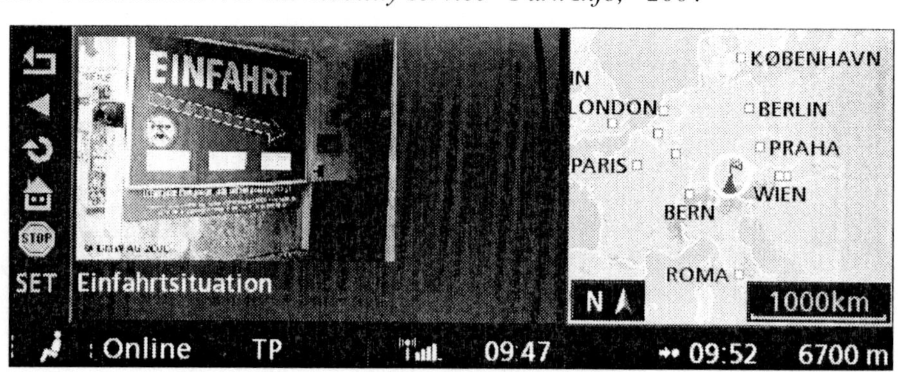

Figure 6. Becker's Traffic Assist Z 201 with 3-D renderings of buildings and landmarks, 2008

Figure 7. Photorealistic three-dimensional map display from Siemens-VDO (now Continental), 2005

Figure 8. Internet-linked Google Earth navigation system from Volkswagen, 2005

altitudinal profiles, to facilitate the planning and locating of off-road routes for drivers of cross-country vehicles.

However, the **GPS** receiver does not capture an objective representation, any more than does travel photography, but creates a visual display that can activate memories of subjective perspectives, perceptions linked to particular locations (Parks, 2005). Although the **GPS** representation of a trip consists only of a series of lines and points (of interest) and the self appears as a set of trajectories, it is principally the movements of the body that transform the map from an omniscient territorial record to an individualistic medium of expression. So the tracking of the personal turns the cartographic discourse into embodied practice (Rose, 1993, p. 7).[3]

In order to permit car users to find their way through this maze of information, the positioning of a subject acquires a whole new meaning through technical features such as dual-view car navigation displays (see figure 9). Thanks to the two-way viewing angle LCDs, passengers can enjoy films uninterrupted, while the driver chauffeurs them through cities or the countryside using the navigation system. The first driver assistance system with dual-view display was built into an *Opel Vectra* station wagon (General Motors Europe, 2005), simultaneously showing different image information depending on the viewing angle.

The speed of the information society not only requires new theories on the visual, but also new forms of subjectivity. Like the shifts in perspective that have marked the great ages of Western civilization, what **Paul Virilio** (1997, p. 24) calls the "trajective state", a speedy state oscillating between the subjective and objective which maps movement from here to there, is a dynamic kind

Figure 9. GM Opel driver assistance system with dual-view display, 2005

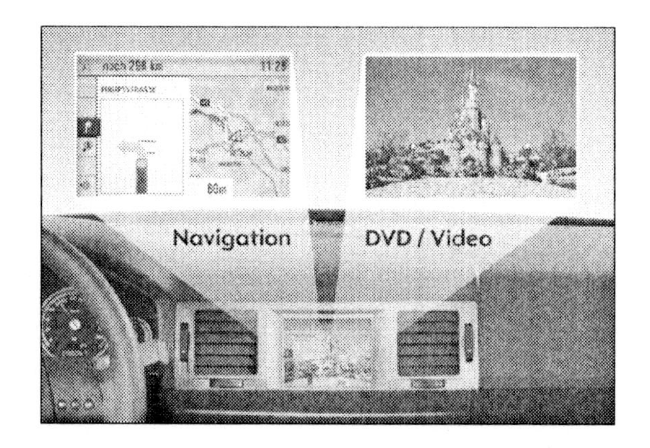

of **augmented** or "fractal **subjectivity**" (Haraway, 1991, p. 193).

Even though over 20 years ago **Paul Virilio** described how vehicles as perceptual machines were ideal candidates for artificial realities in many of their aspects, he most certainly would never have dreamt of such a vision. In view of this development, **Virilio**'s thesis that the modern world had lost an origin of its spatial and temporal living conditions through the (vehicular) acceleration of transmission, does not seem quite so simplistic, as alleged, among others, by Harvey (1989).

In spite of all the criticism of **Virilio**'s conjectures shaped by the rhetoric of loss, "it is thanks to his seismographic feeling for invisible figurations within the visible, that his view has led to the revealing of construction rules in technical media other than those of their simple functional evidence", as Tholen (2002, p. 103) quite rightly notes. **Virilio**'s telematic media perception is enlightening in the sense that it can be applied well to navigation systems.

Virilio already noted that the passenger fixed in the car seat is offered a double view that increases the perceptual effect of his vectorial velocity—the puncture of space. **Virilio**'s language stylizes this into the art of the dashboard (Virilio, 1998, p. 12). The driver is fixed between the dashboard and the "screen" presented by the windshield, on which a reality that has become unreal in the real sense of the word is taking place like a film. The proximity of the video game to driving is obvious. While the landscape disappears of its own accord on the velocity vector, the function of the driver is concentrated purely on maintaining the vector and avoiding obstacles.

Virilio demonstrates this approach in *Open Sky* (1997), using the car windscreen as a paradigm for a cinema screen that is interpreted as an artificial projection of car and driver. The apocalyptic vision of "the aesthetics of disappearance" is closely linked to this aesthetic core zone. Humans and landscape disappear more and more from the digital semiotic scene, as velocity becomes an end in itself that is devoid of meaning. This poststructuralist absence is to **Virilio** an absence of the human being who is lost in the 'non-location' of velocity.

Two vectors or movements form the axis of abscissas, in the net of which the configurations of the velocity factory are apparent: acceleration and normalization. **Virilio** calls their geneology, 'dromology'. According to this, the car's success is the normalization of acceleration, the normality of speed is outside human perception and control. The homely sphere of the inner space in cars is one factor that contributes substantially to this. The driver feels safe in the comfort of the seats and surrounded by security systems, control panels, mirrors and windscreen. High speeds can become normal in this secure space. The dynamic localization also contributes considerably to the normalization of acceleration, in that the navigation display indicates a correct functioning of technology through the constant approaching of the destination.

Means of transport are disorienting. Navigation systems can counteract the "disappearance of detail in the world through the shimmer of speed" (Virilio, 1992, p. 52, own translation), whereby, paradoxically, their orientation performance would not even be possible without movement. In the age of digital technology, movement is an essential prerequisite to orientation in space. This is because, in contrast to analogue maps, the route-dependent orientation of the digital maps can only occur when the navigation system or its user has moved through space (Pammer and Radoczky, 2002, p. 123).

The **Global Positioning System** cannot determine an exact position in the absence of movement. In other words, only movement increases the visual acuity of navigation systems. Indeed, this is a rather generic yet important statement: The act of movement that requires more information to be carried out successfully, produces even more information for the moving perceiving subject. More generally humanity was only able to form spatial concepts through movement (Wunderlich, 1985).

While the user of an analogue map automatically orients the representation of the roads on the

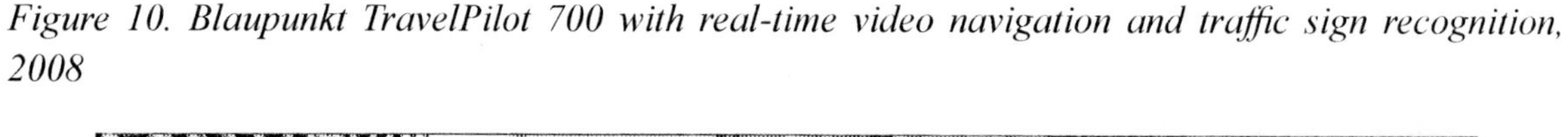

Figure 10. Blaupunkt TravelPilot 700 with real-time video navigation and traffic sign recognition, 2008

map in the direction of the road on which they are located, no solution has yet been found for spatial orientation in digital mobile cartography (Lopau, 2003). Traditional maps therefore have a substantial advantage. They provide much better orientation in the mental representation of the spatial environment. In this process, the cognitive system "as it were, overlays the real road with the representation of the road on the map" (Lopau, 2003, p. 65, author's translation). As such, an 'augmented reality' navigation system no longer differs from that of cognitive mapping in this particular point.

AUGMENTED REALITY NAVIGATION

Even though **Virilio**'s (1997, p. 65) necrology for geography may appear somewhat premature, it is a lasting legacy of his dromological perspective that locations and signs of the digital upheaval in media history were discovered (Tholen, 2002, p. 110). **Virilio**'s "trajectography" (1997, p. 129) questions the prevailing metageographical discourses in cartography, since 'mashuped' location representations are operating with greatly reduced symbolic sign systems (Gordon, 2007). For example, the next generation of Blaupunkt and Continental navigation systems (see figures 10 and 11) dispense with traditional abstract map illustrations, since these may distract and confuse the driver, particularly in complex traffic situations (Wagner, 2005).

Instead, we are once again approaching the vision put forward by Aspen Movie Map. The real-time video navigation system Blaupunkt TravelPilot 700, for instance, presented at the IFA 2008, will represent the environment true to reality. This will be accomplished with an integrated video camera on the back of the device that takes live footage of the road ahead. Blaupunkt's Safe Drive series compares the captured video images with stored street information and incorporates instructions on direction and speed limits—also identified by traffic sign recognition—into the camera-produced pictures of the surrounding area. This process, called '**augmented reality**' (AR),

Figure 11. Continental's 'Augmented Guidance' with head-up display, 2008

gives the driver the same view of the world on the navigation display as they have when looking at the road. The conversion of the abstract map image is no longer necessary. In addition, **augmented reality** in combination with a head-up display can ensure that drivers are looking straight at their route. It is expected that projection onto the windscreen of an integrated video image containing route information will be available for serial installation as early as 2010 (Wagner, 2005).

The constitutional moment of this medium is in fact the constant oscillation between environmental space and two-dimensional projection space. Using the words of Walter Benjamin (1980, p. 18), one could also speak of a "transparent translation of the world that should not obscure the original". On the other hand, a navigation system of this type in combination with a head-up display has the 'advantage' that information which is irrelevant for travel navigation purposes can be masked in its entirety according to the driver's wishes, leaving only pre-defined points of interest visible. From within the car, the world could therefore be perceived only as a map.

Even if car navigation systems no longer use cartographic symbols, the possibility for dis-orientation inherent in these systems remains an ongoing geographic characteristic. Such de-mapping is not unlikely, as cars have access to an interface via navigation systems that permits hackers not only to display data garbage on navigation displays, but also to access the on-board electronics and to cause accidents (Grell, 2005, p. 81).

The new generation of navigation systems that project the travel route onto the windscreen may further result in concrete manifestations of what Edward Soja (1999) describes as the "rise of the perspective of a third space". The driver is himself in the first space and through the windscreen sees a first space that can be experienced physically. Via the head-up display, a second space is simultaneously projected before his eyes as a mental concept of space. These spaces, when overlaid and integrated into each other, represent something like a 'both/and' instead of an 'either/or' through this hybridity, mobility and simultaneity. Such a complex understanding of space opens up new spaces for exploration.

CONCLUSION: AUTOMATIC PRODUCTION OF AUGMENTED SPACES

One of the main effects of '**travel scouting**', '**connected navigation**' and '**augmented reality** navigation systems' is that they create new socio-spatial arrangements—they automatically produce space by actively shaping urban environments, vehicle handling, and driver behaviour inherently as a grammar of action (Dodge & Kitchin, 2006). Here, space is constantly brought into being through navigation practices. Space is endlessly re-created in the moment, it is "a continuous process of matter and meaning taking form as divergent realities [...] constantly come into contact to create new conditions" (Dodge & Kitchin, 2005, p. 178).

Thus, navigation systems make a difference because they alternatively modulate the form, function and meaning of space, producing **augmented spaces** in new ways: They constitute what Graham (2005) refers to as "software-sorted geographies". Navigation systems regulate the access of vehicles to certain spaces. On the one hand they perform as personal navigation aids augmenting the driver's knowledge to find the most efficient route between locations or the least deficient point of interest. On the other, they are disciplining devices leading to a standardization of the tourist gaze. In both cases, they influence how space is brought into being by changing how the 'experience of the city' is performed.[4]

The significantly augmented 'spatial turn' of car navigation systems is overlaying material space with dynamically changing immaterial information. The novelty is the real-time alteration of data, the convergence of different media contents and its personalization. Even if one believes that route guidance can be re-invented with "**connected navigation**" (ABI Research, 2007) or "**travel scouting**" (iPublish, 2006, 2007), it should be remembered that applications for matching address fields to coordinates, for showing tourist attractions and local businesses, for enhanced *Yellow Pages* as well as aids to marketing, field asset management and travel itinerary functions were already integrated into the very first car navigation systems, such as the *ETAK Navigator* (Zavoli & Honey, 1986, p. 362)

However, the 'new' **augmented map** shifts from a means of controlling networks to a means of controlling life immersed in networks (Gordon, 2007). Navigation systems are intimately involved in the '**remediations**' of place- and space-based social worlds. Far from being separated domains, their augmented perspectives underline that the coded worlds of the 'virtual' work to continually constitute the place-based practice of the physical world. Castells (1996, p. 373) calls this the shift from 'virtual reality' to a 'real virtuality'. The guided vehicle in motion is no longer an extension of the body, but the very embodied sensation.

REFERENCES

ABI Research (Ed.). (2007): *Consumer Navigation Devices and Systems. The Changing Dynamics of OEM, Aftermarket, PND, and Handset-based Navigation*. Brochure, New York. Retrieved February 15, 2008, from http://www.abiresearch.com/products/market_research/Consumer_Navigation_Systems_and_Devices

Amin, A. & Thrift, N. (2002). *Cities: Reimagining the Urban*. Cambridge: Polity Press.

Arikawa, M., Konomi, S. & Ohnishi, K. (2007). Navitime: Supporting Pedestrian Navigation in the Real World. *IEEE Pervasive Computing*, 6(3), 21-29.

Benjamin, W. (1980). Die Aufgabe des Übersetzers (1921). In: W. Benjamin, *Gesammelte Schriften, Bd. IV.1* (pp. 9-21), edited by R. Tiedemann. Frankfurt/M: Suhrkamp.

Berland, J. (1996). Mapping Space: Imaging Technologies and the Planetary Body. In: S. Aronowitz & B. Martinsons (Eds.), *Technoscience and Cyberculture* (pp. 123-138). New York: Routledge.

Bolt, R. A. (1984). *The Human Interface: Where People and Computers Meet*. Belmont, CA: Lifetime Learning Publications.

Brand, S. (1988). *The Media Lab. Inventing the Future at MIT*. London: Penguin Books.

Brown, B. & O'Hara, K. (2003). Place as a Practical Concern of Mobile Workers. *Environment and Planning A*, 35(9), 1565-1587.

Castells, M. (1996). *The Rise of the Network Society*. Oxford: Blackwell.

Crang, M. (1998). *Cultural Geography*. London: Routledge.

Dodge, M. & Kitchin, R. (2005). Code and the Transduction of Space. *Annals of the Association of American Geographers*, 95(1), 162-180.

Dodge, M. & Kitchin, R. (2006). *Code, Vehicles and Governmentality: The Automatic Production of Driving Spaces*. NIRSA Working Paper, No. 29. Maynooth: National University of Ireland.

French, R. L. (2006). Maps on Wheels: The Evolution of Intelligent Automobile Navigation. In: J. R. Akerman (Ed.), *Cartographies of Travel and Navigation* (pp. 260-290). Chicago: University of Chicago Press.

Gordon, E. (2007). Mapping Digital Networks: From Cyberspace to Google. *Information, Communication & Society*, 10(6), 885-901.

Graham, S. (2004). Introduction: From Dreams of Transcendence to the 'Remediation' of Urban Space. In: S. Graham (Ed.), *The Cybercities Reader* (pp. 1-29). London: Routledge.

Graham, S. (2005). Software-sorted Geographies. *Progress in Human Geography*, 29(5), 562-580.

Grell, D. (2005, April 4). Tacho-Tüfteln. Was tun gegen die Kilometerstandfälscher. *C't*, 23(8), 78-81.

Haraway, D. (1991). *Simians, Cyborgs and Women*. New York: Routledge.

Harvey, D. (1989). *The Condition of Postmodernity: An Enquiry into the Origins of Cultural Change*. Oxford: Blackwell.

iPublish (Ed.). (2006). *Ziele, Wege, Attraktionen. Das kleine Zauberbuch der virtuellen Reiseführung*. Booklet, München: Ganske Interactive Publishing.

iPublish (Ed.). (2007). *Merian scout Navigator. Innovation by Transformation*. Booklet, München: Ganske Interactive Publishing. Retrieved February 15, 2008, from http://www.merian.de/merianscout/navigationssystem/pta/merianscoutnavigator_brochure_e.pdf

Kant, I. (2003). *Kritik der reinen Vernunft*. Stuttgart: Fourier [Original (1781). Riga: Hartknoch].

Kilian, J. (2008, May 14). The emotional side of design: Merian scout navigator. Presentation, *Where 2.0 Conference*. Burlingame, CA. Retrieved September 15, 2008, from http://assets.en.oreilly.com/1/event/4/Merian%20Case%20Study%20Presentation.ppt

Latour, B. (2005). *Reassembling the Social: An Introduction to Actor-Network-Theory*. Oxford University Press: Oxford/New York.

Laurier, E. (2004). Doing Office Work on the Motorway. *Theory, Culture & Society*, 21(4/5), 261-277.

Lopau, W. (2003). Die räumliche Orientierung von mobilen Displaygeräten. *Kartographische Nachrichten*, 53(2), 64-67.

Manovich, L. (2001). *The Language of New Media*. Cambridge, MA: MIT Press.

Manovich, L. (2006). The Poetics of Augmented Space. *Visual Communications*, 5(2), 219-240.

Mays, E. (2006, February 3). VW to Use Google Earth for the Best in Navigation Technology. Retrieved February 15, 2008, from http://www.autoblog.com/2006/02/03/vw-to-use-google-earth-for-the-best-in-navigation-technology/

Naimark, M. (1997). 3D Moviemap and a 3D Panorama. *SPIE Proceedings*, 3012, 297-305, Retrieved February 15, 2008, from http://www.naimark.net/writing/spie97.html

Pammer, A. & Radoczky, V. (2002). Multimediale Konzepte für mobile kartenbasierte Fußgängernavigationssysteme. In: A. Zipf & J. Strobl (Eds.), *Geoinformation mobil* (pp. 117-126). Heidelberg: Wichmann.

Parks, L. (2005). *Cultures in Orbit. Satellites and the Televisual*. Durham/London: Duke University Press.

Rose, G. (1993). *Feminism and Geography. The Limits of Geographical Knowledge*. Minneapolis, MN: University of Minnesota Press.

Siemens VDO Automotive (Ed.). (2005, December 6): A New Dimension of Navigation: Photo-quality Three-dimensional Map Display from Siemens VDO. Press Release, Schwalbach/Shanghai.

Soja, E. (1999). Thirdspace: Expanding the Scope of the Geographical Imagination. In: D. Massey, J. Allen & P. Sarre (Eds.), *Human Geography Today* (pp. 260-278). Cambridge: Polity Press.

Telematics Research Group (Ed.). (2006, July 24). Is Live Traffic a Telematics Killer App! Press Release, Minnetonka, MN. Retrieved February 15, 2008, from http://www.telematicsresearch.com/PDFs/TRGpress072406.pdf

Thielmann, T. (2008). Der ETAK Navigator. *Tour de Latour* durch die Mediengeschichte der Autonavigationssysteme. In: G. Kneer, M. Schroer & E. Schüttpelz (Eds.), *Bruno Latours Kollektive. Kontroversen zur Entgrenzung des Sozialen* (pp. 180-218). Frankfurt/M: Suhrkamp.

Tholen, G. C. (2002). *Die Zäsur der Medien*. Frankfurt/M.: Suhrkamp.

Thrift, N. & French, S. (2002). The Automatic Production of Space. *Transactions of the Institute of British Geographers*, 27(4), 309-335.

Thrift, N. (2004). Driving in the City. *Theory, Culture and Society*, 21, 41-59.

Urry, J. (1990). *The Tourist Gaze. Leisure and Travel in Contemporary Societies*. London: Sage Publications.

Virilio, P. (1992). Fahrzeug. In: K. Barck, P. Gente, H. Paris & S. Richter (Eds.), Aisthesis. Wahrnehmung heute oder Perspektiven einer anderen Ästhetik (pp. 47-72). Leipzig: Reclam. [Original: Virilio, P. (1975). Véhiculaire. In: J. Berque et al. (Eds.). *Nomades et vagabonds*. Paris: Union générale d'éditions.]

Virilio, P. (1997): *Open Sky*. London: Verso.

Virilio, P. (1998). Dromoscopy, or The Ecstasy of Enormities. *Wide Angle*, 20(3), 11-22.

Wagner, F. (2005, January 30): Navigationsgeräte können künftig sehen. Retrieved February 15, 2008, from http://www.handelsblatt.com/news/printpage.aspx?_p=205913&_t=ftprint&_b=853116

Wunderlich, D. (1985). Raumkonzepte. Zur Semantik der lokalen Präpositionen. In: T. T. Ballmer & R. Posener (Eds.), *Nach-Chomskysche Linguistik* (pp. 340-351). Berlin/New York: de Gruyter.

Zavoli, W. B. & Honey, S. K. (1986). Map Matching Augmented Dead Reckoning. *IEEE Vehicular Technology Conference Proceedings*, 36, 359-362.

General Motors Europe (Ed.). (2005, August 18): Opel Presents Innovations for Tomorrow's Mobility. Press Release, Glattbrugg/Rüsselsheim.

KEY TERMS

Augmented Reality: Is a field of information technology research which represents the possibility of illustrating real and virtual images together. The goal of augmented reality is to add information and meaning to a real object or place. Unlike virtual reality, augmented reality does not create a simulation of reality. Augmented reality is an environment that includes both virtual reality and real-world elements.

Cognitive Mapping: The term 'cognitive mapping' used in cognitive psychology refers to the process of perception of spatial orders in human consciousness. Cognitive maps reflect the world in the way a human thinks it is, without any pretence to correctness. The perspective of the observer is responsible for any aberrations.

Dromology: Dromology is derived from the Greek 'dromos': avenue or race course. The theory of dromology interprets the world and reality as a resultant of velocity. In Paul Virilio's 1977 essay entitled "Speed and Politics", the french philosopher makes a compelling case for an interpretation of history, politics and society in the context of speed. Extending the definition of "dromomaniacs", Virilio argues that speed became the sole agent and measure of progress. He contends, that "there was no 'industrial revolution', only 'dromocratic revolution'; there is no democracy, only dromocracy; there is no strategy, only dromology."

GPS: The Global Positioning System (GPS) is a satellite-based navigation system made up of a network of 24 Navstar satellites placed into orbit by the U.S. Department of Defense. GPS satellite launches started in 1978, and second-generation satellites were launched beginning in 1989. The system became fully operational in 1995, with a signal for military users and a less-accurate signal for civilians.

Head-Up Display: A head-up display is an optical system that superimposes a synthetic display providing navigational information on a driver's or pilot's field of view. Although they were initially developed for military aviation, head-up displays are now used in commercial aircraft, automobiles, and other applications.

Navigation System: A navigation system is a device that has the capability of knowing your current position, and allows you to determine your destination. Today's navigation systems use Global Positioning Satellites (GPS) to pinpoint people's or vehicle's location, compare it with the sought-after destination and guide along the selected route. Before GPS, car navigation was provided by dead reckoning. In these systems the measurements from wheel and compass sensors were combined to determine a sequence of positions which in turn was compared to a map database.

Travel Scouting: Terms the modern way of traveling using a virtual travel guide that identifies its specific location and provides corresponding information on tourist offers, sights and insider tips relevant to the area. The program adjusts itself to the user's situation (road journey, city tour, castle visit etc.) and requirements (information, background reports, recommendations etc.).

ENDNOTES

1. See the 1981 video of the *Aspen Movie Map*. The demo is available under the link: http://www.media.mit.edu/speech/sig_videos.html (retrieved February 15, 2008). Even though the *Aspen Movie Map* was created as a simulator, the navigation of which imitated a real car drive, it was not intended as a prototype for car navigation systems. The project was rather more a by-product of the interest in an airport setup true to the original, in which the Israelis had practised freeing the Entebbe airplane hostages in 1973. The Pentagon had asked whether it was possible to simulate such "experimental mapping" on a computer (Brand, 1988:141).
2. The image produced by a head-up display is reflected onto the windscreen. Just like when looking in the mirror, the driver does not see the image on the surface of the windshield, but floating freely over the hood, at a distance of approximately two meters.
3. Kant's transcendental philosophy (1781/2003) already points out that „Medialität" (the properties and forces generated by a medium), space and physicality are inseparably connected to each other.
4. Driving can be understand as a reconfiguration of de Certeau's consideration of the action and art of walking that configures a certain experience of the city (Thrift 2004).

Chapter XVII
QyoroView:
Creating a Large-Scale Street View as User-Generated Content

Daisuke Tamada
Osaka University, Japan

Hideyuki Nakanishi
Osaka University, Japan

ABSTRACT

A lot of street view services, which present views of urban landscapes, have recently appeared. The conventional method for making street views requires on-vehicle cameras. We propose a new method, which relies on people who voluntarily take photos of an urban landscape. We have developed a system called QyoroView. The system receives photos from users, adjusts the photos' position and orientation, and finally synthesizes them to generate a street view. We conducted two experiments in which the subjects generated a street view using our system. We also observed and interviewed the subjects who participated in order to learn their impression of the system.

INTRODUCTION

Satellite or aerial photos can show users only rough geographical information. Street views can provide detailed geographical information. Street views are panoramic images of urban landscapes that are made from movies or a collection of photos taken along the streets of a city. The very first interactive street view is 'Aspen Movie Map' (Lippman,1980). Recently, various kinds of Websites that provide street views have appeared, such as Google Maps Street View (maps.google.com/help/maps/streetview) and Microsoft Photosynth (labs.live.com/photosynth). Street views can be used for urban navigation, real estate property purchases, land surveying.

Cameras mounted onto cars are generally used to record movies or collect photos to create these street views. This makes it necessary to drive the cars for long distances. Since special equipment and considerable labor are required to produce street views, only large enterprises are usually able to offer large-scale street views.

The amount of user-generated content (UGC), that is, media content created by end users, is rapidly increasing. UGC has become vital in today's daily lifestyles (Wunsch-Vincent, & Vickery, 2007, p. 9). Even if each user creates only a small piece of content, the quantity of the total amount of content sometimes exceeds that created by professionals (Lih, 2004, p. 5).

We developed a system called 'QyoroView,' that can produce a large-scale street view as UGC. QyoroView can synthesize street views from mobile phone photos that contain position data. We assume that typical QyoroView users will be pedestrians who have a GPS camera phone. In Japan, government required "All 3G mobile phones released after 2007 must, in principle, be equipped with GPS." When pedestrians take and upload photos of buildings along streets, QyoroView updates the street view images so they can be viewed using a Web browser. Since QyoroView requires no extra special equipment, it is able to produce street views as UGC.

BACKGROUND

There are several approaches to producing street views. The basic style of this content is a hand-crafted panoramic image (Chen, 1995). An enormous amount of effort is required to construct wide-area panoramic images (Hartono et al., 2006). Using this approach, it is possible to produce street views of some spots, but it is nearly impossible to produce views along streets.

Past studies proposed several approaches to produce street views automatically using on-vehicle equipment. A common technique for collecting panoramic images of many spots along streets has been to use on-vehicle omnidirectional cameras (Koizumi, & Ishiguro, 2005; maps.google.com/help/maps/streetview; preview.local.live.com). A previous study used on-vehicle laser scanners to obtain 3D models of buildings along streets (Fruh, & Zakhor, 2001). Another previous study used on-vehicle cameras mounted to the side of a car to capture panoramic images of buildings along streets (Roman, Garg, & Levoy, 2004). These approaches require special equipment and long-distance driving. Thus, it is costly to produce wide-area street views.

'Photo Tourism' is a system that can make street views from a collection of photos that are freely available in photo sharing sites (Snavely, Seitz, & Szeliski, 2006; labs.live.com/photosynth). So, it requires no equipment or labor to produce street views. However, a sufficient number of photos recording the same location are needed to concatenate photos using image processing technology. Many sightseers take photos at tourist attractions and some of them upload their photos to photo sharing sites, so this approach is applicable to the production of street views of such places. It is, however, not a good idea to apply the approach to the production of street views of featureless urban areas due to the lack of motivation for users to take photos.

'OpenStreetMap' is a UGC system that makes freely editable maps. The map data in the system has mostly been built by users who upload data from their phones or PDAs equipped with GPS devices (openstreetmap.com)]. Recently, Yahoo! allowed OpenStreetMap to use their vertical aerial imagery. So, users can create vector-based maps using online editing tools that can overlay aerial images. The system produces only vector-based maps of urban streets.

There are many services, including 'Panoramio', that have a photo map function that places photos uploaded from end users onto a map (www.panoramio.com). Some projects have tried to extend this basic concept. 'Balog' is a Web-based diary system combined with a photo map function (Uematsu et al., 2004). Some researchers used multimedia projectors to display a large photo map on the floor as installation art (Nakanishi, Motoe, & Matsukawa, 2004). The 'Degree Confluence' project is trying to create a world-wide photo map by collecting photos of integer degree intersections for latitudes and longitudes around the world (www.confluence.org). None of the photo map systems, including those we have described here, produce panoramic images.

QyoroView can produce a large-scale street view as UGC. As discussed above, no conventional approach can do that. On-vehicle equipment enables the production of wide-area street views, but such equipment is not very affordable for end users. The Photo Tourism system can produce street views as UGC, but the system is not applicable to wide-area street views. Photo map systems can collect photos for a wide area, but they do not produce panoramic images.

SYSTEM DESIGN

QyoroView synthesizes street views from mobile phone photos. To make street views, the system needs position data and orientation data for the photos. Some recent mobile phones are equipped with GPS. GpsOne is one of the most accurate positioning services for mobile phones. Although the error rate of GpsOne is five to ten meters even when the device catches signals from more than three satellites (Wang et al., 2004, p. 342), it is still difficult to detect the side of the street on which a user is standing. Some mobile phones are also equipped with a magnetic compass. Magnetic compasses are affected by surrounding magnetic fields. Mobile phones themselves disturb magnetic compasses because the phones generate a magnetic field. Therefore, magnetic compasses are too inaccurate to orient photos. With QyoroView, users modify the positions and the system of QyoroView modifies the orientations of photos. This section explains how QyoroView collects and concatenates mobile phone photos, and how it was designed and implemented.

Collecting Mobile Phone Photos

Figure 1 shows how to take photos based on the standard user interface that is available on almost all of the GPS camera phones released in Japan. We assume that typical users of QyoroView will be pedestrians who have a GPS camera phone. In the figure, the user's phone obtains the current position and displays a map around it. The icon, which indicates the position sent to the server side system of QyoroView, is initially located at the center of the map (Step 1 in Figure). The first task for the user is to adjust the icon's location. The user moves the icon to a target on the other side of the street, which is usually a building (Step 2). Then, the user takes a photo of the target (Step 3).

If the user intends to upload only a single photo, the final task is to send the system an e-mail message that contains the photo and the position data. We have also prepared a more efficient way. The user can upload several photos at once, as long as the street is not very curved. As shown in Figure 1, after taking the first photo, the user walks along the street for a certain distance (Step 4) and takes a second photo (Step 5). This time the user does not need to adjust the location of the icon. The user can take a third, fourth, and more photos in the same way. Only when the user takes the last photo does he or she have to move the icon to the target on the other side of the street in the same way as when taking the first photo. Finally, the user sends the system an e-mail message that contains the photos and the data of the first and last positions. The positions of the intermediate photos between the first and last photos will be calculated by the system.

The user needs to be aware of two things to help the system smoothly concatenate photos. The first thing is that the user should not tilt the phone. Different angles of the phone result in different

Figure 1. How to take photos

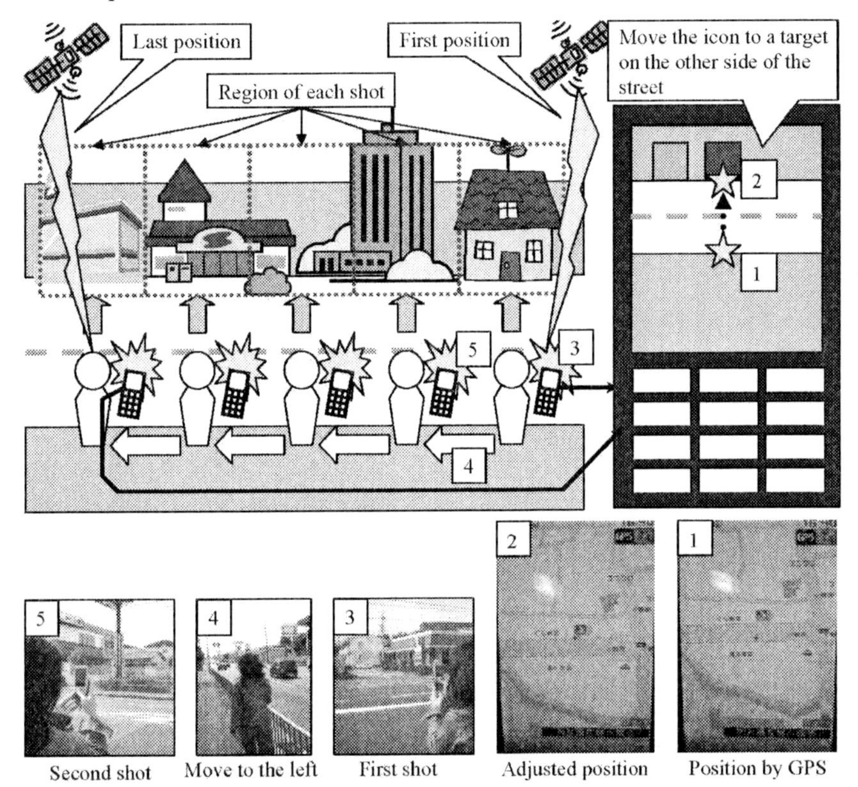

altitudes of the horizon in the photos. To smoothly concatenate photos taken by different users, all users need to keep the phone at the same angle, which is perpendicular to the horizontal plane in the usual case. Another requirement is that the walking distance between each shot should not be too long. If the distance is too long, a missing area will appear between adjacent photos.

Synthesizing Street Views

When the server side system of QyoroView receives the e-mail message that contains the photos and the data of the first and last positions, the system calculates the positions of the intermediate photos between the first and last photos based on a linear interpolation method. As a result, every photo is accompanied by its position. At this time, the photos do not have orientation, and may not stand along the street very precisely due to the user's inaccurate adjustment of the positions, and also due to the curving of the street. Therefore, as shown in Figure 2, the system adjusts the position and determines the orientation of each photo based on a vector map, which contains the border and halfway vectors of streets.

1. Adjusting the Position

The system finds the border vector that is closest to the photo's original position (Step 1 in Figure 2). Then, the system changes the photo's position to a new position that is on the border vector and that is also closest to the original position (Step 2).

2. Determining the Orientation

Since the orientation of the photo must be perpendicular to the border vector on which the photo is placed, there are only two choices for the orienta-

Figure 2. Placing module

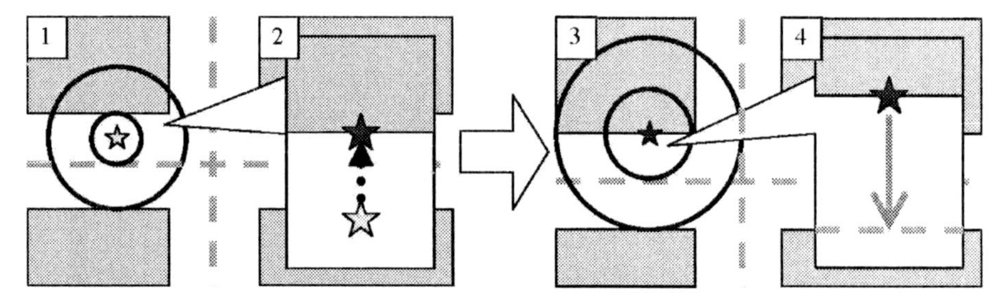

tion. The system finds the halfway vector that is closest to the photo's new position (Step 3). Then, the system picks up the orientation that aims the photo toward the halfway vector (Step 4).

Figure 3 illustrates how uploaded photos are concatenated and pasted along the street. Currently, the street view produced by the system is just a collection of photos pasted on a monochromatic street map that is drawn based only on the vector map.

Figure 4 shows the snapshots of the client side system of QyoroView, which is an interactive map running on a Web browser. You can scroll the map by drag operation, and zoom in and out by rotating the mouse wheel. The figure shows the map in three zoom levels. These functions are similar to conventional two-dimensional map services. You can also rotate the map. Clicking the right mouse button rotates the map ninety degrees counterclockwise. Usually, only three-dimensional map services have this kind of function. QyoroView is basically a two-dimensional map service, but the rotating function is essential, because it is not easy to recognize scenes of upside-down photos.

System Architecture

Figure 5 depicts the architecture of QyoroView. The server side system of QyoroView consists of four modules, three databases, and two data folders. Each module works as follows.

1. Receiving Module

The receiving module parses the e-mail messages sent from the user's mobile phone to retrieve the photos and position data. Then, the module stores the photo images in the photo data folder and stores the indices to the photo images, the position data, the sender's name, and a time stamp in the e-mail database. The e-mail software of mobile phones usually has some limitations. Examples include a maximum number of files and position data that can be attached to a single message. So, the system allows users to send a series of data, which are the photos and the data of the first and last positions, via multiple e-mail messages. To enable the reconstruction of this series of data, the module stores the sender's name and a timestamp for each e-mail message.

2. Positioning Module

The positioning module reads the position data and the indices to the photo images of the same sender from the e-mail database. Then the module reconstructs the series of data, which are the ordered photo indices and the data of the first and last positions, according to the timestamps. After the reconstruction, as explained in the previous subsection, the module calculates the positions of the intermediate photos between the first and last photos based on a linear interpolation method. Finally, in the position database, the module stores the photo indices, each of which is accompanied by its position data.

Figure 3. Drawing module

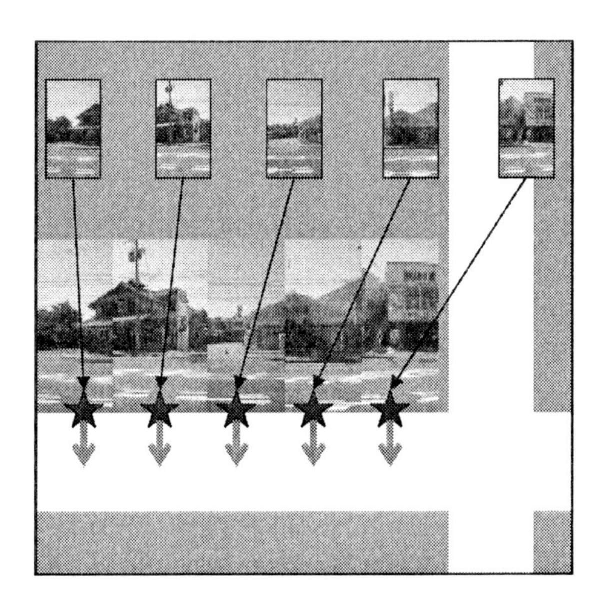

Figure 4. Client side system

Figure 5. Architecture

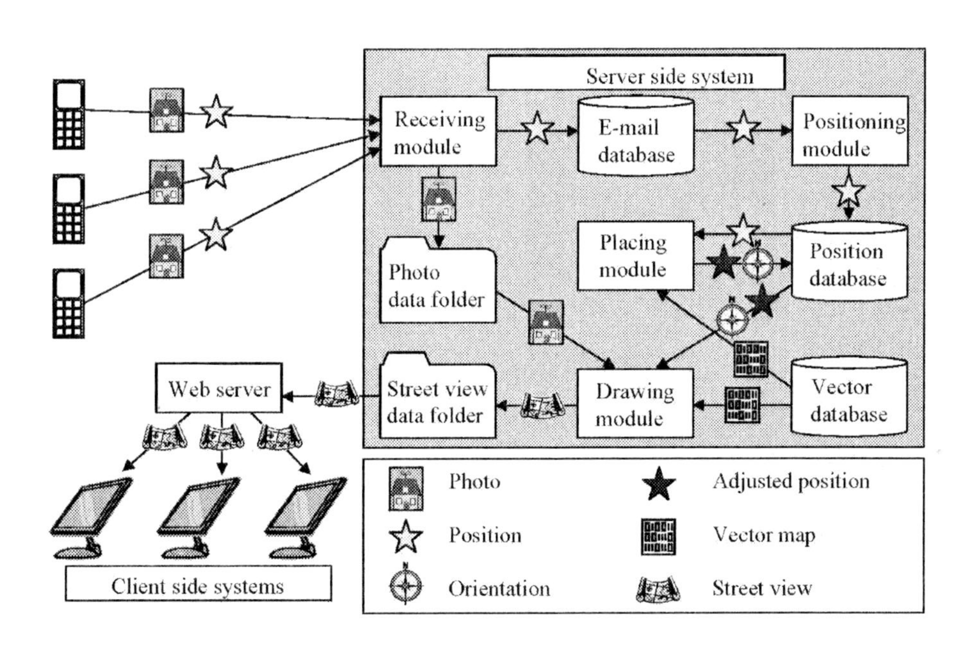

3. Placing Module

The placing module reads the position data which have been added to the position database by the positioning module. The module also reads a vector map around the positions from the vector map database. Then, as shown in Figure 2 and detailed in the previous section, the module adjusts the position and determines the orientation of each photo based on the vector map. Finally, the module stores the adjusted position data and the orientation data in the position database.

4. Drawing Module

In the street view data folder, the street view image is divided into blocks of a fixed size (e.g., 500 pixels square) and stored. When the drawing module updates the street view, the module again produces the blocks that are changed by the new photos. First, the module reads the vector map from the vector map database and redraws the monochromatic street map of the blocks based on the vector map. Next, the module reads the photo images from the photo data folder, and reads the photos' adjusted position data and orientation data from the position database. Then, as shown in Figure 3, the module pastes the photo images on the blocks according to the position and orientation data. Finally, the module overwrites the old blocks in the street view data folder with the produced blocks.

As explained above, the module does not just paste the new photos onto the old blocks, but produces new blocks, because the images of the blocks are stored in JPEG format to keep the file size small. If the module were to read the blocks from the data folder, paste new photos on the blocks, and store the blocks in the data folder again, the quality of the blocks' images would be degraded.

The client side system of QyoroView requests the street view blocks that are necessary to display the map from the Web server of the QyoroView service. The parameters sent to the Web server contain the location, zoom ratio, and rotation degree of the map. Mouse operation on the map determines these parameters and causes a new request to be sent to the Web server.

System Implementation

The client side system of QyoroView is an Ajax-based application. The server side system of QyoroView runs on Debian GNU/Linux 4.0. The modules were implemented in Java 1.5. The databases were implemented in PostgreSQL 7.4. Apache 2.2 is used as the Web server.

Digital Map 2500 (Spatial Data Framework) published by the Geographical Survey Institute of Japan is used as the vector map. The '2500' means that the vector data are in scale of 1/2500.

Currently, the receiving module can deal with e-mail messages sent from KDDI's 'au' mobile phones with the 'EZ Navi Walk' service, since KDDI's phones make up the majority of GPS phones in Japan. KDDI's 'au' is the second largest mobile phone network in Japan and about ninety percent of 'au' users have mobile phones that are equipped with GPS. We are also working on adapting the module to other mobile phone networks, since GPS phones for other networks are appearing.

EXPERIMENT 1

We compared the two methods: the standard method of uploading only a single photo (one-shot capture) and the continuous capture method of uploading several photos at once. We asked three students of our research group to collect photos from around a residential area close to our university campus. The subjects collected photos by the one-shot and continuous methods.

Results 1

Table 1 shows how long it took to take photos with both methods. The table shows that the continuous capture method is more than twice as fast as the one-shot capture method. Figure 6 shows a part of street views produced by the continuous capture method. It took about three and a half hours to collect the photos included in the figure's area.

It is apparently not common to take pictures of buildings from the other side of the street. The subjects reported that it was especially awkward to take pictures of personal houses, elementary schools, and kindergartens. Interestingly, they reported that it was not awkward to take pictures of strange-looking buildings that did not fit with the surrounding buildings.

In this experiment, subjects discussed at laboratory to divide areas where they should take photos. They seemed to think it would be better to cooperate with each other. In the real case of creating UGC, users may need interaction with others. So, we conducted second experiment.

Table 1. Comparing photo-taking times

	One-shot	Continuous
Total number of pictures	29	428
Total time	50 min	5 hour 31 min
Average time per image	1 min 44 sec	46.5 sec

Figure 6. A part of street views

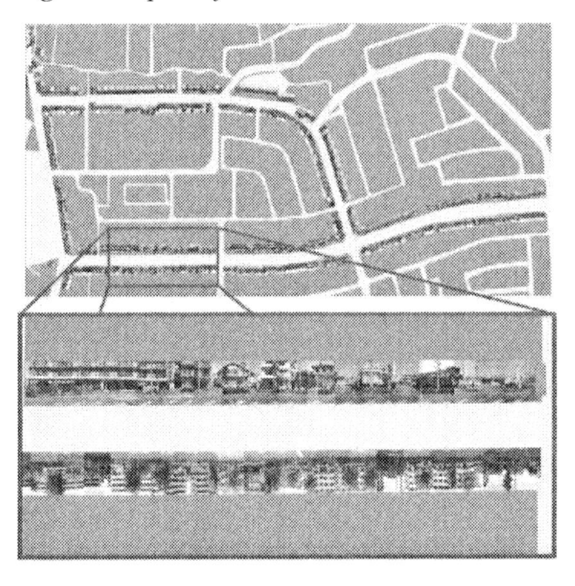

EXPERIMENT 2

We conducted the second experiment to observe how subjects participated in making street views. In this observation we tried to determine how we should support interaction among users to facilitate the production of street views. As listed in Table 2, the subjects consisted of eight undergraduate students (four male students and four female students). They did not begin taking photos on the same day. Two subjects began in August. One of them retired from the experiment on August 5. The left map of Figure 7 shows the location of the two campuses of our university. Five students lived near the west campus, and the other three students lived near the east campus. The subjects constructed street views of a loop road that includes the two campuses. As a simple incentive, we paid a fixed amount of money for each uploaded photo. We also paid them for transportation expenses to visit the area to take photos, and communication expenses to upload photos. We told them that we could not pay for duplicated photos, meaning that they could not earn money if they took photos at a part of the street where other subjects had already taken photos.

Table 2. Subjects of the second experiment

Subject	Gender	Campus	Photos	Begin	End
A	Male	West	861	7/24	8/5
B	Male	West	1898	7/25	8/30
C	Male	East	320	8/12	8/30
D	Female	East	153	7/24	8/30
E	Female	East	139	8/13	8/30
F	Female	West	116	7/27	8/30
G	Female	West	718	7/23	8/30
H	Male	West	241	7/25	8/30

Figure 7. Street view constructed in the experiment

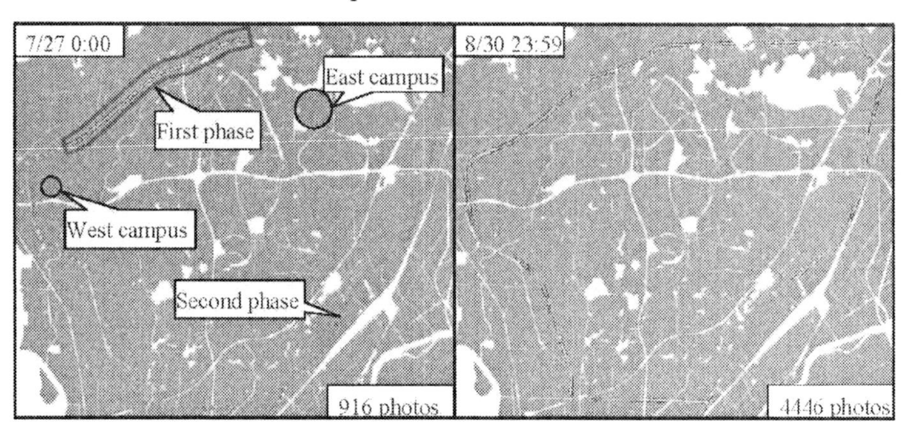

The experiment consisted of two phases. In the first phase the subjects were asked to construct street views of only the north part of the loop road (See the left map of Figure 7). The length of this part was about 5.5 kilometers. In the second phase they were asked to construct street views of the entire loop road, the length of which is about 32 kilometers. The first phase began on July 23, 2007. The first phase ended and the second phase began on July 27, 2007. The second phase ended on August 30, 2007. Since the experiment was conducted in summer, it was very hot and bright around the loop road. So, the subjects tended to take photos during the morning.

We prepared a mailing list for communicating with other subjects. The message log of the mailing list was a resource to tell us what kind of information sharing was needed to construct street views.

We interviewed the subjects on August 30 when the experiment finished. The purpose of this interview was to explore ways to improve QyoroView.

Results 2

This subsection describes the results of analyzing the three kinds of data described above.

System Log

In Figure 7 you can see how the photo taking of the loop road proceeded during the experiment. The figure shows the street view and the number of uploaded photos on July 27 and August 30. July 27 was the day when the message log (described later) was obtained. August 30 was the day when the experiment finished.

The experiment was successful. A total of 4,446 photos were taken by all of the subjects, which covered about three-fourths of the approximately 32-kilometer loop road. However, we learned that we should have been more aware of the distances between the subjects' living areas and the targeted area, i.e., the loop road. It seemed much better to involve subjects whose living areas were spread over the targeted areas.

Message Log

The following exchange of messages occurred on July 27, which was the first day of the second phase. The names of streets, intersections, and stations have been replaced with alphabetical letters.

Message 1 sent from subject G at 9:37 a.m.

I am now at intersection P. I will walk down street Q toward the south. I'm going to take the photos of the east side of street Q. Let me know if you are taking the photos of that part.

Message 2 sent from subject F at 9:40 a.m.

I joined the experiment today. I'm going to begin taking photos near station R and end up taking photos around the city office. I'll proceed toward the south more if I have time.

Message 3 sent from subject H at 11:09 a.m.

I'm going to take photos of the part of street S between station T and intersection U. Let's everybody take many photos, although it's hot again today.

Message 4 sent from subject G at 11:14 a.m.

I finished taking photos of the part of street Q's east side between intersection P and intersection V. Traffic is very congested, so I'm giving up taking photos now. Good luck to everybody who's beginning now!

Message 5 sent from subject F at 11:21 a.m.

I finished taking photos of the part of street Q's east side between station R and the post office. I have to stop because I need to prepare for term-end exams. Good luck to everybody who's beginning now!

Message 6 sent from subject H at 15:03 p.m.

I finished taking photos of the part of street S between intersection U and intersection W. That part includes a crossing with an overpass where I did not take photos.

As shown in messages 1, 2, and 3, the mailing list was used to notify the part of the road where the subjects were going to take photos. After the experiment we asked subjects F, G, and H about how they decided the area to take photos on July 27. In the interview we found that the notifications were able to successfully coordinate the subjects' work. Subject F had planned to begin at intersection P, proceed to the south, and end around station X. But she decided to take photos at the southern area due to message 1, which she received on her mobile phone just after leaving her home. She moved to station R, which was south of station X, and began taking photos. Subject H found that subjects F and G were taking photos of the area around his home when he read messages 1 and 2. So, he decided to take photos at an area that was a little distant from his home.

As shown in messages 4, 5, and 6, the mailing list was also used to announce the part of the road where the subjects had finished taking photos. In the interview they all reported that they carefully chose the spot to end their photo taking. They thought it would be easy for others to take photos after their work if the area they finished was clearly recognizable, e.g., the area between the two large intersections.

There were many exchanges similar to those described above. The mailing list was used to share the plans and results of the subjects' work. The information about the plans was useful for parallel work being done simultaneously. The information about the results was helpful for the work that occurred later.

Interview Results

We selected subject answers that could be categorized as navigation and situations, each of which represents an aspect that may lead to a more efficient production of street views.

1. Navigation

Q. "What should the experimenter have done to facilitate taking photos?"

A. "I think I would have taken more photos if the experimenter let us know which part of the street view of the road had not yet been constructed."

A. "I think I would have been more eager to take photos if the experimenter taught us techniques to take them efficiently."

These answers indicate that the experimenter should have guided the subjects more, i.e., by recommending areas to take photos, and by training the subjects in photo-taking.

2. Situations

Q. "What kind of area was easy to take photos in?"

A. "Areas with a lot of shade made by buildings."

A. "Areas that were less congested with traffic."

Q. "Which periods of time were easy to take photos in?"

A. "Periods when it was cool and there were few pedestrians and cars."

These answers implied that photo-taking could have been facilitated by sharing the knowledge of the area and time when it was easy to take photos.

DISCUSSION

The message log suggests that collaborative support tailored to geographical content creation is a key to the successful creation of geographical UGC such as street views. Collaborative support functions are already common in some UGC systems (Lih, 2004). Collaborative support peculiar to geographical content creation has, however, not yet been clarified. There are several previous studies on geographical UGC (Cheverst et al., 2001; Espinoza et al., 2001; Riva, & Toivonen, 2006), but few of them proposed functions to support collaborative content creation. Below, we discuss functions that should be provided by systems for geographical UGC.

The most basic and important collaborative support seems to be a function to avoid conflicts in the area where each participant creates media content. The system should inform participants of areas that have been filled with content, areas that are now under construction or that will enter construction soon, and areas that are still untouched. Furthermore, as described in the navigation category of the interview results, a function to actively allocate areas for participants to work in may be effective. As the system log showed, there should be a function to take the location of each participant's living area into consideration.

As described in the situations category of the interview results, we think that sharing knowledge of the areas and times that are easy to work in would facilitate geographical content creation. We think conventional geographical UGC systems (Cheverst et al., 2001; Espinoza et al., 2001; Riva, & Toivonen, 2006) can be used for such knowledge sharing.

CONCLUSION

We have proposed and implemented a system that can produce large-scale street views as UGC. The system uses vector maps to concatenate photos sent from pedestrians' GPS phones.

We also presented the results of two experiments. In the first experiment, we compared two methods. The bottleneck in our 'one-shot capture' method was the difficulty of users having to take a lot of pictures. We suggested a 'continuous capture' method and confirmed that a 'continuous capture' method could reduce the time needed to collect photos along streets. In this experiment, subjects seemed to cooperate with others. So we conducted the second experiment to determine how we should support interaction among users to facilitate the production of street views. We presented the results of this experiment that suggests collaborative support functions for geographical content creation. Sharing the plans, current status, and achievements of users who are moving around seemed to be effective for coordinating users.

ACKNOWLEDGMENT

The development of QyoroView was supported by Exploratory Software Project of Information-technology Promotion Agency(IPA), Japan. We would like to thank Marcus Foth and three anonymous reviewers for their valuable comments.

REFERENCES

Chen, S.E. (1995). QuickTime VR: An Image-Based Approach to Virtual Environment Navigation, Proc. SIGGRAPH95 (pp. 29-38.). New York, NY: ACM Press.

Cheverst, K., Smith, G., Mitchell, K., Friday, A., & Davies, N. (2001). The Role of Shared Context in Supporting Cooperation between City Visitors. Computers & Graphics, 25(4). 555-562.

Degree Confluence Project: http://www.confluence.org/

Espinoza, F., Persson, P., Sandin, A., Nystrom, H., Cacciatore, E, & Bylund, M. (2001). GeoNotes: Social and Navigational Aspects of Location-Based Information Systems. LNCS 2201 (Ubicomp2001). (pp.2-17).

Fruh, C., & Zakhor, A. (2001) 3D Model Generation for Cities Using Aerial Photographs and Ground Level Laser Scans. Proc. CVPR2001: Vol. 2. (pp. 31-38).

Google Maps Street View: http://maps.google.com/help/maps/streetview/

Hartono, P., Kawasima, T., Nagai, H., Tsuzuki, M., Yatoku, H., Hayasaka, K., Ishigaki, M., Karuki, K, Kikuchi, Y., Saito, K., Sasakawa, S., Takahashi, Y., Tani, M., & Kakita, S. (2006). GigaPixel-Photo Project: A Success Case of Project Learning in Future University-Hakodate. IEICE Technical Report (Institute of Electronics, Information and Communication Engineers), 106(100), 55-59.

Kaasinen, E. (2003). User Needs for Location-Aware Mobile Services. Personal and Ubiquitous Computing, 7(1). 70-79.

Koizumi, S., & Ishiguro, H. (2005). Town Digitizing: Omnidirectional Image-Based Virtual Space, Digital Cities 2001: Vol.3081. Lecture Notes in Computer Science (pp. 247-258). Heidelberg: Springer Berlin.

Lih, A. (2004). Wikipedia as Participatory Journalism: Reliable Sources? Metrics for Evaluating Collaborative Media as a News Resource. Prof. International Symposium on Online Journalism.

Lippman, A. (1980). Movie maps: An application of the optical videodisc to computer graphics. ACM SIGGRAPH Computer Graphics, 14(3). 32-43.

Microsoft Photosynth: http:// labs.live.com/Photosynth/

Nakanishi, Y., Motoe, M., & Matsukawa, S. (2004). JIKUKAN-POEMER: Geographic Information System Using Camera Phone Equipped with GPS, and Its Exhibition on a Street, MobileHCI2004. Vol. 3160. Lecture Notes in Computer Science (pp. 486-490). Heidelberg: Springer Berlin

OpenStreetMap: http://openstreetmap.com/

Panoramio: http://www.panoramio.com/

Riva, O., & Toivonen, S. (2006). A Hybrid Model of Context-Aware Service Provisioning Implemented on Smart Phones. Proc. ICPS2006. (pp. 47-56).

Roman, A., Garg, G., & Levoy, M. (2004). Interactive Design of Multi-Perspective Images for Visualizing Urban Landscapes. Proc. Visualization2004 (pp. 537-544). Washington, DC: IEEE Computer Society.

Snavely, N., Seitz, S.M., & Szeliski, R. (2006). Photo Tourism: Exploring Photo Collections in 3D. ACM Transactions on Graphics, 25(3). 835-846.

Uematsu, H., Numa, K., Tokunaga, T., Ohmukai, I., & Takeda, H. (2004). Balog: Location-Based Information Aggregation System. Poster Proc. ISWC2004.

Wang, K., Yan, L., Wen, H., &He, K. (2004). GpsOne: A new solution to vehicle navigation. Position Location and Navigation Symposium (PLANS2004). (pp.341-346)

Windows Live Technology Preview: http://preview.local.live.com/

Wunsch-Vincent, S., & Vickery, G. (2007). Participative Web: User-Created Content. OECD Work on Digital Content.

KEY TERMS

Border Vector: Digital data of Border line between a road and a City block

Continuous Capture Method: A method of uploading several photos at once

Halfway Vector: Digital data of Center line of a road

One-Shot Capture Method: A method of uploading only a single photo

Photo Map: A map onto which are placed photos uploaded from end users

Street View: Panoramic images of urban landscapes that are made from movies or a collection of photos taken along the streets of a city

User-Generated Content: Media content created by end users

Vector Map: Digital data of maps that is consisted of X-Y coordinates. If you draw lines and polygons with these coordinates, you can create maps

Chapter XVIII
Virtual Cities for Simulating Smart Urban Public Spaces

Hideyuki Nakanishi
Osaka University, Japan

Toru Ishida
Kyoto University, Japan

Satoshi Koizumi
Osaka University, Japan

ABSTRACT

Many research projects have studied various aspects of smart environments including smart rooms, home, and offices. Few projects, however, have studied smart urban public spaces such as smart railway stations and airports due to the lack of an experimental environment. We propose virtual cities as a testbed for examining the design of smart urban public spaces. We developed an intelligent emergency guidance system for subway stations and used the virtual subway station platform to analyze the effects of the system. This experience allows us to argue that simulations in virtual cities are useful to pre-test the design of smart urban public spaces and estimate the possible outcome of real-life scenarios.

INTRODUCTION

Virtual cities are three-dimensional graphical representations of digital cities (Ishida, 2002a). This chapter describes virtual cities for testing smart environments installed in large-scale crowded urban public spaces such as airports and railway stations. Smart environments are living spaces with embedded abilities to perceive what their inhabitants are doing and support their lives. Since living spaces vary with respect to scale, smart environments of various sizes have been developed: smart rooms (Bobick et al., 1999), smart classrooms (Brotherton & Abowd, 2004), smart homes (Kidd et al., 1999), smart offices (Addlesee et al., 2001), and smart conference sites (Sumi & Mase, 2001). There have been few attempts, however, to develop smart environments in urban public spaces such as airports and railway stations, even though they are an essential part of our everyday life. A manifest reason for this is the sheer vastness of such spaces. Many researchers have built their own room (Bobick et al., 1999) and home (Kidd et al., 1999) in which to conduct their projects, but it is almost impossible to build urban public spaces in research laboratories. Conference sites are too large to be purpose-built as laboratories, too. However, there is no need to do so, since researchers prefer to take advantage of actual conference events as testbeds for evaluating their systems (Sumi & Mase, 2001). Such deployment is preferred also for testing smart classrooms (Brotherton & Abowd, 2004) and smart offices (Addlesee et al., 2001). Deployment of smart urban public spaces is challenging, because it is difficult not only to attach sensors to the spaces but also to involve the visitors in situ; it can be awkward or prohibited to ask them to participate in experiments. Furthermore, it is usually impossible to shut visitors out of the space and occupy it in order to conduct experiments with study participants. We propose virtual cities as a solution. We present a user testing method that utilizes virtual cities populated with scenario-controlled software agents developed by us (Ishida, 2002b).

Figure 1. Evacuation simulation in the virtual Kyoto station

More than 300,000 passengers pass through Kyoto station, the main railway station in Kyoto City, every day. In this station we installed a guidance system that tracks passengers to help their navigation based on their current positions (Nakanishi et al., 2004). Beyond conventional navigation systems, which passively present route information, our system proactively sends instructions to the individuals' mobile phones to control their routes and avoid congestion. The system's primary application is crowd control in emergency situations. Fortunately, we were permitted to attach positioning sensors to the station's subway platform and install the system, though we were not allowed to conduct experiments that would employ many subjects playing the role of an escaping crowd. To conduct experiments without occupying urban public spaces, we developed a virtual city simulator integrating a large number of software agents and humans into the same crowd. This simulator can produce complex group behaviors such as escaping crowds. This simulator enables the agent-based user testing described in this chapter.

The next section explains how agents, humans, and avatars are integrated in the agent-based user testing. The third section presents an experiment, which was conducted to see how the user testing method can work on our guidance system. In the fourth section we discuss implications obtained from the experiment. The fifth section summarizes related work. The sixth section concludes this chapter.

AGENT-BASED USER TESTING

Augmented Experiments by Agents and Humans

We developed an AR (Augmented Reality) based user interface for the virtual city simulator. Even in a crowded urban public space, it is not difficult to test smart environments that support individuals (e.g. normal pedestrian navigation systems (Abowd et al, 1997), since an experiment in which just one person or a group is taking part does not disturb the space. In contrast, it is extremely

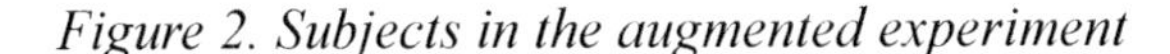

Figure 2. Subjects in the augmented experiment

intrusive to test smart environments that support crowds (e.g. crowd navigation systems such as our emergency guidance system). To solve this problem, we contrived multi-agent crowd simulations that can be overlaid onto physical spaces. A large number of agents in the simulation augment an on-site small-scale experiment. We call this kind of experiment an "augmented experiment" (Ishida et al., 2007). Augmented experiments enable us to conduct large-scale experiments in an urban public space with minimum interference with its daily operation. In the experiment performed on the subway platform at Kyoto Station we overlaid an evacuation simulation in which a hundred agents escaped from the platform to the upstairs concourse through the central staircase. Figure 1 presents a couple of screenshots of this simulation visualized in the virtual Kyoto Station. To avoid disrupting the station's operation, only three subjects escaped in each evacuation trial. The AR based user interface was necessary for the subjects to experience such a simulation on the physical platform.

The AR based user interface was a mobile phone which displays four symbols: a cross, a triangle, a circle, and a double circle. See-through head-mounted displays are not suitable for presenting the simulation of augmented experiments, since it is unsafe to mask the field of view of a walking person with a wide-field image such as a virtual crowd. Semitransparent images are safer but it is harder for users to recognize features. As Figure 2 shows, we used mobile phones, because there are always people looking at their phone's screen, reading and writing text messages while walking around in crowded places. Mobile phones are a simple means of enabling AR. Since small images of a 3D virtual space are difficult to understand, instead of displaying visual simulations, the mobile phones displayed symbols that directly represented what the subjects of the experiment needed to recognize. To produce a situation in which the subjects were often blocked by a surrounding virtual crowd and could not advance freely, we used four symbols representing the following four different degrees of density and slowness of the crowd: When the subjects were facing a crowd too dense to advance through and they had to stop walking immediately, a cross was presented on the screen. If they found a triangle or a circle on the screen, they could move ahead slowly for one or two meters, as they were approaching the crowd. A double circle was presented whenever it was possible for them to walk freely. Before starting each evacuation trial we asked the subjects to keep to these rules. Note that their mobile phones displayed a symbol and also a guidance message (described later). Which symbol and which message a subject's phone displayed was determined by the simulation and his or her position, which was tracked by the guidance system. The direction of subjects' move was estimated by the predefined evacuation route in the experiment.

Participatory Simulations by Agents and Avatars

Since it is conceivable that augmented experiments are not the best solution, we formed another group of subjects, who sat in a laboratory room and controlled their avatars in Figure 1's virtual Kyoto Station. This kind of human-in-the-loop multi-agent simulation is called a "participatory simulation" (Guyot et al., 2005). As Figure 3 shows, the subjects held a gamepad in one hand to control their avatars and a mobile phone in the other hand to receive the guidance messages (they did not receive the symbols described in the previous section). We designed our participatory simulations (VR mode) to contrast with the augmented experiments (AR mode): AR subjects walked around a physical space, whereas VR subjects controlled their avatars just like they would play a videogame. AR subjects received the symbols representing the current situation, while VR subjects could see the visual simulation from their avatars' viewpoint.

Figure 3. Subjects in the participatory simulation

Figure 4. Agent-based user testing

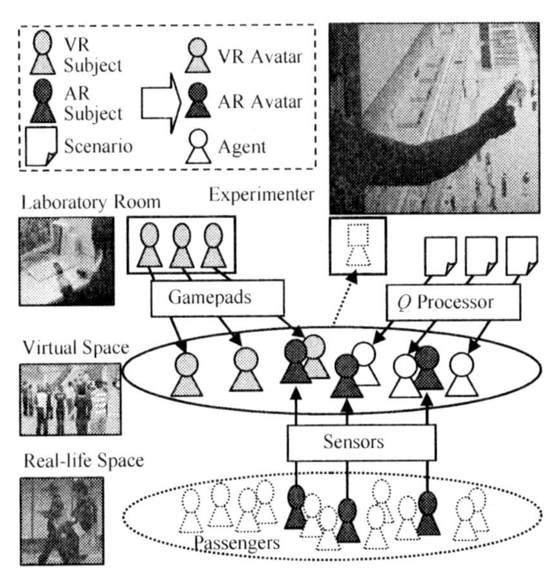

User Testing by Agents, Humans, and Avatars

In general, multi-agent simulations of group behaviors (Noda & Stone, 2003) show a different movement in each trial. If the AR and VR subjects could share the same simulations, we could exclude unnecessary variance in the analysis results of questionnaires, recorded data, and so on. Thus, we designed our simulator to be able to create both symbols and 3D animations from the same simulation, and to use both the sensors and the gamepads to control avatars. Figure 4 depicts agent-based user testing, which is a combination of augmented experiments and participatory simulations. The virtual space contains the VR and AR subjects' avatars, and the agents. The AR subjects and also passengers inhabit the real-life space. The laboratory room includes the VR subjects and an experimenter who monitors the simulation displayed on a large screen and administrates the experiment.

VR and AR subjects, and agents, have different input and output means in order to take part in the simulations. To integrate these heterogeneous participants, we attached a wrapper to the simulator's walking and seeing modules. VR subjects manipulate their gamepads to input the direction which they wish to advance. AR subjects' movements are tracked by sensors that inform the system of their current positions. Agents are controlled by their assigned simulation scenarios, in which their next destinations are specified. The wrapper converts these heterogeneous input data into changes in velocity and orientation. According to these changes, the walking module determines the following positions based on the pedestrian model (Okazaki & Matsushita, 1993) and the gait model (Tsutsuguchi et al, 2000). The wrapper also presents the same data managed by the seeing module in different forms. The data are positions and other parameters of the agents and the avatars. Agents can directly access the data to perceive the situation within the limitation of their visual power and visual field. VR subjects can see the 3D animations drawn according to the data, while AR subjects use their mobile phones to receive notifications of situational changes.

USER TESTING IN KYOTO STATION

The goal of the experiment was a trial use of the agent-based user testing. We tried to confirm that our method was useful to pre-test the design of smart urban public spaces and estimate the possible outcome of real-life scenarios.

Smart Environment Tested in the Experiment

General emergency guidance is usually offered through public address systems which announce general information that is meaningful to the whole crowd. An example of this general guidance is: "There is a fire. Please use the nearest staircase to exit." Our emergency guidance delivered through mobile phones disseminates site-specific information that is suitable for each person. An example of this location-based guidance would be: "Please do not use the nearest staircase because it is too crowded." Location-based guidance systems need sensors to know the location of the addressed person and also the movement of the surrounding crowd.

We attached a vision sensor network to the station. We attached twelve sensors to the concourse and sixteen sensors to the platform. Figure 5(a) is the floor plan, on which the black dots show the sensors' positions, and Figure 5(b) shows how they have been installed. The vision sensor network can track passengers between the platform and the ticket gate. In Figure 5(c), you can see a CCD camera and a reflector with a special shape (Nakamura & Ishiguro, 2002). If we could expand the field of view (FOV) of each camera, we could reduce the number of required cameras. However, a widened FOV causes minus (barrel) distortion in the images taken by conventional cameras. The reflector of our vision sensor can eliminate such distortion. The shape of the reflector can tailor a plane that perpendicularly intersects the optical axis of the camera to be projected perspectively to the camera plane. As shown in Figure 5(d), this optical contrivance makes it possible to have a large FOV without distortion. From the images taken by the cameras, the regions of moving objects are extracted using the background subtraction technique. The position of each moving object is determined based on geographical knowledge, including the position of the cameras, the occlusion edges in the views of the cameras, and the boundaries of walkable areas. Figure 5(e) shows AR avatars synchronized with the retrieved positions of AR subjects.

Figure 5. Positioning sensors installed in Kyoto Station

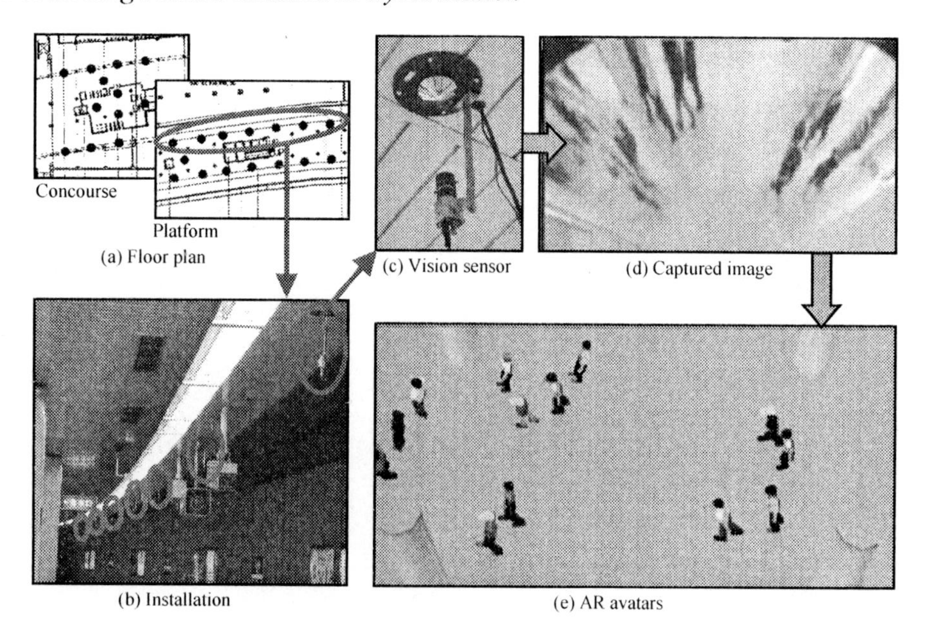

In Figure 2, the AR subjects wore a cap with a halogen lamp attached. The lamp and infrared filters covering the sensors were necessary to avoid errors in tracking the movements of the subjects on the platform, which was crowded with passengers. Passengers did not keep their distance from the subjects because they apparently ignored our experiment. Our location-based guidance system can work without this trick when the platform is sparsely populated.

The system needs the subjects' email addresses to send them guidance messages. In the experiment we registered the addresses before the evacuation began. We suppose that a real-life system would automatically register the addresses via the smart cards that people use to pass through the ticket gates. The system cannot work unless delay in email delivery is short. In the experiment the delay was about several seconds, which was short enough. We kept the email messages short to avoid distracting the subjects. Our previous prototype (Nakanishi et al, 2004) supported vocal instructions instead of guiding email messages, and we did not evaluate ways of conveying the guidance but did evaluate the psychological responses the guidance induces. Thus, the results of this experiment can be applied to both previous and current prototypes.

Hypotheses

Among other things, emergency guidance must be trustworthy in order to safely lead escaping people, who tend to lose their composure and recklessly follow others around them (Helbing et al., 2000; Sugiman & Misumi, 1988). A study on interpersonal trust in remote communication suggests that there are two independent factors in trust (Greenspan et al., 2000). One is the emotion-oriented attribute, which builds supportive impressions. In the case of emergency guidance, this is the degree to which the guide seems to be willing to help people to escape safely. The other factor is the cognitive-oriented attribute, which increases confidence that a task will be successfully completed. In other words, the guide is believed to be able to grasp the situation perfectly and manage the evacuation efficiently. We supposed that location-based guidance is better than general guidance in terms of cognitive-oriented trust. Since location-based guidance takes into account the guided person's position and his/her surrounding situation, we expected that the guidance could give a feeling that the system is monitoring the scene of evacuation and issuing optimal instructions, with the guided person finally feeling at ease. Our hypotheses are that location-based guidance: 1) would be perceived as trustworthy; 2) would be useful as a navigation aid; and 3) might induce more calmness, which is important to prevent panic. We did not pressure the subjects into a panic, since even just a few people could cause serious trouble if they were to panic in a crowded station. Instead, we merely asked the subjects to arrive at the destination as soon as they could. The previous research showed that this sort of moderate pressure does work to a certain extent for investigating evacuations (Sugiman & Misumi, 1988).

Procedure

We recruited nine VR and eight AR subjects for a total of seventeen subjects. All of them were undergraduate students. We paid them for their participation. In each evacuation trial, three VR and three AR subjects escaped together with a hundred agents. To make the evacuation as simple as possible, the system guided everyone to the central staircase that soon became crowded as shown in Figure 1. In an actual evacuation on the subway platform more than a thousand of people would escape. A hundred agents, however, were enough to produce a crowded situation around a single staircase.

The subjects took part in an evacuation where they were guided with either the location-based method or the general way, and answered a questionnaire about the trustworthiness and usefulness of experienced guidance and calmness during the evacuation. Each subject repeated this twice in random order to experience both guidance methods. We analyzed the data of eight VR and five AR subjects, because one VR and three AR subjects were unable to experience one of the two kinds of guidance due to system problems.

In both location-based and general guidance evacuation, the system sent the subjects five guidance messages based on each subject's current location as follows. The first message was sent when the subject started escaping. The second, third, and fourth messages were sent when the subject passed through the spots that are fifteen, ten, and five meters away from the staircase respectively. Delay in email delivery did not become a problem since the subjects walked according to the messages that told them to move ahead slowly or stop walking. The final message was sent when the subject began climbing the staircase.

The messages used in the location-based guidance included information on the direction in which to proceed and the crowdedness around the staircase. At the moment the evacuation started, the subjects received the message "Please escape through the front staircase." Then, after they began walking and had advanced a certain distance, they were told to "Please keep going toward the staircase." When they arrived at the tail of the crowd jostling at the bottom of the staircase, they were guided with "Please use this staircase even though it is crowded." When they were about to finish passing through the crowd, they were told "Please do not hurry because you will pass the crowd soon." Finally, they were advised to "Please go up the staircase calmly," when halfway up the staircase to the concourse. The general guidance messages, on the other hand, were the plain emergency announcements: "Please escape through a staircase close to you," "Please do not hurry when fleeing from here," "Please choose the staircase nearest to you," "Please keep calm during the evacuation," and, "Please use a staircase nearby for evacuation."

Results

After each evacuation, all the VR and AR subjects answered the same questionnaire, which contained three items about trustworthiness, three items about usefulness, six items related to the calmness, and twelve items prepared for hindering the subjects' interpretation of the experimenter's intention. All the items were measured using a nine-point Likert scale ranging from 1 to 9. We analyzed the VR subjects' data and the AR subjects' data separately. We used a two-sided paired t-test to analyze differences in the impressions of the two guiding methods.

From the VR subjects' data, we found that location-based guidance was more trustworthy ($t(7)=3.1$, $p<.05$). This is the result of comparing the means of the "TRUST" index (general: 13.1, location-based: 19.6, 3 was the lowest, 27 was the highest). We obtained this index by summing the scores of these three items: "How much were you willing to follow the guidance?"; "How trustworthy was the guidance?"; and "How persuasive was the guidance?" (Cronbach's α: .84). We also found that location-based guidance was more useful ($t(7)=2.8$, $p<.05$). This was obtained from the "USEFUL" index (general: 15.3, location-based: 20.2, 3 was the lowest, 27 was the highest), which was made from these three items: "How useful were the guidance messages?"; "How easily could you understand the messages?"; and "How kind was the guidance service?" (Cronbach's α: .70). Figure 6 summarizes these results.

In the AR subjects' data, the TRUST and USEFUL indexes did not show significant differences, but we instead observed that the subjects were calmer under location-based guidance ($t(4)=3.1$, $p<.05$). This comes from the "CALM" index (general: 9.8, location-based: 14.6, 2 was the lowest, 18 was the highest), which was the sum of two items: "How unhurriedly did you escape?" and "How calm did you feel during the evacuation?" (Cronbach's α: .91). Figure 7 shows this finding. In the VR subjects' data there was no significant difference in the CALM index.

Note that all the items that had significant difference are presented in the above analysis. We could not find any result that indicated superiority of the general guidance.

IMPLICATIONS

Analyzing the Findings of the Experiment

Interestingly, the location-based guidance elicited different responses from the VR and the AR subjects. The VR subjects appraised the method—the location-based guidance was trustworthy and useful, and the AR subjects became aware of their feelings—the location-based guidance kept them calm. Through analysis of the recorded videos, we explored what caused these different reactions.

First, we found that the VR subjects felt that the location-based guidance system was context-aware and tracking the current situation to make decisions. The VR subjects were watching the PC's screen drawing a visual simulation from their avatars' viewpoint. When the guidance's message informed them of the crowded staircase, they could actually see it. On the other hand, the AR subjects were just signaled to slow down or stop when they received the same message. It was not easy for them to know whether the system was context-aware or not.

Next, we observed that the AR subjects physically experienced the evacuation. The physical experience means that the subjects actually moved their bodies on the platform. When the virtual crowd around the staircase forced the AR subjects to stop walking and wait for the cross symbol—the "pause" instruction—to disappear, they received a different message for each guidance method. The location-based message, "Please do not hurry because you will pass the crowd soon," was predictive, which might have helped to keep the subjects calm. In contrast, the general message, "Please keep calm during the evacuation," did not predict anything, thus the subjects could not estimate how long they had to keep waiting, something that might have made them uneasy. If the AR subjects could see the virtual crowd as the VR subjects could do, they would have felt more uneasy and consequently the 'calm' effect of the location-based message could be observed more clearly. On the other hand, the VR subjects held down the "up" key to keep going forward before reaching the crowd and also after colliding with it. They did not need to release the key, since their avatars would automatically stop walking once the avatars hit other agents or avatars. This kind of operation is also observed in videogame playing. The difference in guid-

Figure 6. VR subjects' response

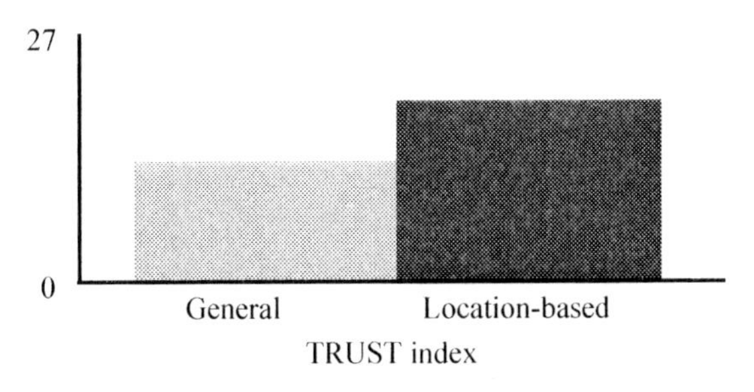

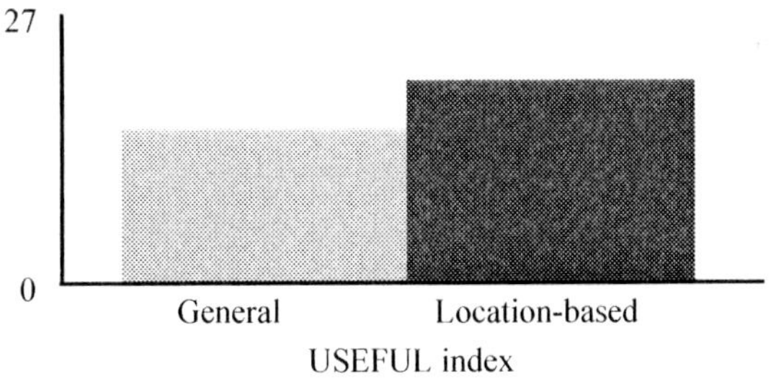

Figure 7. AR subjects' response

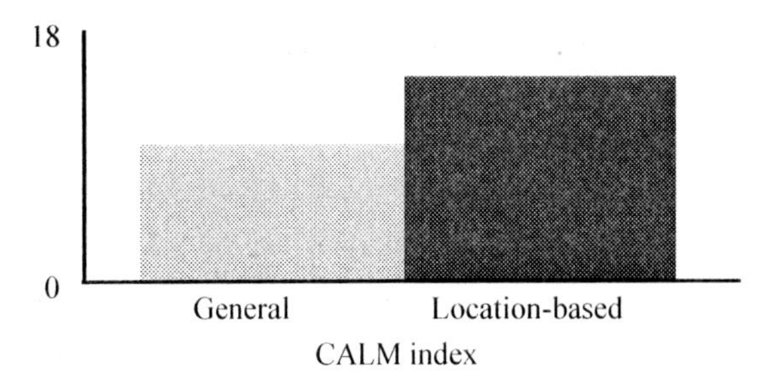

ance messages barely affected the VR subjects because whether they were walking or pausing did not matter to them.

Understanding Agent-Based User Testing

A lack of visual representations prevented the AR subjects from recognizing the crowd of agents and avatars around the staircase, so they were unable to notice the context-awareness of the guidance system. However, physical reality like what they felt when they were blocked by a surrounding crowd led them to indirectly evaluate the context-aware guidance system highly. Since the VR subjects' experience was deliberately differed from that of the AR subjects, it may be possible to generalize this as follows: In the physical space, software agents have difficulty in appearing visually, but they can interact with humans via mobile notification devices in terms of physical movement, even though they do not have a physical body. In a virtual space the agents can easily display their appearance and simulate social interaction with humans by means of 3D computer animations but can barely make them feel physical interaction. The kind of results that can be obtained from agent-based user testing depends significantly on the user interface for interacting with the agents. The VR and the AR interfaces, then, could provide results based on social and physical interactions. The results were different but both of them showed the advantages of the location-based guidance as we hypothesized in the previous section. If we had used a user interface that combined the two kinds of interactions, we might have different results. The development of such a user interface is a future work.

RELATED WORKS

Toward the goal of substituting agents for humans, researchers have been trying to develop various social agents (Nagao & Takeuchi, 1994), such as sales agents (Cassell et al., 1999) and trainer agents (Rickel et al., 2002). These agents are, however, still far from being used in practice. Meanwhile, agents have been used as movie extras (Macavinta, 2002). To realize the dream, it seems to be a much shorter way to develop agents with the ability to follow a scenario describing what to do than agents with complete autonomy. As described in this paper, the user testing of smart environments exemplified how such scenario-controlled agents could substitute for humans.

Our virtual city simulations appear to be similar to SpaceTag (Tarumi et al., 2000), which is location-based informational objects overlaid on the real world. Our simulator is, however, endowed with the gait (Tsutsuguchi et al, 2000), pedestrian (Okazaki & Matsushita, 1993), and multi-agent interaction models (Murakami et al., 2003) so that it can overlay complex group behaviors. Moreover, subjects can take part in our simulations from both mobile and desktop environments, whereas the SpaceTag system is an information-accessing mechanism for mobile users and has little meaning for desktop users. Social mixed reality systems (Crabtree et al., 2004; Okada et al, 2001) can enable mobile and desktop users share the same physical space, though they are not sufficient for the agent-based user testing. Our simulator, on the other hand, is equipped with a wrapper for integrating mobile and desktop users, and software agents. Furthermore, they can become the subjects of experiments, since the simulator is connected to the implemented smart environment, which is our guidance system.

When occupation and replication of the environment are impossible, there is an alternative to our method: exploitation of a similar environment. For example, one study uses a university building as a structural copy of a shopping mall to evaluate a shopping guide system (Bohnenberger et al., 2002). To use this method, one has to be satisfied with a rough mock-up of the actual space unless an existing physical space is found whose structure imitates the space precisely. In our method, it is possible to use the actual space if the effort is made to construct a 3D model of the space and describe the scenario for group behavior. Our simulator FreeWalk/*Q* (Nakanishi and Ishida, 2004) is a combination of a virtual city "FreeWalk" and a scenario description language "*Q*" (Ishida, 2002b). *Q* is based on Scheme, which is a dialect of the Lisp programming language invented by Guy Lewis Steele Jr. and Gerald Jay Sussman. *Q*'s model is an extended finite state machine. According to the scenario written in *Q*, agents can walk, speak, and gesture in FreeWalk's virtual space. Since FreeWalk/*Q* can simulate many kinds of group behaviors including evacuation, our method can be applied to various smart environments installed in crowded places.

CONCLUSION

We proposed agent-based user testing, that is a method for conducting experiments to test smart environments installed in large-scale crowded urban public spaces such as airports and central railway stations, since it is difficult to replicate such environments in a laboratory. Such places do not allow us to ask visitors to participate in our experiments or prevent visitors from entering so that we can conduct our experiments with selected subjects. For this method, we developed a virtual city simulator integrating agents, humans, and avatars. In testing the emergency guidance system installed in Kyoto's central railway station, the simulator overlaid a virtual crowd consisting of a hundred agents, three humans, and three avatars on the station platform so that the experiment could be carried out with little interference to the station's daily operation. The result of the experiment supported our hypotheses that our location-based guidance system installed in Kyoto station was superior in trustworthiness, usefulness, and inducing calmness. This motivated us to conduct a totally physical experiment to confirm these effects.

ACKNOWLEDGMENT

We express our thanks to the Municipal Transportation Bureau and General Planning Bureau of Kyoto city for their cooperation. This work would have been impossible without the invaluable participation of Toyokazu Itakura, Ryo Watanabe, Shinji Konishi, Shunsuke Tanizuka, Takatoshi Oishi, and Armando Rubio Torroella. We thank Shigeyuki Okazaki, Toshio Sugiman, Ken Tsutsuguchi, CRC Solutions, Mathematical Systems, and CAD Center for their support in the development of the guidance system and the social interaction platform. The platform is available at http://www.ai.soc.i.kyoto-u.ac.jp/freewalk/ and http://www.ai.soc.i.kyoto-u.ac.jp/Q/.

REFERENCES

Abowd, G. D., Atkeson, C. G., Hong, J. I., Long, S., Kooper, R., & Pinkerton, M. (1997). Cyberguide: A Mobile Context-aware Tour Guide. *Wireless Networks*, 3(5), 421-433.

Addlesee, M., Curwen, R., Hodges, S., Newman, J., Steggles, P., Ward, A., & Hopper, A. (2001). Implementing a Sentient Computing System. *IEEE Computer*, 34(8), 50-56.

Bobick, A. F., Intille, S. S., Davis, J. W., Baird, F., Pinhanez, C. S., Campbell, L. W., Ivanov, Y. A., Schutte A., & Wilson, A. (1999). The KidsRoom: A Perceptually-Based Interactive and Immersive Story Environment. *Presence: Teleoperators and Virtual Environments*, 8(4), 369-393.

Bohnenberger, T., Jameson, A., Kruger, A., & Butz, A. (2002). Location-Aware Shopping Assistance: Evaluation of a Decision-Theoretic Approach. *Proceedings of International Conference on Human-Computer Interaction with Mobile Devices and Services*, LNCS2411, 155-169.

Brotherton, J. A., & Abowd, G. D. (2004). Lessons Learned from eClass: Assessing Automated Capture and Access in the Classroom. *ACM Transactions on Computer-Human Interaction*, 11(2), 121-155.

Cassell, J., Bickmore, T., Billinghurst, M., Campbell, L., Chang, K., Vilhjalmsson, H., & Yan, H. (1999). Embodiment in Conversational Interfaces: Rea, *Proceedings of International Conference on Human Factors in Computing Systems*, 520-527.

Crabtree, A., Benford, S., Rodden, T., Greenhalgh, C., Flintham, M., Anastasi, R., Drozd, A., Adams, M., Row-Farr, J., Tandavanitj, N., & Steed, A. (2004). Orchestrating a Mixed Reality Game 'On the Ground'. *Proceedings of International Conference on Human Factors in Computing Systems*, 391-398.

Greenspan, S., Goldberg, D., Weimer, D., & Basso, A. (2000). Interpersonal Trust and Common Ground in Electronically Mediated Communication. *Proceedings of Computer Supported Cooperative Work*, 251-260.

Guyot, P., Drogoul, A., & Lemaitre, C. (2005). Using Emergence in Participatory Simulations to Design Multiagent Systems, *Proceedings of International Conference on Autonomous Agents and Multiagent Systems*, 199-203.

Helbing, D., Farkas, I. J., & Vicsek, T. (2000). Simulating Dynamical Features of Escape Panic. *Nature*, 407(6803), 487-490.

Ishida, T., Nakajima, Y., Murakami, Y., & Nakanishi, H. (2007). Augmented Experiment: Participatory Design with Multiagent Simulation. *Proceedings of International Joint Conference on Artificial Intelligence*, 1341-1346.

Ishida, T. (2002a). Digital City Kyoto: Social Information Infrastructure for Everyday Life. *Communications of the ACM*, 45(7), 76–81.

Ishida, T. (2002b). *Q*: A Scenario Description Language for Interactive Agents. *IEEE Computer*, 35(11), 54-59.

Kidd, C. D., Orr, R., Abowd, G. D., Atkeson, C. G., Essa, I. A., MacIntyre, B., Mynatt, E. D., Starner, T., & Newstetter, W. (1999). The Aware Home: A Living Laboratory for Ubiquitous Computing Research. *Proceedings of International Workshop on Cooperative Buildings*, 191-198.

Macavinta, C. Digital Actors in Rings Can Think. http://www.wired.com/entertainment/music/news/2002/12/56778, 2002.

Murakami, Y., Ishida, T., Kawasoe, T., & Hishiyama, R. (2003). Scenario Description for Multiagent Simulation. *Proceedings of International Conference on Autonomous Agents and Multiagent Systems*, 369-376.

Nagao, K., & Takeuchi, A. (1994). Social Interaction: Multimodal Conversation with Social Agents. *Proceedings of AAAI Conference on Artificial Intelligence*, 22-28.

Nakamura, T., & Ishiguro, H. (2002). Automatic 2D Map Construction using a Special Catadioptric Sensor. *Proceedings of IEEE/RSJ International Conference on Intelligent Robots and Systems*, 196-201.

Nakanishi, H., Koizumi, S., Ishida, T., & Ito, H. (2004). Transcendent Communication: Location-Based Guidance for Large-Scale Public Spaces. *Proceedings of International Conference on Human Factors in Computing Systems*, 655-662.

Nakanishi, H., & Ishida, T. (2004). FreeWalk/*Q*: Social Interaction Platform in Virtual Space. *Proceedings of ACM Symposium on Virtual Reality Software and Technology*, 97-104.

Noda, I., & Stone, P. (2003). The RoboCup Soccer Server and CMUnited Clients: Implemented Infrastructure for MAS Research, *Autonomous Agents and Multi-Agent Systems*, 7(1-2), 101-120.

Okada, M., Tarumi, H., Yoshimura, T., & Moriya, K. (2001). Collaborative Environmental Education Using Distributed Virtual Environment Accessible from Real and Virtual Worlds. *ACM SIGAPP Applied Computing Review*, 9(1), 15-21.

Okazaki, S., & Matsushita, S. (1993). A Study of Simulation Model for Pedestrian Movement with Evacuation and Queuing. *Proceedings of International Conference on Engineering for Crowd Safety*, 271-280.

Rickel, J., Marsella, S., Gratch, J., Hill, R., Traum, D. R., & Swartout, W. R. (2002). Toward a New Generation of Virtual Humans for Interactive Experiences, *IEEE Intelligent Systems*, 17(4), 32-38.

Sugiman T., & Misumi J. (1988). Development of a New Evacuation Method for Emergencies: Control of Collective Behavior by Emergent Small Groups. *Journal of Applied Psychology*, 73(1), 3-10.

Sumi, Y., & Mase, K. (2001). Digital Assistant for Supporting Conference Participants: An Attempt to Combine Mobile, Ubiquitous and Web Computing. *Proceedings of International Conference on Ubiquitous Computing*, LNCS2201, 156-175.

Tarumi, H., Morishita, K., Ito Y., & Kambayashi, Y. (2000). Communication through Virtual Active Objects Overlaid onto the Real World. *Proceedings of International Conference on Collaborative Virtual Environments*, 155-164.

Tsutsuguchi, K., Shimada, S., Suenaga, Y. Sonehara, N., & Ohtsuka, S. (2000). Human Walking Animation based on Foot Reaction Force in the Three-dimensional Virtual World. *Journal of Visualization and Computer Animation*, 11(1), 3-16.

KEY TERMS

Augmented Reality: Overlapping virtual objects with a physical environment in order to provide additional information to people in the environment

Avatar: A graphically or physically embodied representation of a human user for virtual or augmented environments

Augmented Experiment: An experiment in which virtual subjects participate

Gait Model: A model for moving human legs

Multi-Agent Interaction Model: A model for deciding next behavior based on perceptual information of other agents

Participatory Simulation: A simulation which interacts with human participants

Pedestrian Model: A model for determining the direction and velocity of human walking

Chapter XIX
The Neogeography of Virtual Cities: Digital Mirrors into a Recursive World

Andrew Hudson-Smith
University College London, UK

Richard Milton
University College London, UK

Joel Dearden
University College London, UK

Michael Batty
University College London, UK

ABSTRACT

Digital cities are moving well beyond their original conceptions as entities representing the way computers and communications are hard wired into the fabric of the city itself or as being embodied in software so the real city might be manipulated in silico for professional purposes. As cities have become more "computable," capable of manipulation through their digital content, large areas of social life are migrating to the web, becoming online so-to-speak. Here, we focus on the virtual city in software, presenting our speculations about how such cities are moving beyond the desktop to the point where they are rapidly becoming the desktop itself. But what emerges is a desktop with a difference, a desktop that is part of the web, characterized by a new generation of interactivity between users located at any time in any place. We first outline the state of the art in virtual city building drawing on the concept of mirror worlds and then comment on the emergence of Web 2.0 and the interactivity that it presumes. We characterize these developments in terms of virtual cities through the virtual world of Second Life, showing how such worlds are moving to the point where serious scientific content and dialogue is characterizing their use often through the metaphor of the city itself.

INTRODUCTION

The idea of the 'computable city' is one that stretches back to a time when the convergence of computers and communications first began to make an impact on the way cities functioned. New forms of electronic interaction began to display themselves in the need for wired infrastructures to support everything from smart buildings to new kinds of information industry (Batty, 1997). The notion that the city through its hardware might become 'intelligent' is something that has been with us since the 1980s. But during this time a somewhat different prospect has emerged with the city itself and its many functions being encapsulated and articulated in non-physical terms, in virtual space rather than real space. At first the impact of the Internet was largely in terms of cities advertising their services to 'virtual tourists' who browsed or shopped the web through simple passive browsing. The early web site *Virtual Bologna* represented the portal to urban services and information about the Italian town of Bologna which become a favourite example of early commentators on the power of the web.

Virtual Bologna was typical of its time with its iconic representation of the city as a gateway to real urban information but what is now happening is that these many technologies which display and transmit information in somewhat passive terms through the web are beginning to take on new forms of interactivity. Increasingly cities and city-like media are being captured on the web and disseminated not as passive web pages but through virtual worlds where the user enters a digital space that is in many ways akin to a real space and engages in interactions which mirror what happens in real space. Virtual cities are being built and inhabited using systems such as *Second Life*, with millions of users making rapid decisions thus shifting these virtual realities minute by minute into new manifestations of digital urban form.

The concept of the 'computable city' is still alive and well in the city itself as more and more computable devices exists within our physical environment. We have not quite reached the stage where such devices are embedded into themselves but all this is becoming routine. It is in terms of what is happening within the computer itself that now marks the cutting edge. The circle has turned completely: computers in cities exist in abundance of course, but it is cities inside computers that now define the digital frontier. This notion of the 'city inside the computer' changes rather remarkably our vision of how one can build virtual cities. Rather than being based on any single real place, they increasingly embody a mix of fiction and reality, digital cities linked together in a virtual urban sprawl, forming part of the 'metaverse' so eloquently anticipated by Neil Stephenson and William Gibson, that genre of science fiction writers that based their visions of the near future on ways in which the physical and virtual merge.

VIRTUAL SPACE

There is a never ending debate about whether or not our knowledge of space is hard wired into each of us or whether it is acquired from early childhood through our senses. However insubstantial and invisible space might appear from an analytic perspective, space is somehow everywhere around us. For most of us, space most hovers between ordinary, physical existence and something that is imposed on us. It alternates in our minds between the analysable and the absolutely given (Benedikt, 1996). In terms of our interpretation of it and the resulting all-important sense of location and place that it inspires, it has a profound influence on our perceptions of reality and of course on the digital worlds that we might create based on such perceptions. Indeed space strongly conditions the way we represent a variety of phenomena, the way we present information, the way we act, and behave in general and it is clear that when we fashion information in the digital world, the metaphor of real space powerfully conditions what we do. Yet it is also clear that because of the digital world, our conception of space is changing. The digital world that beckons, forces us to revise our view of the absolute nature of space. In the virtual world, the constraints of real space, of machine space, and the idea of iconic cities, can be massively relaxed. Virtual space can be nested into itself as many times as one likes, in recursive fashion as we gain the power to embed any digital representation into any other but more specifically into the very digital object that forms the focus in the first place. In this sense, the digital world acts as a mirror, enabling us to scale and transform any object into any other but through processes of embedding an object into itself. It is this that profoundly changes the way we are able to interact with each other in virtual space. These ideas of mirroring realities through recursion with digital media is deeply embedded in contemporary computing and we are but at the beginning of ways in which we might exploit it, as we have begun to explore elsewhere (Batty and Hudson-Smith, 2007). One of the purposes of this chapter is to take these ideas further.

Virtual cities began as digital representations of real cities essentially mirroring their physical form in the most superficial way. They were initially designed so that professionals such as architects and engineers might create environments that could be rapidly and effectively communicated to others for purposes of architectural design urban planning, and a host of serious tasks that defined what cities are about and how they might function better. Traditional digital cities are focused on how to create, represent and communicate place and space on some computerised device, originally made available on some graphics output linked to a digital computer. The type of device has always been central to the nature of such simulations. Once three dimensional representations were limited to high-end mainframe machines but now they have proliferated to the domain of the standard desktop/laptop, the portable hand-held device, GPS-enabled mobile phones and in-car satellite navigation consoles. Doubtless digital cities of this kind which represent icons of the real city can be displayed on any digital device one might imagine. In these terms they have barely moved beyond an obvious representation of the real thing but in digital space. It is an open question as to whether or not these types of cities might be called virtual. In this chapter, we will show how true virtual cities are moving well beyond these initial conceptions.

MACHINE SPACE

There are two central ideas in developing virtual cities into forms where they can be endlessly manipulated in digital terms. First is the idea of the 'Mirror World' first promoted by David Gelernter (1991) in his seminal book *Mirror Worlds: or the Day Software Puts the Universe in a Shoebox.* Gelernter (1991) defines 'mirror worlds' as software models of some chunk of reality, some piece of the real world going on 'outside your window' which can be represented digitally and then rescaled again and again into a form which you can enter and manipulate. However a mirror world is grounded in some real space and its power comes from the way we manipulate the reality. Gelernter (1991) predicted that a 'software model of your city, once set up, will be available (like a public park) to however many people are interested ... it will sustain a million different views ... each visitor will zoom in and pan around and roam through the model as he chooses' (Roush, 2007). In short, mirror worlds are a version of reality existing in the machine, a 'machine space' which in turn can be defined as the 'ParaVerse', or ' ... a parallel virtual world geographically linked to the planet earth or other bodies in the physical universe...' (http://en.wikipedia.org/wiki/Virtual_world). Our view of mirror worlds in city terms is many-fold but all relate back to the physical reality of the real city: as a city that represents the real world inside the computer, as computable space, or as a 'city in the computer'.

Virtual worlds, as distinct from mirror worlds, are worlds which may resemble in many sense the real world but which in essence are worlds created without importing any iconic representation which is tuned to match a real world. This the definition given by the authors of the *Metaverse Roadmap* (Smart, Cascio, and Paffendorf, 2007) who make the distinction between mirror and virtual worlds as one which relates to the source of the media. However as they imply, virtual worlds are unlikely to exist in pure form and increasingly worlds such as *Second Life* are full of material that represents digital icons from the real world; indeed as we will show, it is possible to embed digital representations of the real world—digital cities as mirror worlds—into virtual worlds, thus changing their definition and vastly muddying the digital waters through this kind of intersection. In short virtual worlds are now emerging that we might refer to as virtual mirror worlds which contain both real and fictional media. It is this ability to blend both that marks the way in which virtual worlds are now being used.

To take this argument much further, we must define what we mean by space in a little more detail. Bell (1996) identifies three different kinds of space: visual, informational and perceptual. Visual space is real three dimensional space around us and is defined in terms of all that a normal person can see. It is the array of objects that surrounds us, which we can create collectively, and which we take to be our environment. Each of the objects that comprises this environment has a multitude of different physical attributes, from variations in light and colour to reflectivity. These objects create reality, a fully immersive environment in Cartesian space that can be interrupted and explored by us directly in its three dimensions. In formal geometric terms, if these objects are broken down to singular levels, then each object can be viewed as being made up of a combination of primitives. Primitives are a collection of graphic tokens such as points, lines, and polygons, forming a two-dimensional or three-dimensional arrangement, and it is convenient to think of visual space as being populated by these tokens (Mitchell, 1994). If these points, lines and polygons can be recreated in digital space along with their attributes, then digital space becomes iconic, mimicking and simulating the physical reality, thus creating a mirror of the real world, a 'mirror world' existing in a digital space.

Informational space can be defined as an overlay to visual space as the space in which we communicate and receive information, from urban signage to oral communication. In the digital realm, information is rarely set up in a separate space but becomes an additional attribute of any digital icon defined with reference to its physical space. Digital information takes the form of an embedding of data within digital space or the enabling of communication within a digitally generated environment. Information can illuminate, transform, or displace reality (Borgmann, 1999). With the addition of communication to convey informational space, overlaps occur between the third form of space, that of social or perceptual space. Social space defines the user's identity and role in relation to other users in the visual environment. In digital space, the user's identity is again an additional attribute, explored later in terms of its embodiment and presence in virtual environments. Thus the combination of visual, informational, and social space influences the individual's perception of reality, be it in the real or digital environment, and this is what we define as perceptual space which is key to the digital representation of the built environment. Using digital technologies, reality cannot only be modelled and displayed on the computer screen in the form of points, lines and polygons, but it can also be augmented, manipulated, violated and transformed into environments that convolute the original representations into the wildest of fantasies.

Benedikt (1996) argues that because virtual worlds are not real in the material sense, many of the axioms of topology and geometry so compellingly observed to be an integral part of nature can therefore be violated or reinvented as can many of the laws of physics. It is this reinvention that allows attributes to be enhanced and emphasized, and the laws of gravity, density and weight to be excluded, allowing buildings to be moved with the click of a mouse or allowing the user to fly above the environment. Reality can thus be made virtual and at the same time the virtual can be recursed back and forth into and out of the reality, augmenting it, changing it. But before we explore such concepts, it is useful to take a brief look at how we create this digital space.

VIRTUAL CITIES AS NEW DIGITAL SPACES

The first step on the road to creating a virtual city, a city where bricks and mortar, buildings and their materials are represented as polygons and textures, is digital data. Data is key to our knowledge and understanding of the form of the city but its geometry must be distinguished from its other more substantive attributes which might be both physical and social. The geometry is the raw material comprising the skeleton of streets and buildings, natural vegetation, terrain and so on that provides the physical form used to tag other physical and social attributes. The geometry thus represents the geo-coordinates of the system to which other data can be tagged. Such data is often represented as layers to differentiate and classify different types and in principle, an infinite number of layers can be placed into the cityscape representing the real and/or fictional icons of the world in question. Data thus drives the formation of virtual cities in their mirror worlds and it is the wide array of possible data types that have become available for real cities that is aiding new visualisations and understandings in virtual space.

Our current model which provides a geometric data base for tagging extensive attribute data about Greater London, evolved from a simple model of central London using in the first instance 3D-GIS (geographic information systems) technologies. *Virtual London*, as it is currently called, was then extended to some 3.6 million building blocks covering the 33 boroughs comprising Greater London, an areal extent of some 1600 square kilometres. (Batty and Hudson-Smith, 2005). The model has been tagged with air pollution data, land use, retail data in surface form, it has been flooded as part of our quest to understand issues of climate change, and it has been used for various kinds of simple viewshed analysis involving the impact of high buildings. It is currently developed in *ArcScene* (which is part of *ArcGIS*) but freely ported to other CAD packages, particularly *3D Studio Max* from which movies are made and into which other media such as panoramas, still photographs and fixed animations can be embedded. We show some images from the current model in Figure 1.

Building *Virtual London* in a virtual world however relaxes the constraints we have adopted on developing the model quite considerably. The way virtual worlds operate with free entry of visitors as well as a considerable cadre of members who have rights over what and where to build, makes a focussed virtual city of the kind that comprises *Virtual London* almost impossible to create. Apart from the fact that construction is slow, individualistic and somewhat uncoordinated in comparison to the geometric strictures necessary for digital construction in professional VR, CAD and GIS software systems, the notion of letting the geometry flow differently in such worlds is a central feature. We have experimented with such worlds quite widely beginning with early versions such as *Blaxxun* and *Active Worlds* where the focus was not on real data *per se* (see Hudson-Smith, 2002), moving to more structured forms in *Adobe Atmosphere* where we built virtual exhibition spaces to house our iconic simulations,

Figure 1. Virtual London as a mirror world

a) The Geometric Skeleton b) the Digital Block Model, c) Flooding the Model with a 3 Metre Rise in the River Thames, and d) Layering an Air Pollution Map (NOx) on the Model

importing whole city blocks from *Virtual London*. Even so, our ability to produce realistic renderings and data layers as we do in *Virtual London* and employ the media for the same professional uses in property analysis, urban design and transport planning, is limited. Some of these early experiments are shown in Figure 2.

However the current generation of virtual worlds software enables users to generate much greater realism and many more users to experience this content. The key difference between mirror worlds and virtual worlds is the way interaction with users is enabled. Mirror worlds as *Virtual London* are usually constructed for single user use, for professional use where at most a set of users coordinates their use of the model. It is rare to find several users using the same model as a tool in which to structure their negotiations over design proposals, for example, although this is possible. Much more likely is the use of the models pictured in Figure 1 as tools to enable one-off rather focussed assessments of the future form of

Figure 2. The evolution of virtual worlds

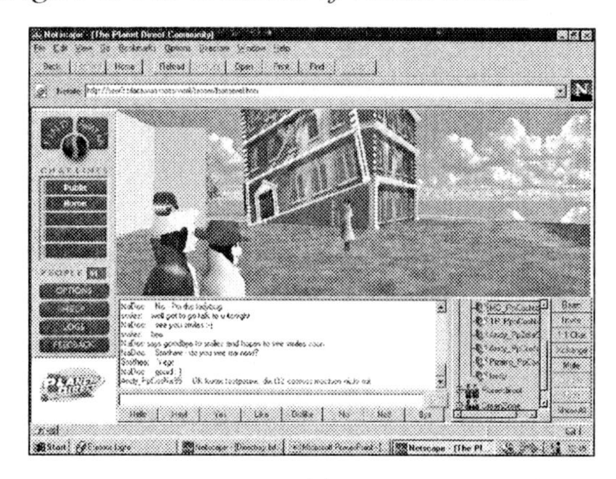

(a)

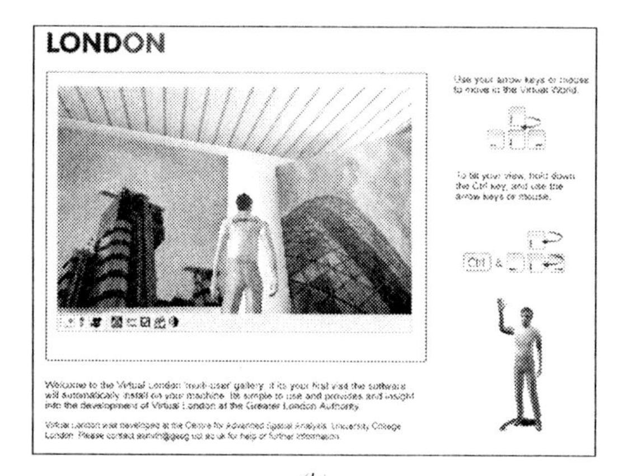

(b)

(c)

a) An Early Rendition of Building in Blaxxun b) A Virtual Gallery c) Virtual London in Adobe Atmosphere

cities rather than as playgrounds for widespread experimentation. Moreover virtual worlds engage the community of users through the web which opens their use to whoever is connected (within the obvious limits of membership and censorship). This ability of many to engage and interact is the key feature that defines Web 2.0 where interaction is the key and where most access is currently achieved through graphical user interfaces. Virtual worlds take this visualisation to the point where users can freely experiment in interacting through real or fictional environments. It is quite rare, for example, to see environments which are entirely one or the other. Users do not yet have the power to easily import entire city blocks but more to the point, there is more limited control over content than in the mirror world. Yet what is happening as we alluded to earlier is that virtual worlds are being populated by mirror worlds, implying a recursion of digital content that is clear from the early examples and is progressing rapidly in newer worlds such as *Second Life*. In

Figure 3. Virtual Amsterdam, a mirror world displayed in the virtual world of Second Life

Figure 3, we show Digital Amsterdam (or some blocks representing that city) as they have been rendered in *Second Life*. Such applications clearly point up the message that these worlds can potentially engage users in many different pursuits, not only in leisure but also in serious science. Who knows? This may be the way of much science in the medium term future, and it is certainly the challenge of Web 2.0.

WEB 2.0 AND NEOGEOGRAPHY

It is now quite clear that the connectivity produced through the Internet enables us to interact across time and space in ways that our ancestors could only dream about. This is based largely on the convergence of computers and communications that two or more generations ago were largely unforeseen even by those who were working with network interfaces to computers themselves. Once these networks were put in place by the late 1990s, then the prospect of using them to compute gradually began to dawn. Sun's old adage and advertising slogan (circa 1992) "The network is the computer" promised a taste of things to come. Now much if not most digital media is being communicated across the Internet.

Only quite recently and certainly since the Millennium has the prospect of using the interactivity of the net become significant, and only now does it appear that in the future this will be the net's main focus. In short, the notion of people communicating and manipulating digital content together and in concert or using it against one another for less benign reasons is the prospect that awaits us. As this kind of interactivity which is sometimes called social networking, gathers pace, then for those of us immersed in notions about building and using the digital city, the prospect of a global community of users who would exploit, extend and develop this digital metaphor in ways we have never anticipated, is gaining pace.

The key to all this is location, geography. We will argue here that location and space this now represents a third force in information technology besides computers and communications. Tagging not only the type of information but where such information is produced, who uses it and at what time it is generated is fast becoming the killer application that roots information about interactivity generated across the web to systems that users can easily access and use in their own interactions with others. GPS (Geo-Positioning Satellite) technologies are at the forefront of this revolution but it is their universal dissemination—first through in-car devices—and now just about through mobile phones—while in the future being embedded in multiple objects that can be carried on the person or in a transport, that is driving this revolution in tagging. Already much is being accomplished and mapping systems such as *Google Maps* are simply the vanguard of a whole series of software systems and virtual worlds that promise to bring geo-location to the fore, and of course to everyone.

This re-emergence of the importance of geography in the Web 2.0 world is becoming known as 'Neogeography'. This is the geography of the everyday person using Web 2.0 techniques to create and overlay their own locational and related information on and into systems that mirror the real world. The term derives from Eisnor (2006) one of the founders of www.platial.com where she defines it (Neogeography) as: "...a diverse set of practices that operate outside, or alongside, or in the manner of, the practices of professional geographers. Rather than making claims on scientific standards, methodologies of Neogeography tend towards the intuitive, expressive, personal, absurd, and/or artistic, but may just be idiosyncratic applications of 'real' geographic techniques. This is not to say that these practices are of no use to the cartographic/geographic sciences, but that they just usually don't conform to the protocols of professional practice". Turner (2006) expands the definition considerably in his pamphlet on the various techniques which non-professional users now have at their disposal. He says: ".... a Neogeographer uses a mapping API like *Google Maps*, talks about GPX versus KML, and geotags his photos to make a map of his summer vacation. Essentially, Neogeography is about people using and creating their own maps, on their own terms and by combining elements of an existing toolset".

The city has thus become a focal point for such visualisations where locational information is added either collectively but mostly individually to some web site or web application that enables the user to tag him or herself in space and time. As the majority of users of these systems currently live in cities or at least urban areas, it is not surprising that the city is one of the key metaphors for Web 2.0. The ways locational information is added to these applications where a website or application combines content from more than one source into an integrated experience, is known collectively as a 'mashup'. In many ways, Neogeography and mashups go hand in hand. Our *Virtual London* model is mashup as the illustrations in Figures 1 and 2 imply. Kopomaa (2000) states that the wonder which virtual spaces awaken in people wandering in the electronic labyrinths of information networks, may also be exploited to revitalize our physical cities if cities are placed in such worlds which exploit the sensitivities and sensibilities of their members and visitors. Indeed a primary aim of our model has always been to inform the public and professionals alike about the future of the city. One of the prospects of Web 2.0 for virtual cities is that as these cities develop over time, new software, new data sources and new ways of digital building will become available. The fluidity that Web 2.0 enables with its focus on individualised updating of information in the locational sense augurs well for highly responsive and timely interventions in real cities. Indeed it is the fluidity of the city that is key to Neogeography as a whole.

Without question the most important innovation in the development of the digital city, its Neogeography, and the mashups that accompany this, is the concept of the digital earth. *Google Earth* and to an increasing extent Microsoft's *Virtual Earth* and NASA's *World Wind* have produced digital cities at a speed and resolution that was unimaginable only a few years ago. These cities act as the base layers for information, a rich canvas onto and into which information can be inserted and extracted at will over the network. In essence they act as our 'space in the machine', a space which can be iconic, photorealistic or multifaceted depending on the user's preference. It is into this space that spatial analysis systems such as space syntax operate, software for analyzing space. Indeed Hillier (1992) actually defines 'space as the machine' and this mirrors a traditional professional usage which is a starting point. But once we grasp the notion that we can put space into the machine, we can then put the machine into the space, digitally, in recursive fashion where the machine *is* space, the space.

In this chapter, we will now explore the way neogeographic systems are being developed to influence both the development of mirror and virtual worlds. *Google Earth* is the example *par excellence* which was born out of Keyhole, a company founded by John Hawke with the aim of creating a 3D program called *Earth* (Roush, 2007). Of note is Hawke's inspiration from Neal Stephenson's (1992) science fiction novel *Snow Crash* which describes a virtual earth created by the Central Intelligence Corporation (CIC). In this context, it is worth quoting Stephenson: "There is something new: A globe about the size of a grapefruit, a perfectly detailed rendition of Planet Earth, hanging in space at arm's length in front of his eyes. Hiro has heard about this but never seen it. It is a piece of CIC software called, simply, Earth. It is the user interface that CIC uses to keep track of every bit of spatial information that it owns - all the maps, weather data, architectural plans, and satellite surveillance stuff. Hiro has been thinking that in a few years, if he does really well in the intel biz, maybe he will make enough money to subscribe to Earth and get this thing in his office. Now it is suddenly here, free of charge".

Indeed the rudiments are now free of charge and *Google Earth* is being used already to store

Figure 4. A framework for neogeography: Google Earth and Virtual London

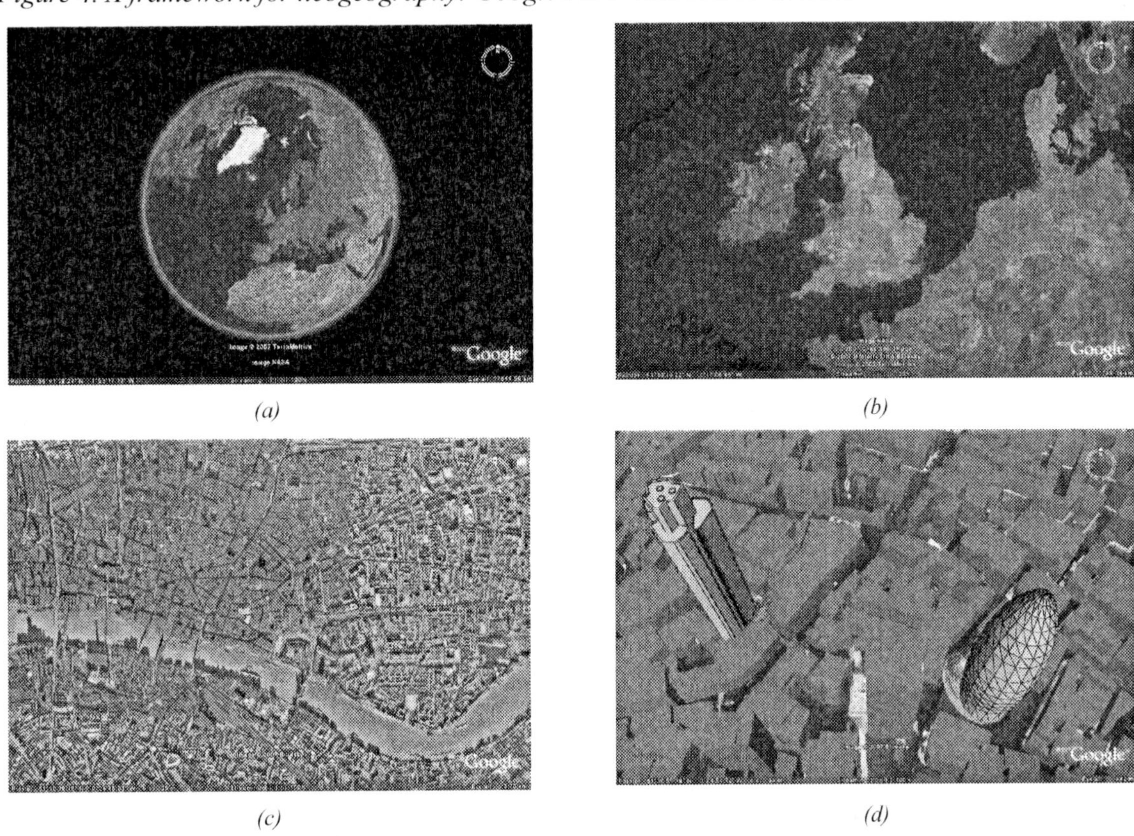

all the information that Hiro describes in this piece. We have ported out own *Virtual London* to *Google Earth* for it provides a 'free' software platform for many professional users who do not have the commercial software to run and explore our product (see www.digitalurban.blogspot.com). This is not yet Web 2.0 but as we will show, it is an essential first step (or rather second step once the mirror world model has been created) in moving in this direction. Figure 4(a) illustrates an opening shot of *Google Earth* and then content can be loaded into the earth which is tagged in such a way that the user can zoom directly (see Figures 4(b) and (c)) to the scale and place where the content is displayed. In Figure 4(d) we see part of central London—the financial quarter—as 3D building blocks.

Google Earth, released in 2005 is important on three levels to visualization and ultimately to simulation. It is simple to use because it is navigable on the x, y and z axis and thus provides a real world geographic area on which to place data. In order to make *Google Earth* represent the Earth, Google have licensed swathes of data from around the world and made it available to view free of charge. This is a notable change, especially in terms of professional spatial and urban analysis, for it provides access to high-resolution aerial imagery that is essentially free and thus challenges the power and authority of many data suppliers who charge for their data unlike Google and many other Web 2.0 companies. The resolution of the system changes according to location of course with 'Googleplex' (the Google Campus complex) currently providing the highest current level at 2.54cm per pixel. In general the highest resolutions of imagery are focused on urban areas and geographic landmarks. *Google Earth* can thus be seen as our first 'universal' glimpse into the mirror, and although it is by no means a Mirror World in and of itself, it does provide a basis on which to build as our example in Figure 4 reveals. Of particular note is the ability to import three-dimensional objects and data with a time dimension into the world, thus transforming the structure into an interface that supports 4D. Three-dimensional cities sit on top of the high resolution aerial imagery, streamed in from numerous sources, either direct from the Google server, the 3D Warehouse which is a repository of user created models, or direct from a user's machine or server (as in Figure 4).

User created content is central to such systems for no software company has either the money or man power to build a complete mirror world. The power of Web 2.0 is that it provides users with the tools and access to create such content and as such, Google released a free version of their *SketchUp* 3D modelling tool in 2006, opening up the ability (and indeed requests from Google) for users to model their own worlds or versions of the real world. This style of modelling extends far beyond the traditional CAD-based view of the world as data can be attached to high-resolution imagery providing the possibility that built environment composed giga-pixel imagery will eventually be produced.

In creating such cities halfway to virtual worlds but very much part of Web 2.0, there are still technological issues involving ground-based capture of imagery and geometry. It is a slow and semi-professional task to photograph the city and turn it into points, lines and the primitives needed for digital geometric content. Microsoft in their *Virtual Earth* have predominately taken the non-Web 2.0 route, by the building cities themselves using aerial based LiDAR (Light Imagery Detection And Ranging) and photogrammetric techniques. These digital cities are amongst the best renditions of cityscapes anywhere and to date, Microsoft have made available 62 cities throughout the US and Europe with a further 500 planned over the next 12 months. Figure 5 illustrates the kind of content with a view of New York in *Virtual Earth* but it is unlikely that this method will prevail for to model the world will require all the world's resources to be mobilised, and that is the power of Web 2.0.

Figure 5. Microsoft's New York in virtual earth

TOWARDS A SOCIAL SPACE: BUILDING VIRTUAL CITIES IN VIRTUAL WORLDS

Web 2.0 is sometimes defined in terms of social networks and social space. Thus we might add to Bell's (1996) classification of visual, informational and perceptual space by including social space which is an elaboration that broadens the context to include collectivities and groups. Social space is all important. In essence it is the network that binds Web 2.0 and Neogeography to an ability to communicate and share information through simple, freely available tools that can be learnt quickly and effectively without immersion in professional activities. As such, these tools and the way they can be used is redefining the very disciplines that traditionally have made sense of such phenomena—sociology and geography—just as economics is being redefined in Web 2.0 as Wikinomics (Tapscott and Williams, 2006). One such example is *Twitter*, a social network based on text communication which provides both an iconic and recursive view of the city as a whole. Text-based messaging is now part of everyday life. The first text based message was sent in 1992, while SMS (short messaging service) was launched commercially for the first time in 1995 (Wilson, 2005). Text-based messaging is very much part of city life. To give an indication of numbers, 1.2 billion SMS message are sent every week in the UK (2007) while in Malaysia 3.2 billion SMS messages were sent in 2006 (Kamal, 2007). Text-based messaging is synchronous and creates a social space. When SMS needs to be shared via a larger network, it becomes one-to-many, and this is what defines *Twitter* space.

Twitter is indicative of the rise of social networking sites which allow people to connect and communicate, and as such it is central to our theory of machine space. Where *Twitter* differs from others such as *MySpace* or *FaceBook* is that it is purely text-based in the SMS format of

Figure 6. Tweets West of London in South East England at 10-20am, November 27th 2007

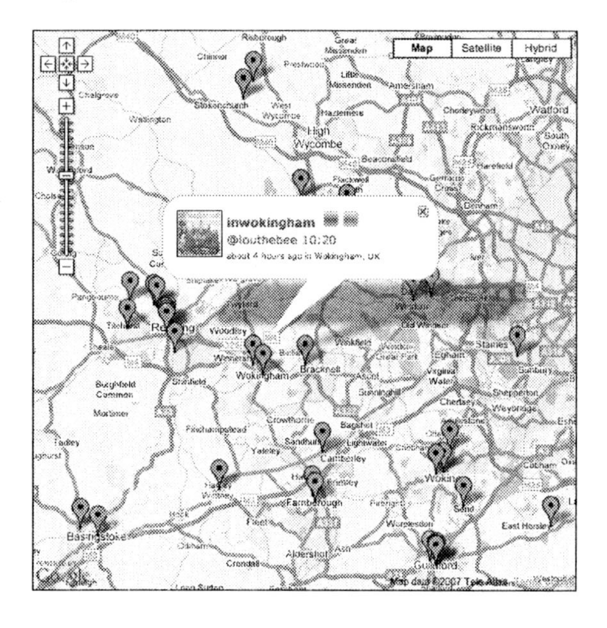

140 characters with a text entry box asking the simple question 'What are you doing?'. Based on SMS, *Twitter* is useable via a mobile phone with messages geo-located by typing in your location after the message. As such the data can be mashed and reused an infinite number of times through the Web or visualized within a digital city in real-time. Messages sent via *Twitter* are known as Tweets, and we can illustrate these with the location of Tweets over a 5 minute period in the South of England in Figure 5 in late November 2007 using the *Google Maps API.*

Tweets can also be visualized in *Second Life*, the most complete of the current generation of virtual worlds that combines visual, social, informational and perceptual space, recursing the city and which is slowly but surely creating perhaps the first true example of a mirror within a virtual world. But we only use the example of Tweets as one of many relating to streaming real time data in general from the real into the mirror world and thence into the virtual world. Much of this data is and will be locational as individuals become equipped with GPS on their phones and other devices such as PDAs, i-Pods and so on. The prospect of enormous quantities of vocational data beckons and it is these mirror and virtual worlds that in their locational-geographical views will be mobilised to make sense of all this, as *Twitter* is beginning to illustrate.

Second Life and its predecessors such as *Active-Worlds* have, in the same way as *Google Earth*, also been strongly influenced by Stephenson's (1992) vision from his novel *Snow Crash* where he first describes the MetaVerse: "As Hiro approaches the Street, he sees two young couples, probably using their parents' computer for a double date in the Metaverse, climbing down out of Port Zero, which is the local port of entry and monorail stop. He is not seeing real people of course. This is all part of the moving illustration drawn by his computer according to the specification coming down the fiber-optic cable. The people are pieces of software called avatars" (p.35).

Avatars are an individual's visual embodiment in a virtual world. They provide an all-important visual and social presence in the digital environment. They are the citizens, the occupants, and the commuters of the digital realm. As such they are also the citizens that can occupy, add data and manipulate the digital built environment. The term avatar—for use in terms of digital environments, that is—was first used by Chip Morningstar, the creator of *Habitat*, the first networked graphical virtual environment developed on the Internet in 1985. The term 'avatar' originates from the Hindu religion as an incarnation of a deity; hence, an embodiment or manifestation of an idea or greater reality. Our presence in virtual worlds is usually through the avatar although it can be any object, and in terms of a mirror world like *Google Maps* as in Figure 6, it is the balloon icon. We have already seen avatars in our early examples of virtual worlds in Figure 2 but here Figure 7 illustrates typical avatars in *Second Life*.

And so to *Second Life* where which we now consider the natural focus for our *Virtual London* model. *Second Life* is a world in which virtual land passes for real dollars and we have been fortu-

Figure 7. Avatars in Second Life

nate in gaining the support of *Nature* magazine who have purchased an island in *Second Life* for demonstrating serious science. We have squatter's rights courtesy of *Nature* on their *Second Nature* Island. What we are doing is porting geographic media in 2D, 3D and through time as streams of online real-time data about the city into this virtual environment. We are fashioning tools to enable us to do this. What *Second Life* provides is the real time context for user engagement with a virtual city through its embedding of mirror worlds. 3D-GIS or CAD software does not provide this content, nor do models embedded in web pages that users can browse and fly through. We need an environment for exploration in which many can interact and fashion the media in diverse ways. We need environments in which we can pose unrelated imagery and content enabling unusual kinds of juxtaposition which users themselves can control and interpret. We need an environment where different kinds of time streams can come together with different kinds of spaces.

What we hear you ask is all this for? Well in our Lab CASA, one of our colleagues is building a tourist information system for Phuket in Thailand using the traditional GIS, planning and decision support which is targeted at decision-makers, planners and tourists themselves. Bringing a great diversity of material together in digital form and co-locating it in a form that resembles the geography of the area is what *Second Life* offers. Moreover it provides an easy entry to space which is attractive and interactive from which users can download material and search for related items of information. This kind of visual space is highly experimental but it offers insights into problems that others may share. We are doing the same for parts of *Virtual London* but we are interfacing this with buildings at different scales and maps which take the scale up to the metropolis itself. Changes of scale are central to an appreciation of cities and *Second Life* enables us to achieve this easily. There is still a major challenge in assembling information coherently and then using it collectively to some purpose but the sheer scale of the environment is such that like *Google Maps* before, millions of users are fashioning a multitude of extensions. We show a piece of our world in Figure 8.

Second Life currently represents the most successful social/visual space on the Internet. Launched in 2003 with little more that a few kilometres of simulated computer space, in May 2007, it covers over 750 square kilometres

Figure 8. Scaling the city: Building Virtual London in Second Life.

(Ondrejka, 2007) which is roughly half of the size of our *Virtual London* model. Of note is the population which is approximately 15,000 residents logged in at any one time, and thus it has a population equivalent to Ilkeston, Derbyshire, or Troutdale, Oregon (Rolph, 2007). *Second Life* is extremely sparsely populated compared to a real city. Vast swathes of the area are devoid of avatars, much of the being a virtual world forming an empty mirror to the real world. But although the density is low, development is intense in the spirit of Wikinomics as defined by Tapscott and Williams (2006): "Today the Net is evolving from a network of websites that enable firms to present information into a computing platform in its own right. Elements of a computer—and elements of a computer program—can be spread out across the Internet and seamlessly combined as necessary. The Internet is becoming a giant computer that anyone can program, providing a global infrastructure for creativity, participation, sharing, and self-organization".

Although Linden Labs, the creators of *Second Life*, developed the program, it is the population of avatars that is creating the hamlets and towns that form its 750 square kilometres and its economy. Millions of Linden Dollars change hands every month for the goods and services residents create and provide. This unit-of-trade may then be bought and sold on LindeX (*Second Life's* official Linden Dollar exchange), or other unaffiliated third party sites where real currency changes hands (Linden Labs, 2007). In these new worlds, the population is in flux as users can 'jack in' and 'jack out', to adopt the terminology of *Snow Crash*. During August 2007, 23 million man hours were spent in *Second Life* time spent by over 974,000 users, an average of 23.6 hours per user. Hof (2006) in *Business Week* states that the as the residents spend "... a total of nearly 23,000 hours a day creating things, it would take a paid 4,100-person software team to do all that. Think of it: the company charges customers anywhere from $6 to thousands of dollars a month for the privilege of doing most of the work ... In other words, your next cubicle could well be inside a virtual world". This is Wikinomics in action, working inside the mirror as a cumulative workforce, something unseen since perhaps the industrial revolution, perhaps never seen before. People as we write, are grouping buildings and forming city plans, beautifying their virtual plots, buying and selling, or just going about their everyday life inside a machine which is increasingly becoming a mirror to the real world. In Gelernter's (1991) terms, the mirror world has entered a virtual world which mirrors the real world in part but only in part and provides a sense of interaction between reality and virtuality which is unprecedented. This is a simulacra in Baudrillard's (1994) terms.

Second Life demonstrates the power of using place within a communications medium, allowing distant participants to leverage real-world metaphors and behaviours to improve collaboration (Ondrejka, 2007). In 1928, Bertrand Russell went on record as saying that "... machines are worshipped because they are beautiful", but our fascination with them has gone far beyond their physical form. Despite the science fiction of it all in terms of man existing in 'the' or 'a' Metaverse, it cannot be denied that people are now existing, trading and communicating inside the machine. Technology acts as a catalyst to change not only what we do but also how we think. It changes our awareness of our self and of one another, of our relationship to and with the world (Turkle, 1984). Perhaps it is ourselves that are recursing into the machine rather than our physical counterparts and containers in the form of the city. Web 2.0 provides the forum on which to engage in such speculations, notwithstanding their apparent far-fetched nature.

Almost a decade ago, Damar (1998) implied that a revolution was on the horizon, the arrive of a 'true cyberspace' that would change the very face of software and our use of computers. Our definition here suggests that the computer is rapidly becoming the most significant of spaces and thus our concept of real geography may indeed no longer be as relevant as in computer space. The notion that we can be anywhere at anytime with anyone changes everything. Ondrejka (2007) calls this the 'collapse of geography' and indeed predicts a redefinition of the nation state with virtual worlds changing the alignment of labour markets and the shapes of large organizations. If real world space no longer matters or matters differently, then reality will indeed recurse into the virtual. Neogeography is set to make the geography of the real world less relevant, and in a sense the Mirror World will be a world where physical location does not matter which is the ultimate recursion.

INFORMATIONAL SPACE: AUGMENTING ICONIC SIMULATION IN THE REAL CITY

When we introduced Bell's (1996) definitions of space, informational spaces were characterised as an overlay to visual space for this is the space in which we communicate and receive information about the city. From urban signage to oral communication, information is communicated in visual space. This is the reality of space, the space that we can overlay with data, augmenting reality and the city with a series of icons. Augmented reality is by no means a new concept. Caudell coined the phrase 'Augmented Reality' in 1990 while at Boeing when helping workers insert and assemble cables into aircraft and we have seen many images of workers augmenting their physical skills through head-mounted displays and eye trackers which deliver pertinent digital information to help them in their physical tasks. Augmented reality contrasts with our mirror worlds we have explored so far for these are synthetic environments while augmented realities refer to situations in which the goal is to supplement a user's perception of the real physical world through the addition of virtual objects (Azuma, 1997).

It is this supplementation, an overlaying of data that mixes realities from the real with the virtual and the perceived that lends itself to iconic simulation. Looking around a city in augmented reality, perhaps via a location-aware portable device, mobile telephone or a head-mounted display, screen information can be overlaid onto the real physical space. For example, looking around a streetscape the device would recognise buildings, transport links, and signage allowing additional data to be streamed in via the network. An example is shown in Figure 9. Such devices, built into light weight glasses are emerging in the market place with mobile telephones being increasingly locationally aware, paving the way for local, augmented reality services. At the heart of the argument is the desire for information, to be part of a wired society and to feel connected to the city: not only on the social and business level but also in terms of our appreciation of environment, architecturally and naturally, combined with the need to know and query what is around us.

The information encoded into the locations around us and used for augmenting reality is defined by Sterling (2007) as 'Hyperlocal'. Sterling states that the databases on Web 2.0 are stuffed with geographical co-ordinates: real positions and real distances. So the bodyware I carry in my pocket and travel bag broadcasts its location to any device within earshot. This data will connect us to the city in a manner that will quickly be taken for granted once it appears and becomes widespread in the same way that *Google Maps* and *Google Earth* are now seen as indispensable. A simple current example is *Mediascape*, freely available software released by Hewlett Packard which allows the development of simple location based information applications (http://www.hpl.hp.com/mediascapes/). It is described as a "series of composed of sounds and images placed outside in your local area". To see these images and hear these sounds, you need a handheld computer or PDA and a pair of headphones. An optional GPS unit can be used to automatically trigger the images and sounds in the right places. To create a *Mediascape*, you start with a digital map of your local area. Using free software, you can attach digital sounds and pictures to places that you choose on the map which we illustrate in Figure 10.

Going outside into the area the map covers, you can experience the mediascape. Using the handheld computer and headphones, you can hear the sounds and see the pictures in the places

Figure 9. Augmenting the real city with digital information

the author of the mediascape has put them. The software is currently in development but provides an insight into how the real world can be easily augmented by users. Move this into Web 2.0 environments as will surely happen and areas of the cities could easily be swamped with media and information. The virtual world will intersect the real world more in the manner sketched a decade or more ago by Batty (1997) in his concept of the computable city. In this way, the virtual world and its mirror gives back to the physical world, completing the loop of recursion in strange and enticing ways. This is then the prospect: of mirror world standing astride both the real and the virtual, of information being recursed into many forms and being made available in diverse ways to people acting as avatars to people acting as themselves but in weird and wonderful environments yet to be invented.

CONCLUSIONS: EMERGENCE ON THE DIGITAL FRONTIER

Future trends are notoriously difficult to predict. **Popular Mechanics** predicted in 1949 that computers in the future would weigh no more than 1.5 tons. The Internet is littered with such comments. The founder of IBM, Thomas Watson stated in 1949 that the world would never need more than half a dozen computers. Bill Gates admitted that when he was asked in 1994 when the first Internet browsers appeared if there would ever be web addresses stencilled on taxis, he told the questioner not to be stupid. With technology moving on at an ever increasing rate, it would be foolhardy to predict beyond a couple of years. We have not seen, nor are we close to a complete Mirror World but the trends are in place, the price of data capture has dropped, and Web 2.0 supplies the man power which is required to populate the world both socially and in terms of this spatial extent. When Gelernter's (1991) book **Mirror Worlds** was reviewed for **Computers & Geosciences** in 1995, John Butler, the Associate Editor, noted that: "the inertia of the web may or may not prevent extensions to pass the initial limitations of design, it may never leave the page-based, one-way link metaphor that is at its root. A useful tool for downloading data from NASA, perhaps, and an online encyclopedia (of dangerously variable quality) to be sure, but not the real-time, rich, and multifaceted infospace that Mirror Worlds could be". We shall see. A

Figure 10. Augmenting the city through the mediascape

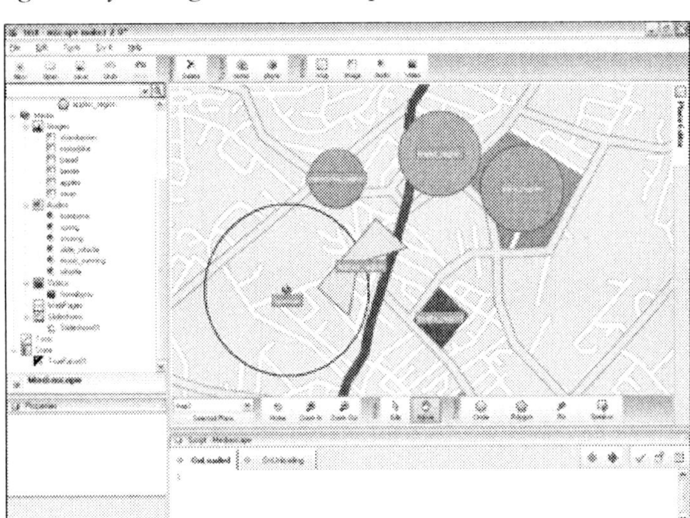

decade on Web 2.0 and innovations like *Second Life* continue to point the way to this cornucopia of rich and multifaceted infospace.

In 2007, NASA has its own *Virtual Earth* in the shape of *World Wind* and Web 2.0 has produced Wikipedia effectively creating an online encyclopedia, already illustrating the unpredictability of this future in terms of the use of technology. Web 2.0 is changing the ways companies work, embracing the consumer, allowing social networks to build content and therefore add value to their system. In many ways, this combination of ideas, work hours, and mass collaborative efforts is like an emergence on the digital frontier, a bottom up model for an interconnected system of relative simple elements which self-organise themselves into a form of intelligent, adaptive behaviour (Johnson, 2001).

Recent talk is of a merger of *Second Life* and *Google Earth*, *Second Earth*, as articulated by Roush (2007) in **Technology Review**. Populating and spawning systems such as *Google Earth* is almost inevitable given the open nature of the net. It is already possible to link Skype and *Google Earth* via avatars with *Unype*, albeit in a crude manner. A populated digital earth is another step closer to a Mirror World and we envisage a number of competing systems coming into the market place within the next year. The prospects for these are not certain: after all, the pull of a virtual world is the ability to build and create. Where do you build in *Second Earth* when the cities are already virtual and space is at a premium: on the green belt or in the deserts? Perhaps the earth will instead recurse itself into a virtual world, complete with all the functionality of zooming and data query but simply as another object in a wider digital environment. We illustrate our own early experiments of this kind of *Second Earth* in Figure 10 with real-time weather data displayed on a digital globe inside *Second Life*.

Embedding a digital earth into a virtual world is perhaps the ultimate recursion. The world will still functions as *Google Earth* but it can be cloned, copied and queried over and over again and rescaled to any size. In essence these are worlds within worlds and worlds that can be scaled according to a user's requirements in computer space. It is in this sense that we began this chapter making the shift from the computer in the city to the city in the computer but this presages a much wider challenge of placing our entire world in terms of our social existence into the machine. Perhaps we have moved from the 'Computable City' to

Figure 11. Recursing the Earth's weather into Second Life

the 'City in the Computer' and now stand at the dawn of the 'Computable Earth'/'Earth in the Computer' with all its components of place and space in an effective and meaningful coupling of the virtual and the physical.

REFERENCES

Azuma, R. (1997) A Survey of Augmented Reality, *Presence: Teleoperators and Virtual Environments*, 6(4), 355-385.

Batty, M. (1997) The Computable City, *International Planning Studies*, 2, 155-173.

Batty, M. and Hudson-Smith, A. (2005) Urban Simulacra: From Real to Virtual Cities, Back and Beyond, *Architectural Design*, 75 (6), 42-47.

Batty, M., and Hudson-Smith, A. (2007) Imagining the Recursive City: Explorations in Urban Simulacra, in H. J. Miller (Editor) *Societies and Cities in the Age of Instant Access*, Springer, Dordrecht, Netherlands, 39-55.

Baudrillard, J. (2004) *Simulacra and Simulation*, The University of Michigan Press, Ann Arbor, MI.

Bell, J. (1996) Architecture of the Virtual Community, Masters Thesis, School of Architecture, University of Wales, Cardiff, UK.

Benedikt, M. (1996) Information in Space is Space of Information, in Michelson, A., and Stjernfelt, F. (Editors) *Images form Afar: Scientific Visualisation—An Anthology*, Akademisk Forlag, Copenhagen, Denmark: 161-171.

Borgmann, A. (1999) *Holding On to Reality: The Nature of Information at the Turn of the Millennium*, University of Chicago Press, Chicago, Il.

Damar, B. (1998) *Avatars! Exploring and Building Virtual Worlds on the Internet*, Peachpit Press, Berkeley, CA.

Eisner, D. (2006) Neogeography, http://www.platial.com, accessed 27th November 2007.

Gelernter, D. (1991) *Mirror Worlds: The Day Software Puts the Universe In a Shoebox ... How It Will Happen and What It Will Mean?*, Oxford University Press, New York.

Hiller, B. (1996) *Space in the Machine*, Cambridge University Press, Cambridge, UK.

Hof, R. (2006) A Virtual Worlds Real Dollars, *Business Week Online*, http://www.businessweek.com/technology/content/mar2006/tc20060328_688225.htm accessed 27th November 2007.

Hudson-Smith, A. (2002) 30 Days in Active-Worlds: Community, Design and Terrorism in a Virtual World, in R. Schroeder (Editor) *The Social Life of Avatars, Presence and Interaction in Shared Virtual Environments*, Springer, Berlin, 77-89.

Johnson, S. (2001) *Emergence: The Connected Lives of Ants, Brains, Cities and Software,* Scribner, New York

Kamal, A. M (2007) 33.2 Billion SMSes Sent Out Last Year, http://www.redorbit.com/news/technology/910859/332_billion_smses_sent_out_last_year/index.html?source=r_technology accessed 27th November 2007.

Kopomaa, T. (2000) *The City in Your Pocket: Birth of the Mobile Information Society*, Gaudeamus, Helsinki, Finland.

Linden Labs (2007) Economic Statistics, http://secondlife.com/whatis/economy_stats.php, accessed 27th November 2007.

Mitchell, W. J. (1995) *City of Bits: Space, Place and the Infobahn*, MIT Press, Cambridge, MA.

Ondrejka, C. (2007) Collapsing Geography, Second Life, Innovation and the Future of National Power, *Innovations: Technology, Governance, Gobalization*, 2, 27-54.

Rolph, S (2007) The Phony Economics of Second Life: What the Business Press Didn't Tell You, *The Register*, http://www.theregister.co.uk/2007/02/20/second_life_analysis/ accessed 27th November 2007.

Smart, J., Cascio, J., and Paffendorf, J. (2007) Metaverse Roadmap Overview, http://www.metaverseroadmap.org/MetaverseRoadmapOverview.pdf accessed 21st November 2007

Stephenson, N. (1992) *Snowcrash,* Bantam Spectra, New York.

Sterling, B. (2007) Dispatches From the Hyperlocal Future, *Wired,* 15.07, 161- 165

Tapscott, D. and Williams, A. D. (2006) *Wikinomics: How Mass Collboration Changes Everything,* Portfolio, New York.

Turkle, S. (1994), *The Second Self: Computers and the Human Spirit*, Granada, London.

Turner, A. (2006) *Introduction to Neogeography*, O'Reilly, PDF Publication, http://www.oreilly.com/catalog/neogeography/ accessed 27th November 2007.

Wilson, F.R, (2005) A History of SMS, http://wwwprismspecturm.blogspot.com/2005_11_01_archive.html, accessed 27th November 2007.

KEY TERMS

Digital Recursion: Is the activity of representing and accessing digital media which is nested in some form within computers and networks.

Mirror Worlds: Are representations of the real world in scaled down simplified form that were originally pictured as working in parallel to the reality itself but with strong interaction both ways between reality and its mirror. The term was first popularised by David Gerlernter.

Neogeography: In its most literal sense this means a new geography but one which is digital From the Platial weblog (http://platial.com/), it is defined as: "Neogeography, as we see it, is a diverse set of practices that operate outside, or alongside, or *in the manner of*, the practices of professional geographers. Rather than making claims on scientific standards, methodologies of neogeography tend toward the intuitive, expressive, personal, absurd, and/or artistic, but may just be idiosyncratic applications of "real" geographic techniques. This is not to say that these practices are of no use to the cartographic/geographic sciences, but that they just usually don't conform to the protocols of professional practice".

Recursive Worlds: Are aspects of reality often represented in literal graphical terms which are captured digitally and recreated within different spaces which can be accessed from each other within the computer and across the net.

Virtual Cities: Are digital representations of city forms which may range from the services of cities embedded in web page through to representations of the geometry of buildings streets and landscapes comprising cities which one can manipulate on the desktop or across the web.

Virtual Worlds: Are representations of reality usually formed in 3D which enable users to enter the world and represents themselves as avatars, thus communicating with other in the world while at the same time transforming the world for educational, leisure or business purposes.

Web 2.0: From Wikipedia (http://en.wikipedia.org/wiki/Web_2): "Web 2.0 is a trend in World Wide Web technology, and web design, a second generation of web-based communities and hosted services such as social-networking sites, wikis, blogs, and folksonomies, which aim to facilitate creativity, collaboration, and sharing among users. The term became notable after the first O'Reilly Media Web 2.0 conference in 2004."

Section V
Wireless and Mobile Culture

Chapter XX
Codespaces:
Community Wireless Networks and the Reconfiguration of Cities

Laura Forlano
Columbia University, USA

ABSTRACT

This chapter introduces the role of community wireless networks (CWNs) in reconfiguring people, places and information in cities. CWNs are important for leading users and innovators of mobile and wireless technologies in their communities. Their identities are geographically-bounded and their networks are imbued with social, political and economic values. While there has been much discussion of the networked, virtual and online implications of the Internet, the material implications in physical spaces have been overlooked. By analyzing the work of CWNs in New York and Berlin, this chapter reconceptualizes the interaction between technologies, spaces and forms of organizing. This chapter introduces the concept of codespaces in order to capture the integration of digital information, networks and interfaces with physical space.

INTRODUCTION

For over ten years—since the mainstream adoption of the Internet with the introduction of the World Wide Web in 1995—researchers, businesspeople and policymakers have conducted studies, launched applications, products and services, and implemented new laws related to the virtual, online, digital and networked properties of the information society. However, in this first decade of the Internet's adoption, the role of physical place has been significantly under-theorized. We are at a turning point. A digital information layer is rapidly expanding throughout the physical spaces of our homes, offices, cities and towns. This digital layer includes mobile and wireless technologies such as WiFi hotspots, municipal wireless networks, cellular networks, Bluetooth headsets, wireless sensors and radio frequency identification (RFID) tags. WiFi hotspots can easily be found in coffee shops—including Starbucks—as well as in parks, airports and other public spaces. And, for the past several years, cities across the country and around the world have been planning to build wireless networks.

This chapter analyzes the people and organizations for whom WiFi networks, and the spaces that they inhibit, play an important role. This chapter draws on a four-year network ethnography (Howard, 2002) of community wireless networks (CWNs) and their role in building, using and innovating local infrastructures in the United States and abroad. Specifically, the chapter draws on participant-observation in NYCwireless, a CWN in New York City, which I have represented as a member of the board of directors since January 2005. Network ethnography is an emerging transdisciplinary method that makes use of a wide variety of network data—using new media including e-mail, websites, log data and social network analysis—in order to study communication in organizations (Howard, 2002). In keeping with network ethnography and following Rogers (2006), I have used Issue Crawler,[1] a network analysis software developed by GovCom.org in order to better understand the ecology of organizations involved in CWNs. I created a list of the urls of the major community wireless organizations from FreeNetworks.org and ran the Issue Crawler to analyze in-links and out-links. The Issue Crawler is a fast way to create a picture of the network by examining aspects such as the centrality and significance of organizations, the domain names of organizations and the linkages between organizations. This chapter addresses the question: What new socio-technical arrangements and forms of organizing are emerging at the intersection of technology and place?

WiFi networks are interesting for a number of reasons. First, they emerged, like the Internet, somewhat by accident. That is to say, the Internet—invented by the Defense Advanced Research Projects Agency (DARPA) as a resilient backup communications network in case of nuclear attack—was not expected to achieve such a widespread commercial success. In a similar way, the technological standard that serves as the basis for WiFi relies on unlicensed electromagnetic spectrum or what is known as the 'junk band' to communicate. Second, they translate digital networks onto physical spaces. Third, they are the domain of a diverse group of volunteers, activists and organizations referred to as CWNs. Fourth, WiFi and related technologies are currently at the center of a number of significant business and policy debates. For example, city governments are struggling to identify sustainable business models for municipal wireless networks. And, policymakers are continuing to set guidelines for issues including spectrum regulation, network neutrality, universal access and community media.

This chapter surveys existing literature and presents key theoretical concepts that are useful in analyzing the people, technologies and places that animate the work of CWNs. First, the global network of CWNs is mapped and the organizational structures through which they are linked are presented. Second, examples of mapping and social network applications are offered in order to build the argument that CWNs are lead users and innovators of wireless technologies. These examples also serve to illustrate the ways in which CWNs are reconfiguring people, communities and spaces. Third, two CWNs, NYCwireless in New York, and Freifunk in Berlin, are described in detail. Finally, this chapter concludes with a discussion of future trends and argues that a new theoretical concept—codespaces—is needed to incorporate the integration of digital networks, information and interfaces in physical space. This chapter concludes with a summary of the main arguments presented.

BACKGROUND

In recent years, there have been a number of studies about CWNs in the United States and Europe

(Bar & Galperin, 2004, 2006; Bar & Park, 2006; Gaved & Foth, 2006; Gaved & Mulholland, 2005; Gillett, 2006; W. Lehr *et al.*, 2004; Longford, 2005; Medosch, 2006; Sandvig, 2004; Werbin, 2006). These studies have documented the emergence of CWNs in the United States (Chang *et al.*, 2005; Forlano, 2006; Meinrath, 2005; Sandvig, 2004), Canada (Longford, 2005; Powell & Shade, 2006), Australia (Jungnickel, 2008) and Europe (Bina & Giaglis, 2006; Medosch, 2006; Priest, 2004). In addition, there have been studies of municipal wireless networks (Fuentes-Bautista & Inagaki, 2006; Lehr *et al.*, 2006; Powell & Shade, 2006; Sandvig, 2006; Sawada *et al.*, 2006; Sirbu *et al.*, 2006; Strover & Mun, 2006; Tapia *et al.*, 2006) as well as the role of urban interfaces for public engagement (Chang *et al.*, 2005). Overall, scholarship in this area tends to focus on the technical, economic or policy aspects of wireless networks rather than the ways in which they are socially constructed and used (Mackenzie, 2005). This aims to fill this gap by describing recent developments in the work of CWNs.

The *social construction of technology* (Bijker et al., 1987) and related concepts including *affordances* (Gibson, 1977; Norman, 1990), *infrastructure* (Star, 1999; Star & Bowker, 2002), *values* (Nissenbaum, 2001) and *disruptive technology* (Christensen, 1997) form the basis of this study of CWNs. WiFi networks are disruptive technologies because they allow multiple people to share the same Internet connection without paying for additional monthly services from the telecommunications company (though doing so may violate their contract. WiFi networks dramatically bring down the cost of Internet access because they greatly remove the need to install cables and wireless underground and throughout buildings. WiFi networks are especially useful in rural settings and developing nations where there is a lack of wired infrastructure. The concepts of *lead users* as innovators and producers, and *user-driven innovation* (Von Hippel, 1978, 2005) are also important in describing CWNs.

Lessig (1999) argued that software regulated behavior in ways similar to that of physical architecture, which he popularized in the mantra "code is law." While more recent legal scholarship (Grimmelmann, 2005; Wu, 2003) has further clarified the ways in which software is similar to, and different from architecture, these discussions do not account for the current convergence of physical and digital spaces. Mobile and wireless technologies complicate this analysis by requiring an explicit discussion of the role of physical space. There are several spatial concepts that are important for the purposes of reconceptualizing the interaction between technologies, spaces and emergent forms of organizing. These are: third places (Oldenburg, 1989), the *space of flows* and the *space of places* (Castells, 1996), *innovation spaces* (Moultrie et al., 2007) and *mediaspaces* (Couldry & McCarthy, 2004). Third places are neutral sites between home and work such as public parks, cafés and libraries where informal, voluntary and playful conversation takes place. Third places are sites of belonging and community and are vital for the functioning of urban social life. Oldenburg fears that third places are rapidly disappearing in the United States (1989). Castells has articulated the tension between the *space of flows*—global networks of technology flows—and the *space of places*—the urban spaces of everyday life. The concept of *innovation spaces* captures the recent interest of firms in designing physical environments that foster innovation and creativity (Moultrie et al., 2007). The concept *mediaspaces* addresses "the ways that media forms shape and are shaped by social space," (Couldry & McCarthy, 2004: 2). However, these spatial concepts are not well suited to capture the integration of digital information, networks and interfaces into physical spaces. In this chapter, I analyze the work of CWNs—groups that are working at the intersection of the digital and the material—and introduce the concept of codespaces to capture this convergence.

MAIN FOCUS OF THE CHAPTER

Beginning in the late 1990s, early adopters of mobile and wireless technologies founded wireless user groups (WUGs), free networks groups (freenets) and CWNs, and began experimenting with, developing software for, and building wireless networks in their cities. For the purposes of this chapter, I define CWNs as volunteer-driven groups, organized around specific geographic communities—cities and towns in the United States (including rural communities and native Indian communities) and internationally—that innovate, build and educate citizens about free, public wireless communications infrastructure. However, this is by no means the only definition of CWNs, which are young organizations with less than a decade of history. Many of the individuals involved in CWNs emerged from the Free Libre / Open Source Software (FLOSS) movement. Some of them developed an interest in the potential of WiFi sharing and community wireless when they found that they could not get high-speed Internet access in their own homes (Hampton & Gupta, forthcoming; Jungnickel, 2008; Longford, 2005; Medosch, 2006; Meinrath, 2005; Powell & Shade, 2006; Sandvig, 2004; Townsend, 2005).

A key component of CWNs around the world is the role of individual volunteers who 'host' free, open wireless networks from their apartments and homes for the use of their neighbors, visitors or passersby. The networks typically reach within a short range of the wireless network (typically 300 to 1000 feet), however, some networks have found ways to reach across significant distances. There is even an annual competition to determine the longest wireless links. These networks are able to grow organically in a decentralized manner with the addition of each new node. This chapter describes the role of CWNs as lead users of WiFi and related technical artifacts including hardware (routers, antennas, solar-powered panels, laptops, WiFi-enabled cell phones and PDAs), software (both open source and proprietary software for captive portals, network management and spectrum mapping) and applications (emerging social network tools that can be accessed on WiFi splash pages). As the following discussion will show, CWNs are motivated by diverse values, ideals and goals, which range from lowering the cost of Internet access and shifting the locus of control over communications infrastructure to community-based initiatives to the thrill of being the first to innovate an open source software to manage WiFi networks. The political nature of their work is evidenced by their efforts to influence pubic policy on a wide range of issues including network neutrality, media ownership, digital inclusion, spectrum management and municipal wireless policy; their participation in activism and protests related to issues such as privacy; and, their ongoing discussions of these issues on listservs, web sites and at face-to-face meetings and summits.

Today, there are thousands of WUGs, freenets and CWNS worldwide in cities including Seattle, New York, Champaign-Urbana, San Francisco, San Diego, Portland, Austin and Boston in the United States, and, internationally in Montreal and Toronto, Canada; London, United Kingdom; Berlin, Germany; Paris, France; Budapest, Hungary; Tallinn, Estonia; Belgrade, Serbia; Johannesburg, South Africa; Jakarta, Indonesia; Montevideo, Uruguay; and Canberra, Australia.[2] The identities and activities of CWNs are linked with global causes while at the same time situated in their local communities. CWNs—composed of activists, entrepreneurs, hackers, researchers and artists—are active in advocating for a number of issues and ideas including network neutrality, digital inclusion and unlicensed spectrum. While wireless networks are largely invisible information layers that blanket towns, cities and spaces thereby going unnoticed, the CWNs are engaged in activism and organizing around specific values, beliefs and principles. This illustrates the social construction of these networks, through their innovators and users (these are often one and the same in the case of CWNs) as explicitly political, economic and socio-cultural.

Despite similarities in the beliefs and values of CWNs, these groups vary considerably in size, membership and activities from country-to-country based on political, economic, legal and socio-cultural factors. For example, while one community wireless organization cultivates the growth of networks in New York's parks and public spaces, another reaches across the Berlin rooftops, and those in Montreal and Budapest center on cafés. In addition, while groups in Seattle, Champaign-Urbana, Montreal and Berlin excel at the development and distribution of open source software, other groups such as the one in New York are more active in policy advocacy, outreach and education. This chapter will introduce and document the work of CWNs, focusing on two of these networks in detail: NYCwireless in New York and Freifunk in Berlin.

Mapping Community Wireless Networks

As indicated briefly above, over the past few years, there have been a number of efforts to connect CWNs worldwide for the purposes of informal interaction, policy-advocacy, knowledge-sharing, collaboration and innovation. These efforts include electronic links such as websites, listservs and audio-conferences as well as face-to-face meetings such as conferences, workshops and week-long camping events. One way to better understand the relationship of CWNs worldwide is through social network analysis. The following diagram illustrates the relationships between the websites of CWNs—understood as incoming and outgoing links on web pages—that are listed on Freenetworks.org. For example, if NYCwireless' website includes a link to Freifunk's website, that would be an outgoing link and if NYCwireless' website is linked to by Freifunk's website, that would be an incoming link.

Figure 1. Social network analysis of community wireless networks (IssueCrawler, November 2006)

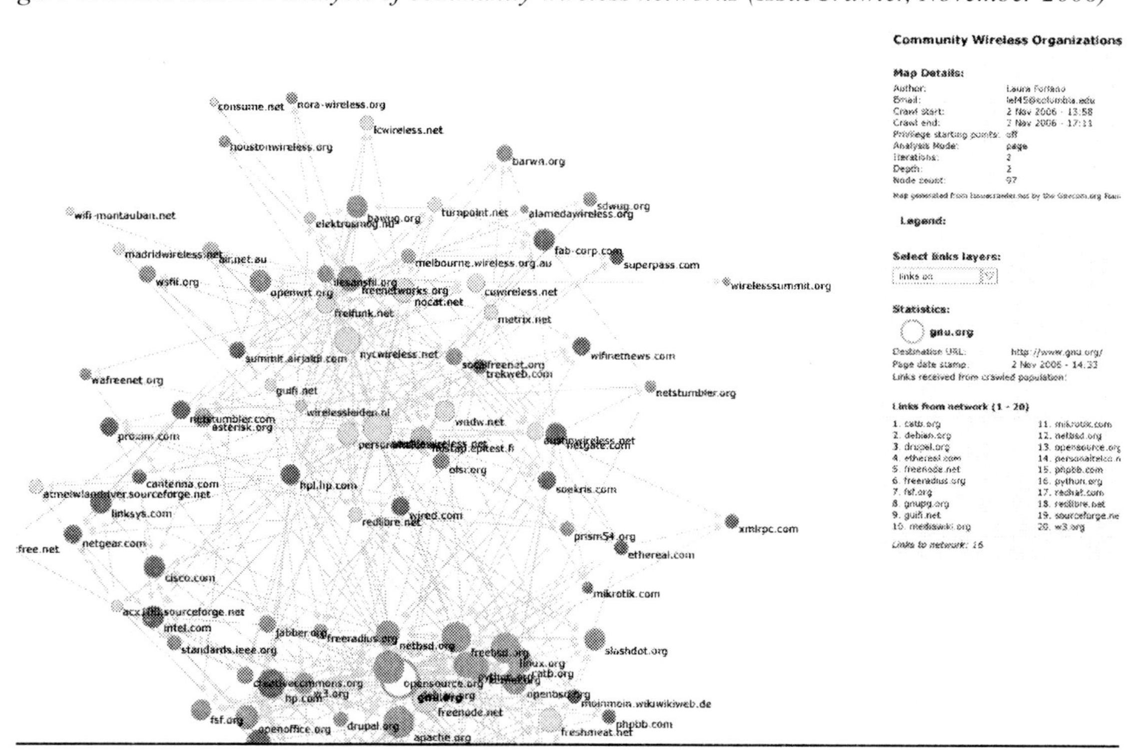

In the diagram, yellow indicates a .net (the majority of CWNs), red indicates a .com, gray indicates a .org, and other colors indicate specific country domains. Larger circles imply a greater number of incoming and outgoing links while smaller circles illustrate a smaller number. More central nodes are of increased importance to the network, meaning that they are well-known organizations that may play a key role in connecting other organizations to one another, while less central nodes are of less importance to the network.

After running the analysis, three distinct clusters emerged: CWNs, open source software and private sector intermediaries. This is important to note because only the urls of CWNs from Freenetworks.org were included in the analysis. Thus, the emergence of non-CWNs and private sector actors is significant. In particular, the diagram illustrates the relationship of CWNs to the open source software community. This makes sense since CWNs are highly interested in using, designing and sharing open source software. In fact, many CWN leaders are open source programmers and a number of CWNs emerged from the open source community.

Among community wireless organizations Seattle Wireless, NYCwireless, CUWiN (Champaign-Urbana Community Wireless Network), Freifunk (Berlin) and Île Sans Fil (Montreal) emerged as some of the most central and most important networks; again, as understood by the number of incoming and outgoing links. Interestingly, the majority of these groups are innovators of open source software, which has been adopted and deployed by other groups. Thus, it can be argued that open source software has an important role in mediating the relationships between CWNs worldwide. In the case of NYCwireless, the network is one of the oldest and most well-known networks because of its location in New York and considerable coverage in the mainstream media.

Figure 2-3. NYCwireless satellite network map showing Madison Square Park, November 2007; (NYCwireless Network Map, March 2008)

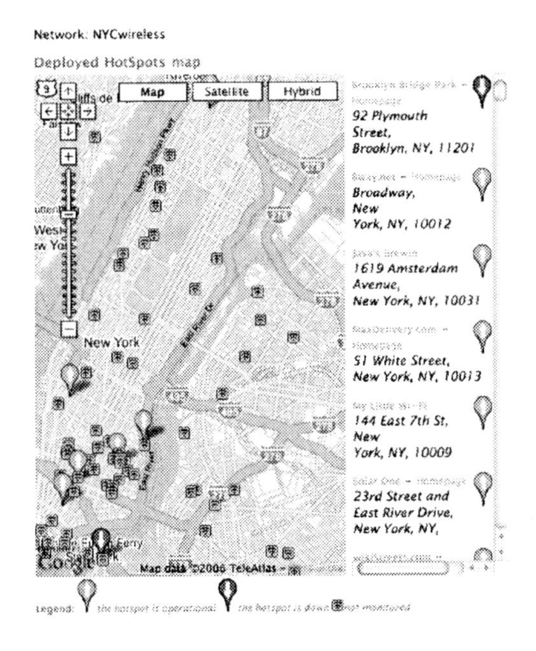

CWNs have also been pioneers in developing and using interfaces such as real-time maps to monitor the status and location of WiFi hotspots. This allows users to reconceptualize cities and spaces as zones of connectivity as well as those of disconnection. The mere use of the term WiFi hotspot to describe specific physical spaces emphasizes the digital rather than the social or spatial aspects of a particular site.

The maps above from the NYCwireless network in New York illustrate the way in which specific hotspots are represented, visualized spatially and linked to one another as CWN members despite being sponsored and supported by different partner organizations such as parks conservancies and business development districts. In this way, CWNs form new relationships and associations between the establishments and spaces that they connect. However, these linkages are not merely within neighborhoods, cities or regions because some CWNs have a national and international presence. Austin Wireless, run by Less Networks, is one network whose presence extends beyond its home city.

CWNs are currently deploying social software applications in order to facilitate communication, social networking and community-building among the users of their networks. For example, the Less Networks "Shout Map" allows network users to make public comments for other network users. The "Recent Users" tab displays the profiles of other people who have used hotspots provided by Less Networks. And, the "Community" tab lists the most recently used hotspot, the profile of a featured user, the total network statistics and the total online users and their locations.

This discussion illustrates the importance of CWNs as lead users and innovators of hardware, software and applications.

Figure 4. Less networks active hotspot map, November 2007

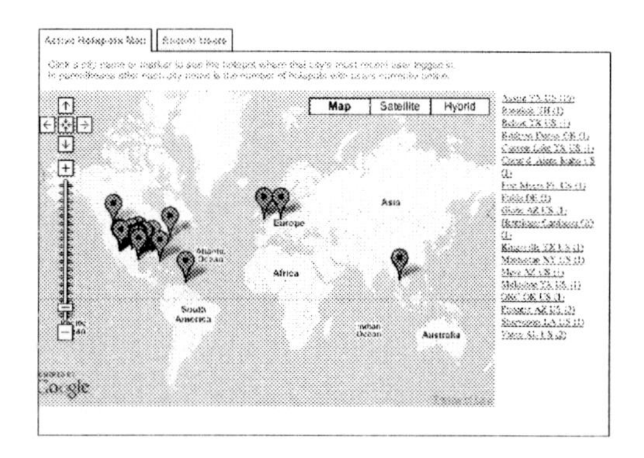

Figure 5-6. Less networks shout map, recent users and community page, November 2007

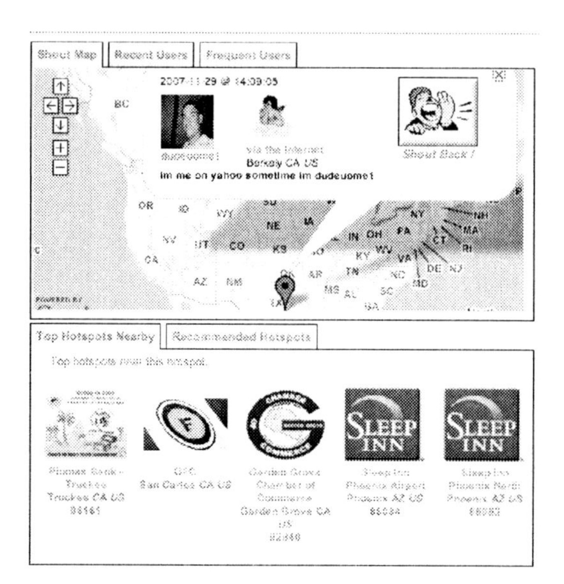

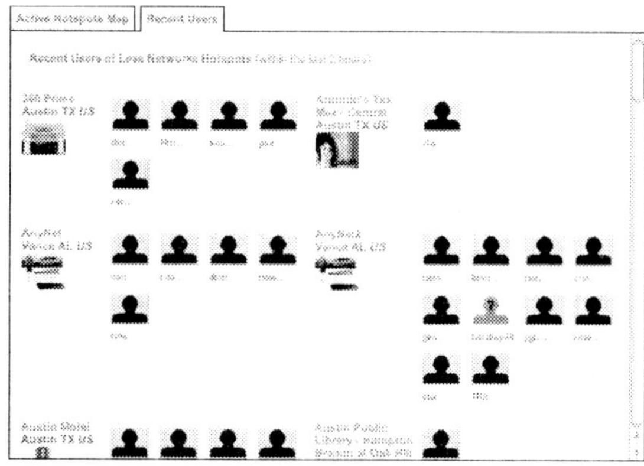

Reconfiguring New York

NYCwireless, a group of volunteers dedicated to building free, public wireless Internet hotspots in parks and public spaces, has been building free, public wireless networks in parks and public spaces in partnership with city parks organizations, business improvement districts and local non-profit organizations since 2001. Specifically, NYCwireless has built hotspots at Bryant Park (in partnership with the Bryant Park Restoration Corporation) and eight locations in Lower Manhattan including City Hall Park, the South Street Seaport, the World Financial Center Winter Garden, and the 60 Wall Street Atrium (in partnership with the Downtown Alliance, a business improvement district). NYCwireless has also worked with Community Access, a non-profit organization that provides affordable housing for people with psychiatric disabilities, to build wireless networks in three residential buildings in Manhattan, Brooklyn and the Bronx. In 2006, NYCwireless built wireless networks in Stuyvesant Cove Park (in partnership with Solar One, an environmental non-profit organization), Brooklyn Bridge Park (in partnership with the DUMBO business improvement district) and Madison Square Park (in partnership with the 34th St. Alliance). NYCwireless has conducted a number of public outreach events including Wireless Park Lab Days (September 2003), New York Live (August 2004), Spectropolis (October 2004), Manchester Live (November 2005), Berlin Live (November 2006). New York Live connected New York and Budapest for five days for five hours a day via videoconference, Manchester Live connected New York and Manchester for a one-hour city hall meeting and Berlin Live connected New York and Berlin for a simultaneous community wireless meeting between NYCwireless in New York and Freifunk in Berlin.

In the summer of 2002, while working in Washington, DC on a summer technology policy fellowship, I first read about NYCwireless in the *New York Times*. "Bryant Park is Manhattan's newest Internet café," I wrote in an August 2002 article, "New York City Goes Wireless," for *Gotham Gazette*, a non-profit news and policy website in New York. This was a year before I became involved in NYCwireless. At the time, there were 70 wireless "hotspots" according to the NodeDB.com The Wireless Database Project. The majority of these were in Manhattan since, at the time, broadband had a relatively low penetration rate. It is necessary to have a broadband connection in order to share the Internet wirelessly. In 2008, New York is still struggling to connect underserved and low-income communities to the high-speed Internet as evidenced by a series of public hearings held by the Broadband Advisory Committee, which was formed in 2007.

Bryant Park, built in partnership with the Bryant Park Corporation (BPC), a not-for-profit, private management company that was established to revitalize the park in 1988, was one of NYCwireless' first major wireless projects. Bryant Park, originally designated as public property in 1686 by New York's colonial governor Thomas Dongan when the land was still wilderness, had faced decline in the 1970s and reopened in 1991. According to BPC, the park is the largest effort in the nation to use private resources—including management and funding—to a public park.[3] Over the past fifteen years, Bryant Park has been transformed into a vibrant public space that is home to a number of privately-sponsored events and activities including movies, music, classes and even a skating rink in the winter. There is also a restaurant and bar, the Bryant Park Grill, which attracts people to the park and is a well-known hang out during the summer.

The installation of the free wireless network, which was sponsored by Intel during the first two years, was part of BPC's strategy to reinvent the park. At the time, for an installation cost of $10,000 and $1,000 per month in ongoing charges for a high-end broadband connection, it was estimated that the network could support 500 users writing e-mails and accessing the Internet. Anthony Townsend, an urban planner, co-founder of NYCwireless, and the author of the forward to this volume, said, "Unwiring our cities will be as important as electrification was at the end of the 19th century...Cities that provide this amenity will thrive. Others will just get left behind," (Forlano, 2002).

One year later, in August 2003, I met Anthony Townsend at the Smart Mobile Workshop held in an auditorium at Roppongi Hills, a brand new residential and retail development in the heart of Tokyo, where he and Howard Rheingold had been invited to speak. What follows is an account of the early history of NYCwireless as derived from Townsend's keynote address (2003). Rheingold was promoting the Japanese version of *Smart Mobs* (2003), which had been translated by the International University of Japan's Center for Global Communications (GLOCOM), the research center in Tokyo, Japan that was hosting the workshop. I had spent the summer at GLOCOM researching the use of mobile phones among Japanese teens on a grant from the National Science Foundation and the Japan Society for the Promotion of Science.

Townsend's remarks focused on NYCwireless, the organization that he had co-founded with Terry Schmidt to build public wireless infrastructure, educate the public about WiFi technology, advocate for telecommunications deregulation and provide emergency communications for New York following 9/11. He introduced the group's philosophy of open wireless networks, which was fostered by hosting public meetings, building free hotspots and providing a forum for online discussion about WiFi networking. One of the group's goals, according to Townsend, was to claim a wireless domain in public spaces as something that would be free to end-users.

In order to further this goal, in 2001, NYCwireless built its first experimental park network in Washington Square Park, which operated for three months out of Townsend's office. Next, a second experimental park network was built in Tompkins Square Park with support from a local record shop. However, when the record store that hosted the wireless equipment on its roof moved in the summer of 2005, the network was taken down. The group's next two major projects were the Bryant Park network, which was said to have 200 users per day by 2003, and the Lower Manhattan Wireless Network, a network of eight hotspots in the financial district, built by Emenity, a for-profit spin off of NYCwireless, in partnership with the Alliance for Downtown New York. Similarly, Emenity built the Union Square network, which was sponsored by *Wired*. In total, by 2003, New York had over 150 volunteer nodes that were owned and managed by individuals.[4] Since WiFi was still a relatively new technology at that time, the large majority of these nodes were hosted by early adopting broadband subscribers in Manhattan.

Decentralized, volunteer-hosted networks are by definition unstable because there is no central mechanism of control. As a result, in 2006, NYCwireless began using WiFiDog,[5] a wireless captive portal that was developed by Île Sans Fil, a community wireless organization in Montreal, Canada (Longford, 2005). WiFiDog runs on OpenWRT, an open-source software for the Linksys WRT54G router and several other platforms that was developed by Freifunk, a community wireless organization in Berlin, in conjunction with C-base, a technical cooperative in Berlin. WiFiDog allows the real-time monitoring of nodes, the customization of splash pages and the creation of statistical reports on network usage. Currently, the portal or splash pages on these networks are beginning to be used for location-based content including news, discussion, events listings and advertising. Using WiFiDog, NYCwireless is able to move from a completely decentralized network to one where there is more centralized control and monitoring.

Townsend concluded that free WiFi was becoming more and more common, in part due to the efforts of groups like NYCwireless, stating that business districts and hotels were providing wireless Internet access. In addition, at the time, Verizon was extending free wireless access from its phone booths to anyone subscribed to their Internet service. However, this service was first scaled back and then cancelled stating "lack of demand". Shortly after, Verizon launched their EVDO cellular wireless service. Thus, the company had a clear profit incentive to eliminate their free wireless hotspots.

In 2003, according to Townsend, there were dozens of similar groups in Europe and North America.[6] However, there was still a need for a global umbrella organization that could give voice to wireless communities, advocate for more open spectrum, educate the public, demonstrate successful uses of wireless technologies and provide a hub for research and development of wireless technology. At the time, Townsend saw a future of mesh networking communities, which would allow users to both send and receive Internet traffic on behalf of the network and become a bottom-up, people-powered telecommunication infrastructure. This would be possible in cities where urban density would help to achieve a critical mass of people and technology. Finally, Townsend stressed the importance of social and community applications that were being designed by artists and technologists, many of whom were active NYCwireless members.

Later that week, at the invitation of Izumi Aizu, an Internet researcher affiliated with GLOCOM, Rheingold, Townsend and I traveled with a group of about 20 others including Joi Ito, an Internet pioneer and venture capitalist, from Tokyo to Kyushu, the third largest and southernmost island of Japan's four main islands, to attend the Hyper-Network Society conference in Oita and visited a small fishing village without Internet access. On the flight back to Tokyo, Townsend and I discussed the possibility of founding a NYCwireless special interest group (SIG) to focus on the socio-economic aspects of WiFi use.

Upon returning to New York in September 2003, I volunteered at an NYCwireless event, manning a table at Wireless Park Lab Days, "a two-day event that celebrated the availability of open wireless (Wi-Fi) networks in Lower Manhattan and explored their implications for art, community, and shared space" co-sponsored by the Downtown Alliance, in City Hall Park (Spiegel, 2003). At the event, I explained wireless technology to passersby and informed them about the efforts of NYCwireless.

According to NYCwireless, the "wireless community movement" is described as "a group of volunteers who work with local organizations to construct a network of computers to share Internet access over radio connections," (Spiegel, 2003). In addition, their efforts allow, "public spaces [to] become equipped with community-owned and open wireless hotspots," and prompt thinking about how a wireless network, "affects the physical space and how urban Wi-Fi users may influence the notions of cyberspace when it becomes grounded in a specific location," (Spiegel, 2003).

Alongside the table, five artists and technologists working with wireless technologies displayed their projects, which explored the relationship between wireless connectivity and public space, challenging "the notion that wireless networks are about web surfing and email communication only," (Spiegel, 2003). The projects included: Mark Argo and Ahmi Wolf's Bass Station,[7] which transforms a boom box into an open wireless node to allow the sharing of digital files; Yury Gitman's MagicBike,[8] a mobile WiFi hotspot that links wireless infrastructure with bicycle culture; Yury Gitman and Carlos J. Gomez de Llarena's NodeRunner,[9] an urban game produced in conjunction with new media art gallery Eyebeam and NYCwireless, that challenges two teams to log into and photograph as many wireless nodes as they can; Ricardo Miranda Zuñiga's Public Broadcast Cart,[10] a shopping cart that allows passersby to produce their own audiocast; and, Dana Spiegel's Virtual People Watching,[11] which visualizes the activities of WiFi users in an online forum allowing others to see what websites they have been accessing.

After discussing the idea for the SIG with other NYCwireless members at the Wireless

Park Lab Days event and getting approval from the board of NYCwireless, I launched the SIG in late-September 2003. The purpose of the group, which became known as the Social Impact SIG, was to:

focus on the unique social changes resulting from the growing adoption of wireless technology in New York. These include changes to social norms, interpersonal networks and the use of public spaces. The group will focus on the social impact of WiFi hot spots (privately-owned and community-based) and park projects in conjunction with the growing proliferation of cell phones. This is especially important due to the potential for future convergence in wireless technology services and devices. It is vital that the social impact of wireless technology be discussed, brainstormed and researched at the earliest stages of the proliferation of these technologies. Such thinking can inform the next generation of technology design, new business ventures and policy planning in this area (Forlano, 2003).

In the months that followed, the SIG—composed of artists, architects, policy experts, engineers and social researchers—explored a number of pressing issues facing advocates of free, public wireless networks including advertising and signage, messaging and advocacy, and measurement and use.

This description of NYCwireless serves to illustrate the ways in which CWNs, as lead users of WiFi technology, experiment with and innovate hardware, software and applications that embody their socio-economic and political values as well as conduct public outreach in order to raise awareness and spread their values throughout the community. These values include the belief that WiFi should be cheap or free to end-users in public space, the understanding that electromagnetic spectrum is a resource that can be shared and the commitment to connecting underserved communities to the Internet. This narrative about NYCwireless illustrates that CWNs are more than merely technical assistance providers in their communities; in short, they are politically motivated activists participating in an emergent social movement. WiFi sharing suggests an alternate economic model based on cooperation rather than individualism. While WiFi sharing is legal, it may violate a business contract with an Internet service provider. Still, there is much misunderstanding when it comes to the question of whether or not WiFi sharing can be considered stealing and this question has not been sufficiently tested in the courts. It is possible to make the argument that WiFi sharing is a political act of civil disobedience because it sometimes ignores private contracts, which are enforced by government. As advocates for what might be considered a political act, CWNs are activists engaged in the building of communications infrastructure.

Authenticating Berlin

In early October 2004, I arrived in Berlin to spend a month on a grant from the American Council of Germany researching the activities of Freifunk, meaning 'free radio' in German, which is a community wireless network in Germany. On my first day, I had a breakfast meeting with Juergen Neumann, the founder of Freifunk whom I'd met at the Fresh Air / Free Networks Summit in Djursland, and his collaborator Ingo Rau. The two ran their own new media company, Ergomedia. Over coffee, Ingo explained to me the early efforts of Freifunk as well as important details of the German telecommunications landscape. Later that day, I arrived at JewelBox, an artist studio and collaborative workspace in Freidrichshain, where Alex Toland, an artist, Ulf Kypke, a system administrator and the founder of the WlanHain, the wireless local area network in Freidrichshain, and members of the Chaos Computer Club worked. The following description of Freifunk is based on an e-mail correspondence with Neumann in February 2007.

Freifunk was formed shortly after Neumann, a technologist and entrepreneur since 1990, moved to Friedrichshain, a neighborhood in the eastern part of Berlin in 2002. There was no broadband access in the neighborhood because Deutsche Telekom had taken out the old, copper cables and installed fiber optic cables instead. However, they did not offer DSL over fiber to end-consumers at the time and it took a number of years for people in the eastern part of Germany to get broadband. At the time, Neumann was living in a housing cooperative with 35 people who all shared a single ISDN connection for Internet access. After some research, Neumann found an Internet service provider about a kilometer away from his apartment building that would provide symmetric Internet access by wireless local area network (LAN). According to Neumann:

I bought routers and antennas and built two wireless LANs on top of our roof. One was to connect the house and its 35 inhabitants to the provider, the other one was connected to an omni-antenna to spread the signal in a radius of about 500 meters, so that other people in the neighbourhood would also be able to connect to a cheap and fast Internet connection.

Shortly thereafter, in October 2002, Neumann attended the BerLon Wireless Culture Workshop,[12] where he met people from community wireless groups including London's Consume.net and Denmark's Wire.less.dk who he continued to stay in touch with at other regional events including the Copenhagen Interpolation[13] and the Freifunk.net Summer Convention in 2003. He realized that these groups were confronting similar problems—the lack of cheap broadband—for a wide variety of different reasons, and that wireless was one part of the solution. At the workshop, Neumann met a number of other Berliners who had a similar idea; many of them were involved in the C-Base,[14] a computer-culture project in Berlin. In the following weeks, Neumann and others including software developers Sven-Ola Tücke (the inventor of the Freifunk firmware), Elektra (the inventor of B.A.T.M.A.N.), Marek Lindner and Sven (aka c-ven) Wagner. This group formed an initial project, wavelan-berlin, which later became the OLSR Experiment.[15] They initiated regular meetings on Wednesdays, which continue to this day, in order to, "share visions and ideas about how to build wireless networks."

The next step was to create a German-language campaign about wireless networking since the majority of the materials at the time were in English. Together with mindworxs,[16] a team of web developers, Neumann created Freifunk[17] in 2003. Since then, the group has been working to build mesh wireless networks. The group's first website went live in March 2003. According to Neumann, "Freifunk.net and our vision of free and user-owned wireless networks is very well known all over Germany and in many other parts of the world today." For example, a group in South Africa is using Freifunk's free firmware[18] to build local wireless mesh networks.[19]

In order to become an active user of the Freifunk network, an individual must buy an access point and 'flash' it with the free, open source freifunk.firmware. The mesh network that is created by all active users is an intranet or open public local access network (OPLAN).[20] The network allows for the free—free as in no charge as well as in open—exchange of data (including files, instant messages, e-mail, voice over Internet Protocol, etc.) within the network at speeds of up to 20 MB per second using the 802.11g standard. This focus on building networks that are free to the end-user is similar across many CWNs around the world. While there is no question that there is a cost for bandwidth use, CWNs support an economic model based on cooperation and sharing rather than individualism.

The network is run under the terms of the Pico Peering Agreement (PPA).[21] The PPA is an attempt to connect CWNs and formalize their interactions by providing, "the minimum baseline

Figures 7-9. Technical-mapping, aerial view and rooftop view of Freifunk's Berlin Mesh Network, July 2006

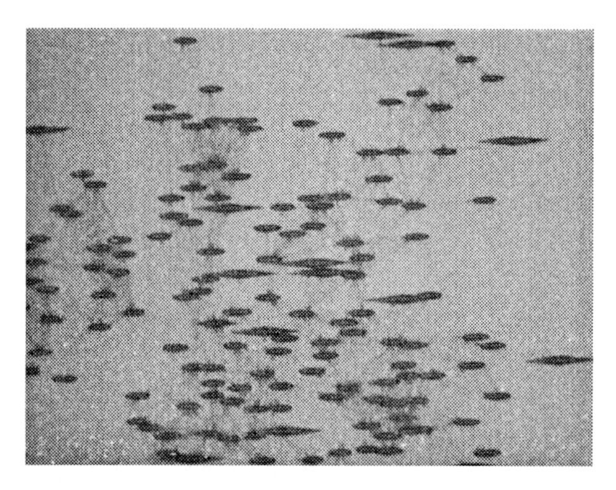

Figure 10. Furnsehturm TV Tower at Alexanderplatz in Berlin decorated for the World Cup Soccer Tournament, July 2006

template for a peering agreement between owners of individual network nodes" in which "Owners of network nodes assert their right of ownership by declaring their willingness to donate the free exchange of data across their networks." The PPA stipulates free transit (including an agreement not to interfere with data that passes through the network), open communication (including publishing information about the network under a free license), no guaranteed level of service, acceptable use, and local amendments.[22] The PPA illustrates that CWNs support a technical model that supports cooperation and sharing in contrast to corporate models. For example, the current debates over network neutrality argue whether or not corporations should be able to charge for differentiated speeds and types of data i.e. e-mails, voice traffic, music files and movie files or block access to certain web sites completely. While this might not be a pressing public policy issue given sufficient competition in the broadband market, the current duopoly of broadband provision has raised concern over what kinds of strategies corporations should be able to use to make a profit and the impact of these strategies on the Internet as a whole.

As the above description of Freifunk illustrates, access to the Internet is only one of the services provided by the network. As long as some of the node operators provide some of their unused bandwidth to the network, other users can connect to the network. In order to access the network, members must download the routing software OLSR to their laptop. OLSR detects the location of the nearest Internet gateway and announces it to the rest of the network. The network is able to grow organically because every new access point will automatically become part of the network and extend the reach of the network. Freifunk's expertise in managing a large network of this kind illustrates their role as lead users and innovators in mesh networking.

Similar to the technological aspects of the network, the Freifunk campaign is able to grow organically because local Freifunk initiatives are encouraged and supported. All website content including logos are published under a Creative Commons license in order to promote the distribution of the idea. Freifunk has expanded to include networks all over Germany, Switzerland and Austria. According to Neumann, Freifunk's Leipzig and Berlin networks are the biggest local, volunteer-run mesh networks in the world. He estimates that the network covers one tenth of Berlin, meaning that, theoretically, 350,000 people could connect to the network. However, some degree of technological literacy is required to join the network due to the need to download the OLSR software in order to use the network and 'flash' the router with freifunk.firmware. In summary, the Freifunk network is decentralized, flexible and growing quickly as a "social initiative but also as a physical infrastructure," wrote Neumann. Like NYCwireless, the Freifunk campaign combines both technical and political values, which underscores the role of CWNs as activists engaged in an emergent social movement. CWNs can be described as socio-technical networks with an important spatial component.

FUTURE TRENDS

While the first decade of the Internet can be characterized by its virtuality; that is, the migration of information, experiences and activities onto digital interfaces, networks and platforms. The next decade of the Internet will be characterized by materiality; or, rather, the integration of digital information into physical spaces. Currently, a palimpsest of digital information has blanketed our homes and offices, cities and towns as well as all of the places in between. In the United States, borders have been outfitted with digital finger print scanners. Passports have been rigged with radio frequency identification (RFID) tags. The hordes of luxury apartment buildings announce security features such as video intercoms and keycard entry. Surveillance cameras are dutifully installed at intersections to ticket speeders and tax peak commuters in city-centers. Once static billboards have been replaced by interactive digital interfaces.

We are unprepared to understand the implications of our rapid adoption of these technologies into our physical environments. Digital information, networks and interfaces may support, enhance, maintain, conflict with or contradict existing ways of shaping social behavior including social norms, laws, markets and physical architecture itself while at the same time being shaped by it. I argue that we require a new theoretical concept—codespaces—to capture the integration of digital technologies into physical space.

CONCLUSION

This chapter introduces the work of CWNs around the world in order to illustrate the ways in which these networks are socially, technically and spatially constructed. CWNs are locally-based but globally-networked. Their identities are closely linked to the geographic communities that they serve. Focusing on the early histories of NYCwireless and Freifunk, this chapter argues that CWNs are lead users and innovators of WiFi technology with diverse strategies and values that have much to offer current discussions of municipal wireless networks around the world. For example, CWNs are at the forefront of experimentation with open source hardware and software as well as the integration of mapping and social network applications. CWNs are important for theorizing about urban informatics because they are deeply engaged in peer production at the intersection of technology and space.

REFERENCES

Bar, F., & Galperin, H. (2004). Building the Wireless Internet Infrastructure: From Cordless Ethernet Archipelagos to Wireless Grids. *Communications and Strategies, 54*(2), 45-68.

---. (2006). Geeks, Cowboys, and Bureaucrats: Deploying Broadband, the Wireless Way. *The Southern African Journal of Information and Communication (SAJIC)*(6).

Bar, F., & Park, N. (2006). Municipal Wi-Fi networks: The Goals, Practices, and Policy Implications of the U. S. Case. *Communications & Stratégies*(61), 107-125.

Bijker, W. E., Hughes, T. P., & Pinch, T. (1987). *The Social Construction of Technological Systems.* Cambridge, MA: The MIT Press.

Bina, M., & Giaglis, G. M. (2006). Unwired collective action: motivations of wireless community participants, *5th International Conference on Mobile Business (ICMB'06).* Copenhagen, Denmark: IEEE Computer Society.

Castells, M. (1996). *The Rise of the Network Society.* Malden, MA: Blackwell Publishers.

Chang, M., Jungnickel, K., Orloff, C., & Shklovski, I. (2005). Engaging the City: Public Interfaces as Civic Intermediary, *CHI '05.* New York, NY: Association for Computing Machinery.

Christensen, C. M. (1997). *The Innovator's Dilemma: When New Technologies Cause Great Firms to Fail*: Harvard Business School Press.

Couldry, N., & McCarthy, A. (2004). *Mediaspace: Place, Scale and Culture in a Media Age.* London: Routledge.

Forlano, L. (2002, August). New York City Goes Wireless. *Gotham Gazette.*

---. (2003). Social Impact Group Description. New York, NY: NYCwireless.

---. (2006). Activist Infrastructures: The Role of Community Wireless Organizations in Authenticating the City. *Eastbound, 1.*

Fuentes-Bautista, M., & Inagaki, N. (2006). Reconfiguring public Internet access in Austin, TX: Wi-Fi's promise and broadband divides. *Government Information Quarterly, 23*(3-4), 404-434.

Gaved, M. B., & Foth, M. (2006). More Than Wires, Pipes and Ducts: Some Lessons from Grassroots Initiated Networked Communities and Master-Planned Neighbourhoods. In R. Meersman, Z. Tari & P. Herrero (Eds.), *Proceedings OTM (OnTheMove) Workshops* (Vol. LNCS 4277, pp. 171-180). Heidelberg, Germany: Springer.

Gaved, M. B., & Mulholland, P. (2005). Ubiquity from the bottom up: Grassroots initiated networked communities. In M. Consalvo & K. O'Riordan (Eds.), *AoIR Internet Research Annual* (Vol. 3). New York, NY: Peter Lang.

Gibson, J. J. (1977). The theory of affordances. *Perceiving, acting and knowing: toward an ecological psychology*, 67–82.

Gillett, S. E. (2006). Municipal Wireless Broadband: Hype or Harbinger? *Southern California Law Review, 79*, 561-594.

Grimmelmann, J. (2005). Regulation By Software. *Yale Law Journal, 114*, 1719-1758.

Hampton, K. N., & Gupta, N. (forthcoming). Community and Social Interaction in the Wireless City: Wi-Fi use in Public and Semi-Public Spaces. *New Media & Society.*

Howard, P. (2002). Network ethnography and the hypermedia organization: new media, new organizations, new methods. *New Media & Society, 4*(4), 550.

Jungnickel, K. (2008). Making WiFi. Retrieved February 17, 2008, from http://www.studioincite.com/makingwifi/

Lehr, W., Sirbu, M., & Gillett, S. (2004). Municipal Wireless Broadband: Policy and Business Implications of Emerging Access Technologies, *Competition in Networking: Wireless and Wireline* (pp. 13-14). London Business School.

---. (2006). Wireless is changing the policy calculus for municipal broadband. *Government Information Quarterly, 23*(3-4), 435-453.

Lessig, L. (1999). *Code and Other Laws of Cyberspace*. New York: Basic Books.

Longford, G. (2005). Community Networking and Civic Participation in Canada: A Background Paper (CRACIN Working Paper No. 2).

Mackenzie, A. (2005). From Cafe to Parkbench: Wi-Fi and Technological Overflows in the City. In M. Sheller (Ed.), *Technological Mobilities*. London & New York: Routledge.

Medosch, A. (2006). On free wavelengths: wireless networks as techno-social models. from http://theoriebild.ung.at/view/Main/FreeWavelength

Meinrath, S. (2005). Community Wireless Networking and Open Spectrum Usage: A Research Agenda to Support Progressive Policy Reform of the Public Airwaves. *The Journal of Community Informatics, 1*(2), 174-179.

Moultrie, J., Nilsson, M., Dissel, M., Haner, U.-E., Janssen, S., & Van Der Lugt, R. (2007). Innovation Spaces: Towards a Framework for Understanding the Role of the Physical Environment in Innovation. *Creativity and Innovation Management, 16*(1), pp. 53-65.

Nissenbaum, H. (2001). How computer systems embody values. *Computer, 34*(3), 120.

Norman, D. A. (1990). *The Design of Everyday Things*. New York: Doubleday.

Oldenburg, R. (1989). *The Great Good Place: Cafés, Coffee Shops, Community Centers, Beauty Parlors, General Stores, Bars, Hangouts and how They Get You Through the Day*. New York: Paragon House.

Powell, A., & Shade, L. R. (2006). Going Wi-Fi in Canada: Municipal and community initiatives. *Government Information Quarterly, 23*(3-4), 381-403.

Priest, J. (2004). The State of Wireless London. from http://informal.org.uk/people/julian/publications/the_state_of_wireless_london

Rheingold, H. (2003). *Smart Mobs*. Cambridge, MA: Perseus Books Group.

Rogers, R. (2006). *Mapping Webspace with Issuecrawler*: GovCom.org.

Sandvig, C. (2004). An Initial Assessment of Cooperative Action in Wi-Fi Networking. *Telecommunications Policy, 28*(7/8), 579-602.

---. (2006). Disorderly infrastructure and the role of government. *Government Information Quarterly, 23*(3-4), 503-506.

Sawada, M., Cossette, D., Wellar, B., & Kurt, T. (2006). Analysis of the urban/rural broadband divide in Canada: Using GIS in planning terrestrial wireless deployment. *Government Information Quarterly, 23*(3-4), 454-479.

Sirbu, M., Lehr, W., & Gillett, S. (2006). Evolving wireless access technologies for municipal broadband. *Government Information Quarterly, 23*(3-4), 480-502.

Spiegel, D. (2003). Wireless Park Lab Days. New York: NYCwireless.

Star, S. L. (1999). The Ethnography of Infrastructure. *American Behavioral Scientist, 43*(3), 377.

Star, S. L., & Bowker, G. C. (2002). How to infrastructure. *Handbook of New Media-Social Shaping and Consequences of ICTs, SAGE Pub., London, UK*, 151-162.

Strover, S., & Mun, S.-H. (2006). Wireless broadband, communities, and the shape of things to come. *Government Information Quarterly, 23*(3-4), 348-358.

Tapia, A., Maitland, C., & Stone, M. (2006). Making IT work for municipalities: Building municipal wireless networks. *Government Information Quarterly, 23*(3-4), 359-380.

Townsend, A. M. (2003). NYCwireless, *Smart Mobile Workshop*. Tokyo: Center for Global Communications (GLOCOM), International University of Japan.

---. (2005). *Wired/Unwired: The Urban Geography of Digital Networks*. Unpublished Unpublished doctoral dissertation, Massachusetts Institute of Technology, Boston.

Von Hippel, E. (1978). Users as Innovators. *Technology Review, 80*(3), 31-39.

---. (2005). *Democratizing Innovation*. Cambridge: MIT Press.

Werbin, K. C. (2006). Where is the 'Community' in Community Networking Initiatives? Stories From the 'Third Spaces' of Connecting Canadians (CRACIN Working Paper No. 11).

Wu, T. (2003). When Code Isn't Law. *Virginia Law Review, 89*(4), 679-751.

KEY TERMS

Codespaces: The integration of digital networks, information and interfaces with physical spaces

Community Wireless Networks: Wireless networks that are initiated, built and maintained by community groups often in partnership with private, government or non-profit organizations

Mesh Networks: A decentralized, flexible and redundant network in which each node is connected to every other node

Municipal Wireless Networks: Wireless networks that are initiated by cities, towns and municipalities but that are often built and maintained by private organizations

Network Ethnography: An emerging transdisciplinary methodology that makes use of a wide variety of network data in order to study communication in organizations

Social Network Analysis: The study of the relationships between people, groups or organizations

WiFi Networks: A network that connects computers to the Internet wirelessly using IEEE 802.11x, which is commonly known as WiFi

ENDNOTES

1. See GovCom.org or IssueCrawler.net for more information. Accessed on May 10, 2007.
2. See FreeNetworks.org for a more comprehensive list. Accessed on May 3, 2007.
3. See Bryantpark.com. Accessed on August 20, 2007.
4. NodeDB.com, The Wireless Database Project. Accessed July 1, 2003.

[5] More information on WiFiDog can be found at http://WifiDog.org, http://OpenWrt.org and http://LinksysInfo.org.

[6] Freenetworks.org. Accessed on July 1, 2003.

[7] For more details, see http://www.bass-station.net.

[8] For more details, see http://www.magicbike.net.

[9] For more details, see http://www.noderunner.com.

[10] For more details, see http://www.ambriente.com/wifi/.

[11] For more details, see http://www.sociableDESIGN.com/nycwchat.

[12] For more details, see http://informal.org.uk/people/julian/publications/the_wireless_event/#berlon.

[13] For more details, see http://www.metamute.org/en/The-Copenhagen-Interpolation.

[14] For more details, see http://c-base.org.

[15] For more details, see http://www.olsrexperiment.de.

[16] For more details, see https://www.mindworxs.org.

[17] For more details, see https://www.freifunk.net.

[18] For more details, see http://freifunk.net/wiki/FreifunkFirmwareEnglish.

[19] For more details, see http://www.balancingact-africa.com/news/back/balancingact_302.html.

[20] For more details, see http://www.oplan.org.

[21] For more details, see http://picopeer.net/PPA-en.html.

[22] See http://picopeer.net/PPA-en.html. Accessed on November 23, 2007.

Chapter XXI
Home is Where the Hub Is?
Wireless Infrastructures and the Nature of Domestic Culture in Australia

Katrina Jungnickel
Goldsmiths College, University of London, UK

Genevieve Bell
Intel Corporation, USA

ABSTRACT

From WiFi (802.11b) with its fixed and mobile high-speed wireless broadband Internet connectivity to WiMAX (802.16e), the newest wireless protocol, extending the reach of WiFi across longer distances and more difficult terrain, new wireless technologies are increasingly thought to impact the ways in which we encounter social spaces in public, civic and commercial sites within large urban centers. This chapter explores how and to what extent these new wireless technologies might also be reconfiguring and reorganizing domestic practice and social relations. Drawing on a year-long ethnographic study of WiFi and WiMax provisioned homes in a major Australian metropolitan center, we argue that new wireless infrastructures are impacting how people imagine and use mobile devices, computers and the Internet in and around the home but not in ways wholly anticipated by commercial Internet service providers.

INTRODUCTION

The commercial rhetoric that surrounds wireless infrastructure proposes to radically alter our everyday lives. A recent Telstra advertising campaign featuring images of a kombi van driving along an open road to the classic tune *I've been everywhere* suggests wireless infrastructure offers the 'freedom' of 'true mobility' (Telstra 2006). Other providers make claims of 'convenience', 'no worries flexibility' and 'always on' connectivity to signal the ease and ubiquity of the service (See Optus 2006; Unwired 2006). On the most basic level WiFi and WiMAX technologies enable the transfer of high-speed data wirelessly and operate via electromagnetic signals broadcast from individually owned modems connected to a broadband connection in the home or accessed via city-wide wireless coverage. This means a person equipped with a wirelessly enabled computer or handheld device can theoretically access the Internet anywhere within a broadcast area. They are no longer restricted to the office desk, the table near the fixed landline in the home, or the Internet café. Contemporary advertising for wireless services promises change to not only *how* we do things but *where* we do them. Indeed, advertising messages from service providers and technology manufacturers, political discourses and governmental agendas, and media coverage all exhort users to release themselves from the constraints of their fixed and sedentary habits and embrace new forms of wireless mobility. For the most part, this mobility is mapped directly onto an urban landscape, which users will now be freer to explore and inhabit. Not only do we reject this notion of a generic, stable, depersonalized user, but there are a number of assumptions implicit in these wireless discourses and imaginings that bear critical scrutiny: firstly, that people find their sedentary habits restrictive and will want to experience the 'freedom' that 'being wireless' offers; secondly, that wireless is 'everywhere' which makes it easy or at least easier than traditional technology; thirdly, that once people gain access to these devices and infrastructures they will use them in new places and in new ways; and finally, that there is a seamless and open terrain in which such access can transpire.

Rather than assuming wireless technology use reflects these kinds of 'anywhere' and 'anytime' imaginings, this paper explores how and in what ways people make sense of wireless infrastructures. By grounding our analysis in cultural and social practices, we offer a snapshot of a specific set of lived experiences that, for the most part, reject the current formulations of 'wirelessness' as a technology of ubiquity. We take as our starting point of analysis, the home as 'hub': here we are playing on the notion of the router as a gateway to wirelessness, and also on the centrality of the home as a unit of analysis for social and cultural practice (Arnold, M 2004; Jungnickel 2006b; Venkatesh 2006; Shepherd C. et al. 2007). We are particularly interested in re-inscribing homes as an important part of the urban computing research agenda. After all for as much as urban spaces are made up of public, civic and commercial sites, they are also composed and comprised of a complicated build out of domestic spaces – from the high-density, multi-family dwellings of Asian cities to the tightly-packed row houses and terraces of many European and British centres, urban spaces are also, already domestic spaces.

By taking the home hub as its theoretical and methodological starting point, this paper explores the concept of 'located mobility'[1] – after all, just because people can use devices everywhere does not mean they necessarily will. Located mobility suggests the existence of rules and boundaries in relation to wireless Internet and computer use in certain spaces, contexts, relationships and periods of time. This conceptual framework allows for critical re-examination of the notions of mobile users in an urban landscape. It directly addresses the rhetoric of 'ease of use' that pervades much technological discourse and problematises the idea that new information and communication technologies (ICTs) streamline and make simple everyday life. Furthermore it also helps inform research questions, specific methods and also field-site selection.

Using an ethnographic approach, this chapter goes behind and beyond the statistics and imaginings of wirelessness and computation to look at how a cluster of new wireless technologies are finding their place in the home. In addition to our own research, we draw on literatures concerned with domestic studies of technology. Whilst many domestic computer studies make clear the importance of the social and environmental contexts, gender issues and construction of personal identity (Livingston 1992; Silverstone, Hirsh and Morley 1992; Nippert-Eng 1996; Lally 2002) there has been less attention on computer use that is mobile and un-tethered from fixed points. This is primarily because wireless modems and services, laptop adoption for home use and residential wireless broadband are relatively recent technological developments. Similarly literature that deals with mobile technology tends to focus upon mobile phones in spaces outside the home (Katz and Aakhus 2002; Ling 2004; Ito et al 2005; Goggin 2006) and with the exception of Goggin's work, much of this type of research is located in Japan, the U.S and Europe. We address these research gaps by asking: What, if anything, happens in Australian homes when broadband Internet becomes unwired? What impacts, if any, does wireless broadband have on ideas of mobility and how does the expectation of 'being wireless' match up to the everyday reality of use? In this chapter we provide a fine-grained, multi-sited ethnographic account of wireless computing and broadband Internet use in urban domestic settings. We consider WiFi and WiMAX not just as urban infrastructures but as technological, social and spatial ones (Star 1999; Dourish and Bell 2007).

BACKGROUND

Despite the remarkable circulation and durability of images of the Australian outback, the reality is that more than 85% of the Australian population lives within fifty kilometers of the coast and the majority in seven key urban centers (ABS 2004) (see Figure 1). Daily life in these Australian cities involves a set of complex and complicated negotiations between home, work and a remarkable array of other social, political, economic, spiritual, commercial and athletic sites. Indeed few Australian urbanscapes are complete without a constellation of pubs, sheds, public pools, parks, beaches, sporting arenas, shops and shopping centers, legalized betting establishments, race tracks, churches, mosques, temples, heritage sites,

Figure 1. Map of Australia showing the location of coastal urban centres

transportation hubs, war memorials, community centers, Returned and Services Leagues Clubs (RSL) halls, the ubiquitous 'bottlos' (licensed alcohol retailers) and 'bowlos' (bowls clubs). Whilst our focus is the domestic sphere we draw attention to a broad array of social and cultural contexts and activities influencing technological practices in and around the home (Bell & Dourish 2007).

Australia has been a strong technology market throughout the last two decades. It routinely leads the Asia Pacific region in technology adoption and uptake – it was an early market for ATMs, cell phones, satellite technology and has recently been a very strong market for laptops. During 2003-04 alone, there was a 110% increase in consumer sales of laptop computers (Sager 2005); three-quarters of the laptops sold during this period came as standard with wireless cards which means the computer can connect to wireless infrastructures. Swiftly falling prices are seen as the prime motivator rather than innovation or design.

Although Australia has very high ICT device uptake, it has lagged significantly behind the region when it comes to Internet connectivity. When we commenced research, of 4.4 million households connected to the Internet, 28% of these using broadband and 69% dialup (ABS 2005a, 2005b). Although the 2007 census highlights the shift from narrowband to broadband with the majority of connections now being broadband (37% out of 58% of households connected), the challenges for Australian users remain the same (ABS 2007). Cripplingly slow speeds and capped download limits still place Australia far behind many other nations. Internet speed per head of population in Australia is reported at just over one megabit per second (Mbps[2]) and has been termed 'fraudband' by many dissatisfied users (SMH 2006). According to some this is because of the Government's 'failure to prevent monopolistic behavior and counter-competitive pricing' which has led to a lack of choice of alternatives, high prices and for some in more rural regions of the country unreliable or non-existent connection to the Internet and a 'less rosy' outlook for the future (Clarke 2004:39). Even with increasing speeds there is still overwhelming concern about Australia's Internet future (Hall 2007), and it was a leitmotif in Australia's most recent federal election cycle. The new Labour Government campaigned on a platform promising to speed up Australians' access to the Internet as a vital component for Australia's economic future.

It is against this backdrop of strong technology ownership and substandard infrastructures that alternative forms of connectivity have flourished. Wireless offers a way for competitors to quickly build out new networks. In 2005, Intel Corporation invested AU$37 million in Australian wireless broadband supplier Unwired to support the installation of WiMAX in major capital cities across the country (Intel Corporation 2005). The service aimed to provide high speed broadband Internet to homes, businesses and mobile wireless networks. By late 2005, Unwired was providing services for 28,000 customers in Sydney. Continued rollout across Brisbane and Melbourne in 2007 has brought their subscription base to a reported 70,000 (Mobile WiMAX 2007).

Sydney, located on the south east coast of Australia in New South Wales (NSW), was identified by Unwired as its first deployment. With nearly a quarter of Australia's population living within the greater Sydney area, it is also considered to be one of the largest cities in the world in terms of urban sprawl and population density. Sydney's 4.2 million residents, reflecting statewide technology trends, represent a strong technology adoption market.[3] In 2005, Intel Corporation with INCITE, a research lab located at the time at the University of Surrey, conducted a year-long, ethnographic study of wireless infrastructures in urban Sydney[4]. Following a multi-sited ethnographic research approach (Marcus 1998), the research data on which this paper draws included observations in select WiFi/WiMAX inhabited locations which are both public and private, participation in getting online in these spaces, qualitative interviews, mapping techniques, engagement with local events and cultural activities, and review of news media and popular literatures related to this area of enquiry. Twenty participants were drawn from the inner Sydney suburbs and nearby commuter communities; they were repeatedly visited over the course of nine months from March to December 2006. The participants represented a cross section of domestic characteristics, Internet usage and social conditions. Interviews were conducted in public places, at work or home and with prior consent, images were taken of people and their spaces and respondents were asked to draw maps indicating use and non-use of technology in certain spaces and at particular times. These maps were then used to guide interviews. Interviews lasted between one and three hours and whenever possible repeat interviews were undertaken to track developments and changes to participants' Internet usage.

Re-Placing Computing Practices in the Home?

Taken less than a month apart, and in neighbouring Sydney suburbs, these two photos represent two kinds of socio-techno ecologies and ensembles (see Figure 2 & 3). Not only is the underlying silicon differently configured and differently demanding but the contexts depicted are carved into different domestic architectures and constituted by objects that suggest distinct forms of leisure and work practices.

In the first image, a laptop sits on a kitchen bench in an open-plan lounge room. A cable snakes down to the sidewall to an electrical outlet shared with kitchen appliances and a wireless modem located in the stereo cabinet in the far wall of the lounge room connects it to the Internet. The laptop shares the bench with a pair of headphones, an MP3 player, a bottle of sunscreen and two lemons in a fruit bowl. A scattering of bills, shopping dockets and a pile of birthday presents attend to the detritus of domestic responsibilities and activities that centralise the kitchen bench firmly in the everyday life of this Sydney family.

The other image presents a fixed desktop computer within the now familiar ecology of the home office. Surrounded by an ensemble of materials and artifacts including keyboards, printers and scanners, pens, discs, paper and post-it notes, it sits on a desk behind which a series of tangled cables run to the phone line and other devices. In many regards, there is little that differentiates this from an office style setup. It is fixed, wired and physically located in one place in the home. People very clearly move to it. It does not move to or with people around the home.

On the surface, these images appear to support the rhetoric that heralds transformative change; wireless infrastructures un-tether conventional, inflexible and desk based computer practices. Our research, however, proposes something altogether different and, at the same time, something fundamentally similar. We agree that wireless presents a significant change in some households, such as those indicated in the photos above, but that these changes are deeply entrenched in the context of each home and bear an uncanny resemblance to existing practices and behaviours. In this chapter we flesh out this argument through an exploration of four themes emerging in the ethnographic research: the enduring features of domestic practice and architecture; technologies of attachment; stubborn domestic and social ar-

Figure 2 & 3. Computing practices and spaces in Australian homes

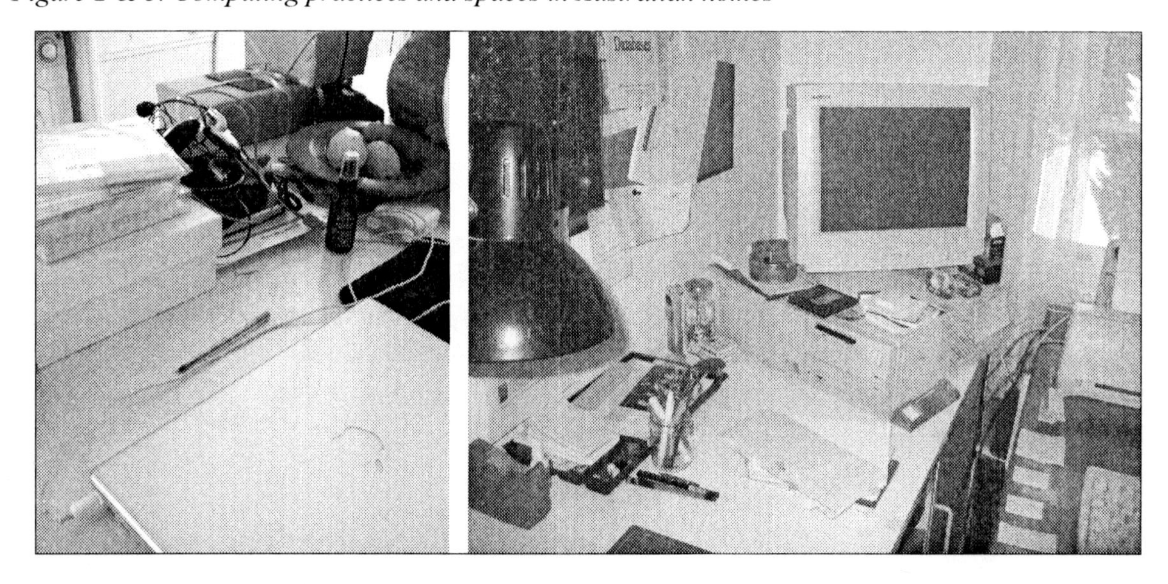

tifacts; social surveillance and forbidden zones within the home.

The Home is an Existing Ecology, Not a Blank Slate

Bernie[5] (36) is an architect at a large Sydney city firm where he manages the interior design department. He lives in the inner suburbs, in an expensive high-design one bedroom flat with his girlfriend Sal (38), a visual artist. Given their shared creative and spatial knowledge, their flat is predictably stylish and well designed. However achieving the same harmony with new wireless technology has been more difficult than Bernie imagined. For instance, he was surprised when he found the new cordless landline phone would not work in the place he wanted to put it.

Figure 4. A layered floorplan drawn by a respondent

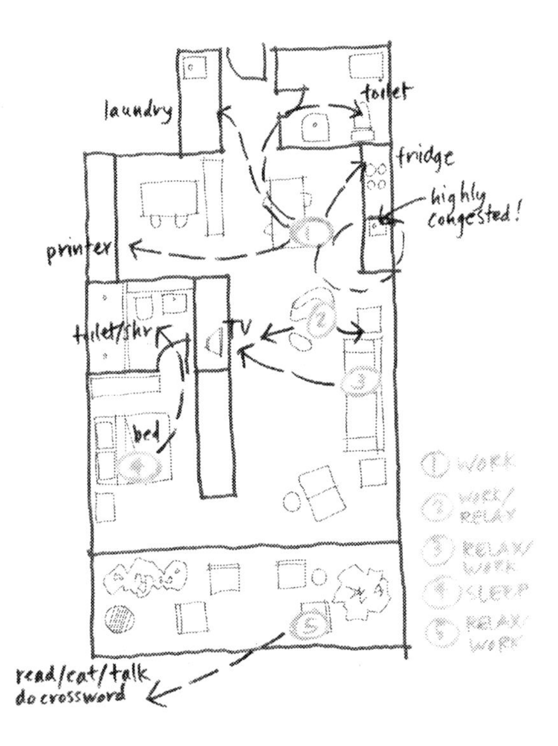

We went to plug the new phone into the technology area [the study] which is where we have our wireless, the printer and everything else, but there was too much interference for the phone so it wouldn't work in its wireless format. It's now in the bedroom, furthest away from other pieces of technology. We are lucky enough to have a one bedroom apartment. If you were living in studio you'd be cactus.

Although the cordless phone should operate anywhere in their small flat, in reality it is restricted to the far edges of their front bedroom. Through trial and error Bernie found a place for it and his mapping of new invisible territories reveals a new design palette that shapes where and how new wireless devices can be used. In addition to rendering visible existing technology and practices, wireless did something else – it collided and resisted. Like Bernie, many of our other research participants told stories about how microwaves interrupted emails, about how baby monitors blocked Internet access and about having to learn about the porosity of internal walls in order to reconfigure their technology use because of signal interference.

Bernie and Sal have moved house three times in the last three years and Sal has made the transition from full time office based architect to freelance artist working from home. All this change means her working practice knits together a variety of domestic and professional, and public and private locations and she relies on her mobile phone and laptop a lot more for both personal and paid work than ever before. When asked to map her domestic space, Sal asked if she could draw two maps because her use of the space and technology alters from day to night. Now that she doesn't have to be plugged into the landline she prefers to work in the kitchen rather than the study. This is fine when she is alone but becomes more complicated at night when Bernie tries to share the space. "Sometimes he tries to iron in the middle of my 'office,'" Sal says laughingly.

Clearly, some spaces become more complicated with wireless at different times. Sal tells of the 'congestion' zone caused by the chameleonic characteristics of the zone between the kitchen table, bench and the lounge. During the day it is her 'work' space, and at night it is her 'relax' space for watching TV, cooking and ironing. Her floor plan clearly reveals the dual nature of artifacts and space within the domestic sphere (see Figure 4). In this case we see how wireless technology interrupts and transforms domestic space, objects and practices as well as the relationships between them, or at the very least suggests complicated fragmentations.

This ethnographic account illustrates ways in which the introduction of wireless technology can complicate everyday life, domestic practices and spatial contexts. Technology is never environmentally neutral and wireless. It is like an introduced species that interrupts, interferes and intervenes with the existing visible and invisible ecology. Much has been written about how the introduction of a new technology draws attention to the ecologies of the home (Silverstone and Hirsh 1992; Livingstone 1992; Tacchi 1998; Lally 2002; Hearn and Foth 2007). In her study of radio, Tacchi (1998) shows how new technologies redefine the role of existing ones in the home, and writing about the introduction of computers into Australian homes, Lally (2002) goes even further to show how existing technologies collaborate or compete with new ones. She describes homes where the computer and television are shared experiences, chairs and tables are subsumed into computing activities and domestic furnishings are used to make computer covers. The homes Lally writes about, and the ecologies she describes, are never fixed or static even though in most cases the computers themselves are. Building on Lally's insights, we argue that a wireless network also cannot be seen as independent object in the home. It relies upon a vast array of objects in which to operate and these objects in turn take on new and unexpected roles in response to these demands, none of which are fixed but rather are in constant malleable and fluid states. As Bernie and Sal's accounts reveal, wireless does not reside in the layout of the house or the location of the furniture rather it lies in the relationship between objects and between users and objects. It is dynamic, in constant negotiation and is socially and materially produced and reproduced in everyday interactions.

If domestic socio-technical ecologies are indeed fluid and regularly reconfigured what might this mean for the study of WiFi and WiMAX? What might it mean to look in from the edges of a wireless space rather than out from the edges of the computer? What might be the consequences for technology deployments of urban density, decreasing household footprints, shared wall residences and radio-signal unfriendly housing materials (such as corrugated iron)? What this suggests is that the home presents new social, material and topographical dimensions than previously encountered with fixed line connections. Because wireless operates on a shared spectrum it leaves unique impressions on existing domestic ecologies in that it impacts both on what can be seen (furniture, wires, electrical outlets) as well as that which cannot (such the efficacy of other broadcasting devices and the porosity of walls and ceilings). It renders important these visible and invisible architectures of the home in that it both governs and interrupts pre-existing activities in particular places.

TECHNOLOGIES OF ATTACHMENT

Raymond (30) is a freelance graphic designer who spends his work time in clients offices and the spare bedroom which he uses as his office at home in the Sydney inner suburbs. He lives with his wife, Ruth (31), who recently gave up a full time job to start her own business as a stylist. This image of their lounge room might look like the corner of almost any small inner Sydney flat, with an ordinary well worn sofa and loose

Figure 5. A laptop kept under the lounge in a Sydney home when not in use

patchwork cover, piles of magazines and nesting tables (see Figure 5). Looking closely, however, reveals a Mac laptop half hidden under the sofa. Raymond explains:

The laptop [...] just lives wherever it lives, but normally somewhere under the couch. My wife's got one as well. They always tend to get shoved under there with all the cat hairs. [...] often I'll sit here and watch television or work on simpler things that don't require the computing power [of his desktop machine] and flick through books and do sort of concept work and lateral sorts of things.

Although Raymond's whole flat is blanketed by a wireless signal, there are very specific places where computing activity takes place. Work and leisure are deeply embedded in these places and in turn space and time code his engagement with the computer. From other research participants, we heard similar narratives. Some read their emails and 'Google' for news in front of the TV whilst others breast-feed while surfing the net. In the kitchen, they look for recipes or talk with friends via IM. In bed they write emails or shop on eBay. What then does it mean to think about these everyday domestic artifacts (sofa, cover, cushions) as constituting highly technical, uniquely personal and deliberately constructed components of domestic wireless infrastructure?

For all the notions of mobility engendered and imagined for wireless technologies, it is interesting to note that place remains pivotal in understanding the role and importance of wireless computer use in the home. There are fixed points (the sofa, a kitchen table, the electricity socket) around which people shape their experience of wireless-ness in the home. What appears unique about wireless infrastructure arises from the paradox that is created. Our research makes clear that instead of diminishing the importance of place, the invisible and pervasiveness nature of wireless infrastructure actually renders it more important. What this highlights is that although computing is much more mobile than in the past, it remains deeply anchored in place in terms of modes of use, firmly influenced by social and cultural boundaries.

Technologies of attachment are more than just technical devices. They constitute much larger ensembles of things and patterns of behavior. In our research, people recounted using an array of objects and spaces to access the Internet just as much as they use their computers. What the secret life of domestic objects serves to illustrate is that wireless infrastructure is neither everywhere or anywhere, as its invisible and pervasive nature along with much commercial rhetoric tends to suggest. It is firmly and tightly anchored in key domestic spaces within the social and material context of everyday life. Wireless technology offers new perspectives on the home; rather than just thinking about the computer in the home, we need to look at the home around the computer. Therefore to understand how wireless works in the home is to understand how it works *with* the home. Uncovering the secret life of domestic space offers a new dimension for exploring the tangible and intangible nature of wireless infrastructures in Sydney homes.

Stubborn Artifacts

Wires, light and the need for power have not gone away. Wireless devices remain reliant on electricity points, power cords, issues of light and weather, and social relations about where and when computing is acceptable. In terms of wire chaos, there is very little to differentiate wired homes and wireless ones (see Figure 6 & 7). Although we may be physically less dependent on ethernet wires and power cords, many wireless households still reflect traditional computer patterns of use. All participants talked about having to contend with the existing infrastructure of the home; the number and location of electricity points, sunlight, walls and furniture. Stories about the persistent need for power, and the chaos caused by the lines of its transmission, recurred over and over again in our interviews. In fact many participants had become experts in the art of wire management and concealment. Regardless of the apparent freedom and flexibility of wireless modes of transmission many artifacts and practices stubbornly remain.

Jed (30) is an artist and graphic designer who works part time as a curator at an inner city photography gallery. He rides his skateboard to the gallery three days a week and spends the rest of

Figure 6. A home with wireless infrastructure

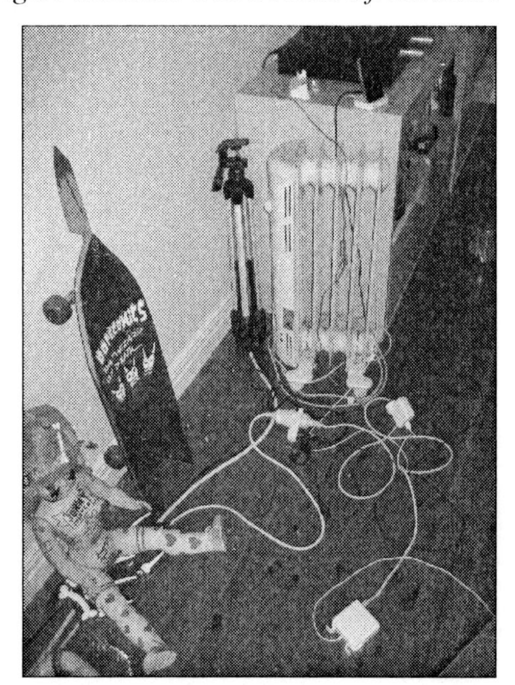

Figure 7. A home with fixed line infrastructure

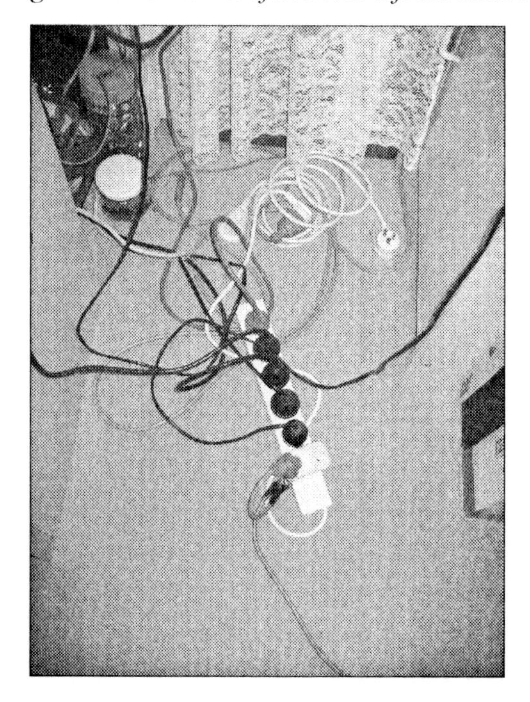

his time working on his own projects at his rented two bedroom flat or in clients offices. Although the WiMAX Internet signal in the gallery is good he has been struggling with getting a similar connection in his house only ten minutes ride away. In talking of his experiences of finding an open wireless signal inside his flat, he reflects,

It was really weird because you could almost channel it. We used to have a table set up here; so this is where it would be best [and] that's a window. I'd have my computer and I would hold it up really high against the window to get a signal, and once the signal was there, then I'd be able to put it on the table and that signal would still remain. It was pretty weird trying to channel it

Jed's WiMAX modem is connected to a wireless router at home but the signal is weak in his area. He thinks it has something to do with the weather and the height of his building. When he needs to send or receive large files he looks for overlapping signals in his flat and 'channels' it as he explains or walks to his friend house and sits on his steps to pick it up. Much like Bernie's process of trial and error, Jed's account reveals a complicated patchwork of strong signals and dead-zones.

Reid (34) is a landscape gardener for the local council and lives in a two bedroom rented house near the beach. He used to be an IT consultant but decided to take some "time out", to try something different, get outside and have more time to spend with his wife Sarah (32) and son Ben (8 months). This does not mean he has stopped thinking about technology, rather he has more time to try to get the WiMAX service working in the way he imagines it should. However it's not been easy. He explains,

If it was raining it was no good. If it was cloudy it was no good. It wasn't consistent. I couldn't monitor the bandwidth and get a constant pattern. We are in a bit of a dip here, but not massively and if it is not line-of-sight then I don't see why a little dip should make a lot of difference. [...] The modem was good only if you got it in the right position. For a while I had it next door in a polystyrene box up on the balcony.

Explaining how he dealt with these challenges Reid talks about the weather, the software on his computer, the hills around his house and his location within them, the modem, even the polystyrene box he built to protect it from the weather and where he placed it on his next door neighbor's balcony. His mapping of the home provides further illustration of the many places he has tried the modem in order to make it work. He eventually passed on his subscription to a friend in the city because he couldn't get a strong signal in his suburb.

Jed and Reid's experiences tell of a technology that is unpredictable and somewhat wild. They do their best to tame and coerce an ensemble of compatible and incompatible aspects of domestic context and external factors together to make it work. Using this apparently seamless and ubiquitous wireless technology, our research participants find themselves contending with a whole new palette of environmental characteristics, social contexts and creative workarounds. In these cases participants resorted to 'hacking the home' to deal with previously un-encountered technological tantrums (Jungnickel 2006).

Traditionally, establishing a 'set-up' for the computer meant creating a fixed place for a desktop, the cables, wires, printer, paper, discs and anything else in the home. Yet using laptops and wireless infrastructure meant some pre-existing notions of computer set-ups no longer applied and new ones need to be addressed. As a result, many struggled to adjust to new spatial and temporal dimensions that being wireless promises. In most homes, after the initial excitement of using the computer in a variety of places around the home, the reality set in and a series of place-based practices are then adhered to. For some this was a surprising, sometimes disappointing and often challenging series of experiences.

SOCIAL SURVEILLANCE AND FORBIDDEN ZONES

Narelle (28) normally works part time at the local fruit shop and delicatessen, but she has recently started maternity-leave as her third child is due in a few weeks. She lives with her husband Mike (28), a site manager for a construction firm, and their two daughters, Rachel (6) and Monique (4) in a small two-bedroom house owned by a relative. This is only an interim house, as they recently sold their two-bedroom house and are looking for something larger to accommodate their growing family. Narelle hates computers and the mess that accompanies them so much that she refuses to have one in her home. The problem is she still wants her children to have access to one. So, she has 'gone halves' with her mum and it will live in the spare room at her mum's house, a 45 minute drive away. Narelle's response is her way of dealing with the pressure to have a computer in her family life but not have it pollute her home. Her response derives not only from how it will be used in the home but its physical presence. It is not an interface that she is prepared to integrate into her everyday life. So resistant is she that she is willing to take her family to the computer rather than bring a computer into their lives. Her account presents her home as a 'No-Go zone' for computing, of any sort, whether it is fixed in place or wireless.

Other homes are more open to computing but still have rules in place. Marcus (34) is the editor of a travel magazine and runs a studio of five designers/writers in the middle of the city. He lives with his wife, Liz (34), in a rented two bedroom flat near the beach, an hour out of the city. They have wireless at home and share it with residents of the apartment block in exchange for an occasional slab of beer. This is Marcus's response to seeing someone working on their computer in a café:

I'm kind of sad [...] that they can't stop working to have lunch. But then again you have a meeting, you might as well have it in a nice environment. I just like to think we all work a little bit less and just stop a little bit more. But it's not true, it's not true of me. But that would be ideal. It upsets me that we work so hard.

Asked if he ever had the occasion to use his laptop in a café or other public space, Marcus reluctantly acknowledges he has but he didn't enjoy it and would avoid it in the future if possible.

Yes, only a couple of days ago someone asked me to go for coffee and although I had too much work on I felt I needed to because they were in a personal dilemma. This is really awful, but it kind of meant that maybe I wouldn't have even been there at all; I took the computer and had to send this email and just kind of started sending it while we were having coffee. The computer was open and things were moving. But I could just leave it there and start talking. [...] I felt rude, yes, because I was always looking over to make sure it was going.

In many ways his attitude towards computer use in public spaces mirrors his mobile phone use.

I hate answering my phone when I'm with other people and I don't like when people do it to me either really. When we're mid-conversation, mid-chat, it kind of annoys me a bit.

Just because wireless infrastructure is available does not mean it will be used and just because it is new does not mean it will generate new practices. The social etiquette of public computing was a recurring theme in our interviews. People's ready anecdotes of inappropriate mobile and computer use, and abuse and of how they personally respond hint at the strength of feeling that surrounds the rights and wrongs of public and private computing and the broader social and cultural framework that shapes this behavior. To use a laptop at the beach in Sydney, for instance, may provoke uncomfortable glances, appear rude and even lead to comments such as 'there's no need for that' or 'time to switch off'. Australian cultural norms, spatial rules and social etiquette define use and non-use of specific technologies in certain places and suggest that wireless is situated, consumed and rejected through larger social dynamics: we are tempted to suggest that this is another reworking of the Australian 'tall-poppy syndrome' where utilizing a laptop in public places might be read as putting on airs (Elder 2007). In interviews, people talked about the ways in which they clearly, tightly and uniformly allocate time and space to computers and computing in and outside the home. They might be able to use their computers and computer devices all the time ('always on' broadband) or have wireless in home but they did not always use it. Invariably there were highly patrolled rules and codes of practice in place in the house – some of which were bound by physical infrastructures and some of which are socially derived.

Spaces code the things that happen in them. Sport spaces, home spaces, work spaces, public spaces – all determine and govern the social activities that abide within them. What this means is that there will be places and times in which wireless infrastructures, irrespective of quality, strength or cost, will only occasionally be utilized and in fact some may never be used at all. Computing, one might argue, much like the consumption of alcohol in a public space, is a socially surveilled practice in Australia. However, unlike public drinking which courts fines for disorderly behavior in clearly marked 'Dry Zones' increasingly being implemented in many community parks and civic squares, there are no helpful signs to guide the public computer user. Instead technology users rely on what is deemed socially and culturally appropriate rather than what is possible, available or indeed legally tolerated. To this end public/urban computing, in some spaces, elicits emotions that run from surprise, to confusion, to pity, and, in some cases, to annoyance. It is for this reason that gaining a sense of the use and adoption of wireless infrastructure in and around Australia homes also requires an understanding of the social and cultural nuances of public interactions and the coding of urban and public spaces.

CONCLUSION

Much has changed over the last five years for Rose (75) since her husband Gilbert (76) had a cerebral hemorrhage and suffered partial paralysis. They had to retire from their printing business, move from their farm to a single storey retirement villa and buy a new car. Caring for Gilbert means Rose has very little personal time now. Although her physical mobility is limited, her social contact with people via the computer is not. Her day starts and ends with writing and reading emails and surfing the Internet. In total she spends at least three hours a day on the computer. She explains,

An email is never just a straight answer. I like to put in background music, animations, whatever. I like to make them look good. If I am using someone else's script then I will go into the source and change it.

The couple has two computers (one is ten years old and can only be used for games now), speakers, external CD drive, a printer and a webcam on a series of desks along a wall between the lounge room and the kitchen. Gilbert uses the computer as well, though it is more for card games now than for emailing his many online 'pen friends'. Together, they have downloaded over 800 games which they both like to play. Although Rose says she would like a wireless broadband connection because she hates 'all the wires everywhere' she much prefers slow speeds and no download caps than limit how she uses the Internet.

As our ethnographic examples illustrate, new wireless infrastructure does not reside in a single location, have a fixed or stable meaning, nor is there, as Rose's account makes clear, a prototypical user. WiFi and WiMAX occupy the space around furniture and walls, they interact with existing infrastructures (both invisible and invisible) and are resilient, porous and repellent to conditions inside and outside the home – one is tempted to describe them as a feral technology. Feral connotes a particularly grounded Australian understanding of the tensions and anxieties of wild and domesticated worlds. Like feral animals, wireless infrastructures often escape domestication, resorting to an unruly condition, requiring of the user, an acute and flexible awareness of the fragile ecologies in which they reside. They only become visible to us at points of attachment with specific devices, at interruptions or interferences with other domiciled artifacts or when they are entirely absent. In no sense can they be considered stable or homogenous. To engage with them predicates an understanding of what they attract and repel, where and how these overlaps interact, what is displaced and what is revealed. Occupying theoretically larger spaces than their fixed line predecessors, wireless connectivity is characterized by located mobility whereby place is still important in how and where devices are used. However, wireless does not find a single place in the home rather it finds *places*. Our findings suggest that domestic culture plays an increasingly pivotal role in the development and innovation of wireless computing practices, and in turn influence how, when and where certain computing activities take place inside *and* outside the home. What this means is that the home is not simply just another site for 'ubiquitous computing' but rather is central to understanding the social, material and symbolic elements that shape daily mobile computing practices. Furthermore the Australian context presents unique social and cultural nuances that indelibly shape local computing practices. These findings represent a fundamental shift away from thinking of the office or work place as the nucleus of computing and the home as a minor satellite point.

REFERENCES

ABS. (2004). Year Book Australia, Population Article: How many people live in Australia's coastal areas? (No. 1301.0). Canberra A.C.T., Australian Bureau of Statistics.

ABS. (2005). March Internet Activity, (No. 8153.0). Canberra A.C.T., Australian Bureau of Statistics.

ABS. (2005b). Household Use of Information Technology, Australia 2004-05, (No. 8146.0). Canberra A.C.T., Australian Bureau of Statistics.

ABS. (2007). 2006 Census of Population and Housing, (No. 2914.0.55.002), Canberra A.C.T., Australian Bureau of Statistics.

Arnold, M. (2004). The Connected Home: probing the effects and affects of domesticated ICTs. In A. Bond, A. Clement, F. de Cindio, D. Schuler & van den Besselaar , P. (Eds.), *Proceedings of the Participatory Design Conference, Toronto, Canada, July 27-31* (Vol. 2). Palo Alto, CA: CPSR.

Bell, G., & Dourish, P. (2007). Back to the Shed: Gendered visions of technology & domesticity, *Personal & Ubiquitous Computing: At Home with IT.* 11(5), 373-381.

Clarke, R. (2004). The Emergence of the Internet in Australia, In G, Goggin (Ed). *Virtual Nation: The Internet in Australia*, (pp. 30-43). Sydney, Australia: UNSW Press.

Dourish, P., & Bell, G. (2007). The Infrastructure of Experience and the Experience of Infrastructure: Meaning and Structure in Everyday Encounters with Space, *Environment and Planning B: Planning and Design,* 34(3), 414 - 430.

Elder, C. (2007). *Being Australian: Narratives of National Identity*, Allen & Unwin.

Gant, D., & Kiesler, S. (2001). Blurring the Boundaries: Cell Phones, Mobility and the Line between Work and Personal Life, In Brown, B., Green, N., & Harper, R. (Eds). *Wireless World: Social and International Aspects of the Mobile Age*, (pp. 121-132). London: Springer-Verlag.

Goggin, G. (2006). *Cell Phone Culture: Mobile Technology in Everyday Life*, Routledge.

Green, L., Holloway, D., & Quinn, R. (2004). @ home: Australian Family Life and the Internet, In G, Goggin. (Ed). *Virtual Nation: The Internet in Australia*, (pp. 88-101). Sydney: University of New South Wales Press.

Haddon, L. (1992). Explaining ICT Consumption; The case of the home computer, In Silverstone, R., & Hirsh, E. (Eds). *Consuming Technologies; Media and Information in Domestic Spaces*, (pp. 82-96). London: Routledge.

Hall, A. (2007). Australian broadband may drag down standard of living: expert, *PM - ABC Local Radio*, October 1st, 2007, Retrieved June 12th 2007 from http://www.abc.net.au/pm/content/2007/s1949401.htm

Heath, S., & Cleaver, E (2004). Mapping the Spatial in Shared Household Life, In Knowles, C., & Sweetman, P. (Eds). *Picturing the Social Landscape: Visual methods and the Sociological Imagination*, London: Routledge.

Hearn, G., & Foth, M. (Eds.). (2007). Communicative Ecologies. Special issue of the Electronic Journal of Communication, 17(1-2). New York: Communication Institute for Online Scholarship (CIOS).

Intel Corporation. (2005). Unwired Australia to Receive AUD$37 Million Investment From Intel Capital, *Intel News Release*, Retrieved November 1st 2005 from http://www.intel.com/pressroom/archive/releases/20050824corp_b.htm

Ito, M., Okabe, D., & Matsua, M. (Eds). (2005). *Personal, Portable, Pedestrian: Mobile Phones in Japanese Life,* Cambridge: MIT Press.

Jungnickel, K. (2006). Hacking the home: Technological tantrums and wireless workarounds in domestic culture, Paper presented at the *Wireless Cultures & Technologies Workshop*, University of Sydney, NSW, Australia.

Jungnickel, K. (2006b). Home is where the hub is?: Domestic cultures and wireless infrastructures in urban Australia, Paper presented at *TASA [The Australian Sociology Association] Conference*, University of Western Australia, Perth, WA, Australia.

Katz, J, E., & Aakhus, M. (Eds). (2002). *Perpetual Contact: Mobile Communication, Private Talk, Public Performance*, Cambridge University Press.

Lally, E. (2002). *At Home with Computers*, Oxford: Berg.

Ling, R. (2004) *The Mobile Connection: The Cell Phone's Impact on Society (Interactive Technologies),* Morgan Kaufmann.

Livingstone, S. (1992) The Meaning of Domestic Technologies: A personal construct analysis of familial gender relations, In Silverstone, R., & Hirsh, E. (Eds). *Consuming Technologies: Media and Information in Domestic Spaces*, (pp. 113-130). London: Routledge.

Marcus, G. E. (1998). *Ethnography Through Thick and Thin*, New Jersey: Princeton University Press.

Mobile WiMAX. (2007). *Unwired Reports 70,000 Subscribers*, retrieved 1st October 2007 from http://www.wimaxday.net/site/2007/07/27/unwired-reports-70000-subscribers/

Nippert-Eng, C. (1996). *Home and Work: Negotiating Boundaries through Everyday Life*, Chicago: The University of Chicago Press.

Optus (2006). Retrieved February 10th 2006 from http://www.optus.com.au

Sager, M. (2005). *Consumer Notebook sales Increased Dramatically in 2004, Finds IDC, Press Release 3rd March*, IDC Australia: IT&T Market Intelligence, Retrieved February 1st, 2005 from http://www.idc.com.au/press/release.asp?release_id=146

Shepherd, C., Arnold, M., Bellamy, C., & Gibbs, M. (2007). The material ecologies of domestic ICTs. *Electronic Journal of Communication* (EJC), 17(1-2). http://www.cios.org/www/ejc/v17n12.htm

Silverstone, R., Hirsh, E., & D, Morely, D. (1992). Information and Communication Technologies and the Moral Economy of the Household, In Silverstone, R., & Hirsh, E. (Eds). *Consuming Technologies; Media and Information in Domestic Spaces*, (pp. 15-31). London: Routledge.

SMH. (2006). *Australian Web access running at a crawl*, 9th March, Retrieved April 10th 2006 from http://www.smh.com.au/news/National/Australian-Web-access-running-at-a-crawl/2006/03/09/1141701635107.html

Star, S. L. (1999). The Ethnography of Infrastructure, In P. Lyman., & Wakeford, N. (Eds). *Analysing Virtual Societies: New Directions in Methodology*, American Behavioural Scientist, 43(3): 377-391.

Tacchi, J. (1998). Radio Texture: Between Self and Others, In Miller, D (Ed). *Material Cultures: Why Some Things Matter*, (pp. 25-46). Chicago: University of Chicago Press.

Telstra (2006) Retrieved February 10th 2006 from http://video.vividas.com/media/4411_BigPond/web/

Unwired. (2006). Retrieved February 10th 2006 from http://www.unwired.com.au

Venkatesh, A. (2006). Introduction to the Special Issue on "ICT in Everyday Life: Home and Personal Environments". *The Information Society, 22*(4), 191-194.

KEY TERMS

Cultural Practice: Objects, events, activities, social groupings and language that participants use, produce and reproduce in the context of making meaning in everyday life.

Ethnography: A qualitative approach that produces a thick description of the ways people live their everyday lives and involves participating in everyday activities, observing what goes on and developing relationships with people in these settings.

Feral Technologies: A particularly grounded Australian understanding of the tensions and anxieties located at the intersection of partially wild and domesticated technologies.

Located Mobility: A theoretical instrument that examines the notion of 'ease of use' that pervades much technological discourse and problematises the idea that new ICTs streamline and make simple everyday life, focusing specifically on the existence of rules and boundaries in relation to wireless Internet and computer use in certain spaces, contexts, relationships and periods of time.

WiFi: Wireless Fidelity (WiFi) is based on 802.11b standards and provides networking capabilities for computers in a localised area to transfer high-speed data wirelessly. It is most often associated with Internet use though it can be used for file sharing, voice over the Internet Protocol (VoIP) and multi-player gaming.

WiMAX: Wireless Microwave Access (WiMAX) is based on a compatible 802.16e standard and covers up to 50 kilometers without direct line of sight.

Wireless Infrastructure: A constellation of technical and social practices.

ENDNOTES

[1] Located mobility is a term initially developed by Nina Wakeford, Director of INCITE, in the design of this research project.

[2] Megabits per second is a measure of bandwidth or a unit of data for the total information flow over a given time.

[3] NSW is the largest populated city in Australia with the second highest incidence of Internet connection at home (62%). Only the ACT ranks higher with 71% (ABS 2007).

[4] Jungnickel conducted the vast bulk of the research, and analysis, with Bell serving as a stake-holder, sometimes collaborator and research partner. A report on the research was delivered to Intel stakeholders in 2007. This paper draws on that report, and additional research analysis conducted by Bell and Jungnickel over the last year. INCITE is now part of the Goldsmiths College in London.

[5] In keeping with sociological and anthropological ethics and practice all names have been changed.

Chapter XXII
Mapping the MIT Campus in Real Time Using WiFi

Andres Sevtsuk
Massachusetts Institute of Technology, USA

Sonya Huang
Massachusetts Institute of Technology, USA

Francesco Calabrese
Massachusetts Institute of Technology, USA

Carlo Ratti
Massachusetts Institute of Technology, USA

ABSTRACT

This chapter presents the iSPOTS project, which collects and maps data of WiFi usage on the Massachusetts Institute of Technology campus in Cambridge, Boston. Instead of simply mapping the locations of WiFi availability, the project is possibly the first to use and analyze log files from the Institute's Internet service provider and to produce spatial visualizations of the observed activity in real time. The aim is to create a better understanding of the daily working and living patterns of the MIT academic community, which changes due to the emergence of WiFi itself. The MIT wireless IEEE 802.11 network, consisting of 3,000 access points (one of the largest of its kind) offers a privileged environment for this research and, in perspective, can provide a test bed for entire cities.

INTRODUCTION

Recent years have witnessed a great increase in wireless Internet access points (WiFi hotspots) in cities around the world. As of the end of 2007, there were over 67,000 public hotspots already available in the U.S. (JWire, 2007), roughly doubling every year. Several forward-looking cities like Boston, MA, San Francisco, CA, and Philadelphia, PA, have launched projects to provide city-wide wireless Internet access for all citizens (cf. Forlano in this volume), WiFi is becoming as common in urban areas as traditional public utilities, such as electricity and land-line phones. The popularity of WiFi is further enhanced by its capacity to communicate multiple types of media over the same protocol: text, voice, images and video can all be streamed over wireless networks instantaneously and globally. As ubiquitous WiFi coverage might be appearing in many cities[1], we see an urgent need to explore the spatial impact of this powerful new communication network from the point of view of an urban planner or architect.

A number of studies have been done to describe WiFi signal availability and intensity in geographic context (see for instance Skyhook, http://www.skyhookwireless.com/). A culture of so-called WiFi 'sniffing' has developed in recent years, which is often related to the mapping of public wireless networks on web pages (e.g., JWire, 2007, the global hotspots finder) 'wardriving' (mapping wireless networks by driving in a car equipped with a sniffer device) and 'warchalking': the drawing of symbols in public places to advertise open wireless Internet networks. Several computer science and engineering studies have used wireless log information to analyze and quantify network traffic to answer questions about network optimization, load balance, and the like (Kotz & Essien 2005). However, there have been few attempts to analyze spatial patterns of traffic on large WiFi networks through log information from Internet Service Providers (ISPs). The lack of such studies can possibly be attributed to the difficulties of accessing raw Internet traffic data and combining it with geo-spatial databases. In the iSPOTS project, carried out by the SENSEable City Laboratory at MIT in collaboration with MIT Information Services and Technology (IS&T), we have had the opportunity to access such data and to provide on-line visualizations of its spatial distribution to the public. A real-time system was set up to gather, process, and visualize the data on the campus map, allowing the MIT community to view and act upon the information instantaneously. A description of the project, including its architecture and preliminary results, is presented below.

The remainder of this chapter is structured as follows. Section 2 provides the background and some key pieces of work related to our project. Section 3 provides background on the MIT campus and network environment. Section 4 describes the data infrastructure, and Section 5 describes our data processing methods. We discuss the weaknesses of our current setup and prescribe directions for future work in Section 6 before concluding.

CONTEXT

A series of campus-wide WiFi studies during early 2000 has paralleled the ongoing transition from fixed wire accessibility to ubiquitous WiFi environment. Some of the most comprehensive of these studies were done in Dartmouth College (Henderson, Kotz & Abyzov 2004, Kotz & Essien 2002, Kotz & Essien 2005). Within the past years the usage of campus WiFi has increased as more people have adopted WiFi-enabled laptops, as well as other WiFi clients such as PDAs and VoIP devices. However, the proportion of WiFi users at popular buildings, in libraries and classrooms appeared to be consistent from the years 2001 to 2004. A similar pattern of preferred WiFi location usage was observed at the University of Saskatchewan in Canada (Schwab & Bunt 2004).

The evidence from these studies was inconclusive regarding types of Internet based activities the users engaged in at these locations. An earlier study at Stanford (Tang & Baker 2002) found that the activities performed while connected to the wireless network varied from person to person, involving both work and leisure communication activities (email and instant messaging). At the University of North Carolina, Chinchilla, Lindsey & Papadopouli (2004) found that users accessed similar web content regardless of their location.

The main purpose of the studies above was to characterize network performance and WiFi users' individual exploitation of the network. Another relevant group of studies revolves around the technologies for indoor positioning. WiFi-based positioning, sometimes called WPS for WiFi Positioning System (Schilit et al. 2003), have been explored by Balazinska & Castro (2003) for locating users by building. Other location systems based on indoor sensor networks include Cricket, an economical yet highly accurate system for indoor navigation (Priyantha et al. 2003), and wireless motion sensors deployed in the Mitsubishi Electronic Research Laboratories building with more intelligent algorithms to determine what movements actually correspond to human presence in spaces (Wren et al. 2006).

The iSPOTS system is applied over the entire MIT campus (with a few exceptions described in the following section) covering thousands of individual room locations. The large amount of data collected, characterized by a predetermined distribution of access points based on the original assigned use of space (e.g. classroom, hallway, residence, etc.) can inform space planners about the efficiency of space usage as evidenced by the location patterns of laptop users. Yet making visualizations of the data available to the public in real time opens up opportunities for more applications that are directly relevant ot users of the space.

THE MIT CAMPUS

Our test environment—the MIT campus—can be regarded as a miniature version of an urban neighbourhood. 10,320 students and 9,414 total employees attend the campus, which consists of more than 190 buildings covering a considerable portion of the city of Cambridge, MA. In the year 2000, when laptops were still expensive and wireless Internet new, MIT decided to undertake a vast operation of building a campus-wide wireless network. Since October 2005, this 168 acre campus has over 3,000 active wireless access points providing full coverage of WiFi in all academic and residential buildings. Besides providing a valuable experiment for examining the spatial patterns of IEEE 802.11 WiFi networks, the analysis of such a large wireless network has also offered us helpful insight for studying other types of communication systems, such as mobile phone networks[2].

The MIT wireless network infrastructure currently uses the IEEE 802.11 protocol exclusively. All access points run by IS&T share the same 'MIT' network name, which permits wireless cards on people's devices to roam seamlessly from one access point to another. The IS&T network division is currently using three different types of wireless access points in the campus-wide wireless network: Avaya Ap-3, Proxim AP3000 and Enterasys AP-3000, with a signal radius from 130 to 350 feet indoors. This allows each access point to serve one or part of a room, as well as neighbouring rooms. shows the relative positions of the analyzed antennae on campus and illustrates the 'ideal signal availability' in the given set of access points without taking into account physical barriers, such as walls and floor plates, which in reality decrease signal propagation.

In the data that has been made available to us so far, we have observed wireless traffic in up to 2659 unique access points in 134 buildings on MIT's Cambridge campus. Data about some access points are not available to us, as they belong to

Figure 1. Schematic map of the MIT campus

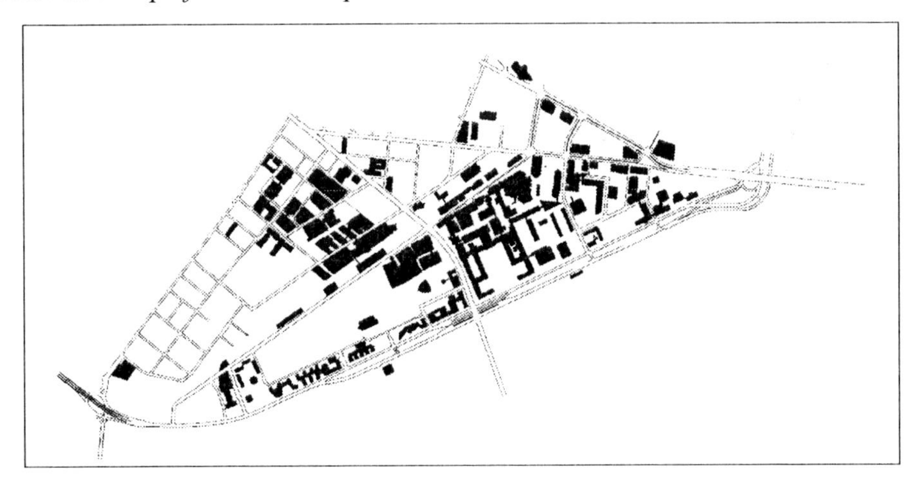

Figure 2. A subset of access points at their locations on the MIT campus

Figure 3. The theoretical availability of the MIT wireless on campus

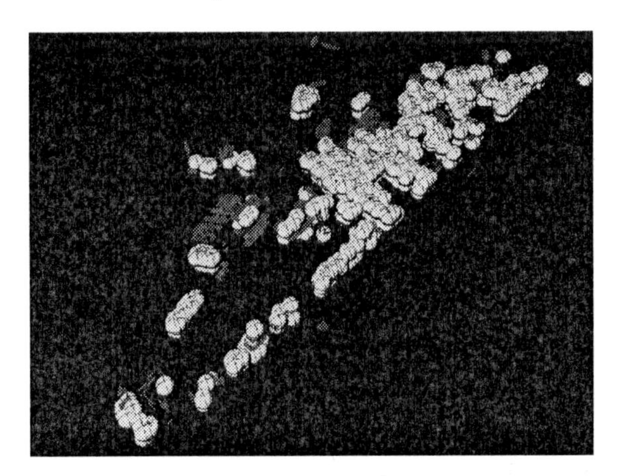

networks operated privately by individual departments. Others we are not able to map, because the GIS data we have about the campus does not yet include some recently constructed buildings. We hope to be able to update that information soon.

There are also other wireless networks on campus maintained independently within labs, departments, and schools through which one is able to connect to the MIT wireless network (MITnet); however, IS&T does not provide support to the MIT community for these networks (for more information on the specifications of MITnet, see MIT IS&T, 2007). Two large independent networks whose data we still lack belong to the MIT Media Laboratory and the Computer Science and Artificial Intelligence Laboratory (CSAIL). We plan to obtain their data in the near future and include the two buildings in the overall maps.

While full WiFi coverage is an amenity of many university campuses in the U.S., Dartmouth College being exemplary (see for instance Kotz & Essien, 2005), the number of access points at MIT is manifold greater. Also, a large majority of MIT students, especially in the graduate community, own laptop computers with WiFi capability, allowing them to freely connect to the wireless network, and the ownership rate has been increasing every semester. According to a study by Dal Fiore, Goldman, and Hwang in 2006, 73% of students bring their laptops either every day or some days of the week to campus. As a result, we have begun to empirically notice changes in the ways that people use the campus facilities for living and working. The intensive evening hours at multiple libraries and the infamous 'Athena clusters' — computing labs consisting of networked terminals that any MIT community member may log in — are giving way to heavy wireless traffic at the student dormitories. Similarly, many laboratories, which until a couple of years ago were bustling with people, now have students scattered in nearby cafeterias, collaborative study rooms and lounge-spaces equipped with WiFi.

As part of the iSPOTS project, we created a digital infrastructure for quantifying such changes. However, as data only visualizes wireless activity patterns, there is a lack of understanding about how non-laptop users exploit the campus. At this point it is difficult and perhaps too early to draw clear conclusions about the impact of WiFi on people's spatial preferences, but work by Dal Fiore, Goldman, and Hwang is beginning to shed some light on the topic.

THE DATA

The spatial analysis of WiFi on campus required data from two sources. The first, obtained from the MIT Department of Facilities, was a geospatial database of all buildings, rooms, and their respective uses as of fall 2005 (buildings and rooms under construction, such as the new student lounge in 10-108 and Building 46, the new Brain and Cognitive Sciences building, are not included in our database). Our second source was MIT IS&T, from which we still keep receiving two constant streams of data. These are data on the number of users per access point, and bytes transferred per access point. Both are measured as totals in 15-minute intervals. A schematic of the streaming data is provided in Table 1.

Table 1. Schematic for streaming data

Users	Transfers
Access point identifier	Access point identifier
Number of users	Number of bytes transferred
Unix timestamp	Unix timestamp

The overall architecture of the system is shown in Figure 4. Each access point of MITnet is georeferenced using coordinates of the centroid of its containing or nearest room. IS&T runs a program that records the number of times a connection is made to the Internet through one of the monitored access points. Each record is refreshed every 15 minutes. These records are transmitted to a MySQL database on the SENSEable City Laboratory server. Users are allowed to view this data through an interactive widget that displays the number of users over time in a chosen space in the past seven days. The chosen space may be the entire campus, a section of campus (e.g. East Campus, West Campus), a building, floor of a building, or a room.

In all public displays of data, the SENSEable City Laboratory follows general guidelines about users' privacy, under which all statistics we present are in aggregate form and only concern network activity. In other words, no data on individual users' locations[3] is obtained by iSPOTS.

DATA PROCESSING

Real-Time Maps of WiFi Usage

As a holistic means of visualizing the spatio-temporal patterns of WiFi usage on campus, we used an ArcGIS script to generate maps of the campus showing the total number of users for

Figure 4. iSPOTS real time data transfer system

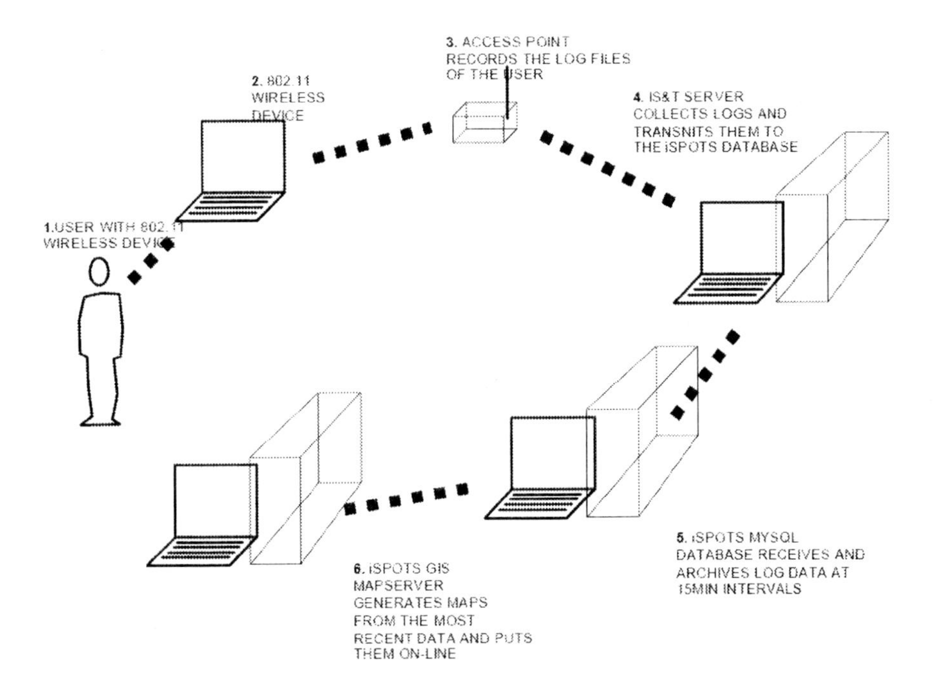

each 15-minute time interval. We further used Macromedia Flash to animate the maps of the latest 24 hours. A sequential visualization of the maps allows users to view how the centres of activity shift between various parts of the campus at different hours of the day. For instance, one can clearly see how the Main Campus (the section of campus housing most of the classrooms) is highly populated during work hours between 10AM and 5PM, while the buildings on West Campus, where most dormitories are located, absorb most of the activity during late evening hours.

Figure 5 shows an example of campus WiFi usage over a period of 24 hours. The original data from access points is in attributed point format. In order to create two dimensional maps, the punctual data was interpolated using standard GIS functions. The activity of WiFi users on campus, as indicated in the maps, is fairly typical of students who attend classes or work on a weekday schedule slightly later than nine-to-five. The growing spots in the centre and east campus between 8:00 AM and 10:00 AM show the increase of users logging on WiFi in academic buildings during the start of the work day. Similarly, usage in academic buildings decreases between 7:00 PM and 9:00 PM. The hours between 12:00 AM and 8:00 AM are relatively quiet, though the row of residences on west campus shows users logged on quite consistently.

TIME GRAPHS OF WIFI USAGE

While the maps described above give viewers a holistic sense of user activity on campus, a more accurate picture is revealed with graphs of WiFi use over time for specific spaces. By mapping the number of WiFi users on the Y axis and a time period on the X axis, a unique signature graph is constructed for each access point's activity distribution over a chosen time period. The selection of spaces available to the user included the whole campus, any section of campus (e.g. east

Figure 5. iSPOTS maps over 24 hours on Tuesday, December 6, 2005. Brighter areas indicate a larger number of users, while black areas indicate no users connected to MITnet via wireless

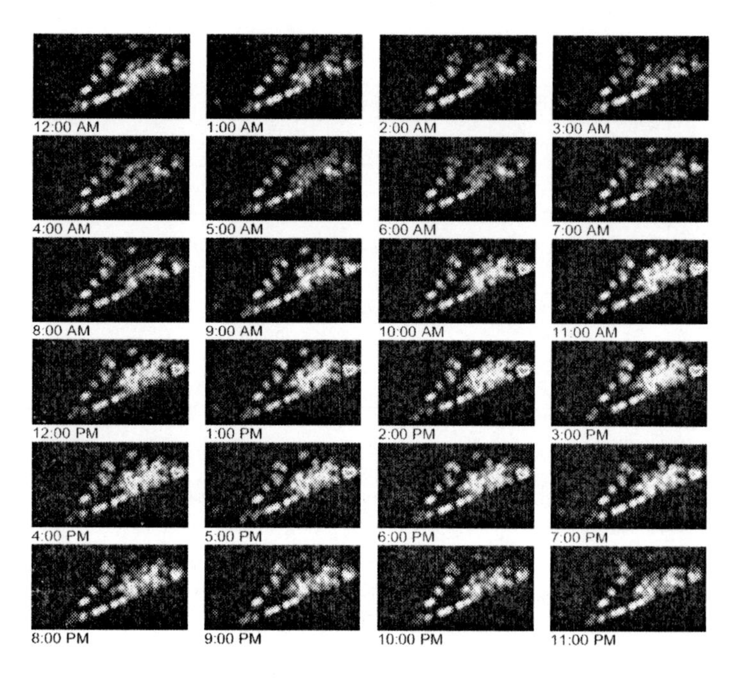

campus, west campus), any building on campus, any floor of a building, and any room with a WiFi access point. The default interface creates graphs of WiFi use over the past week, whereas users can also set custom query periods on a separate statistics web page.

Separating academic, residential, and service facilities over a one-week period shows a fairly predictable pattern of WiFi activity on campus, consistent with the maps above. A one-week graph of WiFi use over the whole campus includes daily peaks typically around 11:00 AM on weekdays, followed by secondary peaks around 8:00 PM. Weekends seem to peak at the same times but with much lower intensity. The first peak is indicative of the high rate of connections made in the morning in academic buildings; a typical academic space witnesses WiFi usage centred around an 11:00 AM peak. The second peak corresponds to the high rate of connections in residences in the evening. Figure 6 shows example graphs of the whole campus, an academic space, a residential space, and a service space.

The double-peaked pattern of WiFi usage we observe over each day appears to be quite common in usage patterns of wireless devices in more general settings. Time graphs of wireless PDA usage on the University California San Diego campus (McNett & Voelker 2003), and bluetooth signals at a city center pub in London (O'Neill et al. 2006) also show a high morning peaks and low afternoon peaks, although the San Diego peaks are a few hours later, possibly because all their subjects are students.

DISCUSSION AND FUTURE WORK

The iSPOTS data mapped above has a precise meaning: it shows the amount of wireless devices connected to the network access points at different locations on the MIT campus in real time. During some of the discussions so far, we have made an implicit assumption that WiFi usage information could be used as a proxy for activity and people.

Figure 6. WiFi usage graphs from top: all of campus, fifth floor of Building 9 (Department of Urban Studies and Planning offices), Sidney-Pacific Residence (largest residence on campus), Student Centre

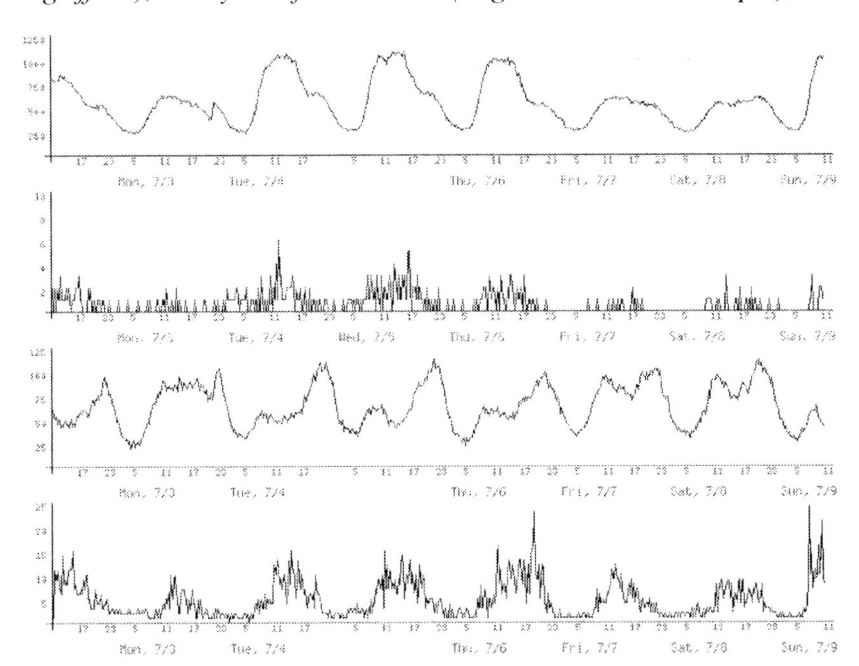

While this assumption probably holds in general terms, a number of biases, both geographical and social, need to be mentioned. Like most behavioural studies, WiFi usage is not likely to exhibit an accurate and periodically repetitive picture of an individual's spatial presence. Both the data and its sampling are too crude for this. In that regard, the spatial analysis of network traffic clearly departs from the more traditional, highly qualitative and in depth studies of social science. However, despite its poorer individual concern, the quantitative network activity based approach allows a far greater sample size and spatial distribution to be analyzed. In addition, once a system is set up, data can be collected over extensive periods of time with virtually no effort involved. A study of its biases and probabilistic representation bounds through data mining, surveys and observations is part of our future work.

One way in which WiFi usage is a biased measurement of actual user activity is clearly, the bias toward spaces where laptop owners are more likely to be present, and where people with laptops are more likely to connect to the network with their laptops. Classes that restrict laptop usage and paper-based exams are just a few examples of situations where our system would not detect the presence of many people at work.

A second source of bias in the interpretation of the data could be the uneven distribution of laptop ownership: while we are quickly moving towards uniform and saturated ownership, graduate students are currently the segment in the MIT community with the highest proportion of laptop ownership. In addition, laptop usage is likely to vary among different departments, employment types as well as temporal habits. In general, as of spring 2006, IS&T registered an average of 5,373 unique WiFi users per day, about a quarter of approximately 20,000 MITnet users (MIT, 2007).

Third, we realize that laptops are often left connected to wireless Internet when not in use and the iSPOTS system directly measures wireless devices and not people. If this information is superimposed onto the data transfer at each access point, it is possible to distinguish between the amount of active and inactive users. However, some biases could still occur (e.g., people may download files when their computer is idle or the machine may be automatically downloading/uploading program updates). It is possible that such activities occur randomly across all computers, and the aggregate quality of the data is thus not significantly hampered, but graveness of the issue remains to be verified. Recent cluster analysis of the aggregate data has shown that the observed WiFi usage patterns in different types of rooms can quite accurately distinguish the academic, residential and service spaces on campus.

An interesting question that our data can address is: given the increased mobility of users, are traditional land use/space use categories still useful ways to classify space? Or do the overlapping uses such as cafes as workspaces and residences as workspaces impily that we should rethink space classification? Jull and Ratti's (forthcoming) analysis provides some evidence that building signatures still correspond to basic space use designations, though we may still expect this to change.

Finally, we recognize that the university campus differs systematically from general urban environments in ways that would be reflected in WiFi usage patterns. For example, the characteristic late-night usage in residence halls is probably a trait endemic to students, who comprise half the university population. More studies in diverse settings would contribute to our knowledge of behavior in urban environments, but campuses may provide the most comprehensive urban sub-setting to the extent that they provide many regular urban functions, including facilities for work, residence, recreation, and other support services.

While we continue working to address the accuracy issues highlighted above, the iSPOTS data can already be used in a qualitative way for campus planning purposes. The data reveals the dominant trends in people's mobility and may be used to substantiate with numerical evidence observations that have been simply anecdotal so far. The Campus Planning Committee of MIT, for instance, could take notice of the emerging spatial changes of life and work environments, and redirect their efforts to support the new trends. A good example is the recently completed 'Steam Café' on the fourth floor of the architecture building, which underwent a complete remodelling during January 2005 (http://steamcafe.mit.edu). The design and execution was left in the hands of architecture students, who not only used the opportunity to redesign the café, but also re-conceptualize the café's image and menu. Before its conversion, the area that is now Steam Café was used by few and only for a limited time each day—the lunch break. Now, thanks to the overlapping of different activities, the presence of WiFi, and a new concept in design (not to mention better food), it is active around the clock. Sales have increased three-fold and the Campus Planning Committee is beginning to recognize that a more efficient environment could emerge by extending similar concepts to the whole campus.

It is also possible to imagine future scenarios, based on the iSPOTS system:

Scenario 1. When the MIT Campus Planning Committee engages in negotiations with a new evening snack cafeteria to open on campus, they can use the iSPOTS archive to aid their decision in site selection. With a few simple queries, they may find, for example, residence areas where WiFi usage shows large numbers of users during afternoon and evening hours. They might even find that the best opening hours for the cafeteria are not standard 9-to-5 working hours, but rather, when most wireless usage occurs in the vicinity.

Scenario 2. A more sobering yet still realistic application is emergency situations. For example, if a security alert, such as a fire alarm or a toxic gas leak occurs in a large building, then security officers could easily check the current status of

WiFi usage in that building and make an intelligent guess about how many people might be inside the building. The predictive capacity of the system depends on calibrating the models with empirically observed usage patters. Even though not everyone uses WiFi and as most people only use it for a limited period during the day, corrections could be applied to the data by using statistics that have been observed over longer periods of time. For instance, with an accurate estimate of the percentage of laptop users in a building and among those, the percentage of them that are WiFi connected, one may calculate the total predicted number of people in the building with fair probability (cf. Nakanishi, Ishida, & Koizumi in this volume).

Scenario 3. Most certainly, the real-time on-line database could become a useful and fun information tool for MIT students on a daily basis. For instance, an architecture student, who is working at home, might be wondering if it would be better to work in his/her studio at school, because other friends might be working there too. In this case he/she could go on-line to the iSPOTS map and run a query for the given studio space and see how many people are currently using wireless Internet in that space. If the network turns out to be busy, then he/she might want to move and join the others. A prototype application "StudioBRIDGE" for the department of Architecture was shown to be useful for this purpose: users did find it useful to check for the presence (without identity) of other users in studios before entering (Yee and Park 2005). For a more general application, it would be most useful to add contextual information to the data, describing how the current presence of users in a certain space compares to the historical average presence of users in the same space and similar time of the day. Furthermore, based on the recent data from the same space, it would be useful to indicate whether the amount of users is currently stable, growing or diminishing.

There are many other examples of possible uses for the real-time analysis capacities of the iSPOTS project. The availability of real time views of the campus activity can allow the creation of feedback mechanisms, which prompt reactions similar to those observed in real time control systems (for a general introduction to control systems see Shinner, 1998; for an exploratory discussion of what that might mean in the urban context see Mitchell, 2005, and Calabrese, Kloeckl and Ratti, 2007). A similar implementation might in the near future apply to neighbourhoods or even cities at large, where urban processes could be observed in real-time from the broadest flux in the city to the highly specific queries about single buildings. As urban GIS data are expanding world-wide, greater opportunities are created for urban analysis. In addition, GIS data exchange over the Internet also enables information to be shared from multiple other databases in the city or around the world. In short, the ever-increasing amount of urban data that are recorded every day with painstaking precision can find many uses in an on-line environment, which enables either all or selected users to find vast amounts of accurate real-time information about the city around them.

The implications of a real-time mapping exercise include not only a new tool for mapping, but also a changing perception of the campus or city as a whole (Ratti and Berry, 2007). We acknowledge that the goals and representations of such mapping are clearly different from traditional urban mapping, and we do not want to contest the value of such maps. Rather, we hope to enrich the palette of urban mapping by introducing a new tool, which can help us visualize the city as a set of processes and broaden our perspective on the complex interrelationships of its elements. If the image of a map changes from static to dynamic and acquires different layers of real-time information, then the map is no longer a fixed reference, representing the durable objects and spaces of the city. A real-time map becomes as lively as the urban environment it represents, and is literally shaped by the users of the environment. In a real-time map, not only urban elements, but also processes are spatially represented. A public on-line distribution of the map allows large numbers of people to monitor the urban flux simultaneously, thus raising the public awareness of the dynamism of the contemporary city through simple cartographic evidence.

Finally, visualizing aggregate people's movement trough secondary sources such as WiFi, is of course not the same as understanding the movements and the causes behind them. Spatial analysis of wireless network traffic merely creates an opportunity for urban scholars to perceive the interactions between people and the built environment. As two dimensional mappings of thousands of users in actual spaces, the iSPOTS maps collapse a great deal of complex information into single images. We acknowledge the challenges and dangers of omitting valuable qualitative data from such representations and focusing perhaps too much on a single chosen variable: the spatial presence of people, which can blind one from a more nuanced understanding of spatial occupancy. The untangling of the complex causal and correlational relationships between physical spaces and their human use is still waiting for its thorough scholarly study, and we look forward to participating in such future work.

CONCLUSION

This chapter reviews ongoing research on the iSPOTS project at the MIT campus in Cambridge, MA. The aim of the project is to analyze usage of the wireless Internet network in order to describe occupancy patterns and movements of its users. Interim results seem to suggest that this type of analysis is powerful and could have many applications—whose relevance could extend to entire cities in future years when they become wireless.

iSPOTS takes real-time data about usage on WiFi access points as a measure of how people use space on the MIT campus. We created interfaces to retrieve data in the form of colour-coded maps and time graphs, for users who may be interested in analyzing the use of specific spaces. We found that data on individual user locations can be useful to users themselves through a peer-to-peer social networking and location-sharing applet. In the next few years, as wireless computing continues to expand into wider urban areas and programs that depend on ubiquitous wireless computing, our system may inform urban planners and administrators of these environments and programs of likely user responses to spatial conditions.

Regarding the MIT campus, we hope that iSPOTS data will soon shed light on a number of changes that are happening due to pervasive wireless accessibility. In particular, we would like to validate or disprove a number of hypotheses. For instance: are we really witnessing a switch towards increasing mobility in individual working patterns? Is it true that traditional classifications of space do not hold anymore, as people are changing their working patterns due to the introduction of wireless communication? And finally, one of the most important questions for architects and planners: if you can be at work anywhere, where would you like to be? What are the architectural qualities of spaces that people vote for with their feet?

ACKNOWLEDGMENT

First, we would like to thank the other members of the iSPOTS team, without whom this project would have not been possible: Daniel Gutierrez, David Lee, Xiongjiu Liao and Jia Lou. Our acknowledgments also go to Jerry Grochow, Vijay Kumar, Teresa Regan, Mark Silis and Dennis Baron, who granted us access to the data on the usage of the MIT wireless Infrastructure and helped us with interpreting them. Vijay Kumar and the MIT Direction of Academic Computing also provided generous seed funds for launching the iSPOTS project. We are also indebted to people at the Massachusetts Institute of Technology for their feedback and for providing an extremely stimulating research environment. In particular, we would like to thank Dennis Frenchman, William Mitchell and Lawrence Vale. We also thank Marcus Foth and four anonymous referees for their helpful suggestions. Janet Owers provided editorial guidance. Of course, any shortcomings are our sole responsibility.

REFERENCES

Balazinska, M. & Castro, P. (2003). Characterizing mobility and network usage in a corporate wireless local-area network. in *MobySys '03.*

Calabrese, F., Kloeckl, K. & Ratti., C. (2007). Urban Computing and Mobile Devices, WikiCity: Real- time urban environments. *IEEE Pervasive Computing*, July–Sept., pp. 52–57. 2007.

Chinchilla, F., Lindsey, M. & Papadopouli, M. (2004). Analysis of wireless information locality and association patterns in a campus. in *IEEE Infocom 04.*

Dal Fiore, F., Goldman E., Hwang E. (2006). Does Laptop Usage Affect Individual Mobility?. Technical report, SENSEable City Laboratory, MIT, Cambridge, MA.

Henderson, T., Kotz, D. & Abyzov, I. (2004). The changing usage of a mature campus-wide wireless network, in *MobiCom '04: Proceedings of the 10th annual international conference on Mobile computing and networking*, ACM Press, New York, NY, USA, pp. 187–201.

JWire (2007). http://www.jiwire.com/search-hotspot-locations.htm, accessed on 12 November 2007.

Jull, M., Ratti, C. (forthcoming) *Urban Ritual: A Study of Building and City Wireless Signatures*

Kotz, D. & Essien, K. (2002). Analysis of a campus-wide wireless network. Technical Report TR2002-432, Dept. of Computer Science, Dartmouth College.

Kotz, D. & Essien, K. (2005). Analysis of a campus-wide wireless network. *Wireless Networks* 11, 115–133.

McNett, M. & Voelker, G.M. (2003). Access and Mobility of wireless PDA users. *ACM SIGMOBILE Computing and Communications Review*, 9(2): 40-55.

MIT (2007). http://web.mit.edu/measures, accessed on 12 November 2007.

MIT IS&T (2007). http://web.mit.edu/ist, accessed on 12 November 2007.

Mitchell, W. (2005). The Real Time City *Domus*, October 2005, 885 pp. 94-99.

O'Neill, E., Kostakos, V., Kindberg, T., Fatah gen. Schiek, A., Penn, A., Stanton Fraser, D. and Jones, T. (2006). Instrumenting the city: developing methods for observing and understanding the digital cityscape. In proceedings of UbiComp 2006, 315-332.

Priyantha, N.B., Chakraborty, A., and Balakrishnan, H. (2000). The Cricket Location-Support System. Proceedings of the 6th annual international conference on Mobile computing and networking. Boston, MA, USA, 32-43.

Ratti, C., Berry, D. (2007). Sense of the City: Wireless and the Emergence of Real-Time Urban Systems. in *Chatelet V. (editor), Interactive Cities*, Editions XYZ, Paris.

Schilit, B., LaMarca, A., Borriello, G., Griswold, W.G., McDonald, D., Lazowska, E., Balachandran, A., Hong, J., and Iverson, V. (2003). Challenge: Ubiquitous Location-Aware Computing and the 'Place Lab' Initiative. *Proceedings of the 1st ACM international workshop on Wireless mobile applications and services on WLAN.* San Diego, CA, USA, 29-35

Schwab, D. & Bunt, R. (2004). Characterizing the use of a campus wireless network. in *IEEE Infocom 04.*

Shinners, S.M. (1998), *Modern Control System Theory and Design.* Wiley-IEEE.

Tang, D. & Baker, M. (2002). Analysis of a metropolitan-area wireless network. *Wireless Networks* 8(2–3), 107–120.

Wren, C.R., Minnen, D.C., and Rao, S.G. (2006). Similarity-Based Analysis for Large Networks of Ultra-Low Resolution Sensors. *Pattern Recognition,* 39(10): 1918-1931.

Yee, S. and Park, K.S. (2005) StudioBRIDGE: using group, location, and event information to bridge online and offline encounters for co-located learning groups. *Proceedings of the SIGCHI conference on Human factors in computing systems,* pp. 551-560

KEY TERMS

Real-Time Map: Instantaneous and dynamic visualization of an environment.

Wireless Mapping: Graphic representation of digital activities in a wireless communication network.

GIS: Geographic Information Systems

Urban Dynamics: A description of the changing movements of people, objects and information in a city.

WiFi Hotspot: An area around an Internet access point where people can connect to the Internet through their wireless communication devices.

Spatial Occupancy Study: An analysis describing how people use certain built spaces over time

MIT: Massachusetts Institute of Technology

ENDNOTES

1. Some of the pilot cities like Chicago have, however, run into financial and technical difficulty in setting up city-wide free WiFi, which has set the trend in question.
2. For instance the Real-Time-Rome project: http://senseable.mit.edu/realtimerome/
3. The voluntary sharing of individual information, on a peer-to-peer opt-in platform is the focus of our subsequent project called iFIND (http://ifind.mit.edu).

Chapter XXIII
Supporting Community with Location-Sensitive Mobile Applications

John M. Carroll
The Pennsylvania State University, USA

Craig H. Ganoe
The Pennsylvania State University, USA

ABSTRACT

We discuss the vision, plan, and status of a research project investigating community-oriented services and applications, comprising a wireless community network, in State College, Pennsylvania, USA. Our project specifically investigates new possibilities afforded by mobile and location-sensitive wireless networking access with respect to community engagement and informal learning, as well as broader changes in community attitudes and behaviors associated with the deployment of this new infrastructure.

INTRODUCTION AND BACKGROUND

During the next several years, many American communities (cites, towns and other relatively populated areas) will consider investing in pervasive wireless networking infrastructures, with the intent of providing better broadband coverage for their citizens and also acting as an information technologies (IT) resource for municipal services. These infrastructures could broaden the currently-typical "Starbuck's scenario" (accessing one's email over a cup of coffee) to include ubiquitous interactions, that is, continuous and location-sensitive Internet access through personal devices. But why should municipalities and citizens do this? More specifically, what are there *civic* rationales for developing these infrastructures? Can these infrastructures serve as more than just a way to access the broader Internet, and additionally provide a means for civic engagement and civic action on a local scale? If so, what are examples of effective civic applications of public wireless infrastructures?

This area is rapidly gathering momentum both as research and as development. Much of the work in this area has focused on creating opportunities for interaction between friends (Burak & Sharon, 2004), classmates (Schilit, et. al,2003), distant strangers (Davis & Karahalios, 2005, Paay, 2005), game players (Chang & Goodman, 2004; Crabtree, et. al, 2005; Vogiazou et. al, 2005), and even dating partners (e.g. match.com offers mobile services) through matching and notification services. Other work focuses on providing users with information through mobile guides and the ability to provide users with information in educational settings such as museums (Kjeldskov, et. al, 2005; Sumi & Mase, 2000).

A few researchers have drafted scenarios that relate to community capacity building. Ananny et. al. (2003) describe a system used by residents of a housing complex who used a text messaging service to add captions to photos of their community as a way to deal with neighborhood change. Lane (2003) described a system in which users could access and add location-specific content to a place (museums, libraries, schools) providing a sense of community memory. Leimeister et. al, (2004) describe a scenario in which information is made available to virtual communities of cancer patients via web-based services. On a limited scale, systems such as Neighbornode (http://www.neighbornode.net/) have been created which allow for the creation of neighborhood hubs allowing news and information to be shared in a community.

Although some work has addressed community capacity building, most of it is policy discussion or envisionment (and not actual design, prototyping, and evaluation) and most of it has not focused on mobility and real-time interactions. For example, Gurstein (2002), in the research literature, as well as Intel's (2005) whitepaper "Core Technologies for Developing a Digital Community Framework" and their 2006 "webinar" collaboration with the Knight Foundation and One Economy, emphasize wireless community applications but, in general, conflate applications specifically-enabled by wireless infrastructures (e.g., real-time, mobile interaction) with access to the Internet *at all*.

It is well known that people use technology to support and maintain existing social relationships in their homes, in the workplace, in their neighborhoods, and in community organizations (Center for the Digital Future, 2005; Horrigan, 2001; Wellman & Hampton, 1999; Wellman, et. al, 2001). However, the existing literature is quite desktop bound: It does not address how people might appropriate mobile/location-sensitive network services to maintain their connections to existing community organizations or to develop new connections. We are creating a test bed to explore this.

Ubiquitous infrastructures are chiefly being used now for individual-oriented notifications and information seeking. This is understandable; such applications are low-hanging fruit. More importantly, from our perspective, these applications are priming the adoption of personal Internet devices. As people learn about and make use of individually-oriented mobile and location-sensitive applications, they are collectively putting in place socio-technical infrastructure that can enable civic and community-oriented applications.

We see ubiquitous infrastructures, and the mobile and location-sensing applications they afford as a new frontier for community informatics. The underlying technologies are already being deployed widely, though there are fundamental and troubling issues about what business models are feasible and appropriate (e.g., Quillen, 2008). Many exciting issues are being articulated about what sorts of community-oriented mobile/location-sensing applications will be attractive and useful to citizens, what sorts of participatory design processes will be feasible and appropriate for such applications, what sorts of community-based learning support will be effective and practical, and what sorts of social and community impacts will be entrained. Research issues are of greatest interest when prerequisite socio-technical infrastructures are available for application and dissemination, but not yet universally applied or adopted. This is precisely the case for mobile/location-sensitive applications of muncipal wireless networks.

The basic technology for realizing these benefits is already here. The main challenges are socio-technical: dissemination, content/application development, adoption, learning, appropriation, and adaptation. The history of technology diffusion shows that these challenges are decisive in the ultimate trajectories of new technologies. This is where most of the creative effort is needed now.

Our work attempts to focus upon and to foreground (1) mobile, location-sensitive services and applications, and (2) communitarian constructions of wireless networks. We feel that social networking services (friend finders), and even commercial services (time and place-sensitive instant coupons) can be supportive of socially more profound communitarian goals, but we are not impressed by revisionist views that attempt to merely conflate communitarian goals with these more superficial interactions (e.g. Florida, 2003; Arnold, Gibbs & Wright, 2003). We do not offer a detailed analysis of these revisionist views here (but cf. other chapters in this volume). But it is important for the reader to understand how we are construing our own work.

Our work descends from the "community networking" tradition, which regards communitarian goals, such as social capital formation, community collective efficacy, social support, community engagement and attachment, and strong local democracy, as the most significant potential outcomes of local network infrastructures (Carroll, 2005; Schuler, 1996). Iconic projects in this tradition from the 1970s and 1980s include the Berkeley Community Memory (Farrington & Pine, 1996), Cleveland Freenet (Beamish, 1995), Santa Monica PEN (Rogers, Collins-Jarvis & Schmitz, 1994), and Big Sky Telegraph (Uncapher, 1999). Our vision is that a municipal wireless broadband network can be a *community network,* that it can and should directly support things that make a community rich and worthwhile. Our research hypothesis is that wireless infrastructures can increase awareness of opportunities for civic engagement, and support new kinds of civic activity; they can bring people together in new ways as well as strengthening existing civic groups.

The chapter proceeds as follows: First, we describe the setting for our current study, the borough of State College, Pennsylvania, and its plans for a municipal wireless network. Next, we describe a core set of communitarian application concepts. Then, we describe technology issues and plans regarding the implementation of these applications. Finally, we consider assessment and evaluation of the infrastructure and applications we hope to put in place.

DESIGNING LOCATION SENSITIVE COMMUNITY APPLICATIONS FOR CIVIC ENGAGEMENT

The borough of State College, PA, in cooperation with Penn State University, is planning a pervasive wireless infrastructure. This infrastructure will be deployed initially in State College, and subsequently be extended to adjacent municipalities. The design focuses on core institutional stakeholders, such as the downtown merchants, the borough government, and Penn State. Example application concepts include advertising and announcements sponsored by merchants, electronic government services such as management of parking fees and policing, and anywhere/anytime access to Penn State students, staff, and visiting alumni.

In the context of this planning effort, we approached the municipal government as part of a broad university-community partnership, developed through the course of a 4-year U.S. National Sscience Foundation (NSF) project on sustainable community-based learning about information technology (2003-2007). Through the course of this project, our research group has collaborated with the nonprofit community in Centre County, PA, to increase capacity to use information technology to achieve civic-oriented goals and sustainably manage technical expertise (Merkel et al. 2004, 2005, 2007). We have worked with a diverse set of community partners that represent a range of people and issues: (i) a water quality group, (ii) an environmental preservation group, (iii) a high school learning enrichment program, (iv) the local historical society, (v) a youth services organization, (vi) the local symphony orchestra, (vii) an organization that promotes home ownership for low-income residents, (viii) a food bank, (ix) an organization that develops community leadership, (x) an emergency response group, (xi) our local public access television, among others. We worked with our community partners to develop the Central Pennsylvania Community Information Technology Workshop, an annual meeting of 50-100 community groups, so that the lessons learned in our research could be more widely disseminated (http://learn.centreconnect.org/).

Our NSF project exemplifies a long-term community-based participatory design model we have developed during the past two decades (Carroll, Chin, Rosson, Neale, Dunlap & Isenhour, 2002; Carroll & Rosson, 2007). In this project, our approach leverages trust relationships with community groups and the State College municipal authorities to develop a strong community component to the planned wireless infrastructure. We were able to convince local authorities to include a free content area in the developing conception for the wireless network. This free area would include community content, services, and applications, accessible to citizens anywhere and anytime as they move around in the town.

During the early months of our project, we developed core set of application concepts. These were inspired generally from our four-year collaboration with a variety of community groups, and in some cases draw specifically on application ideas contributed directly by community partners. Subsequently, we have presented these ideas at meetings with individuals and groups, and at several large community meetings (for example, we presented these ideas for discussion during the opening plenary session of the 2007 Community Information Technology Workshop, an annual meeting for municipalities throughout central Pennsylvania that was initiated as part our prior NSF project). The initial core set of application concepts is given immediately below: Civic Smart Mobs and Nonprofit Collaborative Suite are examples of concepts we were already developing through our previous collaborations with State College community groups; Lost State College is an example of a specific idea from one of our partners:

1. **Civic Smart Mobs:** In his book, "Smart Mobs", Howard Rheingold (2002) discusses how mobile connectivity can help to coordinate social action, though most of his examples involve coordinating protests and social gatherings. It is straightforward to recruit this idea to send out a real-time invitation for volunteering at the Senior Center, or participation in a flower planting or cleanup event at a town park.
2. **Place-based Blogging:** This application of interactive maps would allow people or organizations to host a blog "at" a location in town. People could subscribe to the blog in various ways, including being notified when they are at the corresponding physical location. For example, as a construction project begins in town, people walk by the site and may have thoughts about the project

that they could contribute and share with their neighbors. Urban Tapestries (Lane et al. 2005) prototyped a location-based mobile client where community members could annotate a map with text, images and sound (eventually including RSS and web access).

3. **Lost State College:** This idea leverages an on-going community effort to identify the history of various places in State College. We will embed that content in interactive maps integrated with the wireless hotspots so that people could access the history of a place as they stand in it, and be able to move around adjacent places and their histories. These places can also be annotated by community members to include their knowledge and photographs. Prior research (Carroll et al. 1999; Klaebe et al. 2007) has shown that historical storytelling is an effective means of engaging community members.
4. **Nonprofit Collaborative Suite:** The work of nonprofit organizations is often distributed and place-sensitive. We are working with nonprofits using community-oriented participatory design methods comprising a suite of ubiquitous collaborative tools tailored for nonprofits groups (Farooq et al. 2007). This includes calendaring, wikis, messaging, forums, etc. We will use this initial toolbox to investigate the infrastructure needed to support organizations that use a wireless network to achieve civic-oriented goals.

Our many discussions of these concepts with a variety of citizens and community groups has not only validated them, but it has helped to make the very notion of community-oriented information, services and applications as part of a municipal wireless infrastructure far more accessible to our partners. It has made it more possible to engage our partners in concrete discussions about design plans. In this way, the initial set of application concepts directly helped to evoke refinements and further concepts from our partners. We now have a more extensive set of concepts to work with in the future.

Our view is that planning a design in a "frontier" area like community wireless should be regarded as an open-ended project. We regard our application concepts only as a starting point for a community-wide discussion about how wireless networking can support the community. Indeed, it seems both prudent and realistic to regard applications implemented now primary as mechanisms to trigger broader design discussions from which the real application concepts will emerge. In that spirit, our current focus is to implement the set of prototype applications above that can be demonstrated and/or deployed to help provoke broader community discussion.

Technology for Location Sensitive Civic Applications

Implementing the prototype applications requires integrating a number of location related technologies that are currently emerging as standards for commonplace use. Prior work has had similar ambitions to ours (Lane et al. 2005), but adoptability was often limited by the necessity to provide users with devices capable of the location-based services. Those limitations are disappearing with the growing support on mobile devices for: 1) mapping services, 2) location services, 3) positioning standards for data, and 4) integration between location information and existing collaborative tools. Our effort to provide these location-based civic applications is a systhesis of these emerging technologies along with existing approaches. The following sections emphasize how these capabilities and standards have evolved, and how we will integrate these technologies into our prototype applications.

1. **Mapping services:** Maps are an important component to local civic applications. Numerous systems are emerging that provide

the tools for developers to create location-sensitive applications. Other systems provide public mapping APIs: Google Maps (http://maps.google.com), Yahoo! Maps (http://maps.yahoo.com), and others. Google Maps for mobile (http://www.google.com/gmm) is a standalone mobile mapping application which available on many current mobile platforms. We have also developed a Java infrastructure for synchronous geo-collaboration based on shared, interactive maps (Schafer et al. 2007). We are developing a small-client version of this software to support synchronous/asynchronous map pointing, editing, and annotation of areas within the community, and content generated within these tools can be exported live to standard formats such as KML (http://code.google.com/apis/kml) for integration into other mobile tools.

2. **Location services:** Location services provide the mechanism for devices to determine where the user is physically located. These systems generally either use Global Positioning System (satellite based triangulation) or triangulation around land based wireless signals (such as WiFi access points or mobile phone towers) to determine location. Various developer tools for location triangulation based on nearby WiFi access points are available for use (PlaceEngine.com, SkyhookWireless.com, PlaceLab.org, etc.) and mobile devices are integrating these capabilities onboard. These tools provide the means to relate personal location to the location of online community content.
3. **Positioning Standards for Data (RSS, iCalendar):** Standards have emerged for integrating location information with other content. The Internet Calendar and Scheduling specification (iCalendar), RFC-2445 (http://www.ietf.org/rfc/rfc2445.txt), includes a geographic position property to specify the geographical location of events. The GeoRSS community (http://www.georss.org) has provided a mechanism in XML for specifying geospatial points, lines, boxes, and polygon as part of RSS and Atom feeds. In our own research, we have investigated the use of RSS feeds for managing software development projects and digital libraries (Farooq et al.,2007; Hylton et al., 2005). We are extending this to be location sensitive in two senses, first, notifying clients within a geo-spatial zone that documents pertaining to locations within that zone have been created/modified. Second, this capability will also allow us to investigate location-sensitive RSS in the sense of events posted to a location-specific RSS feed. Many of our community non-profit partners are working toward calendar solutions that include iCalendar support, our goal is to provide an aggregator for their calendar data which includes location-based support.
4. **Desktop/Off-the-Desktop Integration:** The location-sensitive/mobile techniques and applications we develop will be integrated with place-based desktop client views (Carroll et al., 2001). This integration will allow users to virtually navigate to locations within the municipal wireless cloud for editing and browsing when they are not at these locations. Our suite of BRIDGE Tools (http://bridgetools.sourceforge.net/) provides both a synchronous cross-platform Java client for real-time creation and editing of shared content as well as an asynchronous web interface for lightweight interactions with the content Many editing types are supported, such as: wiki pages, calendars, whiteboards and spreadsheets. Chat and discussion tools are provided for conversation, and awareness features show who is online as well as who is working with which of the tools. We have examined mechanisms and visualizations for combining those tools within workspaces for long-term project

work (Ganoe et al. 2003, 2004), and have developed techniques to evaluate the use of collaborative tools for long-term activity that can combine both field and in lab studies (Neale et al. 2004; Convertino et al. 2004). We are integrating location support into the content objects (as well as previously mentioned RSS and iCalendar location support). We are also investigating the addition of similar location features to other open source content management system based blog and wiki engines.

We have also prototyped a Java Micro Edition client for BRIDGE, and investigated designs and evaluation techniques for collaborative mobile problem solving (Massimi et al. 2007). Another community-oriented area for collaborative mobile problem solving would be education. Many educational middle and high school courses ranging from physics to government involve activities outside the classroom. While there has been work on the use of mobile computing for data gathering (e.g., recording pH levels in streams) and sites such as www.sensorbase.org provide places to share location-based sensor data, there has been little work on sharing data and solving problems *while mobile,* yet this would seem to be an attractive application area and directly addresses current educational practices. Some of the applications we will develop and assess in this project (e.g. "Lost State College" described earlier) can easily be adapted for educational projects.

Our goal in this research is to develop a model infrastructure populated with example applications, and documented by empirical studies of people using the infrastructure in daily life. This will be a definitive case study of evolving community attitudes toward and practices with respect to a new class of communications infrastructure. It is a next-generation Blacksburg Electronic Village; in that earlier project, we were able to concretely assess the utility of home Ethernet connectivity years before it was generally available. This test bed will also be a living laboratory: a valuable resource for investigating how types of wireless networks (WiFi, G3, WiMAX, etc.) and mobile network devices (smart phone, ultra-mobile PC, etc.) might modulate community applications and civic effects.

More academically, such an infrastructure provides the opportunity for advancing theoretical concepts in community informatics. Community informatics is concerned with finding ways to apply information technologies in the civic arena in ways that are appropriate for a given geographic community and knowledge domain (Gurstein, 2000). Little is known about the requirements for a location-sensitive infrastructure designed to support civic and social engagement or the specific mobile applications that will achieve these goals, or about how community organizations might use these applications and the specific opportunities and challenges that drive wireless network adoption in this sector. The design of this research is to fill that gap.

A practical outcome of this work will be a better understanding of the infrastructure and applications necessary to improve civic engagement and community life in State College. Our goal is for these approaches to serve as a model for other communities who wish to encourage civic engagement through their own municipal wireless initiatives. Developing community informatics application models for wireless civic infrastructures provides a direct driver for computing services, software, and equipment, and one that is largely independent of entertainment, personal information management, and business/government computing. Making the case that wireless public infrastructure can have effective civic-oriented applications will facilitate the acceptance of these technologies by communities throughout the US and the world.

Evaluation

One of the chronic issues in community technology is evaluation. A key reason for this is that community technology projects are typically carried out with an inadequate budget (certainly so when compared to commercial software development!). But another reason is that community technology projects—and particularly those with strong communitarian objectives (Carroll, 2005; Schuler, 1996)—are just more difficult to evaluate. Thus, a commercial product intended to support Web-page authoring can be assessed by sitting a person down and asking him/her to use it to author a simple Web page. Community networks are seeking to improve social capital, social support, community participation, community identity and attachment, and community collective efficacy—rather profound aspects of social-cognitive affect and disposition! However, these complex predicates of human behavior and experience are relevant to some extent in all collaborative interactions. Thus, another way to construct this situation is that community informatics provides an intimate laboratory for studying essential human aspects of all collaboration infrastructures (Carroll, 2001).

Evaluation in this project focuses on three areas:

1. **Application Requirements:** Our design/development process is an iterative one where we use participatory methods to work with users to develop requirements and evaluate our software designs. These requirements analysis and evaluation processes use standard methods and are being documented. This evaluation not only aids in our development of appropriate applications, but the lessons learned could have implications for device design and provide a reference to others developing community-oriented wireless applications.
2. **User Surveys and Cases:** Once in place, we will survey registered users of the software and non-profit organizations about their use of the wireless network and the developed applications. We will conduct interviews with users and other community members as well as analyze generated public content within the applications to develop more in-depth cases of use.
3. **Context Statistics:** We will capture anonymous usage statistics for the applications, such as relatively where and when different types of applications are being used (along with other typical browser-like statistics: application, OS,etc.). Those statistics can then be combined with community land use data and demographics to help understand the context of use. Initial results from this data may then also be used to aid in location-specific surveys for areas of public use around State College.

We are gathering survey and interview data on community attitudes toward municipal wireless and means of civic engagement at the start of the project, and then at set intervals throughout the project (Carroll et al., 2005). We will also use a case study framework to evaluate expectations and experiences of users as we deploy and introduce applications, as people try out, adopt, appropriate, and adapt those applications, and after the infrastructure and applications has been in place for a period of time (Neale et al., 2004; Merkel et al., 2005, 2007). Our long-term participatory design process will provide us with a lens into the community throughout the project (Carroll et al., 2002; Carroll & Rosson, 2007). We will also use server logging to track usage patterns (Carroll et al., submitted). This project will be a model for civic-oriented applications of wireless networks so it will also be useful to gather practical hands-on information about installation, training, maintenance, potential bottlenecks, and so forth.

Along with our participatory design, development and evaluation approach (described above), the key management element in our plan is to be and remain central players in the design, rollout, and operation of the State College downtown wireless infrastructure. Part of the opportunity in our case is that there is broad local participation in this plan—by the Borough of State College, by the Downtown Merchants, and by Penn State (whose campus is adjacent to one of the primary downtown streets and whose campus wireless infrastructure will seamlessly integrate with the public network). Groundbreaking projects in this area (like Seattle Wireless, http://www.seattlewireless.net/) have been hampered by *not* being broad community initiatives.

One of the most significant impacts on current practices that this project can have is to broaden the types of assumed applications that are taken into consideration in planning public infrastructures. For example, in State College, the borough government immediately began working with police and fire companies and with downtown businesses when they started to design this infrastructure. They were not reluctant to include us and our community group partners, they just did not think of it on their own. This project can help to make civic, cultural, and informally educational applications more salient to public infrastructure planners. A richer set of orienting applications will (obviously) entrain richer and better infrastructures. It could accelerate the adoption and deployment of such infrastructures. We also hope to be able to garner fresh insights on mobile technology usage by community members and thus inform mobility research agendas and design. (see for example, http://softwarecommunity.intel.com/isn/home/Mobility.aspx)

FUTURE TRENDS

Pervasive wireless Internet is potentially much more exciting than just access to the Internet. Wireless means mobile: Access to the Internet while on the move around town. The next generation of cell phones will be Internet-capable personal media devices; their harbinger is the iPhone. Wireless also means location-sensing: In Singapore—today!—one can order a taxi by texting a couple of characters to say "send the closest taxi to where I am standing to pick me up now".

CONCLUSION

Internet access is no longer an arcane indulgence for computer geeks. The Internet presents vital information for healthcare, government, education, shopping, civic activity, and leisure. Internet access needs to be universal, and municipal wireless broadband can contribute to achieving this. But access is only part of the equation, for communities, we argue that as much as universal Internet access is a vehicle to reach out to the global, it can also be a means to connect more deeply with the local.

Municipal wireless Internet can allow people anywhere in our Borough to check municipal information and community news and events. More elaborate and unique services are easily imagined. Wireless Internet access can make it possible to view the history of locations in State College while the user was standing at those locations. The possibility of experiences like this help more intimately connect people to the places they live. In the future, this mobile/location-sensitive paradigm for Internet access will likely become standard. The citizens of a community have a stake in the planning and deployment of their networks, and in the realization of new possibilities for community networking.

A municipal wireless broadband network can be a *community network*: It should directly support things that make a community rich and worthwhile.

REFERENCES

Ananny, M., Biddick, K., C., and Strohecker, C. (2003). Constructing public discourse with ethnographic/SMS "texts." In *Proceedings of Mobile HCI 2003*,Springer-Verlag Lecture Notes in Computer Science, 368-373.

Arnold, M., Gibbs, M. R., & Wright, P. (2003). Intranets and Local Community: 'Yes, an intranet is all very well, but do we still get free beer and a barbeque?' In M. Huysman, E. Wenger & V. Wulf (Eds.), *Proceedings of the First International Conference on Communities and Technologies* (pp. 185-204). Amsterdam,

Beamish, A. (1995). *Communities On-Line: Community-Based Computer Networks*. Masters Thesis, Department of Urban Studies and Planning, MIT.

Burak, A. and Sharon, T. (2004). Usage patterns of FriendZone: mobile location-based community services. In *Proceedings of the 3rd international Conference onMobile and Ubiquitous Multimedia* (College Park, Maryland, October 27 -29, 2004). MUM '04, vol. 83. ACM Press, New York, NY, 93-100.

Carroll, J.M. (2001). Community computing as human-computer interaction. *Behaviour and Information Technology*, Vol 20, No. 5, pp. 307-314.

Carroll, J.M. (2004). Participatory Design of Community Information Systems:The Designer as Bard, In F. Darses , R. Dieng , C. Simone & M. Zacklad, (Eds.*), Cooperative Systems Design: Scenario-Based Design of Collaborative Systems, Volume 107 Frontiers in Artificial Intelligence and Applications.* Amsterdam: IOS Press, pp. 1-6.

Carroll, J.M. (2005). The Blacksburg Electronic Village: A study in communitycomputing. In P. van den Besselaar & S. Kiozumi (Eds.), *Digital Cities 3: Information Technologies for Social Capital.* Lecture Notes in Computer Science, Volume 3081. Berlin: Springer-Verlag, pp. 43-65.

Carroll, J.M. & Bishop, A.P. (2006). Introduction to Special Section on "Learning in Communities". *Journal of Community Informatics, 2-2,* 116-121 (Special Section is pp. 116-133).

Carroll, J.M., Chin, G., Rosson, M.B., Neale, D.C., Dunlap, D.R. & Isenhour, P.L. (2002). Building educational technology partnerships through participatory design. In J. Lazar (Ed.), *Managing IT/Community Partnerships in the 21st Century.* Hershey, PA: Idea Group Publishing, 88-115.

Carroll, J.M. & Farooq, U. (2007). Patterns as a paradigm for community-based learning. *International Journal of Computer-Supported Collaborative Learning, 2(1),* 41-59.

Carroll, J.M. & Rosson M.B. (1996). Developing the Blacksburg Electronic Village. *Communications of the ACM, 39 (12),* 69-74.

Carroll, J.M. & Rosson, M.B. (2003). A trajectory for community networks. *The Information Society, 19 (5),* 381-393.

Carroll, J.M. & Rosson, M.B. (2007). Participatory design in community informatics. *Design Studies, 28,* "Special Issue on Participatory Design", 243-261.

Carroll, J.M., Rosson, M.B., Isenhour, P. L., Van Metre, C., Schaefer, W.A. and Ganoe, C.H. (2001). MOOsburg: multi-user domain support for a community network. *Internet Research, 11 (1)*, 65-73.

Carroll, J.M., Rosson, M.B., Isenhour, P.L., Ganoe, C.H., Dunlap, D., Fogarty, J., Schafer, W., & Van Metre, C. (2001). Designing our town: MOOsburg. *International Journal of Human-Computer Studies, 54,* 725-751.

Carroll, J.M., Rosson, M.B., Van Metre, C., Kengeri, R.R., Kelso, J. and M. Darshani. (1999). Blacksburg Nostalgia: A Community History

Archive. *Proceedings of INTERACT '99.* IOS Press.

Carroll, J.M., Rosson, M.B. & Zhou, J. (2005). Collective efficacy as a measure of community. *Proceedings of CHI 2005: Human Factors in Computing Systems.* (Portland, OR, April 2-7), pp 1-10.

Carroll, J.M., Snook, J. Isenhour, P.L. submitted. Logging home use of theInternet in the Blacksburg Electronic Village. *International Journal of Advanced Media and Communication*

Center for the Digital Future. (2005). *2005 Digital Future Report.* USC Annenberg School, Los Angeles, CA. URL:http://www.digitalcenter.org/pages/current_report.asp?intGlobalId=19

Chang, M. and Goodman, E. (2004). FIASCO: game interface for location-based play. In *Proceedings of the 2004 Conference on Designing interactive Systems:Processes, Practices, Methods, and Techniques* (Cambridge, MA, USA, August 01 04, 2004). DIS '04. ACM Press, New York, NY, 329-332

Chewar, C., McCrickard, D.S. & Carroll, J.M. (2005). Analyzing the social capital value chain in community network interfaces. *Internet Research, 15(3),* 262-280. Won Outstanding Paper Award.

Convertino, G., Neale, D. C., Hobby, L., Carroll, J. M., and Rosson, M. B. (2004). A laboratory method for studying activity awareness. In *Proceedings of the Third Nordic Conference on Human-Computer Interaction* (Tampere, Finland, October 23 27, 2004). NordiCHI '04, vol. 82. ACM Press, New York, NY, 313-322.

Crabtree, A., Rodden, T., and Benford, S. (2005). Moving with the times: IT research and the boundaries of CSCW. *CSCW, 14 (*3), 217—251.

Davis, B. and Karahalios, K. (2005). Telelogs: a social communication space for urban environments. In *Proceedings of the 7th international Conference on HumanComputer interaction with Mobile Devices & Services* (Salzburg, Austria, September 19 -22, 2005). MobileHCI '05, vol. 111. ACM Press, New York, NY, 231-234.

Farrington, C. & Pine, E. (1996). Community memory: A case study in community communication, P. Agre, and D. Schuler, Eds., *Reinventing technology, rediscovering community.* Ablex.

Farooq, U. & Carroll, J.M. (2003). Mobilizing community networks. *Proceedings ofHOIT 2003: Home Oriented Informatics and Telematics, The Networked Home and theHome of the Future.* http://www.crito.uci.edu/noah/HOIT/2003papers.htm.

Farooq, U., Ganoe, C.H., Carroll, J.M. & Giles, L. (2007). Supporting distributed scientific collaboration: Implications for designing the CiteSeer collaboratory. *Proceedings of HICSS 40: Hawaii International Conference on Systems Science,* (Hilton Waikoloa Village, January 3-6).

Farooq, U., Ganoe, C. H., Xiao, L., Merkel, C, Rosson, M.B. and Carroll, J. M., Supporting community-based learning: Case study of a geographical community organization designing their web site, *Behaviour & Information Technology: Special Issue on Computer-Support for Learning Communities,* Volume 26, Number 1 (January-February, 2007), pp. 5-21. London, UK: Taylor & Francis.

Florida, R. L. (2003). Cities and the Creative Class. *City and Community, 2*(1), 3-19.

Ganoe, C. H., Convertino, G., and Carroll, J. M. (2004). The BRIDGE awareness workspace: tools supporting activity awareness for collaborative project work. In

Proceedings of the Third Nordic Conference on Human-Computer interaction (Tampere, Finland, October 23 -27, 2004). NordiCHI '04, vol. 82. ACM Press, New York, NY, 453-454.

Ganoe, C.H., Schafer, W.A., Farooq, U. & Carroll, J.M. (2001). An analysis oflocation models for MOOsburg. In M. Beigl, P. Gray & D. Salber (Eds.),*Proceedings of Workshop on Location Modeling for Ubiquitous Computing*, ACMUbicomp2001 Conference on Ubiquitous Computing (Atlanta, September 30—October 2), http://www.ubicomp.org/, pp. 45-48.

Ganoe, C. H., Somervell, J. P., Neale, D. C., Isenhour, P. L., Carroll, J. M., Rosson, M. B., and McCrickard, D. S. (2003). Classroom BRIDGE: using collaborativepublic and desktop timelines to support activity awareness. In *Proceedings of the16th Annual ACM Symposium on User interface Software and Technology* (Vancouver, Canada, November 02 -05, 2003). UIST '03. ACM Press, New York, NY, 21-30.

Gurstein, M. (2002). Weaving community with community fibre: communityinformatics and the broadband revolution. In *Proceedings of the 35th AnnualHawaii International Conference on System Sciences*. HICSS'02, vol. 8, IEEE Computer Society, Washington, DC, 214.

Gurstein, M. (Ed.) (2000). *Community Informatics: Enabling Communities withInformation and Communications Technologies*. Idea Group Publishing, Hershey, PA.

Horrigan, J. October 31, (2001). *Online Communities: Networks that Nurture Long-distance Relationships and Local Ties*. Pew Internet & American Life Project, Washington, DC. URL: http://www.pewInternet.org/reports/toc.asp?Report=47

Hylton, K., Rosson, M. B., Carroll, J. M., and Ganoe, C. H. (2005). When news is more than what makes headlines, *ACM Crossroads: Human-Computer Interaction*,Winter 2005, 12.2. URL: http://www.acm.org/crossroads/xrds12-2/rss.html

Intel Corporation, (2005). *Core Technologies for Developing a Digital Community Framework*. White Paper. URL:ftp://download.intel.com/business/bss/industry/government/digitalcommunity-framework.pdf

Isenhour, P., Rosson, M.B. & Carroll, J.M. (2001). Supporting interactive collaboration on the Web with CORK. *Interacting with Computers, 13*, 655-676.

Kjeldskov J., Graham C., Pedell S., Vetere F., Howard S., Balbo S. and Davies J.(2005). Evaluating the usability of a mobile guide: the influence of location, participants and re-sources. *Behaviour and Information Technology*, 24 (1), 51-65.

Klaebe, H., Foth, M., Burgess, J., and Bilandzic, M. (2007). Digital Storytelling and History Lines: Community Engagement in a Master-Planned Development. *Proc. 13th Intl Conference on Virtual Systems and Multimedia*. VSMM 2007, Brisbane, Australia, 23-26 Sept 2007.108-120.

Lane, G. (2003). Urban tapestries: Wireless networking, public authoring and social knowledge. *Personal Ubiquitous Computing, 7 (3-4)*, 169-175.

Lane, G. and Thelwall, S. (2006). Urban Tapestries: public authoring, place and mobility. URL: http://socialtapestries.net/outcomes/reports/UT_Report_2006.pdf

Leimeister, J. M., Daum, M., and Krcmar, H. (2004). Towards mobile communities for cancer patients: the case of krebsgemeinschaft.de. *International Journal of Web Based Communities, 1 (1)*, 58–70.

Massimi, M., Ganoe, C. H. and Carroll, J.M. (2007). Scavenger hunt: an empirical tool for collaborative mobile problem solving. *IEEE Pervasive Computing*, January 2007, 6(1), 81-87.

Merkel, C.B., Clitherow, M., Farooq, U., Xiao, L., Ganoe, C.H., Carroll, J.M. &Rosson, M.B. (2005). Sustaining computer use and learning in communitycontexts: Making technology part of "Who they are and what they do". *The Journal of Community Informatics, 1(2)*, 134-150.

Merkel, C.B., Farooq, U., Ganoe, C.H., Rosson, M.B. & Carroll, J.M. (2007). Managing technology use and learning in nonprofit community organizations:Methodological challenges and opportunities. *Proceedings ofCHIMIT 2007: First ACM Symposium on Computer-Human Interaction for Management of Information Technology.* March 30-31, 2007: Cambridge, MA.

Merkel, C.B., Xiao, L., Farooq, U., Ganoe, C.H., Lee, R., Carroll, J.M. & Rosson, M.B. (2004). Participatory design in community computing contexts: Tales from the field. *Proceedings of the Participatory Design Conference* (Toronto, Canada, July 27-31). New York: ACM Press, pp. 1-10.

Neale, D. C., Carroll, J. M., and Rosson, M. B. (2004). Evaluating computer-supported cooperative work: models and frameworks. In *Proceedings of the 2004 ACM Conference on Computer Supported Cooperative Work* (Chicago, Illinois, USA, November 06 -10, 2004). CSCW '04. ACM Press, New York, NY, 112-121.

Paay, J. (2005). "Where we met last time": a study of sociality in the city. In *Proceedings of the 19th Conference of the Computer-Human interaction Special interest Group* (CHISIG) of Australia on Computer-Human interaction: Citizens online: Considerations For Today and the Future (Canberra, Australia, November 21 25, 2005). ACM International Conference Proceeding Series, vol. 122. ComputerHuman Interaction Special Interest Group (CHISIG) of Australia, Narrabundah,Australia, 1-10.

Quillen, K. (2008). Future of Earthlink's municipal wireless business uncertain. The New Orleans Times Picayune, February 15.

Rheingold, H. (2002). *Smart Mobs: The Next Social Revolution*. Perseus, Cambridge, MA.

Rogers, E. M., Collins-Jarvis, L. & Schmitz, J. (1994). The PEN project in Santa Monica: interactive communication, equality, and political action. Journal of the American Society for Information Science, 45, 6, 401-410

Rosson, M.B., Dunlap, D.R., Isenhour, P.L. & Carroll, J.M. (2007). Teacher Bridge: Creating a Community of Teacher Developers. *Proceedings of HICSS 40: Hawaii International Conference on Systems Science,* (Hilton Waikoloa Village, January 3-6).

Schafer, W.A., Ganoe, C.H. and Carroll, J.M. (2007). Supporting CommunityEmergency Management Planning through a Geocollaboration SoftwareArchitecture. *Journal ofComputer-Supported Cooperative Work*, 6 (4-5), 501-537.

Schafer, W. A., Ganoe, C. H., Xiao L., Coch, G., and Carroll, J. M. (2005). Designing the next generation of distributed, geocollaborative tools. *Cartography and Geographic Information Science, 32 (2)*, 81-100.

Schuler, D. (1996). *New Community Networks: Wired for Change*. Addison-Wesley, Reading, MA.

Schilit, B. N., LaMarca, A., Borriello, G., Griswold, W. G., McDonald, D.,Lazowska, E., Balachandran, A., Hong, J., and Iverson, V. (2003). Challenge: ubiquitous location-aware computing and the "place lab" initiative. In *Proceedingsof the 1st ACM international Workshop on Wireless Mobile Applications and Services onWLAN Hotspots* (San Diego, CA, USA, September 19 -19, 2003). WMASH '03. ACM Press, New York, NY, 29-35.

Sumi, Y. and Mase, K. (2000). Supporting awareness of shared interests and experiences in community. *SIGGROUP Bull. 21 (3),* (Dec. 2000), 35-42.

Uncapher, W. (1999). New communities/new communication: Big Sky Telegraph and its community, M. Smith and P. Kollock, Eds., *Communities in Cyberspac.* Routledge.

Vogiazou, Y., Eisenstadt, M., Dzbor, M., and Komzak, J. (2005). From buddyspace to CitiTag: large-scale symbolic presence for community building and spontaneous play. In *Proceedings of the 2005 ACM Symposium on Applied Computing* (Santa Fe, New Mexico, March 13 -17, 2005). L. M. Liebrock, Ed. SAC '05. ACM Press, New York, NY, 1600-1606.

Wellman, B. and Hampton, K. (1999). Living networked on and off line. *Contemporary Sociology, 28 (6)*, 648-654.

Wellman, B., Quan Haase, A., Witte, J., and Hampton, K. (2001). Does the Internet increase, decrease, or supplement social capital? Social networks,participation, and community commitment. *American Behavioral Scientist, 45 (3)*, 436-455.

Xiao, L. & Carroll, J.M. (2003). Wireless Community Support for CommunityNetwork. IWSAWC 2003: 3rd International Workshop on Smart Appliances andWearable Computing (Workshop at the 23rd International Conference on Distributed Computing Systems (ICDCS 2003), May 19-22, 2003, Providence, Rhode Island USA).

KEY TERMS

Civic Applications: Software designed to promote awareness for and engagement by citizens of a local community

Civic Smart Mobs: A real-time gathering of individuals contacted through mobile technology for the purpose of conducting a civic activity , adapted from "smart mobs" (Rheingold, 2002)

Community Network: Computer system designed to support a geographical community by enhancing existing physical entities and social relationships

Location Sensitive Aapplications: Software that is enhanced by features that know and utilize the physical location of the user

Mobile Problem Solving: Sharing and analysis of data while "on the go" in the process of solving a problem task

Participatory Design: Design process which directly engages end users for the purpose of ensuring that the design meets their needs

Place-Based Blogging: An article or story along with follow-up discussion centered around a physical location; possibly as a location sensitive application

Wiki: Web page(s) or content that can be easily and directly edited by Internet users

Chapter XXIV
From Social Butterfly to Urban Citizen:
The Evolution of Mobile Phone Practice

Christine Satchell
The University of Melbourne, Australia

ABSTRACT

Early 21st century societies are evolving into a hybrid of real and synthetic worlds where everyday activities are mediated by technology. The result is a new generation of users extending their everyday experiences into these emerging digital ecologies. However, what happens when users re-create their human identity in these spaces? How do the tools of new technologies such as the mobile phone allow them to capture and share their experiences? In order to address these issues, this chapter presents the findings from a three-year study into mobile phone use in urban culture. The study revealed that for a new generation, the mobile phone was integral in the formation of fluid social interactions and had accelerated urban mobility. Users once restrained by pre-made plans were able to spontaneously traverse the city and suburbs, swarming between friendship groups and activities. Distinct user archetypes emerged from these mobile phone driven sub-cultures whose practices brought about fundamental changes in social mores with respect to engagement and commitment, to notions of fluid time versus fixed time and, ultimately, to urban mobility. Connectivity had become central to what it means to have a social identity and users responded to this by merging bits of data to create their "ideal digital self" through which they communicate socially. Yet, recent developments in mobile phone design reveal the potential for a new generation of people to recontextualize their use in a way that moves beyond "the social" as they utilise sensors and data capturing and sharing functionalities in new mobile devices to augment their "social butterfly" identity with an ideology of a "socially conscious urban citizen."

INTRODUCTION: THE MOBILE PHONE AS A CULTURALLY LOADED ARTEFACT

Ubiquitous, more addictive than cigarettes, empowering, disruptive, the most intimate communications device in the modern world, the new car; just how should we define the mobile phone and the mobile phone user? The discourses surrounding the mobile phone herald the artefact as the defining cultural icon for the digital generation, the one item a person can possess to represent their status as a participating member in early 21st century society. "If you want to assure yourself that you belong to the new century, this is the object to have in your hands" (Myerson, 2001, p3). Claims of this nature reveal much about the cultural times in which we are living. The shift from a post-industrial to a digital society has resulted in a culture that is not only obsessed with being in constant contact with each other, but where the idea of connectivity actually defines the culture.

The literature indicates that the many changes brought about by mobile phone use are occurring within a cultural dimension. The implications are that the mobile phone brings with it more than communication, it brings powerful notions of personalization and identity as can be seen in the work of Carroll, et al. (2002); Counts & Fellheimer (2004); Geser (2004); Goggin (2004); Ishii (2004); Ling (2002); Plant (2001); and Taylor & Harper (2002). Therefore, when the study presented in this chapter investigated young people in relation to mobile phone use, the major focus was on how users interacted with new technologies in order to create a sense of identity, both for themselves and the world they lived in.

Although the emerging themes were focused on the process through which mobile phones were engaged to enforce pre-existing social networks and do not encompass advances that happened after the study finished in 2005, the insights are potentially relevant for those designing for new uses of interactions in urban environments because they provide an informed snapshot of the nature of a young, urban mobile user and the environment they inhabit.

Understanding everyday practices is important for urban informatics practioners as was emphasised by Rhinegold (2003) who draws attention to urban informatician Anthony Townsen's prediction that the mobile phone will have as much impact on urban environments as the automobile. Therefore, the first part of the chapter will describe the study design. The second section will present the findings from the study. The third section will examine the relevance of the findings as users, who previously defined themselves through social use and conceived the mobile phone as a private mode of communication, begin to identify themselves through new mobile paradigms such as active members of physically co-located communities (Bilandzic et al., 2008) or environmentally conscious citizens (Paulos, 2008). As Oktay (2002, p.261) notes:

Identity is one of the essential goals for the future of a good environment. People should feel that some part of the environment belongs to them, individually and collectively, some part for which they care and are responsible, whether they own it or not. At the urban level, the environment should be such that it encourages people to express themselves and to become involved.

STUDY DESIGN

The qualitative based study (Lindlof & Taylor, 2002) was conducted for the User Environment arm of The Smart Internet Technology Cooperative Research Centre (SITCRC). Thirty-five technologically competent users, 18-30 years old, living in Melbourne, Australia participated. The open-ended interview technique (Minichello, 1995) was employed to conduct the initial user study. This interview technique provided insights into the culture that was being studied by allowing the researcher and the user to engage with each other. The dialogue helped produce a result where the participants revealed in-depth accounts of their ideas, opinions and experiences. The implementation of this method encouraged participants to share narratives that revealed their uses of new technologies in their everyday lives by focusing on the moment of interaction rather than the adoption of technology to fulfill social and cultural goals.

Grounded Theory techniques were used for the analysis, although the data was not scrutinized in minute, line-by-line detail, or with the strict theoretical agnosticism as specified by Glaser (1992). Rather, cultural theory lenses were engaged to understand the patterns of mobile interactions in the context of the users *and* the settings they were occupying (Satchell, 2008). Once the emerging themes had been identified a matrix was developed to map the main observations.

The first theme relates to distinct user archetypes that emerged from mobile phone driven sub-cultures. These archetypes are the nomad, the iconic, the updater and the resistant user.

The second theme looks at how young people are using their mobile phones for the spontaneous formation of social networks and examines the emerging hybrid of digital/real-time relationships, what Ito and Okabe (2003) call the 'augmented flesh meet'. This behavior has led to the emergence of 'passive scheduling' as a form of social communication and as a means for maintaining presence.

The third theme reveals that the mobile phone is creating a generation of conflicted users trying to balance the need for connectivity with the desire to be at times, uncontactable. This results in users putting mechanisms in place to exercise more control over digitally constructed mobile space. The need to be able to convey meaning without real time interaction was the most commonly cited improvement that users wanted for their mobile phones.

The fourth theme is concerned with how the social dynamic resulting from mobile phone interactions is driving the notion of identity, leading to a youth culture where connectivity itself has become a defining part of what it is to have a social identity. An extension of this is the fear of dis-connectivity which in turn, highlights the empowering nature of the artefact.

FINDINGS

Theme One: Emerging User Archetypes

Four distinct, although not mutually exclusive archetypes of users emerged from the mobile phone driven subcultures of youth culture. They provided useful metaphors for understanding youth and urban cultures and the central role technology plays in them. As Sawhney (1994) points out, "We have to accept the fact that although the use of metaphors is not a particularly elegant or sophisticated technique, it is perhaps the only conceptual tool that we have for understanding the development of a new technology" (p. 293).

Nomads

The 26 nomadic users in the study were characterized by always being on the move between different groups and activities. Furthermore, they were nomadic in that unlike previous generations they did not have centralized meeting places where they could get together. They were disconnected physically, leading fragmented lifestyles as discussed by Carroll et al. (2002) and were often without a consistent home base. However, because they are connected virtually via their mobile phone handsets and networks, they could seamlessly map their own journeys through a continual series of activities and events. This is in keeping with Deleuze and Guattari's ideal of the nomad who is characterized by freedom of movement and is not constrained by time and space: "One can rise up at any point and move to any other" (Deleuze and Guatarri, 1987, p. xiii).

The nomadic users found cohesion in otherwise fragmented lives through their use of mobile phones. A participant noted how they provided control for managing increasingly fragmented everyday activities:

I am constantly in contact with colleagues and peers and friends.... Because of my technology I am always able to be part of what's going on in all aspects of my life. I have more control in a digital world than in a real world, but the digital world helps me control the real world. What I'm saying is that if I didn't have my phone to coordinate it all I would start slipping up and not meeting my social or professional obligations.

It can be seen that for this archetype of user, the handset becomes like a surrogate home base or virtual lounge room from where the nomad can maintain a continual virtual presence, summoning or joining real and virtual groups at will. This represents a paradigm shift in urban mobility not seen since the car liberated a generation of teenagers in the 1950s.

Iconics

Pankraz (2002) points out a 'global youth culture' is being driven by technology, rather than fashion, music, and sport as in previous generations. This was, to varying degrees, true for all the users in the study. For example, 33 of the 35 users customized the look and/or sound of their mobile phones to reflect their taste and style. For these users the mobile phone has transcended its functionality as a communication device and has become an icon or status symbol. Twelve users actually spent more time changing the look and sound of their phone to reflect their current mood than they did talking or texting. This was in keeping with Hulme & Peters (2001) who drew on Baudrillard (1993) to look at the process through which the mobile artefact has become appropriated by youth culture as a defining icon for their generation.

Two users in the study deliberately shunned new developments in mobile phones in favor of 'old school' or 'brick' models. "I don't want my technology shiny and new; I want it organic or at least a brick". Yet, this choice of the mobile phone as the site for an anti-aesthetic statement illustrated that whether it be in celebration or rejection of its form, the mobile has become an artefact for the expression of taste for every user in the study.

Updaters

Plant (2001), likens mobile driven phrases such as 'on my way" or "on the bus" to global dance tracks. She proposes that "Where are you?" is the perfect mobile question and "On the mobile" is the perfect mobile answer (p.29). This question is fundamental to mobile telephony and distinguishes the difference from land-line based communications by providing the possibility that the caller could, indeed, be anywhere! This has not only created a need for users to position and re-position themselves at the start of every phone call, but given way to a culture of use where the dominant part of the message relates purely to the activity or location and no further information is supplied. This is the 'updater'—representing 26 users for whom mobile phone ownership goes hand-in-hand with a need to regularly update others of their actions.

The updater archetype highlights the way mobile technology allows users to maintain a virtual presence in each other's lives, creating a new generation of 'always on' friends. This is also related to Moblogging (mobile blogging) although it is also subtly different in that updaters only broadcast within a strictly walled network of users, while Internet bloggers disseminate the digital account of their lives on the Internet.

The updaters place a lot of importance and derive much pleasure from documenting, circulating and consuming digital accounts of day-to-day experiences. They become active content producers rather than passive consumers of technology and use their own experiences to create new consensual meaning. It could be seen from the 'updaters' in the study that these users want to maintain a digital presence in each other's lives and as mobile artefacts rapidly converge with other technologies and services such as digital cameras, MP3 players, the Internet and 3G networks, 'updaters' are creating increasingly sophisticated home produced multi-media content to do so.

Resistants

The archetype of the resistant user revealed that while 33 users expressed affection, attachment, and identification with their mobile phones this did not translate into an unqualified embrace of a lifestyle generated by mobile technology. In fact, 26 users said that they hate their phones. What then was to be made of this inconsistency? The data was revisited and it could be seen that users are resistant to ubiquitous mobility, desiring connectivity but then not satisfying their need for being unreachable. This was encapsulated by one user's attempts to resist the mobile presence of her friends:

I turn it down so I can't hear it ringing, but then I can see it flashing. I try and look away but I can still see and sense it and I just stop concentrating on what the person (I'm with) is saying. I usually even know who it will be on the other end—and they know that I know. Then they hang up, it stops flashing and I can go back to being part of the conversation. Then 30 seconds later it starts flashing again. After about five missed phone calls I actually start to worry, maybe it's some sort of emergency, so the next time it flashes I look at the screen to see who it is and sure enough it's the person I thought it was—just being demanding and wanting my attention right now.

For two of the participants, resistance to mobile phone ownership was purely financially driven. These two users rarely used their phones for social purposes; rather, connectivity was inextricably linked with security. This resulted in the reluctant financing of a mobile phone as if it were a health or travel insurance policy. This indicates that users are happier to pay for mobile phones when they are used for social interaction than if they are for security purpose

Theme Two: Spontaneous Formation of Social Networks (Swarming)

Mobile phones provide a fluidity of interaction through access to a digital, networked, social world. For 26 users in the study, this resulted in a reality where technology and friendships are inseparable from each other. "Technology has increased my intimacy with my friends. It gives me the power to be part of the group whenever I want. I never turn my phone off". For these users the main reason to own a mobile phone is for the formation and maintenance of social networks. Long term plans were avoided giving way to spontaneous encounters.

Twenty-seven of the participants in the study rarely planned to meet up, rather they gathered spontaneously on a minute-by-minute basis. This was only achievable because of the user's mobile phones. "I couldn't live without it because then I would have no friends. I mean I would have friends but how would I find them?" This sense of always being connected and immediately available has brought about a huge cultural shift. Connectivity frees users from the constraints of the need for a predetermined physical locale as the mobile phone becomes the user's virtual home base. "The one place that you can always find me". The impact of swarming young people on city environments brings about new issues however, and the need for public places to "hang out" takes on a new importance. We still need conventional market squares, central parks and a plethora of cafés. How is a city (government and planners) supposed to respond to the needs of swarming youth?

Blurring the Boundaries between Real and Virtual Interaction (The Augmented Flesh Meets)

For 22 participants, the sense that they are always connected with each other came from the ability to switch seamlessly between real and virtual environments. As one user described: "I will usually send a text saying that I am on my way, even when my friends would know that. Often we start our conversation (via text) before we physically hook up..." Furthermore, a group of friends gathered in real life will frequently be joined, via mobile phone, by someone remote that is known to the group. The incoming call or text then becomes a new focus for the group's conversation. This indicates that the mobile phone lends a new dynamic, opening up the social space to include those not present. Ito and Okabe (2003) call this the 'augmented flesh meet':

Mobile phones have become devices for augmenting the experience and properties of physically co-located encounters rather than simply detracting from them. Teens use mobile phones to bring in the presence of other friends who were not able to make it to the physical gathering, or to access information that is relevant to that particular time and place. The boundaries of a particular physical gathering, or flesh meet, are becoming extended through the use of mobile technologies, before, during, and after the actual encounter. (pp.17-18)

It can be seen that rather than alienating other members of the group who are there physically, the 'augmented flesh meet' can create new sites for enforcing intimacy.

Interaction Occurs Primarily in the Context of Regular Face-to-Face Contact

The high level of comfort users experienced with mobile facilitated interaction exists only in the context of pre-made, real life friendships. This makes mobile phone supported interaction different to that of other virtual worlds, such as gaming or Internet communities where, at least in some cases, users traverse time and space in order connect with people they have never met in real life. Of the 22 users who revealed that there is an almost equal satisfaction to be gotten from mobile phone interaction as there is from face to face communication interaction, all point out that this is only with people that they see regularly in everyday situations:

At least when I'm texting my friends (as opposed to communicating with them on the Internet) my parents know that I'm not talking to people I don't know because nobody would use their mobile phone to talk to strangers.

This conception of the mobile phone as a private mode of communication provides users with a barrier between themselves and those outside their network.

Scheduling: As an Activity in Itself (The Approximeeting)

The nature of mobile technology is such that it lends immediacy to the formation of social networks. The outcome of this is that the act of scheduling itself becomes an important and pleasurable activity. Rather than just meeting a friend, the physical meeting is anticipated with a series of text messages and mobile phone calls:

There is no such thing as an organized get-together, or if there is, the time and location will invariably change. Even something like going to the movies cannot happen without 20 text mes-

sages—three changes of cinema and five different possible movies.

Eighteen participants used what they consider 'dead time' to engage in scheduling. Definitions of dead time included being on public transport (17 users), waiting in queues (8 users), lectures (3 users), and driving (15 users). Rather than finding this type of interaction an intrusion, 22 users found the scheduling based dialogues rewarding:

Even at work I'm constantly organizing to do things with my friends. I go back and forth between windows at work, it's like a work document, e-mail, work document, text message, work document, mobile phone call. Then I leave work, double check my mobile messages, send a few texts.

This intense mobile interaction is creating a new genre of social communication, with five users stating that they prefer the scheduling to meeting:

Sometimes the actual meeting is not as good as the planning—and even when you do meet, you're too busy organizing to meet with another group of friends to pay attention to the people that you are with.

It can be seen that the 'aproximeeting' is significant in that it embodies a huge cultural shift, challenging traditional ideas of the nature of engagement and commitment.

Theme Three: Control—Maintaining Boundaries of Virtual Spaces

Connectivity provides mobile phone users with a sense of reassurance; however, with this comes the consequence of vulnerability due to unwanted calls. Users try to protect themselves against this seemingly contradictory dynamic of openness vs. isolation. Essentially, the mobile phone is creating a generation of conflicted users trying to balance the need for connectivity with the desire to be at times, uncontactable. A participant described how the phone can intrude on his life and the mechanisms he puts in place to try and control it:

Good old caller ID. It helps stop my phone being a total pain. I look at it and know if I want to answer it or not. All the people I want to talk to are listed in my phone so I can see which name comes up and know if I want to answer it or not. But then there are those calls that come in 'number unknown'. I hate that because it's just a lucky dip. It might be someone I want to talk to calling from their home phones because some phones for some reason always seem to come up 'caller unknown'. Of course when I'm at work there is no choice but to answer every call. My job actually pays for my phone and all my calls, even my social ones so it's my obligation to make sure that I answer it during business hours, but when it's past 6pm on the weekend it's another story. One thing I think that is going to be a huge problem is videophones, there are certainly things I wouldn't want to see.

This sentiment was supported by another participant who expressed the need to control where and when communication occurred. "I love caller ID—I talk to certain people at certain times. It does make me have better relationships. I would rather converse when I feel like it, not when I am tired or watching TV." An international student studying full time in Melbourne taking part in the study noted that with e-mail you can have multiple e-mail addresses for friends, family and work, this makes having control over your digital space easier. Unless you have two or more phones though, it is harder to impose these divisions over the digital spaces.

These findings indicate that while users desire the fluidity of social interaction that mobile phones afford them, they also resent their intrusion in their lives and are seeking new ways to exert control over digital space and reduce unnecessary contact.

As shall be discussed in the next section, one of the ways in which they are doing this is to use their mobile phones to convey meaning.

Connectivity vs. Contactability

A re-occurring problem for users in the study was intrusion. Twenty-six participants responded that their own mobile phone was a regular intruder in their lives. Eighteen users responded to this problem by carefully policing the boundaries of the digital world, for example, not answering a call if they did not recognize the caller. It is significant that these complaints about mobile intrusions were mostly in regards to incoming calls; the less intrusive text message was only cited by three users as a source of annoyance, a finding that was noted by Woodruff et al., (2004). This sentiment was best encapsulated by a participant in the study who explained that her ideal interaction with friends was improved through the barrier of technology:

If I have a friend over, we will both sit here doing exactly the same thing except that then I have to make concessions as to what we do. I would prefer to have a friend online, then I could just keep them in a minimized window most of the time.

Conveying Meaning without Communicating

Twenty-four of the participants in the study agreed that even when a phone call is not answered the status of the phone itself reveals a lot about the current availability of the user. For example, a phone that is switched off indicates the person is not in social interaction mode. A phone that is turned off mid-ring indicates either that the person is engaged in an activity where it is not appropriate to talk, or a deliberate rejection of the caller. A phone that rings out or goes to voice mail is seen to mean that the person is willing to be contacted but cannot get to the phone at that particular moment. It is an indication to try again soon. One user stated:

Sometimes it is nice to just go out of circulation. When I want to do that I just turn my phone off and people see that and can tell that I am not interested in catching up. The problem though, is sometimes my phone goes flat and I end up sending the 'out of circulation message' when really I am just out of battery (Lucie, 18, student).

Twenty participants responded that they would like this concept of conveying meaning without having to directly communicate to be developed further, a finding supported by Donner (2007). Currently, the act of turning the ringer down or rejecting a call are unpleasant relative to the message we are trying to convey and there is a need to be able to change fluidly between these states or modes.

Theme Four: Identity Defined Through Connectivity

Unlike notions of digital identity within ubiquitous computing which seek to provide security by withholding information about the person, participants in the study were not concerned with what information was restricted but rather what is revealed. A user stated, "Their synthesized persona is quite revealing." Furthermore, participants described how they were merging bits of data to create their 'ideal digital self' through which they communicate socially. This means that instead of voice or text, users are communicating through a hybrid of still and moving images, sound-bytes, symbols and logos. "I merge lots of different pieces of data to create the one single identity for a specific purpose… Why just be text when you can be multi-media? I have just started becoming multi-medium too because I have got a picture phone and I e-mail from it." This was in keeping with the findings of Counts and Fellheimer (2004) who found that Lightweight photo sharing, particularly via mobile devices, is fast becoming a common communication medium used for maintaining a presence in the lives of friends and family.

Fear of Disconnection

The social dynamic resulting from mobile phone use has driven and redefined the notion of identity with 'connectivity' itself becoming a defining part of what it is to have a social identity. However, for 20 participants, a natural extension of the desire for connectivity was the fear of being disconnected. "I cannot bear to be out of the loop even for a minute" (Lucie). This was supported by a DoCoMo Study (2001) that found the main criticism young people have of their mobile phones is that they are not able to offer truly ubiquitous mobility, they could not call from anywhere and flat batteries could leave them stranded. A participant in our study noted:

My phone makes me feel safer if for example, I leave work late and am going into the underground car park. I know what it is like to have a mobile phone and I know what it's like not to have one and believe me the world is a much safer place for a girl when she's on the phone to someone.

Another participant describes how she feels strongly 'connected' to a digital network, empowered by connectivity and would go to great lengths to maintain it:

When I am on my own I feel safer with my mobile, because it means that I am never on my own. I hate being out of range. I feel very uneasy when I start to see those few little bars of coverage disappear. As soon as I see those little bars on the side start to disappear I start to worry. It's not that I have to talk to people all the time, it's that I need to know that I'm connected to a network. You know things like 911, I mean 000. But its not even that I want to call emergency services although I like to know that they are there. I just need to be connected, I guess.

This fear of being disconnected is significant because it reveals the mobile phone as a personally empowering artefact.

DISCUSSION: EXTENDING MOBILE NETWORKS BEYOND PRE-EXISTING SOCIAL FORMATIONS

The findings presented so far in this chapter focused on the process through which mobile phones were engaged to enforce pre-existing social networks. It could be seen that real life social networks have gone digital, or more specifically, require a digital component to flourish. Connectivity has become central to what it means to have a social identity and the result for the participants in the study was that the mobile phone had become a culturally loaded artefact that both shaped and was representative of their 'social butterfly' lifestyle. Yet recently mobile interactions have extended beyond these prefabricated formations and are affecting other types of activities in urban environments. Mobile phones are being used to construct new digital spaces where interaction can occur and as social networking in these mobile facilitated digital spaces grows, new paradigms of interaction evolve. For example, Dourish (2006) and Foth & Sanders (2008) find these interactions are increasing the significance of place and space in cities while Klamer et al., (2002) note that the mobile phone leads to interactions that would not have occurred otherwise, enhancing movement between networks as well as facilitating personal action.

Bilandzic et al., (2008) explore the way people use a mobile device to socially navigate urban environments with their mobile application CityFlocks and in doing so, enhance the experience of those physically co-located in communities.

The system provided two options for its users to access local knowledge from urban residents about the services and places of a city. They can either use the system to set up a link via phone call or SMS to a local expert resident who would provide direct help with questions about particular urban places, or they could request a list of location based comments and recommendations left behind by local residents.

The research conducted by Paulos, Honicky, & Hooker (this volume) re-conceptualised the mobile phone, transforming it from a communications device to a 'personal instrument' that could help us measure and understand the world around us. This 'Participatory Urbanism' is achieved by embedding mobile phones with sensors that can capture and share information about the urban environment. In doing so the owner of the phone is similarly transformed, moving from a passive consumer to an active citizen who engages the mobile artefact to capture, share and remix data for the benefit of both local and global communities. The desired result is technologies that could lead to an enhanced awareness of the urban environment thus empowering users to actively contribute to healthier cities.

When contextualized in light of these new practices the findings from this study can be helpful, contributing to an understanding of what aspects of the technologies might inhibit or enhance adoption. For example, the user study illustrated how lone females walking through deserted streets at night were empowered by the connectivity of their mobile phones which complements Klamer et al.'s (2002) notion of solo navigation of urban spaces.

CityFlocks and similar mobile phone applications have the potential to bridge virtual and real world social networks. However, the study highlighted ingrained user needs that would have to be appreciated. For example the need for mechanisms to balance connectivity with contactability in a social context mirrors the user needs uncovered during the usability testing of CityFlocks. They established that when introducing mobile technologies into urban environments the asynchronous push / pull communication mode worked better than real time interaction: "Even though those direct channels generally provide a richer form of communication, our test users would not bother using them" (Bilandzic et al., 2008).

Finally, the user archetype of the 'iconic user' for whom the mobile artefact was engaged to express their style, taste and personality, illustrated the potential for the realisation of the vision of Paulos et al. (this volume) as the primarily socially driven individual engages with 'Participatory Urbanism' using their mobile phone to take part in activities that transform them into an environmentally conscious urban citizen.

CONCLUSION: TRANSFORMING MOBILE IDNETITY

In 'Making Meaning of Mobiles—A Theory of Apparatgeist' (2002) Katz and Aakhus explain the phenomenon of mobile use through the theory of the 'apparetgeist' which can be literally translated to mean 'machine spirit'. The apparetgeist suggests "the spirit of the machine that influences both the designs of the technology as well as the initial and subsequent significance accorded by users, non-users, and anti-users" (p. 305). For Katz and Aakhaus, the theory of the apparetgeist is embodied by the mobile phone—a device that is stronger than cultural or geographical differences and able to establish new, similar patterns of 'use', 'design' and 'anti-use' around the globe. In a similar vein, Langdon Winner (1999) argues that artefacts can and do have politics, thus highlighting the potential of the mobile phone to shape ideological practices. The study presented in this chapter confirms this, illustrating the huge role that mobile phones play in helping people establish their sense of identity and achieve their goals. Until recently, these goals have been of a social nature with users embracing the mobile phone as a private mode of communication that provides a barrier between them and those outside their network. However, the research such as that being conducted by Bilandzic et al. (2008) and Paulos et al. (this volume) illustrates the potential for mobile identity to be re-contextualised into a notion that includes involvement in the collective enhancement of one's physical community and the conservation of our natural environment.

Mobile phones have been called the twenty first century cigarette by Stewart (2003) because with cigarette advertising being banned, mobile phone advertising has taken over as the item that most represents a 'lifestyle' as opposed to an actual functional product. As the study presented in this chapter study has shown, the culturally loaded nature of the mobile phone provides the opportunity for an evolution where the mobile phone, while still supporting social butterfly needs, can be recontextualaised as an artefact that encourages users to identify themselves as culturally and civically aware individuals or what Burgess, et al (2006) call 'urban citizens'.

REFERENCES

Baudrillard, J. (1983). The ecstasy of communication. In H. Forster. (Ed.) *The anti aesthetic essays on post modern culture.* Seattle: Bay Press.

Bilandzic, M., Foth, M., & De Luca, A. (2008) CityFlocks: Designing Social Navigation for Urban Mobile Information Systems. *Proceedings ACM Designing Interactive Systems (DIS)*, Cape Town, South Africa.

Burgess, J., Foth, M. & Klaebe, H. (2006) Everyday Creativity as Civic Engagement: A Cultural Citizenship View of New Media. In *Proceedings Communications Policy & Research Forum*, Sydney.

Carroll, J., Howard, S., Vetere, F., Peck, J., & Murphy, J. (2002). Just what do the youth of today want? Technology appropriation by young people. *Proceedings of 35th Hawaii International Conference on System Sciences (HICSS-35).* Maui, Hawaii.

Counts, S., & Fellheimer, E. (2004). Supporting social presence through lightweight photo sharing on and off the desktop. *Proceedings of the 2004 Conference on Human Factors in Computing Systems*, 599-60. Vienna, Austria, April, 2004. ACM Press.

Deleuze, G., & Guattari, F. (1967). *A thousand plateaus: Capitalism and schizophrenia.* Minnesota: Thousand Oaks Press.

DoCoMo. (2001) Report No.10, Current trends in mobile phone usage among adolescents. Retrieved May, 2001, from:http://www.nttdocomo.com

Donner, J. (2007). The rules of beeping: Exchanging messages via intentional "missed calls" on mobile phones. *Journal of Computer-Mediated Communication*, 13(1)

Dourish, P. (2006). Re-Space-ing Place: Place and Space Ten Year On. Proc. ACM Conf. Computer-Supported Cooperative Work CSCW 2006 (Banff, Alberta), 299-308.

Foth, M., & Sanders, P. (2008, forthcoming). Impacts of Social Interaction on the Architecture of Urban Spaces. In A. Aurigi & F. De Cindio (Eds.), Augmented Urban Spaces: Articulating the Physical and Electronic City. Aldershot, UK: Ashgate.

Geser, H. (2004). Towards a sociological theory of the mobile phone. Retrieved January 17, 2005, from http://socio.ch/mobile/t_geser1.pdf

Glaser, B. (1992). *Basics of grounded theory analysis.* Mill Valley, CA: Sociology Press.

Goggin, G. (2004, Jan 12). 'mobile text'. *M/C: A Journal of Media and Culture,* 7, Retrieved

March 18, 2004, from://www.media-culture.org.au/0401/03-goggin.php

Hulme, M., & Peters, S. (2001). Me, my phone and I. The role of the mobile phone. *Proceedings of CHI 2002, Workshop on Mobile Communications.* Seattle, Washington. April, 2001.

Ishii, K. (2004). Internet use via mobile phone in Japan. *Telecommunications Policy, 28*(1), 43-58.

Ito, M., & Okabe, D. (2003). Technosocial situations: Emergent structurings of mobile email use. Retrieved May 27, 2004, from http://www.itofisher.com/PEOPLE/mito/mobileemail.pdf

Katz, J., & Aakhus, M. (Eds.). (2002). *Perpetual contact: Mobile communication, private talk, public performance.* Cambridge: Cambridge University Press.

Klamer, L., Haddon, L. and Ling, R. (2000) *The Qualitative Analysis of ICTs and Mobility, Time Stress and Social Networking,* Report of EURESCOM P-903, Heidelberg.

Lindlof, T., & Taylor, B. (2002). *Qualitative communication research methods.* Thousand Oaks, CA: Sage Publications.

Ling, R. (2002). The social juxtaposition of mobile telephone conversations and public spaces. *Proceedings of the Conference on the Social Consequences of Mobile Telephon,* 3. Chun Chon, Korea. July, 2002. Retrieved October 15, 2002, from www.telenor.no/fou/program/nomadiske/articles/rich/(2002)Juxtaposition.pdf

Minichello, V., Aroni R., Timewell, E., & Alexander, L. (1995). In-depth interviewing: Researching people (2nd ed.). (Chap.4, In-depth interviewing. pp 61-104). Melbourne: Longman.

Myerson, G. (2001). *Heidegger, Habermas and the mobile phone (post modern encounters).* Cambridge: Icon Books Ltd.

Oktay, D. (2002). The quest for urban identity in the changing context of the city: Northern. *Cities, 19* (4), 261-271.

Palen, L. (2002). Mobile telephony in a connected life. Commun. ACM 45, 3 (Mar. 2002), 78-82.

Paulos, E. (2008) Citizen Science: Enabling Participatory Urbanism. *Urban Informatics: Community Integration and Implementation.* Ed Foth, M. IGI Global.

Pankraz, D. (2002). The power of teens online. Retrieved January 27, 2003, from

http://www.brandchannel.com/papers_review.asp?sp_id=102

Plant, S. (2001). On the mobile: The effects of mobile telephones on social and individual life. Motorola. Inc. Retrieved August 27, (2002), from http://www.motorola.com/mot/doc/0/234_MotDoc.pdf

Rheingold, H. (2003) Cities, Swarms, Cell Phones: The Birth of Urban Informatics + Social Currency. *Live Journal.* Retrieved Jan 6th, 2008 from http://community.livejournal.com/unwired/3450.html.

Satchell, C. (2008) Cultural Theory and Real World Design: Dystopian and Utopian Outcomes. *Proceedings of CHI,* Florence, Italy.

Strauss, L., and Corbin, J. (1997). Grounded Theory in Practice. Sage. Stewart and Segars.

Stewart, J. (2003). Mobiles phones: Cigarettes for the 21st century. Retrieved July 15, 2004, from http://homepages.ed.ac.uk/jkstew/work/phonesandfags.html

Taylor, A.S., & Harper, R. (2002). The gift of the gab: A design oriented sociology of young peoples' use of mobiles. Proceedings of the Conference on Human Factors and Computing systems CHI 2002. Minneapolis, Minnesota, April, 2002.

Winner, Langdon. (1999). Do artefacts have politics. In D. MacKenzie, & J. Wajcman. (Eds.). *The Social Shaping of Technology* (2nd ed.). Buckingham: Open University Press.

Woodruff, A. & Aoki, P. M. Push-to-talk social talk. CSCW '04, 409-441.

KEY WORDS

Apparatgeist: The ability of an inanimate machine to possess a spirit.

Digital Augmentation: The enhancement of virtual presence through the use of digital content.

Grounded Theory: ...A technique for analysing data that uses a bottom up approach allowing the researcher to build theory as the themes emerge.

Identity: Relates to the way the person sees themselves either individually or as member of a group.

Mobile Phone: The mobile phone (also: cell phone) is a portable electronic device used for mobile communication. In addition to the voice function, standard mobile phones include SMS for text messaging, email, packet switching for access to the Internet and MMS for sending and receiving photos and video.

Qualitative Research: Qualitative research involves the use of qualitative data, such as interviews, documents and participant observation data, to understand and explain social phenomena. It provides an in-depth analysis by examining phenomena within a social and cultural context.

Swarming: The spontaneous formation of social networks facilitated by mobile technologies.

The Swarm: A patented mobile phone prototype that allows the user to simultaneously represent multiple digital identities and embed their virtual presence with digital content. http://www.pixelshifter.net/client_login/swarm_2007

Section VI
The Not So Distant Future

Chapter XXV
u-City:
The Next Paradigm of Urban Development

Jong-Sung Hwang
National Information Society Agency, Korea

ABSTRACT

u-City is South Korea's answer to urban community challenges leveraging ubiquitous computing technology to deliver state-of-the-art urban services. Korea's experience designing and constructing u-City may be a useful benchmark for other countries. This chapter defines the concept of u-City and analyzes the needs that led Korea to embark on the u-City project ahead of others. It examines the opportunities and challenges that the nation faces in the transition stage. What has enabled Korea to pioneer the u-City concept is the development of IT infrastructure and the saturation of the IT market on the one hand, and the balanced national development strategy on the other hand. Success of u-City requires a national capability of designing forward-looking institutions to enable better cooperation among stakeholders, the establishment of a supportive legal framework and promotion of technology standardization.

INTRODUCTION

u-City can be thought as the next paradigm of urban development that leverages IT to advance urban functions by several levels. Employing IT to enhance urban functions is not a new attempt as many cities are already using IT extensively in a variety of applications such as transportation, disaster control, law enforcement and facility management (ICF, 2007). However, by utilizing new cutting-edge ubiquitous computing technologies, u-City introduces totally new innovative urban functions and socioeconomic paradigms that up until now have never been realized.

Korea is the first country to envision and realize u-City. Right after the IT839 strategy was established by the Korean government in 2003 to accelerate the development of ubiquitous computing technologies and services, the nation began to think about a variety of u-City implementation strategies that involve ubiquitous computing technologies comprehensively and, in just a little more than a year, the first generation of u-City projects kicked off simultaneously in several areas including Busan and Dongtan in 2005. In 2006, u-City found its way into official Korean government policies such as the u-Korea Master Plan and in 2007, although small in scale, u-City test beds were put in place at 6 different locations across the nation (NIA, 2007a).

During the same period, cities in other nations focused on the deployment of their IT infrastructure including broadband or wireless networks (ICF, 2007; Bar & Park, 2005). However, such initiatives were predominantly aimed to reinforce urban infrastructure, and thus differed fundamentally from u-City, which focuses not only on the establishment of infrastructural aspects but more importantly on embedding intelligence into the urban environment. Although there were some projects that also evolved around the notion of an intelligent environment, such as the Cooltown project of Hewlett Packard from the early 1990s (Barton & Kindberg, 2001), such projects were very small in size, thereby failing to evolve beyond the experimental level. Therefore, u-City of Korea deserves the title of being the first in the world to actualize an intelligent city.

This paper introduces the concept of the intelligent city of the future by reviewing current u-City projects in Korea. Notably, it will clarify underlying assumptions and government roles necessary for implementation of intelligent city projects by analyzing why Korea embarked on u-City construction projects ahead of others and what constraints the projects are facing now.

UBIQUITOUS COMPUTING AND u-CITY

Understanding u-City requires an understanding of ubiquitous computing, which is the platform of u-City. As the term u-City itself is a compound word of ubiquitous computing and city, u-City requires the deployment of ubiquitous IT services in an urban framework first and foremost.

Mark Weiser (1993, p. 1), who coined the concept, defined ubiquitous computing as "the method of enhancing computer use by making many computers available throughout the physical environment, but making them effectively invisible to the user." The basic idea behind ubiquitous computing was to make computers "autonomous agents that take on our goals" (Weiser, 1993). In other words, ubiquitous computing means embedding computing technologies in our physical surroundings so that virtual and physical objects may deliver services autonomously without human intervention.

Ubiquitous computing has properties totally different from those of conventional information technologies that we have used to date. Conventional IT, or "legacy IT" creates virtual space that exists only in a computer network and works independent of the real world (Weiser, 91; Schmidt, 2002). Of course, human command is needed to control the virtual space. However, ubiquitous computing infuses computers into the real world and renders the distinction between the real and the virtual world meaningless. As the virtual space communicates with the real world, human beings do not have to give any directions or orders. This is the distinction. Legacy IT essentially makes "digital space separated from the real world" whereas ubiquitous computing makes "digital space integrated with the real world."

With this understanding, ubiquitous computing has three distinct services in comparison with legacy IT. First, it supports real-time service (Fleisch, 2004). As digital space is integrated with the real world, ubiquitous computing can ensure prompt responses to events occurring in the real world. Since conventional IT relies on human input, there is bound to be room for lag and human error, an aspect that is minimal if not non-existent in ubiquitous computing. Second, since ubiquitous computing system is context-aware, it can provide the optimum solution for a given situation (Schmidt, 2002). Third, objects and environments surrounding human beings offer services autonomously. Therefore, not only computers but also services are invisible to users.

Since u-City means the application of ubiquitous computing technology in an urban space, the city itself becomes intelligent. The city is configured to create the most optimum service intelligently based on its real-time capacity as IT is embedded in physical environments that make up the city. This is the most remarkable characteristic of u-City pursued by Korea that sets it apart from IT-centric urban development models of other countries focusing on infrastructural establishment with broadband or Wi-Fi networks. It will surely take a long time to fulfill the vision of u-City to perfection. However, efforts to integrate city and IT into one have already begun. Although still in a primitive form, intelligent, autonomous services that u-City intends to deliver are already found in today's intelligent buildings or intelligent transportation system (ITS).

From a technological perspective, u-City integrates information technologies so seamlessly with each other that the entire city works as if it were a single system, which starkly contrasts with today's information system. Since information systems built in today's urban space are designed to deliver unit services, it is hard to find one developed from ground up with the entire city in mind among them. However, u-City designs IT architecture from the perspective of not a single application but the entire city. Therefore, information systems in u-City do not work in silos but in collaboration with each other to provide functions needed across the entire city in a real time environment.

With such extensive collaborative technological processes, u-City prompts the convergence of services and a revolutionary transformation of social systems. It is true that services are already converging with one other (for example, communication and broadcasting, mobile and payment) in several different domains of society. However, u-City will go even many steps further and tear down boundaries among industries or sectors, turning convergence into the ordinary rather than extraordinary. Notably, the conventional boundary between manufacturing and service industries will disappear (Fleisch, 2004). Automakers will provide drivers with multimedia content, maintenance service, route guidance and many other different services. Construction contractors will sell not only houses but also diverse home services including education, entertainment and healthcare. Service convergence will become a reality as different types of information systems and contents can be combined in u-City as easily as Lego blocks are put together. By its functional structure, u-City essentially involves service convergence.

EVOLUTION OF U-CITY

It will take a very long time to completely bring ubiquitous computing into reality. If ubiquitous computing is defined by applying a very broad interpretation as a state in which IT devices and services are utilized anytime, anywhere in our everyday lives, it can be certainly said that Korea as well as a number of countries with advanced IT capabilities have already entered the ubiquitous computing stage (Bell & Dourish, 2006). However, if ubiquitous computing is defined as ubiquity of IT intelligence that goes beyond ubiquity of just IT devices, as noted above, the current level of IT development is far off from such a vision. It is not a goal that can be reached by undergoing a few innovative technological developments alone. Only after experiencing countless innovations as well as a series of trials and errors over a span of several decades, will we be able to realize the vision of ubiquitous computing in its complete sense.

u-City, too, will undergo the same development process as ubiquitous computing. In this regard, u-City is not a specific project but a vision of the future. It can be understood in the same context as e-government, which at first glance seems to refer to a specific project but in reality denotes a model of government innovation in the era of informatization (Fountain, 2001: OECD, 2003). Typically, a project should have specific plans regarding its goals, schedule and methods. Like e-government, there is a clear vision as well as a sense of direction with regards to u-City, but how to get to the desired destination has yet to be defined. During the process of implementing u-City, projects that are suitable for each era and environment will be carried out in diverse forms. Accordingly, u-City will continue to evolve in the future.

While it is impossible to accurately predict how u-City will evolve, general stages of development can be deduced in a logical manner. That is, u-City will start from real-time data, move on to the stage of context awareness, and finally arrive at autonomous services. These stages of evolution are based on the level of reduction in the effort that users must make in order to utilize urban services. At the real-time data stage, data collection is carried out automatically, but users still need to analyze the data and take necessary measures. While the context-awareness stage takes it one step further from the previous stage to automate interpretation of context and provision of basic services, it still fails to embed intelligence into the entire process of a service. However, in the final stage of autonomous services, not only is the entire process involving a service based on context-awareness but there is a connection amongst the services being provided. Therefore, if a user makes key decisions, the rest is handled by the u-City system that offers the optimum solution to the user's needs.

Let us take an example of a transportation system to better understand the stages of u-City evolution. At the real-time data stage, which is the most primitive form of u-City, a user can have real-time access to various information regarding vehicles or roads. Currently roads that can automatically measure vehicle speed are already commonly used in developed countries, and a lot of vehicles are equipped with a function to check tire condition on a real-time basis. However, in u-City, automatic measurement will be expanded across the entire transportation system and the obtained data will be shared in a more efficient manner. In the context-awareness stage, techniques to analyze real-time data are greatly advanced to offer a larger number of intelligent services. Taking it one step further from simply measuring traffic speed, these services may, for example, calculate how long it will take to get to a destination for each possible route, or assign a different parking spot by identifying when an elderly person or an infant is in the car. The final stage of autonomous services will enable automobiles to handle such tasks as filling up for gas or being able to make repairs by themselves while roads can automatically address changes in their condition, such as an accident or damage.

It is not an easy task to accurately forecast u-City's evolution beyond the context-awareness stage. That is because the concept of context that computers have to cover keeps changing as technology and services continue to advance (Schmidt, 2002). Nevertheless, from this stage onward, the city will have artificial intelligence of some sort. That is, if an abnormal sign is detected in one function or part of the city, the city itself will recognize such abnormality and take the appropriate countermeasures against it, thereby recovering the normal condition (Stalberg, 1994). Intelligent ability is applied not only to urban infrastructure, including transportation, communication, gas, and electricity but also to social services, such as the environment and health care, which will raise the city's value dramatically. It is true that the city will have a greater ability to react (that is, sensitivity to potential or actual changes) than now even at the real-time data stage (Stalberg, 1994).

However, human beings still have to engage in and control the city's responses at this stage. In contrast, it will be the city itself that makes the decisions at the context-awareness stage.

What benefits will human beings gain from the fact that u-City has intelligence? While it is still too early to make a systematic prediction about the benefits u-City will bring about, two advantages can be pointed out as examples. One is that u-City will ensure sustainable urban development. Up until now, cities had no choice but to tear down their own foundation for existence. u-City, on the contrary, will have the ability to protect the foundation on its own. This prediction follows the logic that motor safety will be further enhanced when automobiles themselves have the ability to ensure safety, rather than relying solely on the capabilities of drivers.

Another benefit u-City will offer is personalization of urban services. In ubiquitous computing, the components of context surrounding a user communicate and share information with each other to create solutions (Schmidt, 2002). The further u-City evolves, the wider the range becomes for the components of a city to be combined with each other freely, like Lego blocks, as mentioned. This leads to a greater possibility for citizens to design and customize urban functions in order to satisfy their personal preferences. For instance, some parents may choose to receive notification when their young children get to school safely, while others choose not to use such services for whatever reasons. If a software service model which allows users to utilize others' applications, instead of their own, is referred to as "Software as a Service (SaaS)," then u-City enables "City as a Service (CaaS)," allowing citizens to access and customize urban functions to their likings.

U-CITY DEVELOPMENT IN KOREA

Korea's launch of the u-City project was not planned. The application of ubiquitous computing technology to urban space was considered a mere theoretical possibility when vigorous discussions on ubiquitous computing as a next-generation IT paradigm took place around the year 2003. At the time, the general perception was that an actual project to implement u-City would not be embarked upon in the near future. However, no sooner was the concept of u-City introduced than local autonomous governments and developers began utilizing it as a vision for the respective local development projects that they were promoting. In addition, when it was clear that the u-City vision was rather warmly received by the public as well as the market, u-City became a common goal for almost every new urban development project in Korea.

As of 2007, 22 u-City projects are known to be carried out in Korea (Kim, 2007, p. 64; NIA, 2007a, p. 15). Of these, 14 are being implemented in existing cities while 8 projects involve new cities. Busan, which was the first already-established city to launch the u-City project, is focusing on embedding intelligence in the city's logistic infrastructure to meet its particular needs arising from having Korea's largest container port. Seoul, on the other hand, is concentrating efforts on building intelligent environment in state-of-the-art business complexes (NIA, 2007a). Meanwhile, new cities are committed to laying ubiquitous computing infrastructure across the entire city, rather than adding intelligence to particular functions of the city as pursued by old cities. For instance, a sensor network connecting the entire city is being built across the city of Sejong to which Korea's administrative capital will be relocated. The project, which encompasses underground facilities, roads, structures and parks, is scheduled to be completed in 2011.

Existing cities and new cities take completely different approaches to building u-City. While newly built cities take a holistic approach, old cities take an incremental approach. The new towns establish a comprehensive IT strategy from the outset when formulating urban development plans, whereas existing cities apply ubiquitous computing technology only to geographical areas or functions that require renovation. While the holistic approach employed in new cities offers an advantage of implementing a larger number of advanced services, it requires great investments all at once and also involves a higher risk of failure. Compared to this, although existing cities can provide intelligent services within a very limited range, it is easier to secure necessary investments and has relatively lower risk of failure due to its incremental evolution.

Another comparison can be made regarding the scope of u-City development between one that covers the entire city and another that involves only certain part(s) of the city (Lee & Lee, 2007). While u-City projects are implemented across the entire city in most new towns, old cities undergo the two types of development at once. To revisit the examples cited earlier, Busan's project covered the entire city for a very specific purpose of building intelligent logistic infrastructure, whereas Seoul is implementing a variety of services in a specific area of Sangam where a digital media city is being built. While projects that deal with an entire city more or less have characteristics of public works, certain beneficiary groups are formed from projects that involve specific regions and as such, the benefit principle is applied to such projects (Lee & Lee, 2007).

Korea is implementing a variety of policies to support u-City construction. The greatest attention is being paid to creating a proper organizational framework for u-City projects (Kim, 2007). Because a large number of stakeholders are involved in u-City projects, an effective means to coordinate and adjust their interests is a must for its successful implementation. To this end, the government organized the u-City Forum as a policy network in 2005, which comprises of IT and construction experts. In addition, the government is pursuing the enactment of the Basic Act on u-City Construction in order to establish a uniform system to make policy-related decisions.

On the technological front, test beds have been selected since 2007, which are used to develop standard models for each service to be provided. The most worrisome aspect of Korea's u-City projects is that there is no common foundation for standardization, which might undermine the compatibility between different services. Because many regions are implementing u-City projects simultaneously, a market failure caused by the absence of standardization is anticipated. Therefore, the Korean government has built test beds in 6 regions in 2007 to develop standard models of 6 services (NIA, 2007a) – underground facility management, facility management, riverine ecosystem management, tourism service, pollution management and construction site management.

RATIONALE OF U-CITY

Korea began to focus on u-City far earlier than any other country. As noted before, there were predecessors to u-City such as the Cool Town project of HP that aimed to build an intelligent environment. However, most of them were on the small experimental scale and did not lead to significant investment in the full-scale urban infrastructure like u-City. Osaka, Japan uses the title u-City, but the project is about building a seamless information and communication service network (Kim, 2006), which reveals a fundamental difference from Korean cases.

So, what drove Korea to take the risk and be the first mover in u-City well earlier than other nations? What made Korea move first even if the U.S., Europe and Japan were well ahead of Korea in terms of technological prowess necessary for u-City implementation?

The first clue is found in the outstanding IT infrastructure of Korea which began to take off and dramatically increase in the mid-1990s that led to Korea being ranked first in the world at the beginning of the 2000s. As of the end of 2006,

Figure 1. ICT infrastructure development of Korea (1985~2006) (Source: NIA. Yearbook of Information Society Statistics: Republic of Korea.)

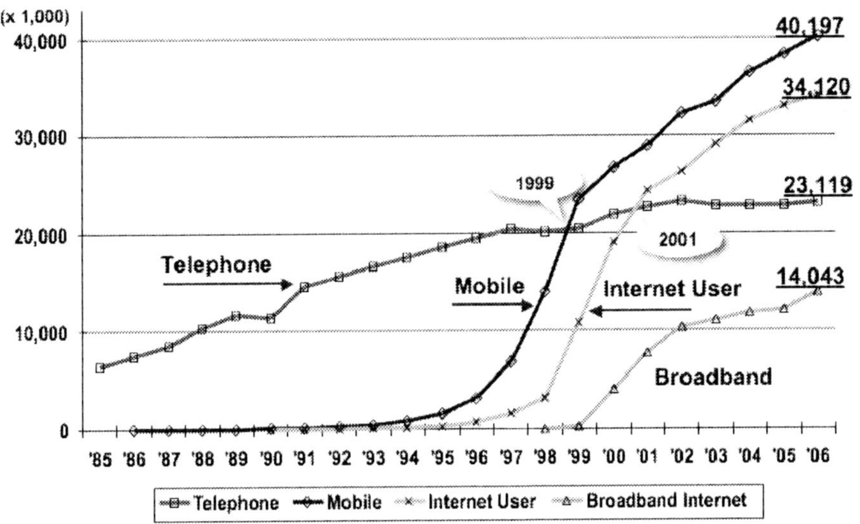

more than 83% of the entire population own mobile phones, 77% or more of the entire households have broadband Internet connections and 74% of the national population are Internet users (NIA, 2007b). In comparison with other countries, Korea follows behind Denmark, the Netherlands and Iceland in terms of a broadband penetration rate (OECD, 2006). However, if considering the quality of broadband service in terms of network speed and subscription fees, Korea is deemed to have the most advanced infrastructure (ITIF, 2007).

In addition, Korea is also active in constructing the next generation ICT infrastructure. Korea launched the world's first 3G mobile communication service in 2000 and has already achieved a high penetration rate of 81.7% in the middle of 2007 (ITU, 2005; NIA, 2007b). In June 2006, Korea also launched the world's first mobile WiMAX service under the name of WiBro enabling users to tap into high-speed Wi-Fi service even while moving fast and has been expanding in that market since then. With broadband networks, 3G and WiBro supporting high-speed Internet connections in wired and wireless environments respectively, Korea has already reached the "Always Connected" state. Furthermore, almost all types of infrastructure necessary for implementation of ubiquitous services such as DMB (Digital Multimedia Broadcasting) service enabling users to watch TV from on mobile devices are already in place.

The cutting-edge IT infrastructure in Korea is the prime driver in defining the differences between Korea and other nations in terms of the approach to incorporating IT with urban development initiatives. Many countries including the U.S. are now keen on establishing mobile Internet infrastructure, and the Wi-Fi City projects in several cities of the U.S., the Hot City project of Spain and the Mobile Taiwan project in Asia represent such efforts (Bar & Park, 2005; NIA, 2007a). However, as wired and wireless access environments are already very mature, Korea is now focusing on intelligent services that can effectively use this advanced network infrastructure.

The relative saturation of the IT market in Korea is also one of the drivers behind Korea's pioneership of u-City programs. As mentioned earlier the penetration ratios of broadband Internet and mobile phone have surpassed 77% and 83% respectively. IT businesses have had to create new markets in order to continue to generate profits. To increase the number of subscribers to new services such as 3G mobile communication or WiBro already launched in the market, they needed to introduce a new breed of intelligent services enabling real time information access anytime and anywhere such as RFID/USN and u-health, rather than using the Web passively as was the case in the past. On top of that, large-scale projects such as u-City were necessary to encourage the penetration of fast-emerging futuristic services by creating demands.

IT companies are advocates of the u-City construction and roll-out in Korea. City developers such as Korea Land Corporation or Korea National Housing Corporation and other construction companies are playing a major role in many u-City projects. Furthermore, IT businesses create momentum and incentives to engage offline businesses in u-City projects. IT businesses in turn play a highly aggressive role in forming consortia with city developers and construction companies, developing u-City service models and building communication networks and RFID/USN platforms in support of such service models, etc. It is also possible for Korea to commence forward-looking u-City projects earlier than others in an effort to ease the saturation of the conventional IT market as the country has promoted IT business quite aggressively as part of its strategy to transform Korea into an advanced information economy.

A state-of-the-art IT infrastructure and the saturation of the IT market are necessary but not sufficient conditions to account for the early adoption of the u-City initiative in Korea. There are many advanced countries that have advanced IT infrastructure and almost saturated IT markets. However, Korea is still the only country that has embarked on u-City development projects on a relatively extensive scale. Therefore, additional reasons can be pinpointed from circumstances specific to Korea.

By 2003, the Korean government has put its highest priority on balanced development across the country. And one of the main strategies was to build competitive new cities in provincial areas, including a new administrative capital and innovative cities. The aim was to attract businesses and populations from the Seoul Metropolitan area (BND, 2007). To support this, it established dedicated organizations, enforced legal frameworks and prepared special financing arrangements. At the same time, the national government has increased considerably its funding for regional development projects initiated by local governments.

This Balanced National Development policy spearheaded by the Korean government from 2003 accounts as one of the reasons for Korea's early u-City construction. In the wake of rapid industrialization since the 1960s, an imbalance in levels of development occurred across different regions in Korea: the gap between Seoul Metropolitan Area and other provincial areas had widened to an alarming extent by the 2000s. In terms of the concentration of national population, the population of Seoul Metropolitan Area jumped dramatically from 20.8% in 1960 to 38.4% in 1980 and to 46.3% in 2000. Considering that the population of the capital area accounts for 31.9% of the national population in Japan, which is famous for the high population density in its capital area, the population concentration around Seoul Metropolitan Area is extremely high and a serious matter.

The u-City concept attained powerful momentum due to the need for new local development models. Pioneers of u-City projects such as Busan and Dongtan promoted u-City aggressively in search of a development strategy to improve the attractiveness of local areas and triggered similar moves quickly in other areas as their projects were met enthusiastically. The ripple effect also spilled over into the private sector development projects, and all large-scale residential district development projects have opted for a ubiquitous model.

POLICY ISSUES

The challenges that Korea faces in the u-City implementation process have significant implications for other countries, which will soon embark on similar projects. Issues raised in relation to u-City also provide excellent opportunities for us to preview the emerging policy agenda of the ubiquitous computing era.

The most pressing issue in u-City implementation is to define new roles for principal players. The conventional role classification in IT, construction and utility service sectors poses challenges to u-City implementations. For example, in case of constructing intelligent underground utility facilities, different stakeholders, such as water suppliers, electricity or gas service providers, argue over conflicting interests. This often results in project delays. Integrating information relevant to city functions, which is available across different sectors, such as traffic management, healthcare and security, also complicates matters due to existing regulations and institutional frameworks. Notably within the government, different ministries must cooperate with each other to support u-City programs. However, a government-wide policy coordination system for u-City has yet to be established.

From an institutional perspective, the existing legal framework constitutes a bottleneck for u-City implementations. It requires a complete overhaul. First, there is no legal basis supporting personal data collection via RFID tags and sensors; second, no clear statutory definition for the scope of remote healthcare service and legal liability is available; and third, mandatory regulation governing standardization of tags, sensors and devices to ensure cross-service compatibility has yet to be put in place. There are so many issues relevant to u-City in the existing legal framework that it is not possible to settle them all at once. The issues need to be resolved incrementally. Nevertheless, it is sensible to establish a powerful steering organization that engages relevant government ministries and stakeholders to ensure the effectiveness of the u-City program and issues get dealt with in an efficient and timely manner.

From a technology perspective, it is necessary to define a standard platform for the variety of technologies utilized in u-City. Standardization is the area that is suffering market failure in relation to u-City. It benefits everyone if all stakeholders cooperate with each other and establish a common standard. However, if each stakeholder maneuvers so as to push for only their gains, this fails to produce a common good for everyone. As u-City involves many different participants and a wide variety of technical requirements, the government has an active role to play in establishing foundations for standardization, providing test-beds for common use by private businesses, and verifying the applicability of standard technologies via pilot projects.

GOVERNMENT'S ROLE

The government plays a pivotal role in compensating for uncertainties and costs related with u-City. It may sound obvious, but it is hard to expect investment or entrepreneurship from private businesses if uncertainties are high and the cost structure is unfavorable. In ordinary circumstances, it is sensible to leave it up to free market mechanisms to reduce uncertainties and improve the cost structure. However, in case of u-City, it is not desirable to purely rely on free market mechanisms due to possible market failure.

To sum up the role of the government in the u-City program, it must construct u-City infrastructure first in order to lower the entry barriers. This enables competent entrants to participate. As an example, even though a company has great growth potential in terms of technological as well as business prowess and has an excellent business plan in u-City, it will have little or no chance of success if it has to build all infrastructure such as communication networks or tag platforms. Furthermore, if the institutional framework such as relevant laws and regulations are not in order, transaction and entry costs will soar significantly because private businesses will have to carry out an overhaul of the legal frameworks.

For that reason government must overhaul the physical and institutional infrastructure as early as possible to reduce uncertainties and costs for private businesses. Even so, it is necessary to clarify role models between the national and local governments. Specifically speaking, both national and local governments must adhere to their own roles within the limitations of their respective authorities. However, it is ultimately the national government that is responsible for furnishing the necessary infrastructure for u-City. As u-City has to grow across the nation with interfaces between different cities, a nationwide infrastructure must be constructed first.

Second, the government must actively encourage demand. In an emerging technology project like u-City network externalities make a significant difference. Initial demand plays a critical role in offsetting a vicious circle whereby weak demands discourage potential customers (Katz & Shapiro, 1985). When the public sector leads investments in service infrastructure in the early days of u-City, demand creation may not prove to be a critical issue. However, if the services keep evolving and the private sector provides most of mainstream services, there should be a variety of channels through which demand can be created. In addition to the public sector being a customer or enforcing the use of u-City service by regulation, it is necessary to develop incentives such as a more favorable cost structure toward u-City users.

Let us take a look at the u-City service cost issue in relation to demand creation. As the case of broadband Internet in Korea has proven, user charges are a highly important variable in the penetration of new service. Service charge schedules based on the cost from the supplier's perspective may backfire, suppressing demand and inviting losses in the end. Hence, it is necessary to determine service charges and enlist alternative mechanisms such as subsidies in consideration of the customer's willingness to make payments.

Lastly, government must play an active role in encouraging convergence across different industries. One of the features characterizing ubiquitous IT services is that several industries converge with each other to provide a bundle of services (Fleisch 2004). In u-City, numerous services and urban functions, for instance, automobile traffic, entertainment, healthcare industry, converge. Government must develop a policy framework encompassing IT, BT (bio technology) and NT (nano technology) across their individual boundaries. In addition, it must also place a priority on convergence in supporting R&D and product development initiatives, and improve the education system to reorient manpower development framework to transcend conventional boundaries between academic sectors.

CONCLUSION

The u-City shows tremendous value as a development strategy of the future. It will be a flagship project spearheading the ubiquitous computing era when things will provide necessary services intelligently and autonomously. As the case of Korea illustrates, u-City provides not only a visionary model for local development but also momentum to accelerate such projects.

However, it is still uncertain whether u-City will thrive in the future. Futuristic city models such as u-City will be the norm sometime in the future, but it is still not certain whether we can build a successful u-City prototype in the near future. The key to the success of u-City will lie in creating a systematic framework optimized to fit the needs of a ubiquitous computing paradigm. The most important success driver for u-City is the ability of the nation to coordinate a variety of stakeholders in a cooperative network and establish laws, institutions and technical standards appropriate for the city of the future.

REFERENCES

Bar, F., & Park, N. (2005). Municipal Wi-Fi Networks: The Goals, Practices, and Policy Implications of the U.S. Case. *Communications & Strategies*, 61, 107-125.

Barton, J., & Kindberg, T. (2001). The Cooltown User Experience. HP Laboratories Palo Alto, HPL-2001-22.

Bell, G., & Dourish, P. (2006). Yesterday's Tomorrows: Notes on Ubiquitous Computing's Dominant Vision. from http://www.ics.uci.edu/~jpd/ubicomp/BellDourish-YesterdaysTomorrows.pdf.

BND (Bureau for Balanced National Development). (2007). *2007 Annual Report on Balanced National Development Plan* (Korean).

Fleisch, E. (2004). Business Impact of Pervasive Technologies: Opportunities and Risks. *Human and Ecological Risk Assessment*, 10 (5), 817-829.

Fountain, J.E. (2001). *Building the Virtual State: Information Technology and Institutional Change*. Washington, D.C.: Brookings Institution Press.

ICF (Intelligent Community Forum). (2007). *The Top Seven Intelligent Communities of 2007*. from http://www.intelligentcommunity.org.

ITIF (Information Technology and Innovation Foundation). (2007). Assessing Broadband in America: OECD and ITIF Broadband Rankings.

ITU (International Telecommunication Union). (2005). *ITU Internet Reports: The Internet of Things*.

Katz, M.L., and Shapiro, C. (1985). Network Externalities: Competition and Compatibility. *The American Economic Review*, 75(3), 424-440.

Kim, B.R. (2006). Current Status and Implications of Ubiquitous City Construction in Japan (Korean). *Weekly Technology Trends*, 1271, 14-26.

Kim, J. (2007). u-City Implementation Policy (Korean). In MIC (ed.). *International Conference on u-City 2007* (pp. 57-75). Proceeding.

Lee, B.C., & Lee, Y.J. (2007). u-City Business Module and u-Service (Korean). *TTA Journal*, 112, 72-82.

NIA (National Information Society Agency). (2007a). *u-City*.

NIA (National Information Society Agency). (2007b). *2007 Informatizaion White Paper: Republic of Korea*.

OECD. (2003). *The e-Government Imperative*.

OECD. (2006). OECD Broadband Statistics to December 2006.

Schmidt, A. (2002). *Ubiquitous Computing: Computing in Context*. Unpublished doctoral dissertation, Lancaster University, U.K.

Stalberg, C. (1994). The Intelligent City and Emergency Management in the 21st Century. from http://www.stalberg.net/cespub2.htm.

Weiser, M. (1991). *The Computer for the 21st Century. Scientific American*, 265(3), 94-104.

Weiser, M. (1993). Some Computer Science Issues in Ubiquitous Computing. *Communications of the ACM*. from http://www.ubiq.com/hypertext/weiser/UbiCACM.html.

KEY TERMS

DMB (Digital Multimedia Broadcasting): DMB is a digital radio transmission system that delivers radio and TV content onto mobile devices like a cellular phone.

RFID (Radio Frequency Identification): RFID can identify objects automatically by attaching tags to objects and reading them remotely with RFID readers. This technology is being used for a very wide range of services, including passport and cashcard.

Ubiquitous Computing: This concept was introduced in the late 1980s and has been developed under various names such as pervasive computing and ambient intelligence. The basic idea is to make computers autonomous agents that take on our goals.

u-City: It is an urban development model in the future that will advance urban functions greatly by utilizing ubiquitous computer technologies extensively and embedding intelligence into the urban environment.

u-Health: It is one of ubiquitous computing services, which deliver intelligent and autonomous services based on context-awareness. With the development of sensor technology, u-health service is focusing on monitoring the conditions of client's health in real time.

USN (Ubiquitous Sensor Network): It is a network that enables the collection of information from all types of sensors wirelessly. It installs various types of sensor nodes and monitors condition of target objects in real time.

Wibro (Wireless Broadband): This technology enables Internet users to tap into high-speed wireless Internet service while moving fast at a speed of more than 60 km/h. The first commercial service was launched in Korea in 2006.

Chapter XXVI
Urban Informatics in China: Exploring the Emergence of the Chinese City 2.0

Dan Shang
France Telecom Research and Development, Beijing, China

Jean-François Doulet
University of Provence (Aix-Marseille 1), France

Michael Keane
Queensland University of Technology, Australia

ABSTRACT

This chapter examines the development of information and communication technologies (ICTs) in urban China, focusing mainly on their impact on social life. The key question raised by this study is how the Internet and mobile technologies are affecting the way people make use of urban space. The chapter begins with some background to China's emergence as a connected nation. It then looks at common use of web-based and mobile phone technologies, particularly bulletin boards, SMS and instant messaging. The chapter then presents findings of recent research that illustrates communitarian relationships that are enabled by mobility and the use of technologies. Finally, these findings are contextualized in the idea of the City 2.0 in China.

INTRODUCTION

Urbanization in mainland China has increased dramatically during the past two decades. The urbanized population of China spiraled from 215 million in 1982 to 410 million in 2000, increasing to 32 percent of the total population (Friedmann 2005: 133 n17; see also Donald and Benewick 2005, 27). According to one authoritative source, by 2014 an estimated 40.2 percent of the Chinese population will be urbanized (Garner et al 2005,

67). New cities have emerged and many existing towns and regions have been reclassified as cities. According to Mars and Hornsby (2008), twenty new cities are constructed per year, a phenomenon that is likely to continue until the year 2020.

Urbanization has brought with it great social change. The last two decades of the twentieth century were characterized by widespread economic reforms and population shifts to cities, in part due to the reclassification of many non-urban household registrations (*hukou*) to urban districts. Many traditional social maintenance structures have subsequently come under stress; people have been forced to acquire new skills in order to survive in a changing urban landscape and to develop strategies to seek out jobs, housing preferences and possessions.

Migration has impacted upon the spatial transformation of cities. Unskilled semi-literate workers from the countryside now provide labor for the construction of high-rise gated apartments that house China's aspiring white collar classes. Bulldozers and wrecking balls relentlessly demolish historic factories and traditional courtyard residences (*hutongs*)—and with this, traditional ways of life (see Zhou 2006). In this radical makeover of urban space overpasses, underpasses, ring roads, technology parks, theme parks, shopping malls and convention centers are the material manifestations of economic development.

This chapter examines the development of information and communication technologies (ICTs) in urban China, focusing mainly on their impact on social life. China provides a contrast with other studies in this book. China is 'in-between', less creative and technologically literate than its neighbors Korea and Japan, but ascending fast in its desire to compete as an equal in the knowledge-based society of the new millennium. In many respects China is also less tolerant of disruptive social behavior than some of the other robust democracies discussed elsewhere in this collection. One of the key questions raised by this chapter therefore is how the Internet and mobile technologies affect the way people use and understand urban space. Furthermore, how do ICTs reconstruct communication networks in ways that allow people to feel a sense of connection and shared identity without causing the ruling regime to fear mass social uprising? Following some background information that acknowledges China's technological 'leap forward', we present findings of research that shows the relationship between urban mobility and the creative use of communication technologies. One of the conventional indicators of urban mobility is the automobile (Doulet and Flonneau 2003). We examine perceptions of urban space, strategies to master city trips through the use of web-based technologies, and community celebration of urban experience.

A MORE CONNECTED URBAN SOCIETY

The extensive rollout of information and communication technology services during the past two decades has given China an unprecedented platform to catch up to the developed West, something it was unable to do under Maoism (cf. Keane 2007). By comparison with revolutionary China, the China of today is open to both ideas and business. Furthermore, the gradualist commodity economy development model espoused by Deng Xiaoping in the early 1990s has adapted to global conditions. With cities absorbing more migrants and more international influences, urban space has been reconstructed, reconfigured and made more productive. Castells ([1993] 2007, 486) argues, 'In the new economy, the productivity and competitiveness of regions and cities is determined by their ability to combine informational capacity, quality of life, and connectivity to the network of major metropolitan centres at the national and international levels'. Castells coined the term 'space of flows' to refer to the system of exchange of information, capital and power. In particular, the communication revolution has had a major impact on social relations and values in a society that formerly lacked mobility. In looking at the social implications of ICTs in urban China, we confirm therefore that they enable the adaptation of spatial behaviour to an ever-changing social context. From this perspective ICTs are tools that allow people to be more flexible. The rapid increase in telecommunication infrastructure, together with the state assisted development of the Internet in China—and more specifically Web 2.0 applications and services—has increased the 'space of flows', and with it, the number of online communities. The result has been a redistribution of social meanings and values.

The global dimension of information technology innovation provides us with four key ideas to contextualize China's 'peaceful rise'—the latter is a felicitous term adopted by the Chinese Communist Party to draw attention to China's increasing influence and positioning in the global economy. First, information technology goods and services are now produced, marketed and absorbed at an ever increasing rate. Second, consumers and workers must upgrade skills to stay abreast of developments (Ruggles 2005). Third, users of new technology are increasingly involved in influencing the design of technologies. Fourth, while information technologies are to be found in all strata of society, a great deal of user innovation is associated with young mobile urbanites.

These developments apply equally well to China. Despite a low per capita income compared to most Western nations, China has experienced high-speed growth in access to new communication technologies[1]. By October 2003, the number of mobile subscribers had already overtaken fixed line telephony. According to the Ministry of Information Industry (MII), by December 2007, there were 412 million mobile users (CNNIC 2007). In actual fact, because many users have more than one phone the actual number of devices registered was 548 million. In big cities, like Beijing and Shanghai, mobile penetration is over 90 percent.

By the end of 2007 China ranked second after the United States with more than 210 million Internet users (CNNIC).[2] However, compared with national mobile phone penetration (30 percent), the Internet reaches only 16 percent of the total population. Penetration in urban areas (27.3 percent) is of course much higher than rural areas (7.10 percent) (CNNIC December 2007). Mobile Internet represents a growth market. At the end of 2007, 50.4 million people, or 24 percent of all Internet users, were using mobile devices to connect to the Internet. Another way of understanding this is that 12.6 percent of the total 412 million mobile phone users accessed the Internet (CNNIC 2007). This is an astounding figure when one considers that third generation (3G) licenses are yet to be rolled out (as of writing). Slow download speeds of the 2.5G system results in high charges as they are calculated on a per data basis. This means that viewing anything on the Internet via a mobile device will cost much more than on a computer. With 3G technology China will enter a new informational age: the costs of connecting to the Internet through the mobile phone will reduce as a direct result of network externalities, that is, there will be many more mobile users.

The total number of SMS (*duanxin*) sent in China grew from just half a billion in 2000 to more than 592 billion in 2007, while the number of Internet café users grew from 60,000 to over 60 millions during this same period. In the rapidly changing urban environment, customized services are available for low-income workers, including SMS, prepaid phone cards, Internet Cafés and little smart mobile phone (*xiaolingtong* / PHS). Indeed, studies have shown that the mobile phone plays an important role in helping migrant workers to develop and extend a large number of weak ties with casual acquaintances or organizations, and exchanging mass job information within such ties (Law, 2005).

*Figure 1. Penetration rate of fixed-line, mobile and Internet in China (1995-2007) Source: MII, CNNIC and National Bureau of Statistics (Beijing) *Note: fixed includes PHS/Xiaolingtong subscribers.*

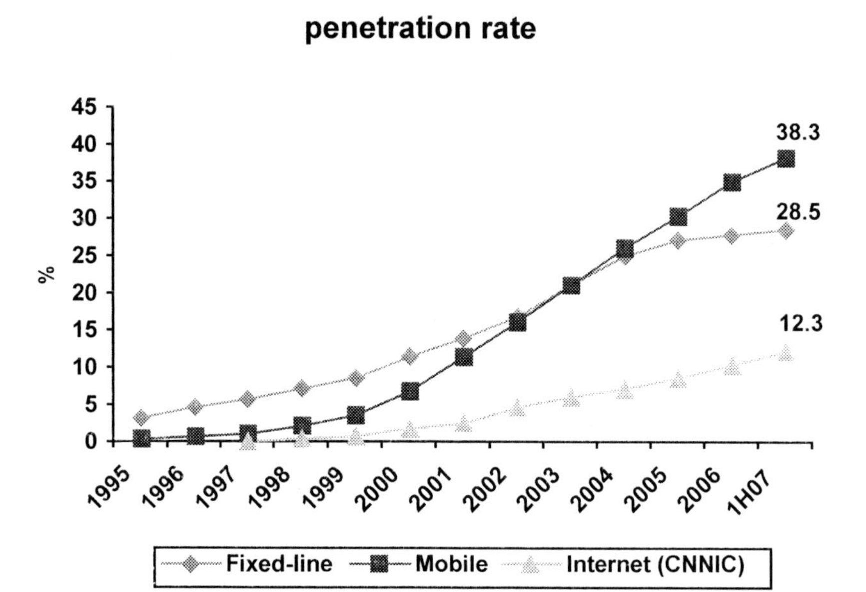

A Fast Adoption of ICT Among Marginalized Groups

According to the China Internet Network Information Center (CNNIC), the 'typical' Chinese Internet user is young, with low income or no fixed income. Indeed, 69 percent of Chinese Internet users are under 30 years old, while some 50.9 percent are below the age of 25 (CNNIC 2007). The most popular online activity is entertainment. Bulletin boards (BBS) and gaming are extremely popular while there is high instant messaging (IM) penetration. IM is more widely used than email. Many users go to Internet cafés to chat over QQ, the leading IM service provider. For youth, instant messaging and game-playing are major obsessions. The leading provider of IM is QQ, which was officially launched by Tencent in February 1999. QQ announced a massive 221.4 million active users in its 2006 report. QQ has built up several value added services such as QQ.com, QQ Game, QQ Zone, QQ Show, and Paipai.com. All of these services have been seamlessly integrated with QQ IM and can be accessed with a user's QQ account. QQ even introduced a virtual currency named Q-coin, which can be used to buy virtual clothes, hairstyles, furniture—and even virtual pet food for virtual pets. In a way, QQ combines aspects of the social networking site MySpace, the video sharing site YouTube and the online virtual world of Second Life.

I play with QQ about three to five hours a day... I usually play QQ games, buy game stuff from the QQ Game and buy decorations for my QQ show.

—A 21-year-old college student in Beijing[3]

There is an interesting social observation to be drawn from these activities. Because of the One Child Policy, which began in the 1980s, China's teenagers don't have sisters and brothers. As a result they rely heavily on sites such as QQ to communicate with their peers, to express their personalities and release pressure from the high expectations of their parents. Many become hooked on Tencent's QQ offerings in high school and stick with it in college. By providing all-in-one packaging of entertainment offerings, QQ has achieved dominance in this youth Internet market.

These data show that Internet based technologies have allowed Chinese netizens to connect across "scale free networks" to a great degree. Scale free networks are those that aggregate users in highly connected nodes surrounded by many less active connections. The point about scale free networks is that they are often vulnerable to epidemics. The Internet itself is a prototypical scale free network. However what is perhaps more interesting than highly connected hubs is the development of many "small world networks". By comparison, small world networks are often characterized by shared identity; sometimes they are constituted by friends of friends or by people who share common interests or pursuits. These networks are inherently more stable: that is, no single individual will influence everyone and there are recognized observed protocols. In the following section we discuss one kind of small world network that has formed around the use of the motor car.

URBAN MOBILITY AND THE INCREASING USE OF TELECOMMUNICATION TECHNOLOGIES

In the developed Western economies new technologies have facilitated a great degree of mobility. The widespread consumer absorption of new products and applications, and the mastery of ICT tools, has allowed greater individual fulfilment together with participation in product re-development. As mentioned above the Chinese society has undergone wide-ranging economic reform during the past two decades, resulting in more mobile urban lifestyles. The high penetration of mobile phones, now linked to the Internet, illustrates how mobility is central to understanding the social dimension of technology in Chinese cities.

China is a transitional state in which modern behaviour competes with the values of traditional society. Increased mobility is challenging the concept of community, particularly in big cities. In other words, the increase in mobility challenges inherited forms of sociability, especially those related to the notion of proximity (Davis, 1995). Likewise the use of the Internet and mobile technologies distances people from traditional face to face relationships. In addition to these direct effects of technological devices, contemporary community-building processes are undergoing radical changes (as we shall see below).

Mobility also has a visible material significance. Since 1992, many urban dwellers in China have invested in the purchase of homes and cars. The acquisition of the home is determined by income but it is increasingly subject to location preferences; for example, some neighborhoods are more fashionable than others. Attention to location displaces the former work-based mobility of the Maoist era. In Maoist China, travel distances were limited because of a strong geographical integration of work and residence. Residences were organized within or close to the work unit. Under the effect of the reforms, the work units are losing this structuring role. Today, companies have less control over the residential location of their employees: work places and residential places are scattered, resulting in longer travel distances. And because of severe traffic congestion, Chinese urban dwellers have to spend more time commuting everyday, hence the growing Chinese love affair with the car.

The transformation of urban space has largely been a result of planned reconstruction, aided by greater social mobility and affluence. In this combination of space and capital individuals have developed a more informed knowledge of the city. For decades, travel was effectively structured around the home-to-work pattern. Today, because of a more individual-based society and the increasing significance of leisure, people are multiplying their urban experiences. Travel for work, education and shopping remains the most common travel motives; however, for an increasing number of individuals, leisure is a necessity rather than just an option.

Up until the mid-1990s, urban mobility depended mainly on bus and bicycle. With improvements in transportation networks, people gained access to faster transportation (taxis, company cars, private cars, subway and light train). On average, every 1.46 household in Beijing has a car[4]. Moreover despite the costs, taking a taxi is very common practice. Considering those contextual elements, it seems that mobility has become a key entry point to analyze the impact of telecommunication technologies on urban life in China. In the following part, we will discuss some of our findings.

"CAR CLUB"

Orange Labs Beijing (The France Telecom group) created a two-year research program with the purpose of: exploring ICT usage in China in order to assess the impact of mobility on the concept of community, investigating urban mobility as a social phenomenon, and examining urban space as a natural environment for interaction. To understand these ideas the lab examined the relationships between different forms of sociability and mobility linked with ICTs through in-depth interviews as well as new "mobile" methods.[5]

The Car Club is a "small world network", an online community of owners of particular brands and models. We analyzed Car Clubs' online and offline activities, their organization as well as their evolution, language, communication patterns and values. In understanding urban mobility as social behavior, we were looking at how new technologies provide a site for creating new forms of collective practice and meaning.

Car ownership continues to grow at a rapid pace in urban China. The number of passenger cars on the road in 2000 was 6 million; today the figure stands at more than 20 million. 28 percent of Beijing residents now own cars. In 2006, there were 3.1 million registered cars in the Beijing municipality, of which about 2.4 million were private vehicles. 30 percent of daily trips are now done by car. Every day, 1,000 new cars (and 500 used ones) are sold in Beijing alone.[6] Because of this boom "car clubs" and "self-driving trips" are popular across the nation. What is particularly interesting, however, and different from most western countries, is that the surge of automobile ownership has occurred synchronously with the rapid adoption of mobile phone and Internet in China. Both play a key role in the construction of a new urban lifestyle and act as symbols of independence.

In China, the car is also about status. Chinese social relations are typified by reciprocal social networks. The Car Club was online at the outset, but now plays a role in organizing real social networks. The coordinated activities of multiple individuals produce larger-scale effects, thus one of the effects produced by this "on the go" car club community is a responsive collective mobile intelligence.

In Beijing 200,000 car owners belong to 79 car clubs. Car fans share their love, knowledge and problems about cars on BBS, which also serve as a conduit for offline activities among car fans.[7] Netizens have developed a system of nicknames and acronyms to refer to their beloved cars and the car clubs they belong to, for example:

Camry: (little Fu) 凯凯 (Kai Kai), KK, KMR, CMR
Focus: 小福, 福福 (Fu Fu), FKS, FCS
Polo: 菠萝 (pineapple)
Peugeot 307: 小狮 (little lion)
Peugeot 206: 小六 (little six)

Audi A4: 小4 (little 4)
Nissan Taida: 达达 (Da Da), DD, QD
Opel: 宝宝 (bao bao/baby), 小欧 (little ou)

The Car Club communities are non-profit sites, formed by car owners independently from car manufacturers. They constitute an open community of car fans, car owners and potential owners. Compared with many online communities, particularly scale free networks, these "small world networks" are selective: social status is linked with certain car brands. Members of different car clubs in turn look for different experiences. Some seek to expand their social networks (e.g. BMW car club). Others pursue leisure activities among singles (Ford Focus car club), find persons with the similar "nomadic" lifestyles (Beijing Jeep car club) and learn how to get a better driving experience by exchanging knowledge among drivers. In a way, people join these car clubs not only for leisure activities or to share common interests, but also for daily mobility solutions in the face of the new urban life.

We noticed among respondents an evolution of attitudes towards the car club and its collective activities, from conspicuous consumption to community building and mobile lifestyle adoption. As one respondent says:

At the beginning, I noticed the online car club because I bought my first car and I was curious about who had the same car as me; then it started being great fun when we drove our cars on the road together in a line, it was very cool and attracted lots of attention. As time goes by, I begin to realize it is part of my life from just reading the post on the BBS, asking questions about how to better use the GPS, joining FB activities, and organizing group trips as the team leader, and helping new comers. I even was on a business trip, I still logged on to our BBS, even just to have a look at the conversations."

—Member of an online car club

TYPICAL LANGUAGES AND ACTIVITIES

Four types of topics are regularly generated on Car Club BBS or QQ groups: these are "asking for help", "sharing experiences", "organizing trips" and "providing information". The following are representative of language and practices we observed in the car clubs.

- **FB** (short for *fubai*)
  Are you searching for a place for FB event! Note: Considering the actual participant numbers will be far beyond the sign-up numbers, if you come without reservation, we will accept your money, but we won't let you in!

The literal meaning of *fubai* is corrupt. It comes from describing the act of corrupt officials freely spending to enjoy life. Within auto forums, the term has been appropriated to refer to netizens spending their own hard-earned money to enjoy life through such FB activities as going out for good food and traveling. Often these offline FB activities are organized online within the forums. Online friends, united by their love of a particular car, can quickly become offline friends.

- DX (short for *daxia*)
  I have customized my car, would love DX to give comments on it. I am uploading pictures now!

  —a netizen

The literal meaning of *daxia* is big shrimp. DX serves as a sort of honorific hero title for a male netizen who says something cool or does something interesting.

Within online car clubs, there is always someone who is highly active at organizing the group's self tour. These people will post the tour proposal either on BBS or in QQ groups, with a detailed description including route, activities, accommodations, budget per person and theme.

The following shows how a group leader organized such a collective activity and the various communication tools he used for the coordination.

1. *I posted a message on the car-mate forum to introduce the activity like a promotion.*
2. *I announced it in my QQ group;*
3. *I waited for people to sign up for participation;*
4. *I counted the number of participants;*
5. *I contacted the place to arrange the detail of accommodations and schedule, and visited the place by myself to fix the route in advance; I contacted participants with QQ (when I was out of office and home, and used mobile QQ) and posted the schedule and expected expense on the forum, asking everybody to bring their mobile phone and interphone.*

With an increasing income and an increased number of vacation days per year, China has entered the society of leisure.[8] Many people are attracted to out of town locations, even for weekend trips. Meanwhile, an increasing number of urban residents now choose to travel alone or search for travel mates online and organize group trips to go somewhere together. The boom in China-focused online travel agencies makes planning travel inside China much easier and cheaper than before. As a result, many Chinese people visit sites like Ctrip.com (a consolidator of hotel rooms and airline tickets in China that started in 1999) and Qunar.com (a price comparison site for travel) to make their holiday plans. In order to organize vacation, people obtain information in advance using websites or discussion groups and make plans collectively.

MOBILITY AS COLLECTIVE BEHAVIOR

The traditional pattern of Chinese sociability relies on kinship and spatial relationships which are locally embedded in specific social structures. As illustrated below today's sociability is opening from small social circles, such as family and close friends, to larger self-designed "small worlds" based on common interests such as car clubs, traveling enthusiasts and cooking-mates.

Being mobile in an opening social circle goes along with a more sharing attitude towards space and collective mobility practices. The prevalence of car clubs is a case in point. In China, a car is not an individual consumption, it's about status. The way Chinese people behave with their cars shows

Figure 2.

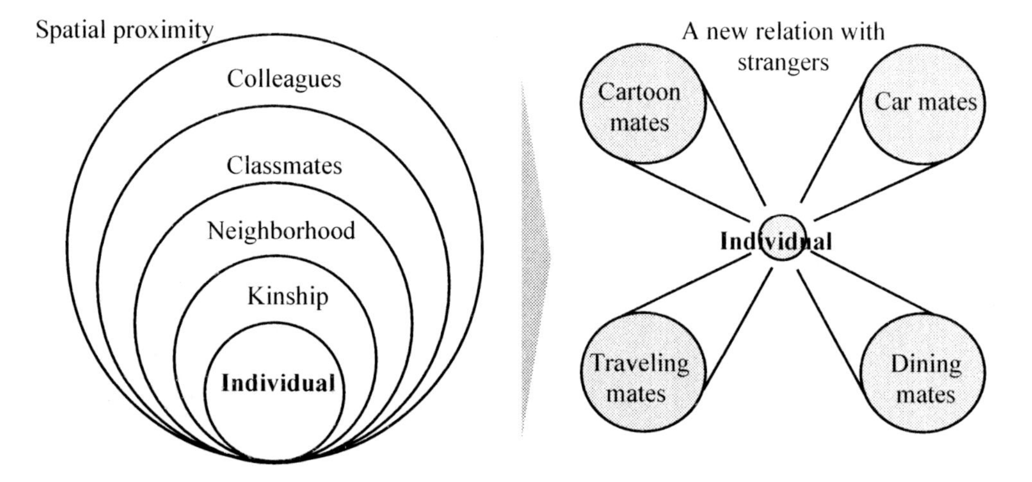

the strategies to face the new urban life with the characteristics of increasing mobility, changing urban landscape, expansion of weak ties and ICT usage. A car club is built online at the beginning to offer a kind of organizational belonging to the segmented urbanites, but later it plays a role in everyday city life as a quite efficient reciprocal and practical social network. One of the effects produced by the "on the go" car club, is the responsive collective mobility intelligence.

Therefore, the significance of "Car Club" and "Travel Mates" is not the club or the online community itself, but the link it creates between online and offline activities, that is, from *collective sociality* to *collaborative spatiality*. Beginning with the concept of mobility, rather than technology, we propose a functional link between people and technology. From this perspective, while urban and mobile technologies are tools for problem-solving, they are more importantly sites at which social and cultural practices are produced and reproduced (see Dourish et al, 2007). We observe that new technologies provide a site for creating new forms of collective practice and meanings. Being mobile therefore is not only the adoption of the car, or certain mobile technologies, but also the changing attitude towards urban life, reinvented sociability patterns and new relations with urban space.

CONCLUSION: CONTEXTUALIZING THE CITY 2.0, CONNECTIVITY, AND MOBILITY

In the past two decades, China has witnessed rapid development of urban infrastructures for telecommunication, a reconstruction of communities, a higher level of social, residential and daily mobility —especially with leisure-oriented mobility and the rapid adoption of emerging technologies and services. Web 2.0 innovations in China increasingly come from areas such as BBS/Online forums, IM services, P2P streaming, Internet TV, Mobile, Online gaming and virtual goods. Considering that the number of mobile users is more than double the number of Internet users in China, the future is about "automating interaction" (Ruggles 2005). Here is the key point. Looking at China from the long view, one may observe the totality of control, the Communist state, the view pushed by China's critics leading up to the Beijing Olympics. Looking close-up, however, we see a great freeing up of interaction: people talking on mobile phones, gamers connecting with distant players, SMS and IM networks continually connecting, and thousands of "small world" community networks like Car Clubs. The future of the Chinese City 2.0 is framed by ever-increasing connectivity, allowing mobile lifestyles that integrate personal mobile devices with use of space. The emerging City 2.0 allows individuals to collect greater intelligence about being mobile, to share knowledge within small world communities and to enable distant others to interact with that knowledge.

As we mentioned in the introduction, important trends in informatics are reflected in urban lifestyles. The first point we noted was that information technology goods and services are produced, marketed and absorbed at an ever increasing rate. In China, wireless and Internet-based innovations have accelerated. Second, consumers and workers are upgrading skills and literacies as the boundary between online social networking and offline social networking blurs. Both small world and scale free networks co-evolve. Third, mobility is providing opportunities connected with culture and sociability. Dourish et al. (2007) have investigated different approaches to dealing with mobility in urban computing. For instance, one classic approach is to view mobility as a "problem" that requires mobilization of static applications like PDAs to resolve disconnection from stable working situations. On the other hand, seeing mobility as opportunity foregrounds interactive opportunities for city travelers.

In recognizing these transitions, moreover, the developers of mobile applications have begun to incorporate lessons from social science to understand ways in which people produce spatial experience. In this inversion of market push, users of new technology become increasingly involved in influencing the design of technologies. This is the vision of the Chinese City 2.0. We hope that this may also be the design for a more open and democratic China. This is the challenge presented by information and communication technologies in transitional societies. On the one hand, a great leap forward for productivity, on the other, a democratization of information and a reshaping of social values.

REFERENCES

Castells, Manuel (2007). European cities the information society, and the global economy. In R LeGates and F. Stout (Eds). *The City Reader* Fourth Edition (pp. 478—488). Milton Park: Routledge: originally in *The Journal of Economic and Social Geography*, Vol. LXXXIV, 4, 1993.

CNNIC (2007). *The 21th Statistical Report on China's Internet Development*, Retrieved March 30 2008 from www.cnnic.cn

Davis, D., ed. (1995). *Urban Spaces in Contemporary China*. Berkeley: University of California Press.

Donald S. H. and Benewick, B. (2005). *The State of China Atlas*. Sydney: UNSW Press.

Donald S. H. and Benewick, B. (2008). *The Pocket China Atlas*. Berkeley:University of California Press.

Doulet J.F and Flonneau, M. (2003). *Paris-Pékin, civiliser l'automobile (Paris-Beijing, Civilizing the Automobile)*. Paris: Descartes et Cie.

Dourish, P.Anderson, K, and Nafus D.. *Cultural Mobilities: Diversity and Agency in Urban Computing*. INTERACT (2) 2007: 100-113.

Friedmann, J. (2005). *China's Urban Transition*. Minneapolis: University of Minnesota Press.

Garner, J. (2005). *The Rise of the Chinese Consumer*. Chichester: John Wiley and Sons.

Keane, M. (2007). *Created in China: the Great New Leap Forward*. London: Routledge

Law, P. (2005). *Mobile Communication, Mobile Relationships, and the Mobility of Migrant Workers in Guangdong*. Paper presented at the conference on Mobile Communication and Asian Modernities, October 20-21, Orange Lab, Beijing

Mars, N., and Hornsby, A. (2008). *The Chinese Dream: a Society under Construction*. Rotterdam: Uitgeverij 010 Publishers.

Rong Zhou (2006). Upon the ruins of utopia. *Volume Special Issue Ubiquitous China*, 2: 44- 47.

Ruggles, M. (2005). *Automating Interaction: Formal and Informal Knowledge in the Digital Network Economy*. New Jersey: Hampton Press. Inc.

KEY TERMS

City 2.0: A new expression designating new approaches to organizing a city along the principles of Web 2.0, inviting participation from citizens to define urban services, a more widely shared type of urbanism.

Collective Mobility: Trips are motivated by a more open sociability and individuals adopting mobility behaviors that are instigated by one or more social groups; for example, a car club self-driving tour.

Collaborative Spatiality: The perception, as well as the experience, of urban spaces is partly shaped by the collaborative activities within groups. For example, through car club BBS, people build a "City-Wiki" including trip routes, cheap parking manuals and a GPS guidebook to share their mobility experience and knowledge.

Sociability Pattern: The way individuals build their social circles and the interactions between these circles.

Urban Mobility: The whole of trips generated daily by the inhabitants of a city, and the methods and conditions associated with such trips (modes of transport selected, length of trip, time spent in transport, etc.)

ENDNOTES

1 The average wage in 2005 was 18,364 yuan ($US2371). See Donald and Benewick 2008, 49

2 Users are those that go online for at least one hour per week.

3 Cited from "Internet Boom in China Is Built on Virtual Fun", By David Barboza, Published: February 5, 2007, New York Times.

4 Beijing Traffic Management Bureau, 2007

5 This research is based on two rounds of fieldwork: The first round was conducted in the summer of 2006, with 40 in-depth interviews and 6 time-space diaries, in Beijing, Shanghai and Guangzhou. Based on the findings of the first round, we moved on to the second round in the summer of 2007, with 16 in-depth interviews (8 in Beijing and 8 in Shanghai), 1 focus group, 1 creative session and 1 participative observation organized by online collaborative travel communities. All the interviewees are "highly mobile urban individuals", defined by a high level of motility (i.e. potential of mobility).

6 Beijing Traffic Management Bureau, 2006

7 CIC, a leading Internet Word of Mouth (IWOM) research and consulting firm in China, collected Car BBS Posts for September: Number of messages:3,826,853; Number of unique posters: 93,720; Average number of posts per day:127,561

8 China has three seven-day holiday periods each year: Lunar New Year or Spring Festival early in the year, Labor Day (May 1-7), and National Day (Oct. 1-7). The three "Golden Weeks" are a welcome break for urbanites. Up until the mid 1990's workers in China used to put in six-day work weeks with only a long weekend or two for leisure during the year. Now if all weekends and holidays are counted workers can enjoy up to 114 non-working days; in other words, about a third of the year is spent on leisure time. Instead of staying at home or visiting relatives and friends within the local area, the holidays see millions of Chinese on the move, shopping, dining and traveling.

Chapter XXVII
WikiCity: Real-Time Location-Sensitive Tools for the City

Francesco Calabrese
Massachusetts Institute of Technology, USA

Kristian Kloeckl
Massachusetts Institute of Technology, USA

Carlo Ratti
Massachusetts Institute of Technology, USA

ABSTRACT

The real-time city is now real! The increasing deployment of sensors and handheld electronic devices in recent years allows for a new approach to the study and exploration of the built environment. The WikiCity project deals with the development of real-time, location-sensitive tools for the city and is concerned with the real-time mapping of city dynamics. This mapping, however, is not limited to representing the city, but also instantly becomes an instrument for city inhabitants to base their actions and decisions upon in a better informed manner, leading to an overall increased efficiency and sustainability in making use of the city environment. While our comprehensive research program considers a larger context, this chapter discusses the WikiCity Rome project, which was the first occasion for implementing some of WikiCity's elements in a public interface—it was presented on a large screen in a public square in Rome.

INTRODUCTION

The WikiCity project at the sense*able* city laboratory at MIT is a multi-year research effort that builds on different research strings, combining under one common vision the aspects of sensing, time- and location-based data structuring, and the input and output articulation of this data within the context of urban environments. These different fields are combined both to offer new perspectives in analyzing a city's dynamic and for the conception and elaboration of novel tools for citizens to make optimal use of their environment (Calabrese, Kloeckl, and Ratti, 2007).

The WikiCity Rome project, on the occasion of La Notte Bianca (White Night, www.lanottebianca.it) in Rome on September 8, 2007, has been an opportunity to present a first glimpse of the more comprehensive WikiCity project to the public. This first implementation allows people access to the real-time data on dynamics that occur where they are at that moment, creating the intriguing situation that the map is drawn on the basis of dynamic elements of which the map itself is an active part. 'How do people react toward this new perspective on their own city while they are determining the city's very own dynamic?' and 'How does having access to real-time data in the context of possible action alter the process of decision making in how to go about different activities?' are our guiding research questions.

The overall WikiCity research program considers such questions in a larger context that includes the active uploading of information by citizens, local authorities, and businesses regarding an ever increasing field of data; an elaborate approach to semantic data structures to enable novel ways of querying the data; and a rich array of multimodal access interfaces for users to interact with the data in a meaningful way.

The remainder of this chapter is structured as follows. Section 2 describes the project, taking into account different aspects of creating an open platform. Section 3 describes WikiCity Rome, first implementation of the WikiCity concept. Section 4 describes relevant work related to the WikiCity concept. Section 5 describes the design concept and different application scenarios, while Section 6 deals with access modalities and interfaces. Section 7 describes the software infrastructure and its main elements. Finally, Section 8 describes directions for future work and conclusions.

WIKICITY

People moving and acting in a city base their decisions on information that is, in most cases, not synchronized with their present time and place when making that decision[1]. How often have you arrived at the airport just to find out that your flight has been delayed, or been surprised by a traffic jam, or found that a product is out of stock or a service operator busy at the moment you need it.

In the same way, a person acting in a city contributes himself to dynamics of which others are not aware when making their decisions. Looked upon in this way, a city resembles what Deleuze and Guattari describe as a *rhizome* (Deleuze and Guattari, 1977). The rhizome is a philosophical network structure where every part is necessarily connected with every other part of the system. There are no preferential connections because every connection alters the overall network structure. As a consequence, the rhizome cannot be plotted since the plotting action itself is part of the rhizome and thus in the very moment of plotting its structure, the structure changes.

The WikiCity project, in a similar way, is concerned with the real-time mapping of city dynamics. This mapping, however, is not limited to representing the city, but instead becomes instantly an instrument upon which city inhabitants can base their actions and decisions in a better-informed manner. In this way the real-time map changes the city context, and this altered context changes the real-time map accordingly, with the ultimate aim of leading to an overall increased efficiency and sustainability in making use of the city environment.

Will such a WikiCity lead to more people attempting to be at the same place at the same time or in an increasing number of different places at different times? Designing a tool to address such a question requires considering whether and how the real-time map is capable of communicating different and context-based information to users in different circumstances and how people's decisions, which were taken on the basis of the real-time information, are fed back into the system.

The project aims to create a common format for interchange of real-time location-based data and a distributed platform able to collect, manage, and provide such data in real-time. This platform is conceived as an open platform, meaning that different users can access, upload, and modify data on this system platform. WikiCity therefore functions as an additional infrastructural layer of a city, similar to electricity, transport, and telecommunications. Two relevant issues emerge from the choice of this platform type: the reliability of data sources, and the understanding of how different data can be constructively combined in response to a user's query.

Information and Trust in an Open Platform

Since the content is to be provided by an open community and is ultimately going to be combined in such a way that users may base decisions on it, the issue of trust (with regard to content data) deserves careful reflection. How do you know you can trust information that is provided from an open community? Whom do you trust, and can you decide whom to trust? (cf. Foth, Odendaal, and Hearn, 2007; and Beer and Burrows, 2007).

In conventional infrastructures this aspect is most often resolved through publicly recognized certifications and recognitions of singular and well-defined agents; this underlying element of trust is part of the reason why people choose to rely on such infrastructures. How can we arrange such a system of trust in WikiCity's open platform, which by definition is open and therefore cannot be based on any single agent that attributes reliability onto other agents?

In an open community platform, trust should come from within the community of users. A strong example is eBay's member feedback history: with each new transaction, the two members involved give each other feedback, and each new piece of feedback is combined with the entirety of the user's feedback history in order to create an overall rating. In this way, feedback provided by a single user is merged with every additional input provided in the future, and the decision whether to trust a single input can be based upon a comparison with the whole feedback history.

Another and perhaps complementary approach leads to the definition about whom I want to trust when accessing data from WikiCity. An open platform might enable me to choose whose data I want to rely on. I might want to consider data trusted by myself, trusted by others whom I trust directly ("first-level trusted data providers"), and those whom *they* trust directly. In this way, we use levels of separation as an indicator of trust.

Both of these approaches involve the addition of elements to any set of data supplied to WikiCity that enable users to evaluate its reliability. Different levels of such trackability shall be considered, giving the end user different possibilities as to which data to consider and which to ignore (e.g., I might be interested in considering data from sources without any history of trust in certain situations, and by my giving feedback on their encountered reliability, a trust record is created).

Making the Users Create the Links

While one of the challenging tasks in structuring a data platform consisting of a diversity of data, which one cannot and does not want to predict or limit up front, is to describe the actual data, another is to create relationships between data sets. For a combined query regarding perfect places to fly a kite, one would be interested in considering weather data (wind, rain), daylight hours, and an area's topology and vegetation. Instead, knowing the electrical resistance of a device present in that location is of no use for this query, but how does the platform know?

An interesting hint can be taken from a project developed by Carnegie Mellon researchers: reCAPTCHA (http://recaptcha.net) puts the 60 million daily descriptions of CAPTCHAs[2] on the World Wide Web to use in resolving words that OCR (optical character recognition) programs cannot read when digitizing books.

In a similar way we can imagine WikiCity collecting different types of data on its open platform, which people can use, through an API (Application programming interface), to custom-code their data-fusion applications. In this way, specific pieces of information can be aggregated and visually represented to end users through an interface. When this happens, all data types involved in this process receive an increase in the level of attractiveness between each other. A human has chosen that it makes some sense to combine these data types for a real-world application. This level of attractiveness can then be used in a more open query of the platform in order to suggest or give priorities to certain types of data to be involved in the query reply.

WIKICITY ROME

Our group takes a bottom-up approach to developing WikiCity: the overall structure is created stepwise through implementations on a reduced scale. In a similar way, the aspect of trust as related to data is approached gradually, combining traditional elements of trust with novel ones as outlined above. We are at this stage working together with established and publicly certified partners such as telecommunications entities, telephone and address directory services, satellite imaging, public transport, and newspaper publishers in order to establish a range of different data sets that can be combined on the WikiCity platform. Next, we aim to enlarge this base group of know-how and content providers as well as open up the system gradually to provide users with direct input and output access on the platform itself.

Figure 1. Photos from the WikiCity Rome project presented at La Notte Bianca, 08-09-2007

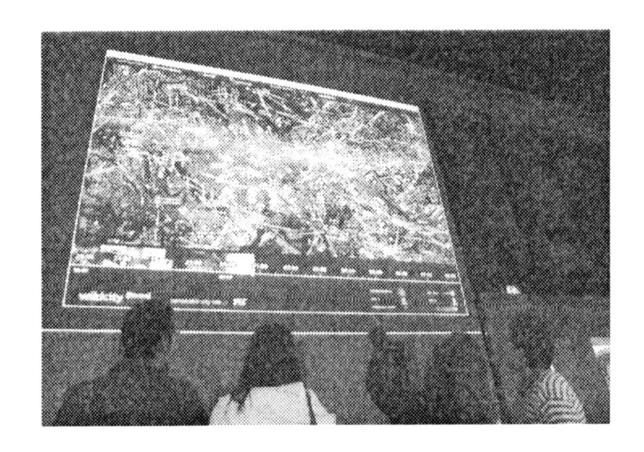

A first implementation of the WikiCity concept was presented in Rome, Italy, during La Notte Bianca on September 8, 2007. This demonstrator (see http://senseable.mit.edu/wikicity/rome) comprised the presentation, on a big screen in a major square of Rome and on the Web through a Web applet, of real-time population distribution by the use of cell-phone data, the location of buses and trains, real-time news feeds from a main Italian newspaper, and its mapping to certain events happening in the city.

RELATED WORK

Recent literature has increasingly used the term 'city' in reference to a real-world space offering social opportunities and cultural landmarks for urban analysis (Thom-Santelli, 2007). Hitherto, implementations have been based on ideal users who perfectly fit a certain application. Recently, social software service implementations try to promote different user-specific interpretations of urban landscapes and by that enable the development of multiple meanings and representations of city use. This also results in a move from representing the city as a structured lineup of buildings to a space of interpretative cultural interaction.

The same direction is envisioned by Bassoili (Bassoli et al., 2007) which states that a paradigm shift in urban computing is just occurring, in the sense that it is not only about the city, but that the city itself and its inhabitants are part of the application and vice versa—i.e., urban computing becomes part of the city. In consequence, the context of the city is growing toward a platform for the collective creation of content and social interactions. Thus, the city itself is increasingly perceived as a source of experience rather than a place of trouble and hassle. The authors mention disconnection, dislocation, and disruption as the three main challenges when designing urban applications.

A general trend in current research is that applications using wirelessly interconnected devices have to integrate seamlessly into everyday life until people don't even recognize their presence anymore. Thus, pervasive information needs can be satisfied through smooth integration with social structures, fast personalized service access, seamless sharing of information, multiple types of interfaces, and ubiquitous networked user services (Trevor and Hilbert, 2007).

DESIGN CONCEPT AND SCENARIOS

WikiCity is about envisioning new application scenarios on the basis of a technology potential involved in location- and time-sensitive information. Different aspects of the concept design are presented in this section. We start identifying the agents the metafunctions and the technology features involved in the system. Then we address the concept of time as applied in the project, and use the real-time control system as a working metaphor. This leads to the elaboration of different scenarios used to conceptualize the applications of WikiCity.

Within the social context of a city, we have identified three main groups of element: the various agents involved, the specific environmental conditions in which they operate, and the potential of new technologies.

1. **Agents:** Private individuals, associations, local authorities, companies, non-profit organizations
2. **Environment:** City architecture, infrastructures, landscape, waterways, climatic conditions
3. **Technology Features:** Positioning, detecting movement and interaction, evaluating density, visualizing, sensing environment values

Figure 2. Application scenarios generated by intersection of agents, environment, and technology features

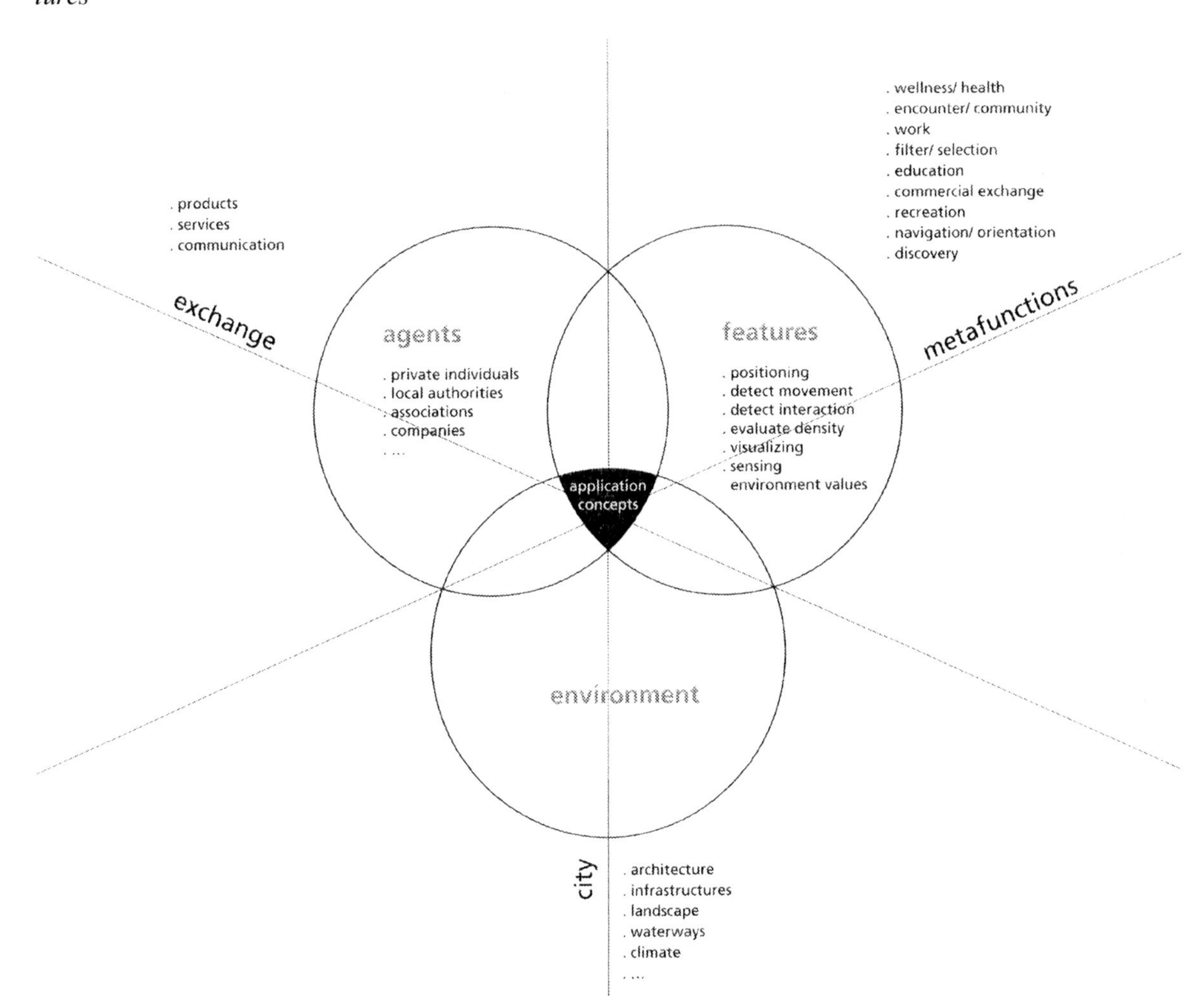

Subsequently we identified the intersections of the capacities and requirements of these three element groups.

Considering people living in urban areas, we distinguish among four main macro needs that can be addressed in the design of services and products for urban dynamics: safety, interaction, comfort, and mobility.

On the basis of these macro needs, we identify metafunctions—different for any specific geographic, cultural, or technological context—that try to accommodate these needs. They are as follows:

- **Wellness / Health:** How do people maintain their physical and mental health and well-being, and what do they consider these to mean?
- **Work:** What do people do to interact constructively with their environment? How and what do they create?
- **Encounter / Community:** How does the pleasure of living together with many others in one city translate into actual encounters? How do people arrange to meet: where and when and whom do they meet, and on which occasions?

- **Filter/Selection:** In the urban environment, people are confronted with a massive amount of events that may attract their attention. What strategies do they use to filter out what is not of interest, and how do they find and focus on events that do match their range of interests?
- **Education:** Assuming that learning is not confined to school buildings: which occasions and places do people identify with enlarging their knowledge, and in what ways do they proceed to do so?
- **Commercial Exchange:** What commodities are included by society within the sphere of commercial exchange? Value patterns are continually changing, and items that are free today might become valuable products or services tomorrow. What products or services are part of the commercial exchange dynamic? Which ones are not a part of that dynamic now but might become a part in the future? How will this affect them?
- **Recreation:** What do people do for enjoyment and pleasure. What constitutes pleasure for people in a city, how can an urban environment foster these activities and situations, and what factors prevent them?
- **Navigation / Orientation:** Moving physically within a city requires knowing where to go and how to get there. How do citizens choose their routes through the urban landscape, and how do they adapt to the changes in that environment? How do they rely on external communication elements for this navigation, and how much do they instead base their orientation on personal knowledge?
- **Discovery:** An urban environment presents risks and opportunities that are known to a citizen, but it contains the potential of unexpected discoveries in all aspects of life. What elements of a city contribute to constructive new discoveries by its citizens? What behavior do people adopt in discovering new aspects of their environment?

Figure 3. Technology features and metafunctions intersect to generate application fields

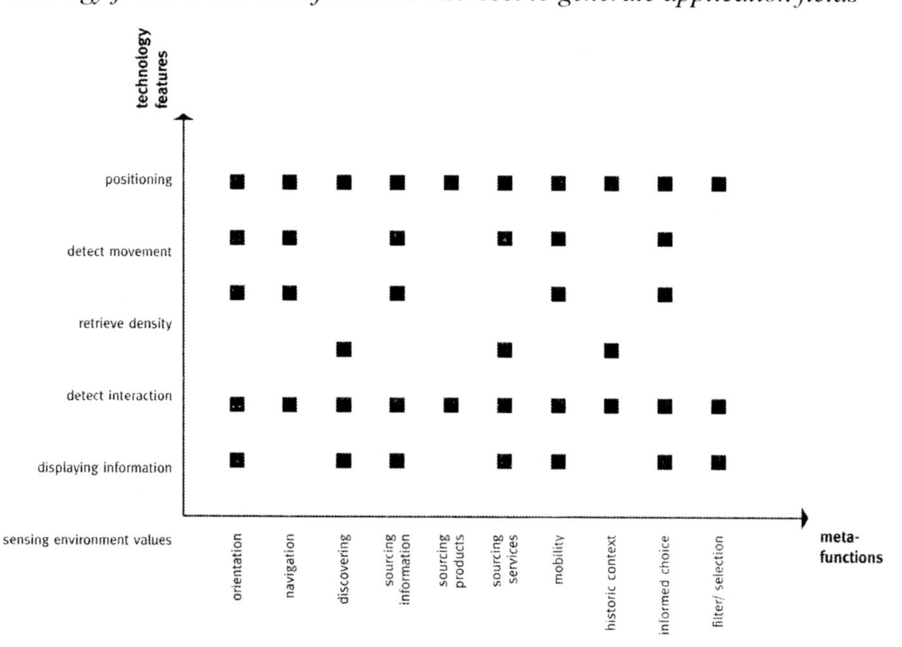

By examining the intersections of metafunctions and potential technological features, we can identify new application concepts for the WikiCity project, as explained in the subsections below.

Time Value

A key characteristic of WikiCity is the circulation of information on a real-time basis. Before proceeding, we should clarify our interpretation of the term *real-time* and the scope of its implications.

Often the term *real-time* refers to a system in which data is processed within a small fraction of time—for example, a sensor that returns a measurement as a signal within a fraction of a second. The difficulty with this definition is that it does not provide a specification of the time interval in question, and this makes it difficult to judge any given system as to whether it does or does not operate in real-time.

A more useful definition refers to real-time as "the actual time during which a process or event occurs" (Oxford, 2007). Consequently a real-time process implies that there is a deadline before which a given piece of data is useful to the system, whereas that same data is not useful—or may even be destructive—to the system thereafter. While the deadline implies a process, identifying the usefulness of respecting such a process's deadline implies the existence of a higher-level mission. Considering now that it is evidently this mission that defines the parameters of the deadline, we end up with an idea of real-time in which there is no stringent necessity to speed up data transfer to arbitrarily defined "very fast" limits but rather to identify reasonable deadlines for data-transmission that are related to specific missions.

Let us consider some examples to illustrate what this implies for WikiCity and its position within the framework of real-time systems. When setting up the data integration for the public transport vehicle position for the WikiCity Rome project, we had initially arranged for a location data feed in five-minute intervals. Visualizing in this way the position of buses around the city gives a good overall impression of the distribution of buses around the city throughout the day. In fact during a previous sense*able* city lab's project, Real-time Rome (see Calabrese, Ratti, 2006 and http://senseable.mit.edu/realtimerome), such a time interval was perfectly sufficient to overlay the information with the cell-phone activity that occurred at the same time in order to examine how transport lines work regarding people's aggregation. In WikiCity Rome, our aim was instead to turn the visualization into an instrument useful for citizens while the information is being processed, suggesting the possibility that people could identify when a bus was about to pass by a stop near their location. Seeing a bus's location of five minutes ago is clearly a case in which such information cannot be considered real-time anymore; it has passed its deadline of usefulness, and the system has failed to deliver information on time to accomplish the mission, which in this case is catching the bus.

On the other hand, when we consider information about upcoming, starting, and running events displayed on a map at their relevant locations, we must opt to visualize this data some minutes before the start of the event (in order for people to be able to attend the event); however, the deadline for this visualization of the event is not critical because the importance of displaying that the event is ongoing, for most missions, decreases as the event approaches its end.

The Real-Time Control System as a Working Metaphor

In order to identify the functional elements needed to construct such an instrument, we chose the real-time control system as an analogy to start with. In the past decades, real-time control systems have been developed for, and deployed in, a variety of engineering applications. In so doing, they have dramatically increased the efficiency of systems through energy savings, self-organization and -repair, regulation of dynamics, increased robustness, and disturbance tolerance.

Now: can a city perform as a real-time control system?

Let us examine the four key components of a real-time control system:

1. entity to be controlled in an environment characterized by uncertainty;
2. sensors able to acquire information about the entity's state in real-time;
3. intelligence capable of evaluating system performance regarding desired outcomes;
4. physical actuators able to act upon the system to realize the control strategy.

A city certainly fits the definition of point 1. Point 2 does not seem to pose problems either: today's deployment of a range of remote sensors in urban areas allows for unprecedented data collection and analysis. As an example, the Real-time Rome project used mobile phones and GPS devices to collect the movement patterns of people and transportation systems a well as their spatial and social usage of streets and neighborhoods. Information regarding further aspects are already collected continually by distinct computing systems that track product and service availability, environmental values, climatic conditions, acoustic values, events, etc.

What about points 3 and 4? How to actuate the city? Although the city already contains several classes of actuators such as traffic lights and remotely updated street signage, their range of use is currently limited. A much more flexible actuator would be the city's own inhabitants: they represent a distributed actuation system in which each person pursues his individual interest in cooperation and competition with others, with the overall behavior of the system governed by the interaction among individuals. People can also form part of the overall intelligence of the control system.

Toward the above goal, the WikiCity project can be thought of as adding further, interaction-oriented layers to a real-time map of the city and making location- and time-sensitive information accessible to users, giving them full control of the database, where they can upload and download data. In this way, people become distributed intelligent actuators and thus become prime actors themselves in improving the efficiency of urban systems.

Scenario Elaboration

Considering the above working analogy of the real-time control system enables us to identify the functional elements of the system making up WikiCity. As a further step in concept and scenario elaboration, the technique of storyboarding is used to visualize dynamic situations enabled by WikiCity in a modeled real-world situation. The technical potential of location- and time-sensitive services becomes apparent by applying it to a coherent scenario. The following figure shows an example of such a storyboard:

Various scenarios have been developed in order to identify potential applications for the WikiCity platform:

- **Jogging Path:** By considering real-time environmental data like air quality, noise, or pollen flow together with the traffic situation and personal health conditions, the WikiCity system can suggest ideal jogging routes, taking these factor into account together with the starting position of the user.
- **Bus-Hopping:** Instead of waiting at the bus stop for a bus that theoretically adheres strictly to a static timetable but in reality deviates from it, having access to real-time information about bus locations promotes a more flexible approach to using the public transport system, allowing the user to arrive at the stop only when the bus is about to turn the corner, or to use an alternate bus line to

Figure 4. Jogging path scenario

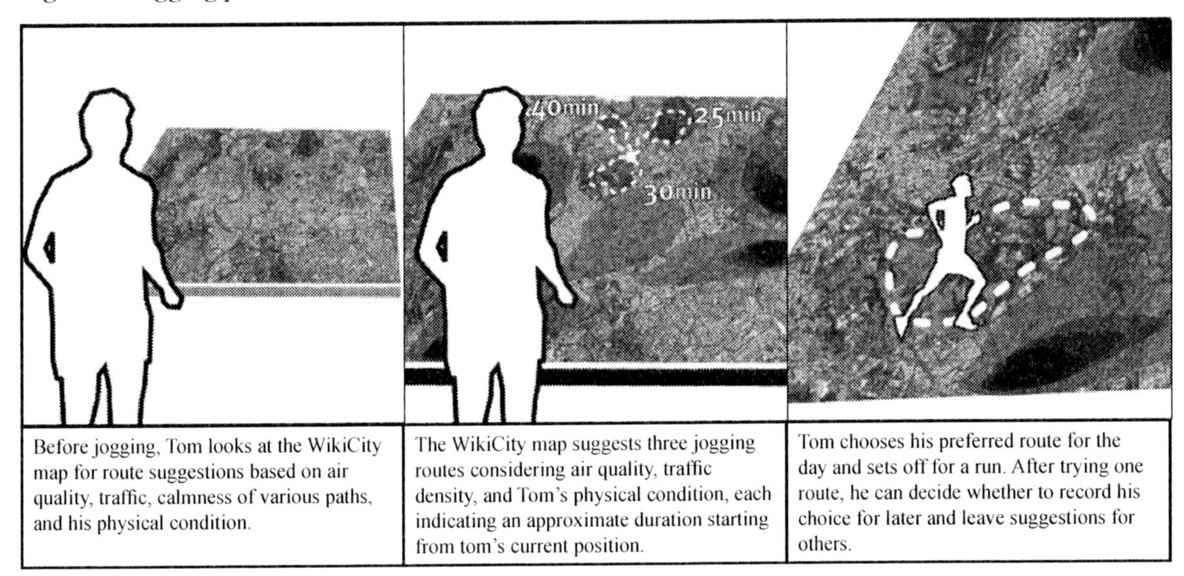

get close enough to the destination to finish the way by foot.

- **Event Spotter:** Walking through the city, one is aware that many events that are both interesting and close by will be missed simply because the typical pedestrian is unaware of them. On the other hand, often programs are planned in advance or at distant locations but are subsequently cancelled without notification to potential attendees. Being able to receive notifications about ongoing or upcoming events within a set radius of the user's current location opens up a happy combination of strolling through the city and discovering events as one comes near them.
- **Sight Density:** Sightseeing in a city often involves finding long queues at popular sites. Knowing in advance about the wait situation at different sites will help the sightseer to organize a visit that allocates more time to actually seeing the sights than to waiting for them. At the same time, because of the increased level of organization resulting from visitors' being better informed, the site's management benefits by being better able to deal with visitor access.
- **Nearest Aid:** When experiencing a minor health issue, one would not want to call an ambulance or even visit a hospital. Instead, knowing about nearby pharmacies, waiting ambulances, or other medical outposts would allow a citizen to stop by and receive adequate advice or treatment without going through a more costly ordeal. Spreading knowledge of such available places can be a simple contribution to people's well-being in a city.
- **Informed Shopping:** It often happens that one knows exactly what product one is looking for but still ends up driving from one place to the other to actually find it. Product-availability information already exists in digital form in logistics databases for most shops. Being able to consider this data together with proximity and store hours would make moving physically through the city in the search of a product or service more efficient.
- **Emergency Support:** Emergency situations may benefit most from a real-time system such as WikiCity since they consist of a drastic alteration of the usual context

> and produce an acute need for up-to-date information on the state of the situation in order to best cope with it. As an example, the location and availability of potable water at any given moment is crucial information to have in areas affected by flooding or earthquake, since collecting it can be a difficult undertaking.

As illustrated in the following figure, in order for these scenarios to feed back into the development process of the WikiCity system architecture, this process helps to identify the attractiveness of certain data types, which enable different applications. This is especially important in the early phase of the project's implementation, during which careful selection of data providers can make the project platform more or less versatile.

Proceeding with the design of application scenarios for the WikiCity platform, we aim to remain true to Deleuze and Guattari's rhizome metaphor. "The rhizome is a centered, non

Figure 5. Seven scenarios

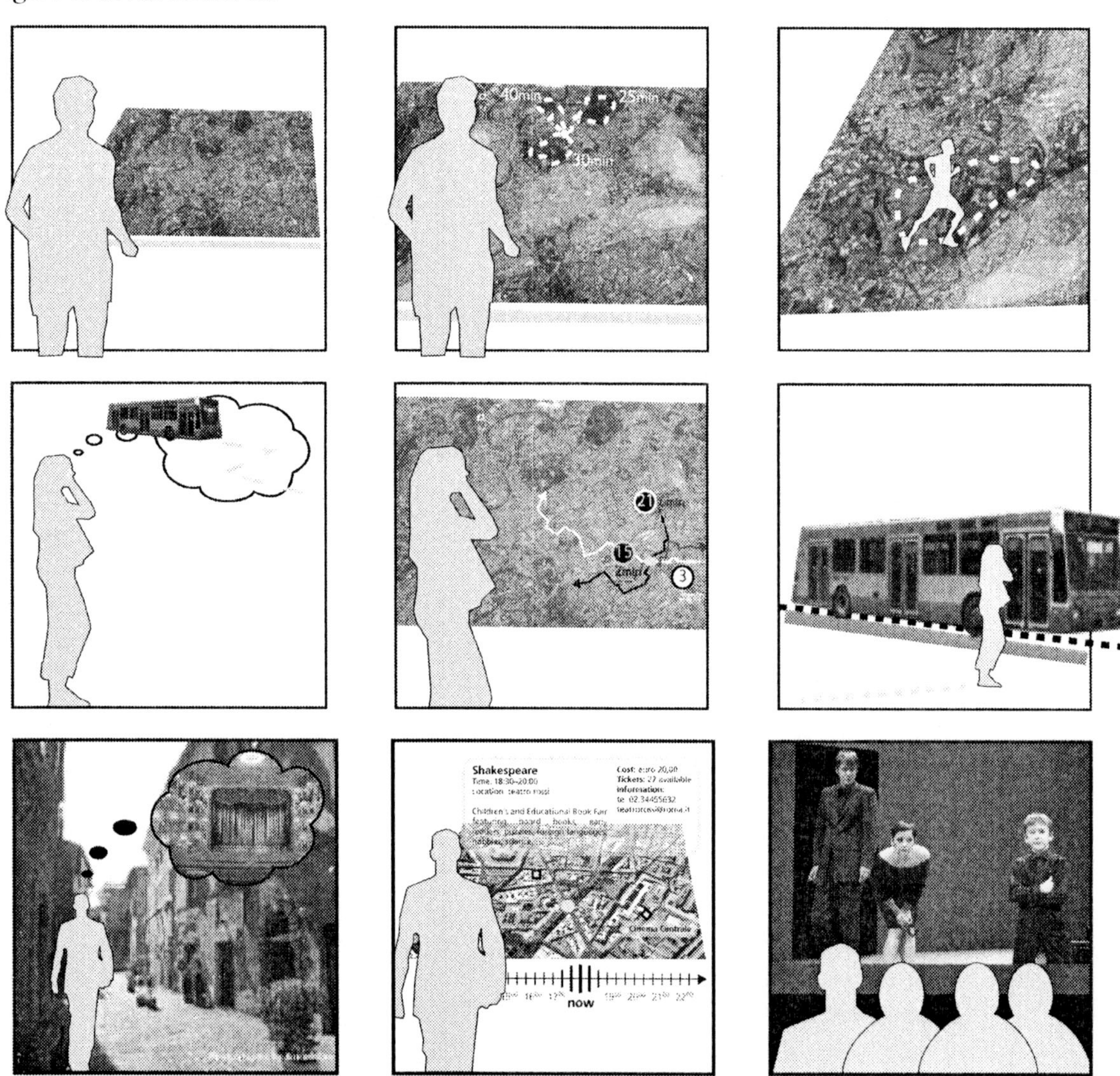

Continued on following page

Figure 5. Continued

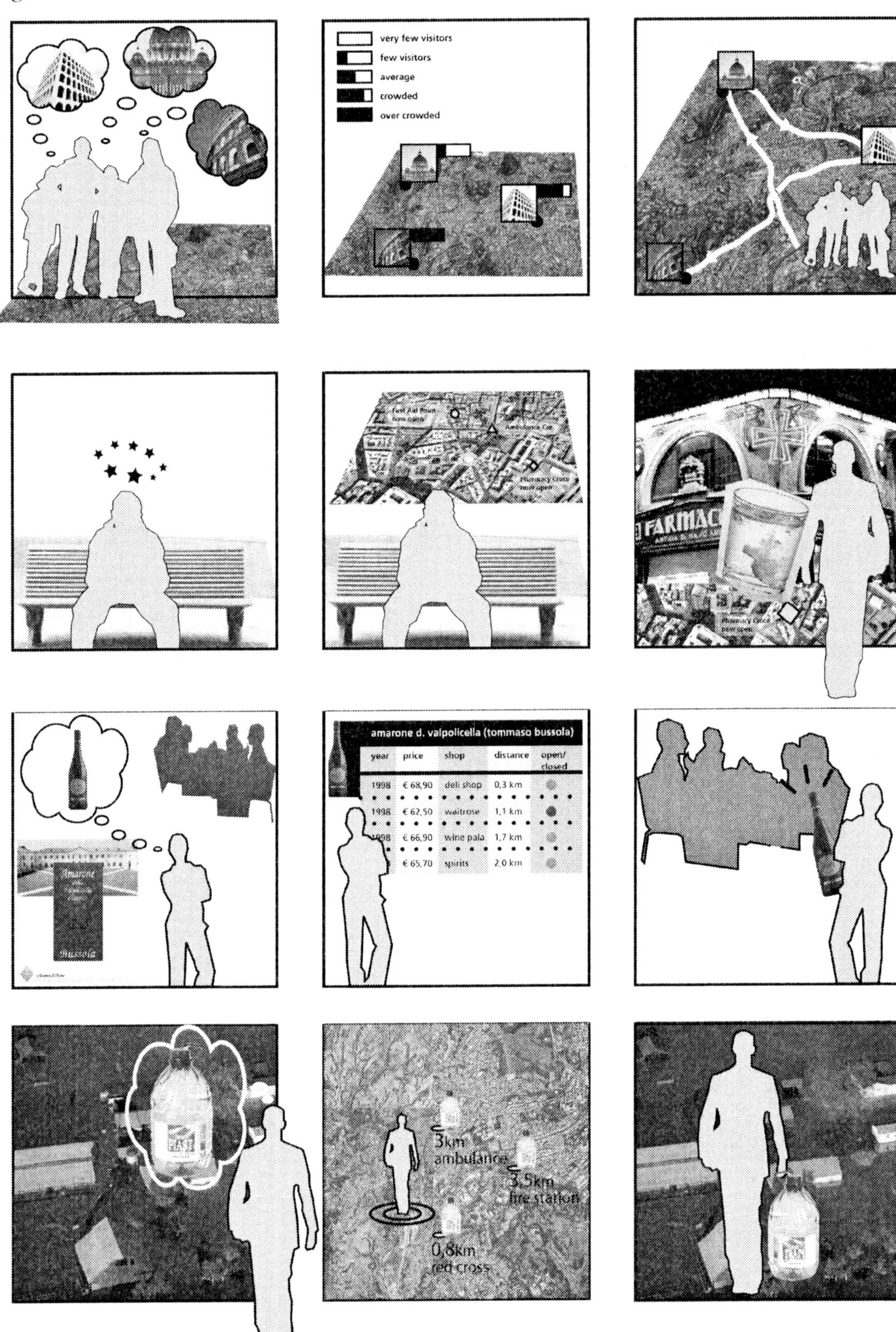

Figure 6. What data enables what scenarios?

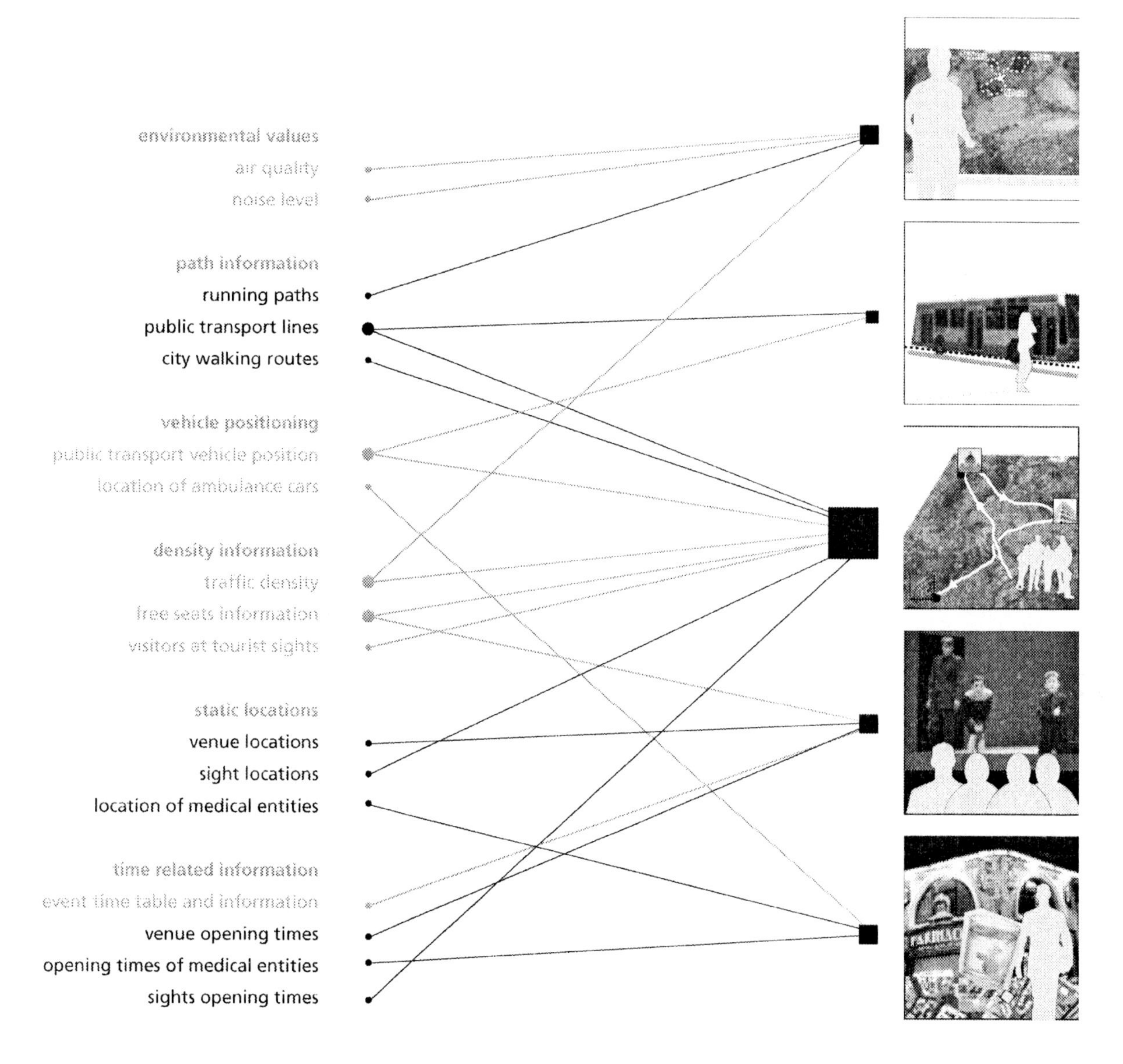

hierarchical, non-signifying system without a General and without an organizing memory or central automation, defined solely by a circulation of states" (Deleuze, Guattari, 1977). What is obtained is utopian freedom—liberation from the constraints of hierarchal systems. The concept of utopian freedom of movement supplied by Deleuze and Guattari has previously been employed by cyber theorists such as Wray (1998) as a means of explaining how users freely traverse digital space. While this analysis is intended to relate to the Internet, it would seem that the concept of the nomadic Internet user extends naturally to explain the city dweller who, through the use of technologies such as WikiCity, enters, traverses, and leaves real and urban spaces.

However, an essential part of the Deleuze-Guattari model of the rhizome is that utopian elements can exist only in tension with dystopian forces. The essence of the rhizome is that it exists in contention with arboreal (and, for Deleuze and Guattari, capitalist) interests that seek to curtail the potential freedom presented by the rhizome, and the outcome is a continual battle of territorialism and re-territorialism.

This dualism of the rhizome/arboreal model might provide a useful framework for understanding the limits or design aspects best avoided in the implementation of WikiCity and help ensure that the outcome is the liberated rather than the disenfranchised citizen.

ACCESS MODALITY: INTERFACES

This section deals with the design of access modalities that allow users to interface with the WikiCity system, enabling the implementation of the scenarios developed in the previous section.

Interfaces for WikiCity are about creating connections between the tangible level of the city that surrounds us and the functional layers of data and their integration. Different kinds of interfaces shall be considered in order to offer a broad range of access modalities, lowering the difficulty of connection and allowing access to this WikiCity platform to be as open as possible.

The application scenario and the involved agents, the type of data, and the environmental context within which the scenario is located determine the different interface modalities. A multimodal interface approach shall be followed, combining different input and output methodologies. For the actual project implementation, however, available technology, project partners, and time constraints will determine a step-by-step approach to integrate the entire range of interface modalities. We shall therefore distinguish two main groups of interface designs for WikiCity: 2-D display interfaces on the one hand and genuinely multimodal interfaces on the other.

Figure 7. Access modality can be positioned more closely to the user, the built environment, or mobile vehicles

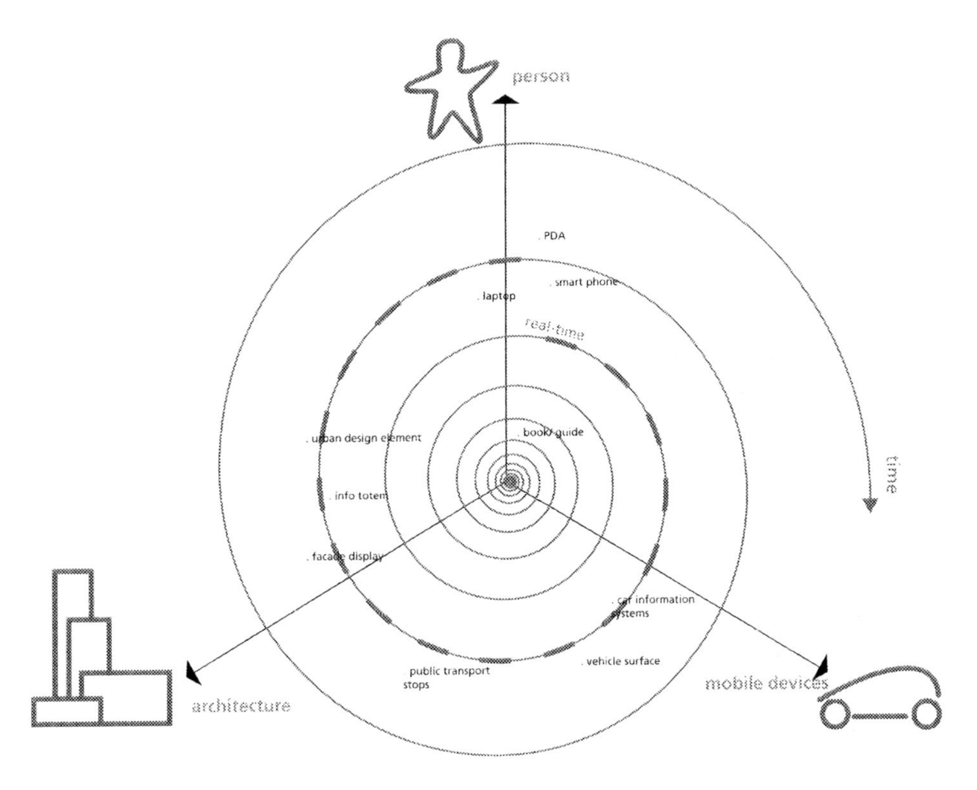

2-D Interfaces: WikiCity Rome

WikiCity Rome has been the first implementation of some of the elements composing the more comprehensive WikiCity research program. For this occasion, a large (10x5 meters) public projection was chosen and positioned on a public square in Rome. During the entire twelve hours from 8pm until 8am, a projection showed a satellite image of Rome with overlays indicating:

- real-time cell-phone activity
- real-time location of public buses
- starting and ongoing event tags at their corresponding location
- live news feeds from journalists at the location where they were reporting from

Certainly this public display as an interface for the WikiCity Rome project implies some limitations such as a static location, a lack of interactivity and personalized queries, and a fixed set of data types. On the other hand, it is interesting to consider some of the less obvious positive points of such a setting:

Low Access Barrier

A public display allows access to the displayed information to anyone passing by the display location. No prior experience or knowledge is required beyond that of reading a map, or better, a satellite view of one's own town (which, as a matter of fact, is not so easy a task, and always presents difficulty when dealing with maps).

Figure 8. WikiCity Rome interface used for a 10x5-meter projection in a public square

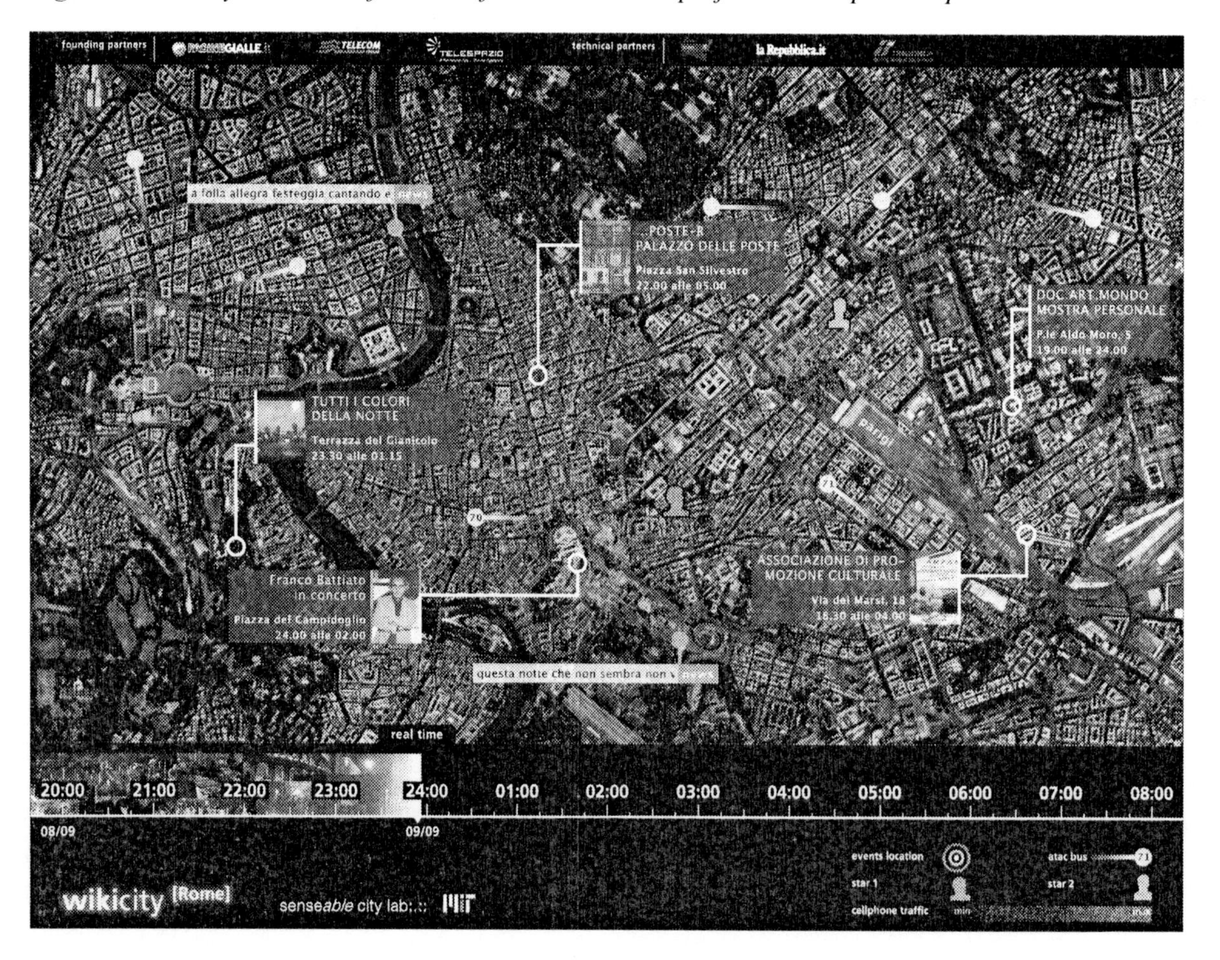

Social Aspect of Consultation

a large public projection allows for a group of people to observe—that is, to access—the WikiCity information, comment on it, and share impressions, ideas, and proposals for actions to be taken on the basis of the information read. Unlike personal devices, such a large-scale interface allows WikiCity to become a tool that can effectively foster social dynamics between users. Additionally, the aspect of difficulties in map-reading mentioned above is alleviated by knowledge exchange among the audience.

Regarding the different interface elements, before describing each in more detail, here are some general observations: Combining different kinds of time- and location-based data brings up the importance of the place of reference, which in the WikiCity Rome case was the location of the public projection, while in portable devices it would be the present location or, similar to a location-independent search, the chosen location of interest. One visualization difficulty encountered in WikiCity Rome was the clash between a big amount of relevant information near the reference location and the small space available that it refers to on the map. On the contrary, there is much map space available farther away from the projection location. Yet, a dense visualization of it is not as relevant. The one element that reflects this consideration in WikiCity Rome is the motion of the buses' visualizations, which show their respective line numbers only within an area around the display location, while transforming into visually less invasive, smaller elements without numbers anywhere outside this area.

While on a personal device, it might be feasible to approach this aspect by zooming in or focusing only on specific locations, in the public projection in Rome this was not an option since both perspectives have to remain intact—that is, the perception of possible action that can be taken on the basis of locally relevant information as well as the perception of the overall dynamic showing the city in its entirety.

Interface Elements and Their Behavior in WikiCity Rome

- **Map:** A satellite image from Rome modified digitally to enhance contrast is used as the base map layer.
- **Cell-Phone Activity:** In WikiCity Rome, cell-phone activity, based on an aggregated and anonymized data set, is used to suggest the distribution of people within the city. Since this is a type of data that tends to change noticeably over a time interval that is longer than the average time a person could be expected to observe the projection during La Notte Bianca, a blue flickering has been chosen to correlate better to what was visualized—that is, people moving.
- **Public Buses:** A yellow dot indicates the real-time position of buses moving around Rome. The number inscribed in the circle indicates the bus number when the bus moves within proximity of the projection's location, allowing for observers to decide whether to take the bus or not. The length of the bus tail indicates the velocity at which the bus travels.
- **Events:** The event label shows up whenever an event is starting and while it is occurring. To avoid the risk of cluttering the map with too many event labels open at a time, we developed an algorithm that would control dynamically the distance of the event labels from the edge and between each other, extending the guide lines accordingly. To avoid clutter and to maintain readability, it also controls the visibility of ongoing events so as not to visualize more than three labels at the same time.
- **News Feed:** Having a major Italian newspaper as technical partner in the project, we arranged to have various journalists report live at short intervals about occurrences in different parts in Rome. On the projection, these location-tagged news feeds were pre-

Figure 9. Satellite map of Rome

Figure 10. Blue flickering indicating cell-phone activity

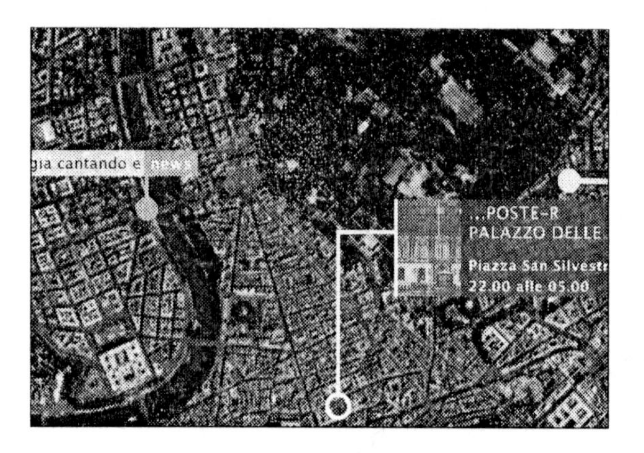

Figure 11. Icon representing the public buses

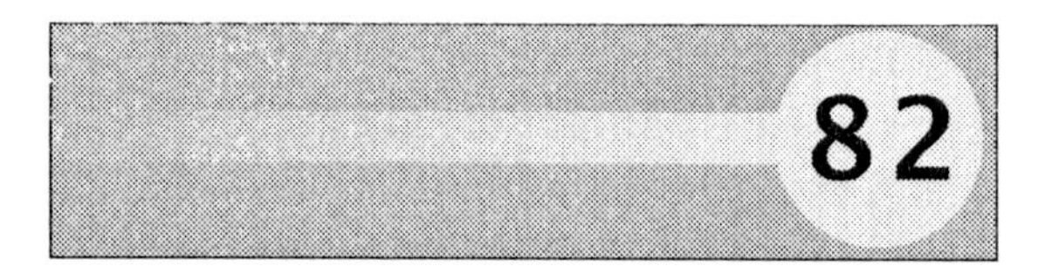

Figure 12. Event label indicating name, place, time, and image for an event

Figure 13. Location-referenced news-feed label

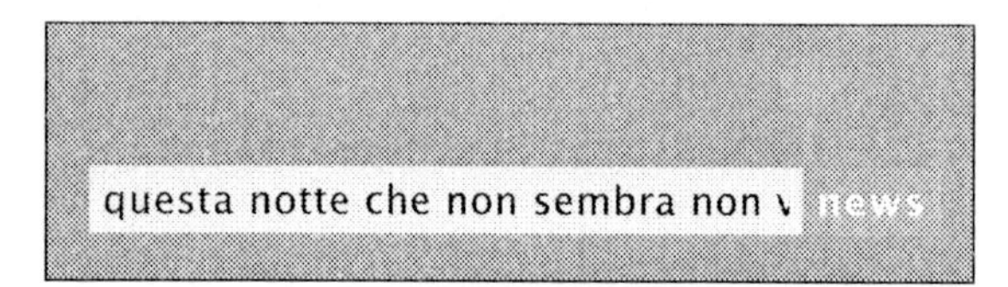

sented as scrolling text at the location where they were reported from, creating a simple but effective way of visually geo-referencing information about city dynamics.

Multimodal Interfaces

In the search for connections between the data realm and the user within the urban space, the second focus for WikiCity interfaces shall be put upon built structures. Throughout the urban environment, inhabitants are surrounded by structures such as urban furniture or built infrastructures, many of which already carry various types of information, even if mainly in a static way.

Information from the virtual realm of WikiCity shall also be accessed through interaction with these structures. Three approaches can be clearly distinguished:

1. Embedded access in existing structures
2. Embedded access in existing structural topologies
3. New structural typologies as a way to access WikiCity

Different from mobile devices, the location is a known constant in all three of these 3-D interfaces. Furthermore, the information type accessible through immobile 3-D interfaces can be limited to a specific mix that results from the interface typology and the environment surrounding it.

We are currently looking into different aspects and research done within the range of "tangible user interfaces" (TUI) to further enhance the WikiCity interface considerations.

SOFTWARE IMPLEMENTATION

In order to realize the WikiCity application able to provide the services described in Section 3, we have identified two crucial aspects that drive the development of the system.

- **Semantic Web Technologies:** The role of semantics for interoperability and integration of heterogeneous data, including geospatial information, is widely recognized (see Berners-Lee, 2001 and W3C, 2007). By the use of semantics, information processing (retrieval or integration) can be based on meaning instead of on mere keywords. The W3C Semantic Web Activity Working Group has been working on a series of standards in this sense. We think that ontologies can be used in order to improve interoperability and sharing of information and knowledge among various data sources and services. Moreover, we are developing contextual, spatial, and temporal ontologies and correlating them to allow analytical processing and reasoning on multimodal data.
- **Decentralized Infrastructure:** WikiCity will not have a single authority running the whole system. It will be composed of a collection of authorities that will participate in the whole system at various scales. So large authorities running large databases, such as telecom operators, will be part of the system as well as small organizations or individuals. As an analogy, it is exactly the same as the Web is today: we have large information systems belonging to large corporations that are part of the Web, as well as people that simply connect to the Internet and run a Web server on their own machine. Decentralized information is more pronounced in Web 2.0 applications (cf. Kolbitsch and Maurer, 2006; Beer and Burrows, 2007), where people can share information on somebody else's system, if allowed and encouraged to do so. Moreover, some specifically addressed peer-to-peer infrastructure will be evaluated in order to make information flow from sensors via (distributed) intelligence to the end users and actuators of the WikiCity. As a result, the WikiCity will be shaped as a multi-centered infrastructure, where many entities providing services exist and are organized as in the Web, and where some entities can self-organize in a peer-to-peer manner for coordinating tasks and pushing information.

Based on these observations, we have designed the software architecture as composed of several interacting elements, listed below.

- **Data Authoring:** WikiCity requires several tools, scripts, and methodologies to extract metadata from syntactically (including unstructured, semi-structured, and structured data) and semantically heterogeneous and multimodal data from diverse sources (wireless sensor networks, mobile phone data, citizens, etc.).
- **Data Acquisition:** The WikiCity application requires an interface for data providers to send the locational data (data stream) and manage the real-time constraint imposed by the application.
- **Data Extraction and Processing:** A new type of browsing across data sources, based on location and time constraints, is required to support the data extraction. We are experimenting with data processing based on Web-services composition and orchestration to achieve this.
- **Interactive visualization:** A critical part of WikiCity is the development of interactive visualizations. This task involves the creation of a variety of interactive tools for browsing, searching, and navigating through and among diverse, distributed collections of information. New browsing methods are being developed (e.g., browsing by timeline or distance range) to support the new type of data available. The User Interface is designed in two ways:
  - **City Level:** It is based on city maps, or other abstract representations, where different

Figure 14. Interfaces in 2-D and 3-D as connections between the tangible and the virtual realm of a city.

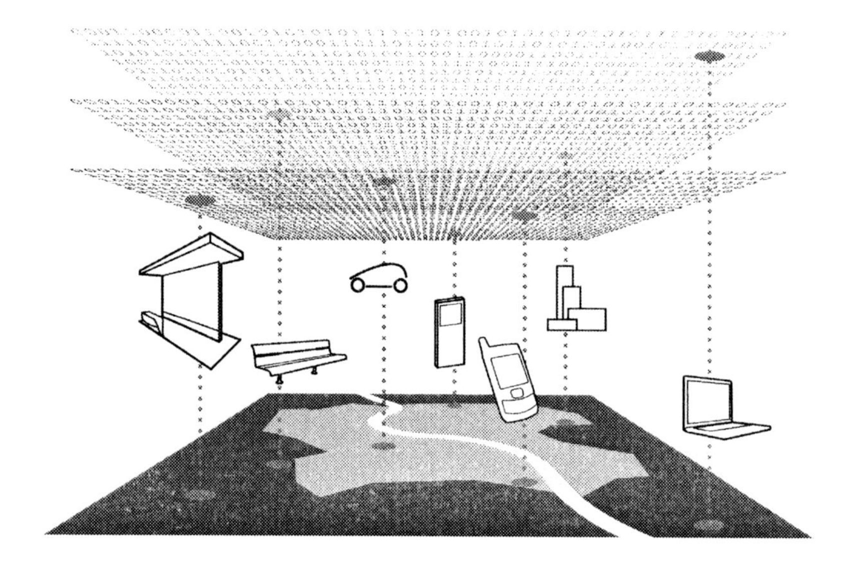

layers can be added. Such layers represent real-time information available in WikiCity. The user views the data as though watching the city using a particular lens.

- ○ **Personalized:** The interface is also based on the user location (retrieved in an automatic way by using location technologies or defined by the user) and a specific query.
- **Management Interfaces:** Many of the components of WikiCity will need interfaces in order to be configured and managed (e.g., setup of sensor networks, interfaces for key-distribution in encrypted P2P connections, etc.).

In the following, some technical aspects about the two most important components of the architecture are addressed.

Data Acquisition

A challenge for WikiCity is the question of how an enormous amount of user-generated data can be exploited to create meaningful and intuitive outputs without causing information overflow. To tackle this problem, as already explicated in the previous section, we make use of semantic Web technologies, and in particular of ontologies and semantic Web services.

The ontology is described by means of the Web Ontology Language (OWL, 2004). Generally, all data in the system are related to one or more categories in the ontology, which is the basis for a structured query system and well-defined data maintenance. The Ontology Service is used to upload and access information about the data within the domain described by the ontology.

To manage the data stream coming from the data providers, we use the Web services technology, which provides a uniform interface able to manage the real-time constraint imposed by the application. The connection between Web services and ontology is obtained by semantic mark-up of Web service (OWL-S, 2004). This is obtained by defining, in the service profile, the output data as individuals of one of the classes defined in the ontology.

Moreover, we allow the definition of processing services, which are semantic Web services that are able to perform some processing on the data described by the system's ontology.

Data Extraction and Processing

A new type of browsing across data sources, based on location and time constraints, is required to support Data Extraction. To this end, we are using a combination of methods of ontology-based information searching (see, for instance, Yu, 2003) and data processing, based on Web services composition and orchestration (see BPEL4WS, 2007).

The services provided by this component can be divided into two categories:

- The Query Service acts as a management service for requests originating from user services, in particular Data Provision Services, which can query data from Semantic Web Services.
- The Composition service is used to create business choreography from a Processing request. Such choreography is then executed by retrieving the relevant data, sending a request to a Processing Service and returning the processed data back to the originating User Service. This service makes use of a BPEL4WS engine, which creates the business process at run-time based on the client's request (Figure 15).

As an example of a business process, Figure 16 shows the BPEL orchestration process created to answer the query of the storyboard described in Section 4.5 (Jogging Path). Such process is composed of 4 services' invocation that allows downloading data about: traffic, air quality, noise level, adequate running path. Then a MergeData service is used to process such data in order to retrieve the best jogging path, corresponding to the location where all the indications give good results, in terms of traffic, air quality, noise level, and adequate paths.

Figure 15. Scheme of the data search

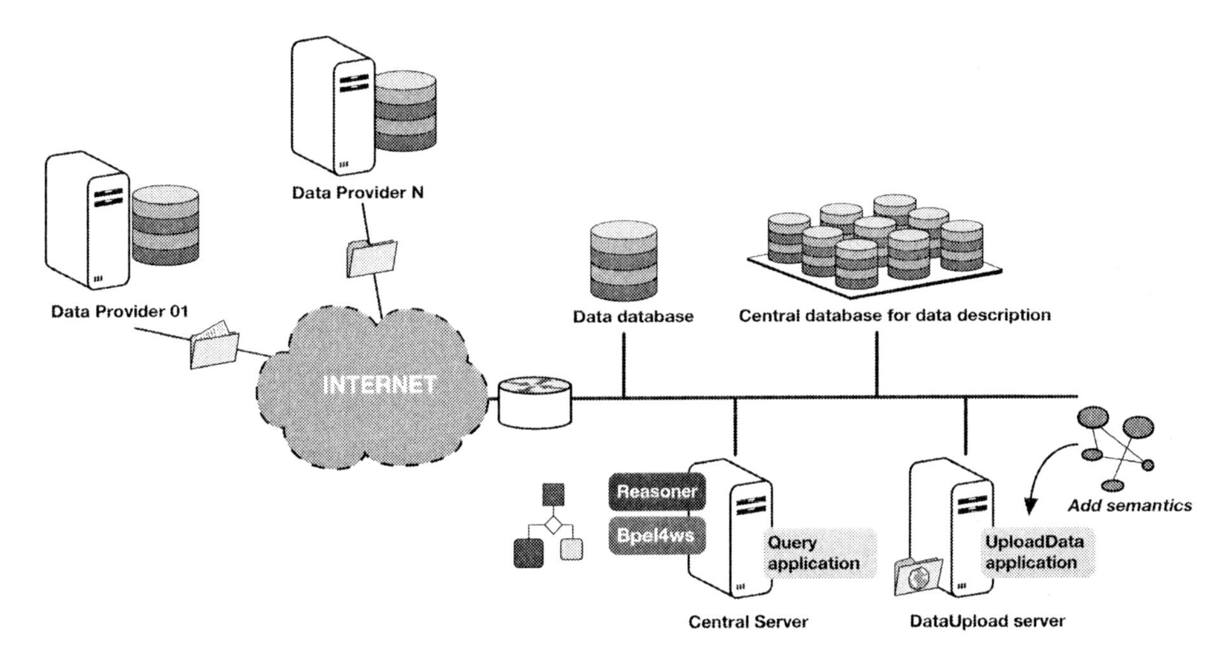

Figure 16. Business process related to the query Jogging Path

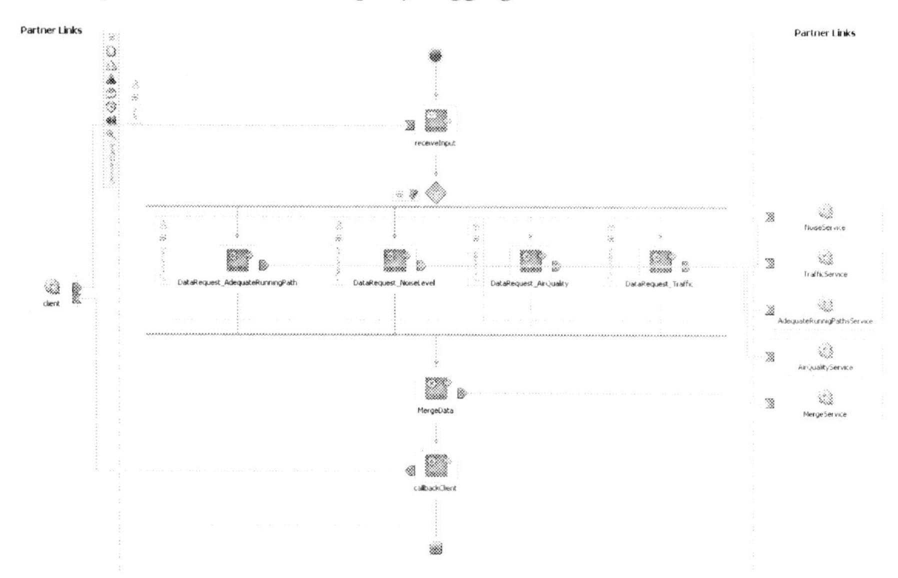

Figure 17. Web-services connection scheme

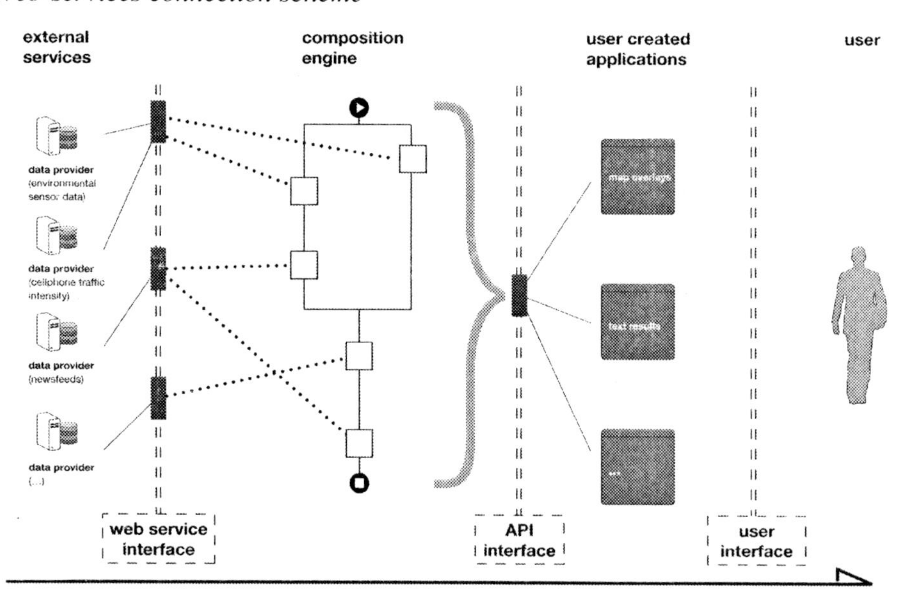

A scheme of the Web-services connection is depicted in Figure 17.

CONCLUSION

From a conceptual analysis, the benefits of real-time location-sensitive information to city inhabitants seem fairly clear, indicating how this could contribute to the efficiency of various real-world situations. Critical aspects have emerged, however, as to how this new form of information may impact some situations in terms of diffusing or concentrating attention of users. Further analysis of such potential situations will feed back into the design of the way real-time location-sensitive information is communicated and made accessible.

While aiming at the construction of a diffused network structure, we have seen that for the initial start-up phase a hybrid system is necessary, which combines the two different approaches of the centralized database and those located within the internal network and in the service providers' servers.

As such a first step, WikiCity Rome has managed to combine a limited number of different kinds of real-time data supplied by established sources, integrating them into an interface that was successfully used by a large number of people during a one-night event. A more in-depth analysis about the results of this pilot project is still being carried out and will be finalized in further work.

Upcoming research steps will focus on distinctive aspects of the overall concept, one of them focusing on the integration of different data types for the application in a multimodal transport system. Without doubt, there is no one ideal means of transport. Rather, different types of trips require different means of transport; or, better, within a constantly changing context, it is often a combination of different means of transport that leads to the most efficient and effective implementation of a trip. Relying in such a way on a combination of different transport systems not only requires the smooth functioning of each of those, but also poses crucial requirements on the interoperability of the systems as well as a coherent and integrated communication of the state of each system in real-time and their combinatorial potential to prospective users.

Another direction follows the adequacy of a WikiCity real-time system for evacuation purposes in an extended urban area (cf. Nakanishi, Ishida, & Koizumi in this volume). In such emergency scenarios, the possibility for a system to know the real-time presence of people to be evacuated is extremely interesting for the delivery of distinct and possibly different instructions supplied for a safe evacuation of large numbers of people, each being considered according to his or her particular circumstances.

Taking the real-time mapping of urban dynamics to a finer granularity brings us to the consideration of not only means of transport and people but also the mapping of objects in general. So far we have been considering mainly one typology of objects for this purpose: cell phones. However considering in the context of WikiCity the possibility of real-time mapping of various types of objects opens up interesting possibilities. When considering the sharing or rental of products, for example (car sharing is a prominent and by now increasingly well established example, but consider also do-it-yourself tools, sports equipment, etc.), knowing the location and state of a product potentially increases the efficiency with which such a product can be put to use by a large user group and at the same time addresses issues of maintenance that may arise from such an intense usage. Products and services can be supplied to whoever needs them in that place and in that moment if their location, availability, and condition is known in real-time, enabling the creation of a system of dynamic resource allocation for end users within the urban context.

In the field of logistics, this approach has been around for a long time. An object's position can be identified throughout entire supply chains, often across large parts of the globe. This capability has led to dramatic changes in the way that technological resources are used by companies throughout the production and distribution processes. How might these systems be useful models for real-time demand-responsive product and service supply schemes in an urban context?

Especially this last consideration of directions toward which to take WikiCity brings up the critical issues involved in this type of project: privacy and control. Mapping the status and location of people and objects in real-time has implications for an individual who is concerned about reserving his privacy. Using a non-connected bus with a generic paper ticket is something quite different from taking a bus whose position is tracked and using an electronic ticket that feeds into a system the user's identity. It alters the visibility of that specific person to a system that can technically be read by other systems and individuals, and at the same time it alters the potential of service improvements for the specific user and for the overall transport system.

A critical consideration comes from looking at the example of the supply-chain management mentioned above. Thomas Friedman illustrates in "The World Is Flat" (p. 584) how all the thousands of parts of his Dell Laptop have come together during production on the basis of a sophisticated globalized supply-chain management and points out how a temporary shortage in the supply of one component such as a 40-gigabyte hard drive leads to an immediate relay to the marketing department which offers a 60-gigabyte hard drive for the same price over the next 2 hours. Active demand shaping is the term used for such a process enabled by the tight tracking of status and location (availability for production in this case) in the production realm. How would this dynamic translate into the urban context and citizens moving within and making use of their city? A crucial aspect of further development in this area will have to be focused on how to ensure that the technology of real-time location-based mapping remains focused on providing better information for people to base their decisions on instead of formulating decisions for the people.

Any decision is based on knowledge and insight in the context in question. The better a situation and the actual dynamics in place are known, the better one is able to interact in an effective way with that situation and open up at best the implicit potential of that circumstance. Understanding urban dynamics with the help of digital technologies that enable real-time and location-based information is a powerful instrument to support just that, and it will be exciting to see how this tool can be used in constructive and inclusive ways for the benefit of a city.

ACKNOWLEDGMENT

We would like to thank Marcus Foth and three anonymous referees for their helpful suggestions.

REFERENCES

Bassoli A., Brewer J., Martin K., Dourish P., Mainwaring S. (2007). Underground Aesthetics: Rethinking Urban Computing, IEEE Pervasive Computing, 6(3), 39--45.

Beer, D., & Burrows, R. (2007). Sociology and, of and in Web 2.0: Some Initial Considerations. Sociological Research Online, 12(5).

Berners-Lee T., Hendler J., Lassila O., (2001). The Semantic Web, Scientific American, May.

BPEL4WS, (2007). Business Process Execution Language for Web Services version 1.1, http://www-128.ibm.com/developerworks/library/specification/ws-bpel/

Calabrese F., Ratti C., (2006). Real-time Rome, Networks and Communication studies, vol. 20, n. 3/4.

Calabrese, F., Kloeckl, K., & Ratti, C. (2007). WikiCity: Real-Time Location-Sensitive tools for the city. IEEE Pervasive Computing, 6(3), 52--53.

Deleuze G., Guattari F., (1977). Rizoma. Pratiche Editrice, Parma-Luca.

Foth, M., Odendaal, N., & Hearn, G. (2007, Oct 15--16). The View from Everywhere: Towards an Epistemology for Urbanites. Paper presented at the 4th International Conference on Intellectual Capital, Knowledge Management and Organisational Learning (ICICKM), Cape Town, South Africa.

Friedman, T. (2007). The world is flat, release 3.0. New York : Picador.

Kolbitsch, J., & Maurer, H. (2006). The Transformation of the Web: How Emerging Communities Shape the Information we Consume. Journal of Universal Computer Science, 12(2), 187--213.

OWL, (2004). Web Ontology Language, http://www.w3.org/TR/owl-features/.

OWL-S, (2004). Semantic Markup for Web Services, http://www.w3.org/Submission/OWL-S/

Oxford (2007). Definition of Real-time, Oxford English Dictionary, Oxford University press.

Sterling, B. (2005). Shaping Things (Mediaworks Pamphlets). The MIT Press.

Thom-Santelli J. (2007). Mobile Social Software: Facilitating Serendipity or Encouraging Homogeneity?, IEEE Pervasive Computing, 6(3), 46--51.

Trevor J., Hilbert D. M. (2007). AnySpot: Pervasive Document Access and Sharing, IEEE Pervasive Computing, 6(3), 76--84.

W3C, (2007). Semantic Web, http://www.w3.org/2001/sw/

Wray, S. (1998). On Electronic Civil Disobedience". Presented at the Socialist Scholars Conference March 20, 21, and 22 New York, NY

KEY TERMS

Multimodal Interfaces: Interfaces that allow user interaction in more than one way such as visual display, sound, touch, and others.

Real-Time System / Real-Time Control System: System for the control of a real entity by means of sensors, intelligence, and actuators

Rhizome: A philosophical network structure, put forward by Gilles Deleuze and Félix Guattari, in which every part is necessarily connected with every other part of the system. There are no preferential connections because every connection alters the overall network structure. The rhizome as a flat network is in contrast to arboreal structures connoted by a hierarchical structure.

Sensor Network: Network of spatially distributed devices using sensors to cooperatively monitor physical or environmental conditions

Semantic Web: Extension of the World Wide Web in which the content is expressed in a way that is readable by software agents.

Tangible User Interfaces: Interface type that allows the user to interact with digital information by acting upon physical elements in the users' environment.

Wiki: A Wiki is a software that allows users to collaboratively compose Web pages whose content is cross-referenced. One of the principles of the Wiki is that all users can actively create and edit the content in a continuous and open process which creates an ever-evolving whole.

ENDNOTES

1. "A SYNCHRONIC SOCIETY synchronizes multiple histories. In a SYNCHRONIC SOCIETY, every object worthy of human or machine consideration generates a small history. These histories are not dusty archives locked away on ink and paper. They are informational resources, manipulable in real-time." (Sterling, 2005)
2. A CAPTCHA is a program that can tell whether its user is a human or a computer. CAPTCHAs are used by many Web sites to prevent abuse from "bots," or automated programs usually written to generate spam (source http://recaptcha.net).

Chapter XXVIII
Citizen Science:
Enabling Participatory Urbanism

Eric Paulos
Intel Research Berkeley, USA

RJ Honicky
University of California, Berkeley, USA

Ben Hooker
Intel Research Berkeley, USA

ABSTRACT

In this chapter, we present an important new shift in mobile phone usage—from communication tool to "networked mobile personal measurement instrument." We explore how these new "personal instruments" enable an entirely novel and empowering genre of mobile computing usage called citizen science. We investigate how such citizen science can be used collectively across neighborhoods and communities to enable individuals to become active participants and stakeholders as they publicly collect, share, and remix measurements of their city that matter most to them. We further demonstrate the impact of this new participatory urbanism by detailing its usage within the scope of environmental awareness. Inspired by a series of field studies, user driven environmental measurements, and interviews, we present the design of a working hardware system that integrates air quality sensing into an existing mobile phone and exposes the citizen authored measurements to the community—empowering people to become true change agents.

Tell me, I forget.
Show me, I remember.
Involve me, I understand.

—Chinese proverb

MOTIVATION

Mobile phones are powerful tools indeed—collapsing space and time by enabling us to reach out to contact others at a distance, to coordinate mico-planning events, and to reschedule activities at a moment's notice. But with all of their abilities they lack the superpower we perhaps need most—the ability to measure and understand the real world around us.

We carry mobile phones with us nearly everywhere we go; yet they sense and tell us little of the world we live in. Look around you right now. How hot is it? Which direction am I facing? Which direction is the wind blowing and how fast? How healthy is the air I'm breathing? What is the pollen count right now? How long can I stay outside without getting sunburned? Is the noise level safe here? Were pesticides used on these fruits? Is this water safe to drink? Are my children's toys free of lead and other toxins? Is my new indoor carpeting emitting volatile organic compounds (VOCs)? Now look to your phone for answers about the environment around you. What is it telling you? For all of its computational power and sophistication it provides us with very little insight into the actual conditions of the atmospheres we traverse with it. In fact the only real-time environmental data it measures onboard and reports to you is a signal to noise value for a narrow slice of the electromagnetic spectrum (Figure 1).

Certainly one could imagine accessing the web or other online resource to find an answer to some of these questions. But much of that online data is calculated and published for general usage, not for you specifically. For example, the civic government may say that the temperature is currently 23°C by taking one measurement at the center of the city or averaging several values from multiple sites across town. But what if you're in the shade by the wind swept waterfront where it is actually 17°C or waiting underground for the subway where it is a muggy 33°C. The measurement that means the most to you is likely to be the one that captures the actual conditions you are currently experiencing, not citywide averages.

Imagine you are deciding between walking to one of two subway stations and could gather live data from the passengers waiting on the platform at each stop about the temperature and humidity

Figure 1. The only real-time environmental data displayed on a mobile phone: a narrow slice of the electromagnetic spectrum with a tiny readout of cell tower signal strength

of each station at that very moment? What if you were one of the 300 million people who suffer from asthma (WHO, 2006b) and could breath easily as you navigated your city with real-time pollen counts collected by your fellow citizens? What if you could not just be told the level of noise pollution in your city but measure and publish your own actual decibel measurements taken in front of your home? What if you were one of the more than 3 billion people, nearly half the world's population, that burned solid fuels, including biomass fuels (wood, dung, agricultural residues) and coal, for their energy, heating, and cooking needs indoors and yet had no way to monitor the health effects of the resulting pollutants on yourself and your family even though nearly 2 million people die annually from indoor air pollution (Ezzati & Kammen, 2002)?

Mobile technology is with us and is indeed allowing us to communicate, buy, sell, connect, and do miraculous things. However, it is time for this technology to empower us to go beyond finding friends, chatting with colleagues, locating hip bars, and buying music. We need to expand our perceptions of our mobile phone as simply a communication tool and celebrate them in their new role as personal measurement instruments capable of sensing our natural environment and empowering collective action through everyday grassroots citizen science across blocks, neighborhoods, cities, and nations. Our goal is to provide our mobile devices with new superpowers and "super-senses" by outfitting them with novel sensors and providing an infrastructure for public sharing and remixing of these personal sensor measurements by experts and non-experts alike. The overall long-term goal is to enable actual and meaningful local and global changes driven by the desires of everyday citizens. Our work is but a small effort towards this end.

INTRODUCTION

We have already seen the early emergence of sensor rich mobile devices such as Apple's Nike+iPod Sport Kit (music player + pedometer), Apple's iPhone (mobile phone + proximity sensor and accelerometer), Nokia's 5500 (mobile phone + pedometer), Samsung's S310 (mobile phone + 6 axis accelerometer), LG Electronics LG-LP4100 (mobile phone + breathalyzer), t+ Diabetes (mobile phone + blood glucose sensor), and Samsung's planned body fat (K. K. Park & Hwang, 2007) and fertility monitoring phone (J. J. Park, Lee, Jung, & Kim, 2007). Similarly, we have seen the "Web 2.0" phenomenon embrace an approach to generating and distributing web content characterized by open communication, decentralization of authority, freedom to share and re-use, and "the market as a conversation" (Beer & Burrows, 2007; Josef & Hermann, 2006; O'Reilly, 2005).

We assert that there are two indisputable facts about our future mobile phones: (1) that they will be equipped with more sensing and processing capabilities and (2) that they will be driven by an architecture of participation and democracy that encourages users to contribute value to their tools and applications as they use them as well as give back collective value to the public. There are countless examples already in existence of such systems from Flickr to Wikipedia to Creative Commons (Lessig, 1994) to open source movements such as FLOSS (Free/Libre/Open-Source Software) where the *de facto* moral etiquette of openly contributing and sharing the collective repository of knowledge is upheld as the foundational and driving principle of the technology. What if we simply enable mobile phones to more easily participate in this emerging computing paradigm? There is an inevitable and powerful intersection of people-centric sensing with the current online remix culture.

More specifically, what happens when individual mobile phones are augmented with novel sensing technologies such as noise pollution, air quality, UV levels, water quality, *etc*? We claim that these new mobile "sensing instruments" will promote everyday citizens to uncover and visualize unseen elements of their own everyday experiences. As networked devices, they reposition individuals as producers, consumers, and remixers of a vast openly shared public data set. By empowering people to easily measure, report, and compare their own personal environment, a new citizen driven model of civic government can emerge, driven by these new networked-mobile-personal-"political artifacts" (Winner, 1999).

Our strategy is to design and deploy a series of networked measurement instruments that are embedded within our everyday places as well as coupled to personal mobile devices to collectively capture a view of our environment. More importantly, our research positions citizens as the driving element for collecting, reporting, interpreting, and collectively improving the health of our natural environment. Our hypothesis is not only that a wealth of novel and important untapped computing interactions exist in this research space, but that such experiences are certain to become a dominant paradigm in our evolving relationship with technology.

The technological debate radically expands from beyond simply how to design a few functional mobile applications that satisfy the needs of thousands of people (such as a location service, a friend finder social networking system, or a mapping overlay tool) to how thousands of mobile individuals can measure, share, and remix publicly sampled data into a wide variety of more personally meaningful mobile experiences and tools. Large-scale services, while tremendously important, often suffer from lowest common denominator effects as they seek to make a single system satisfy the needs of everyone. We see our future technologies as a mixture of large-scale systems and personally customized small tools. In this chapter we are interested in exploring this new model of citizen measuring, public sharing, and personal remixing of the environment driven by personal experiences and measurements. The result is a technological future that hopefully conveys personal meaning to citizens, a more informed and responsive civic government unburdened from its reliance on low resolution, generic, and filtered data driven solutions, and a better place for us all to live. By elevating everyday citizens into the role of data collector, commentator, and policy maker, we hope to directly empowering such individuals to act as change agents within their world.

Outline

In this chapter we define the territory of *citizen science* along with the challenges to its adoption, introduce the mobile phone as a measurement instrument for environmental sensing, and outline a taxonomy of sensor and mobile phone interactions. In the second half of the chapter we explore the measurement of air quality specifically, discuss current techniques in use for measuring and presenting air quality, report results from an air quality awareness survey, present a field study using several human driven atmospheric environmental sensors carried across the capital city of Ghana, and present a specific instantiation of a working system for *citizen science*—personal environmental monitoring with air quality sensors integrated with a mobile phone.

CITIZEN SCIENCE

Citizen Science builds upon a large body of related projects which enable citizens to act as agents of change. There is a long history of such movements from grassroots neighborhood watch campaigns to political revolutions. Some of the more well known movements are the National Audubon Society's Christmas Bird Count (CBC) where a census of birds in the Western Hemisphere is performed annually by citizens since 1900 (LeBaron, 2006).

More recently, the success of online approaches such as SETI@Home and "The Great World Wide Star Count" an international event that encourages everyone to go outside, look skywards after dark, and report the count of stars they see, inform us that there is an immense public interest in such collective movements.

Our work leverages Corburn's "street science" framework, which emphasizes local urban insights to improve scientific inquiry and environmental health policy and decision-making. Corburn underscores the importance of local community knowledge as "the scripts, images, narratives, and understandings we use to make sense of the world in which we live" (Corburn, 2005). Even more emphatically he states that a community's "political power hinges in part on its ability to manipulate knowledge and to challenge evidence presented in support of particular policies" (p. 201). While such local knowledge and community-based practices are sometimes labeled as romantic or populist, Corburn insists that such views overlook the structural and global dimensions of problem solving for urban communities. Corburn believes that "street science" leverages community power imbalances, and can increase agency or decision maker understanding of a community's claims, thereby potentially increasing public trust. He insists that such local knowledge informs environmental health research and environmental policy making in four distinct ways: 1) by making a cognitive contribution by rectifying the tendency towards reductionism; 2) by fostering of a "hybridization" of professional discourse with local experience; 3) by pointing out low-cost and more effective interventions or remedies; and 4) by raising previously unacknowledged distributive justice concerns that disadvantaged communities far too often face.

We also draw from the work of German sociologist Ulrich Beck who postulates that as people become less constrained by social institutions, they are in a position to mold the process of modernization rather than remain passive observers of a system in which they hold no stake (Beck, Giddens, & Lash, 1994). In Beck's world, individuals have the opportunity to become change agents by way of information. For him information is key to the (re)shaping of the social and political world. For us the creation, sharing, and remixing of information is a fundamental component of our citizen science design approach.

Finally, in *The Death and Life of Great American Cities* Jane Jacobs writes (Jacobs, 1961) that to understand cities we need to "reason from the particulars to the general, rather than the reverse [and] to seek 'unaverage' clues involving very small quantities, which reveal the way larger and more 'average' quantities are operating" (p. 440). Jacobs continues, "Quantities of the 'unaverage', which are bound to be relatively small, are indispensable to vital cities" (p. 443). Citizen science attempts to elevate the local expertise of citizens and their personal, small, unusual, local, particular experiences across urban life.

URBAN COMPUTING

Urban Computing captures a unique, synergistic moment - expanding urban populations, rapid adoption of small, powerful, networked, mobile devices, bluetooth radios, tiny ad hoc sensor networks, and the widespread influence of wireless technologies across our growing urban landscapes. According to the United Nations, 2007 marks the first point in human history that over half of the world's population live in urban areas (UN, 2003). In developed nations the number of urban dwellers is even more dramatic, often exceeding 75%. The very essence of person, place, and community are being redefined by personal wireless digital tools that transcend traditional physical constrains of time and space. New metaphors for visualizing, interacting, and interpreting the real-time ebb and flow of urban places are emerging. The large population densities of cities, the widespread proliferation of wireless and digital infrastructure, and the early adoption of mobile and wireless technologies by its inhabitants make such urban landscapes a rich territory for researching emerging social and technological phenomena. Urban Computing strives to expose, deconstruct, and understand the challenges of this newly emerging moment in urban history and its dramatic influence on technology usage and adoption. Several bodies of collected work outline this research frontier (Dave, 2007; Kindberg, Chalmers, & Paulos, 2007; Shklovski & Chang, 2006).

What is not Urban Computing?

Urban Computing focuses on our lifestyles and technologies within the context of public urban spaces. Its research challenges differ from those found within the home where technologies readily intermingle across our intimate relations with friends and family members. It diverges from office and work environments where productivity and efficiency often dominate our computing tools. Urban Computing is **not** a disconnected personal phone application, a domestic networked appliance, a mobile route planning application, an office-scheduling tool, or a social networking service. Urban Computing is also not simply concerned with mobile computing which is actually a misnomer for "nomadic computing" in which people interface with a "computing device" from a finite set of discrete places such as "at the park", "on the bus", and "in the cafe". Rather we are concerned with the continuous needs, opportunities, and styles of interactions that arise not just within the scope of these nomadic places but across the full range of places, transitions, and flows. Finally, while cities are places of dynamic social interactions, Urban Computing is not focused on simply solving the challenges of social networking in public places, but exploring a much wider gamut of urban live from the personal to the social, from the solitary to the crowd, from emphasizing connectivity to celebrating the disconnect, from promoting passivity to inspiring activism, curiosity, and wonderment (Paulos, Jenkins, Joki, & Vora, 2008).

Urban Computing Framework

Urban Computing establishes an important new framework for deconstructing and analyzing technology and urban life across five research themes - *people*, *place*, *infrastructure*, *architecture*, and *flow*. The diversity of these important themes promotes rich interdisciplinary research within the field of Urban Computing. The authors assert the following key themes for Urban Computing:

- **People:** Who are the people we share our city with? How do they influence our urban landscape? Where do we belong in this social space and how do new technologies enable and disrupt feelings of community and belonging?
- **Place:** How do we derive the meaning of various public places? What cues do we use to interpret place and how will urban technologies re-inform and alter our perception of various places? How does technology create new places?
- **Infrastructure:** How will existing urban systems such as power, water, subways, public transportation, signal lights, toll booths, etc be used and re-appropriated by emerging technologies?
- **Architecture:** What new techniques and smart surfaces will emerge for interacting with buildings, public surfaces, sidewalks, benches, and other "street furniture"? What role will new structures, shapes, and forms play?
- **Flow:** What is a path or route through a city using these new urban tools? How will navigation and movement, either throughout an entire city or within a small urban space, be influenced by the introduction of computing technologies?

PARTICIPATORY URBANISM

Inspired directly by citizen science and in the spirit of Urban Computing (Paulos & Jenkins, 2005), *participatory urbanism* is more directly focused on the potential for emerging ubiquitous urban and personal mobile technologies to enable citizen action by allowing open measuring, sharing, and remixing of elements of urban living marked by, requiring, or involving participation, especially affording the opportunity for individual citizen participation, sharing, and voice. Participatory urbanism promotes new styles and methods for individual citizens to become proactive in their involvement with their city, neighborhood, and urban self-reflexivity. Examples of *participatory urbanism* include but are not limited to: providing mobile device centered hardware toolkits for non-experts to become authors of new everyday urban objects, generating individual and collective needs based dialogue tools around the desired usage of urban green spaces, or empowering citizens to collect and share air quality data measured with sensor enabled mobile devices.

A clear research challenge is to understand the roll that emerging *in situ* mobile technologies will play in promoting citizen science. Using only text and picture messaging, citizens have already initiated several significant citizen science themed actions.

- **People Power 2**: a four-day popular revolution that peacefully overthrew Philippine president Joseph Estrada in January 2001 where text messaging played a leading role (Mydans, 2001).
- **Orange Revolution**: a series of protests and political events coordinated using text messaging that took place in the Ukraine in 2004 that exposed massive corruption, voter intimidation, and direct electoral fraud between candidates Viktor Yushchenko and Viktor Yanukovych (Myers & Mydans, 2005).
- **TXTmob**: a open source text messaging system used to coordinate protests during the United States Republican Presidential Convention in 2004 (Justo, 2004).
- **Hollabacknyc.com**: A blog where women "holla back" at harassers by taking their pictures with camera phones and post them online. Inspired by Thao Nguyen's Flickr image of Dan Hoyt indecently exposing himself to her on a New York public subway in 2005 (May, 2007)
- **Parkscan.org**: a system setup in 2003 allowing people voice concerns on park maintenance by uploading information about public park conditions as text and pictures from mobile devices and the web (Farooq, 2006).

Several research projects have also begun to explore technology's role in promoting citizen science such as Equator's Ambient Wood Project (Randell, Phelps, & Rogers, 2003) using PDAs for sampling the environment by children and White's LeafView mobile phone system for capturing, logging, and cataloging plants in the field (White, Feiner, & Kopylec, 2006) by non-scientists. More recently, UCLA's Center for Embedded Network Sensing has setup a research initiative called "Participatory Sensing" that is developing infrastructure and tools to enable individuals and groups to initiate their own public "campaigns" for others to participate in by using networked mobile devices (Burke et al., 2006). Similarly, the MetroSense project outlines an exciting opportunistic "people-centric" approach to mobile phone sensing including several deployments with bicycles (Campbell et al., 2008). As strong advocates of such participatory models, our work complements this research by 1) focusing on an initial capstone application of air quality sensing, 2) emphasizing the measure-share-remix metaphor for "on-the-go" citizen participation, and 3) expanding the integration of new sensors for mobile devices. We have also seen exciting new work that addresses sensor data sharing and remixing with Microsoft's SenseWeb (Santanche, Nath, Liu, Priyantha, & Zhao, 2006), Nokia's SensorPlanet (Balandina & Trossen, 2006), Platial (www.platial.com) and SensorMap (Nath, Liu, Miller, Zhao, & Santanche, 2006), Mappr (www.mappr.com), Swivel (ww.swivel.com), and IBM's ManyEyes (Viégas, Wattenberg, Ham, Kriss, & McKeon, 2007).

MOBILE DEVICE AS MEASUREMENT INSTRUMENT

One major vector of citizen science is the enabling of networked, personal, mobile devices to become easily augmented with novel sensors that empower individuals to personally collect, share, compare, and participate in interpreting the personal measurements of their everyday life. As we stated in the introduction, our future mobile devices will have not only more processing power but also a wealth of new sensing capabilities. These new personal mobile sensors open up a rich new design territory. We envision a taxonomy of sensors and mobile devices as follows:

- **Onboard:** Sensors that are physically built into the mobile device. Examples are the onboard accelerometer in the Apple iPhone and Nokia 5500 and the existing microphone and camera on today's mobile phones. In the future it could include mobile phone capable of measuring carbon monoxide, pollen, radiation, epidemiological viruses or wind speed and direction for a golfer.
- **Worn:** Sensors that are worn on the body or clothing and are wirelessly connected to the user's mobile phone using a personal area network (PAN) such as Bluetooth. Examples, are the shoe worn pedometer in the Nike+iPod Sports Kit, personal heart rate monitors, to perhaps a leg kick sensor for dancers, a hat mounted UV sensor, or a wristwatch sulfur dioxide sensor. We can envision a new genre of environmentally conscious fashion designed around air quality sensing enabled clothing.
- **Left Behind:** In the spirit of sensor networks (Akyildiz, Su, Sankarasubramaniam, & Cayirci, 2002) very low-power and low-cost sensors that are left behind to collect measurements which are retrieved upon the users next visit or encounter with the sensor. Examples include an ambient temperature logger left under a bus seat, a UV light detector left in a park, a soil moisture sensor stuck in a public planter, and a nitrogen dioxide sensor attached to a street sweeper.
- **Temporarily Scattered:** Sensors that are spread out within a nearby area and used by an individual while they are within that given context. Later they are typically collected and taken away by the user. Examples include a hall-effect spoke counter by a bike messenger on her bicycle, a motion sensor used by a group playing soccer to demarcate the goal and side lines, and accelerometers mounted on a skateboard for "recording" the acceleration profile of a "casper stall" trick.
- **Infrastructure:** Sensors, typically powered, highly calibrated, and static in the urban environment that broadcast their sensor data via a global network and/or a personal area network such as Bluetooth or Wibree. Typically these sensors are operated by reliable agencies and used as a public resource for authenticating and calibrating the mobile sensor platforms carried by individual citizens. Examples include a carbon monoxide sensor in a subway station, a temperature sensor at a bus stop, and pollen count logger in a park.

BARRIERS TO CITIZEN SCIENCE

While sensor rich mobile devices usher in a compelling series of new mobile device usage models that place individuals in the position of influence and control over their urban life, there are a number of important barriers to the development and adoption of such systems. These research challenges are presented below:

- **Hardware Extensibility:** Currently, attaching new devices and sensors is a skill reserved for experts and scientists. In order to gain widespread usage non-experts must be able to easily attach new sensors to their mobile phones. Standard connectors, for providing power (i.e. 500mA), and standard protocols for communication with sensors need to be adopted. Most mobile devices provide RS-232 serial, USB, and/or Bluetooth as external communication mechanism to attached sensors. But standards across vendors for pins and power are lacking. Nokia has initiated research into this problem with N-RSA (Nokia Remote Sensing Architecture) (Balandina & Trossen, 2006).
- **Open Platforms:** Developers need access to all of the features and hardware on mobile devices. This does not imply that such open platforms are prone to open hacking and are inherently insecure. In fact recent work on embedded OS device isolation and virtual machines such as J2ME, Parallels, and VMware demonstrate strategies for designing secure, open systems. Mobile phones must adopt these established computing practices to insure open sensor development for citizen science.
- **Software for Sharing:** The need for common file formats and Internet protocols like Google maps, geo-tagged images, XML-schema, and RSS/ATOM for "on-the-go" devices must be established. Microsoft's SenseWeb (Santanche et al., 2006) has outlined an approach to sharing sensor data from fixed sensor deployments. Without common formats, the growth and adoption of grassroots citizen science efforts will be stunted.
- **Power:** The addition of new hardware, new sensors, more data logging, and more radio power for sharing this data, puts extreme demands on power management for these new mobile devices. Creative strategies for opportunistic sampling, sharing, and processing of these new data feeds must be developed. Delay Tolerant Networks (DTN) and similar approaches currently hold much promise (Fall, 2003).
- **Privacy and Anonymity:** Users may desire to participate in this public data collection but not at the expense of publicly disclosing their daily location traces and patterns. We need mechanisms to insure privacy and guarantee a level of anonymity for users and yet enable communities to make connections and foster open debates with their data.
- **Authentication and Trust:** Erroneous or intentionally bogus measurements by an adversary need to be easily detected and flagged as such for removal from the system. The validity and integrity of the entire system are based on insuring this fundamental level of trustworthy data.
- **Calibration:** Citizen science by definition explicitly enables the use of scientific data collection equipment by non-experts. The handling and usage of the sensors and measurement conditions will vary wildly—in and out of elevators, handbags, pockets, subway stations, *etc*. How can we attempt to calibrate these sensors "in the wild"?
- **Sensor Selection:** What are the reasonable set of sensors to use and what conditions make sense to measure? Where should the sensors be mounted and in what contexts and positions are they best sampled?

- **Super Sampling:** Intuitively, millions of mobile sensors should be better than a single fixed sensor. What are the algorithmic techniques that can provide a mechanism to use the sensor data to super sample locations and activities? How can this super sampling enable detection of anomalous or adversarial sensor readings and improve overall sensor calibration.
- **Environmental Impact:** Finally, perhaps of greatest importance, while the vision is to provide millions of sensors to citizens to empower new collective action and inspire environmental awareness by sampling our world, the impact of the production, use, and discarding, of millions of pervasive sensors must be addressed. Does the overall benefit of citizen science enabled by these new devices offset their production, manufacturing, and environmental costs?

MEASURING ENVIRONMENTAL AIR QUALITY

The World Health Organization reports that 2 million people now die each year from the effects of air pollution, two times the number of deaths from automobile accidents (WHO, 2006a). Direct causes of air pollution related deaths include aggravated asthma, bronchitis, emphysema, lung and heart diseases, respiratory allergies, visual impairment, and even Sudden Infant Death Syndrome (SIDS) (Klonoff-Cohen, Lam, & Lewis, 2005).

How is Air Quality Measured and Reported?

The main method of measuring and reporting air quality is the Air Quality Index (AQI). The AQI is a standardized indicator of the air quality in a given location based on United States federal air quality standards for six major pollutants as regulated by the Environmental Protection Agency (EPA). The AQI is primarily an 8-24 hour moving average weighted measurement of sulfur dioxide (SO_2), nitrogen dioxide (NO_2), carbon monoxide (CO), ground-level ozone (O_3), 2.5 micron particulate matter ($PM_{2.5}$), and 10 micron particulate matter (PM_{10}). A non-gas, particulate matter is the term for tiny particles of solid (a smoke) or liquid (an aerosol) suspended in the air and is a major contributor to lung tissue damage, cancer, and premature death. The AQI translates these daily air pollution concentrations into a number on a scale between 0 and 500 which is reported to the public along with a standardized color indicator: 0-50 good (green), 51-100 moderate (yellow), 101-150 unhealthy for sensitive groups (orange), 151-200 unhealthy (red), 201-300 very unhealthy (purple), and 301-500 hazardous (maroon).

Many of these measured pollutants are generated as byproducts of fuel combustion from vehicles, smelting, and other industrial processes. In cities, automobile exhaust alone can contribute to as much as 95 percent of all CO emissions (EPA, 2007). Recent studies have also revealed that 97 percent of European citizens living in urban areas are exposed to pollution levels that exceed EU limits (EU, 2004).

Currently, citizens must defer to a small handful of civic government installed environmental monitoring stations. For example, the entire San Francisco Bay Area in California, the sixth-largest consolidated metropolitan area in the United States, home to more than seven million people, composed of numerous cities, towns, military bases, airports, and associated regional, state, and national parks sprawled over nine counties and connected by a massive network of roads, highways, railroads, and commuter rail contains only 40 measurement stations (Figure 2). However, not every pollutant is measured at every site. For example, the Bay Area Air Quality Management District reports dangerous $PM_{2.5}$ at only 6 locations (Figure 2) and ozone (O_3) at 21 (AirNOW, 2006).

Figure 2. Map of the 40 air quality monitoring sites (left) and air quality map for the 6 locations where (PM2.5) is reported (right) for the San Francisco Bay Area

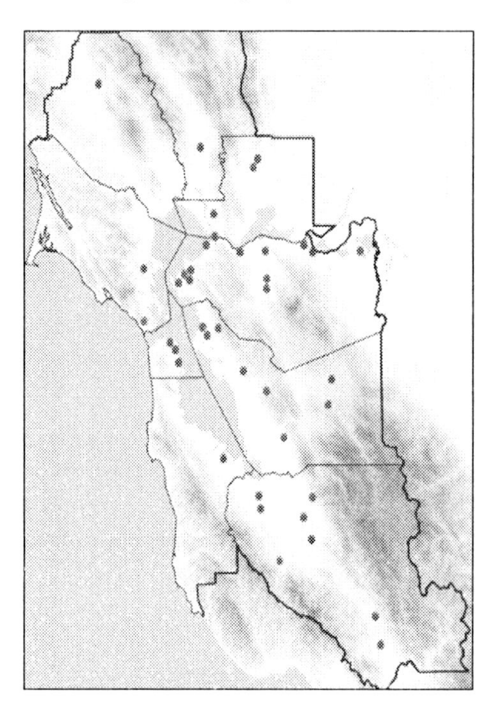

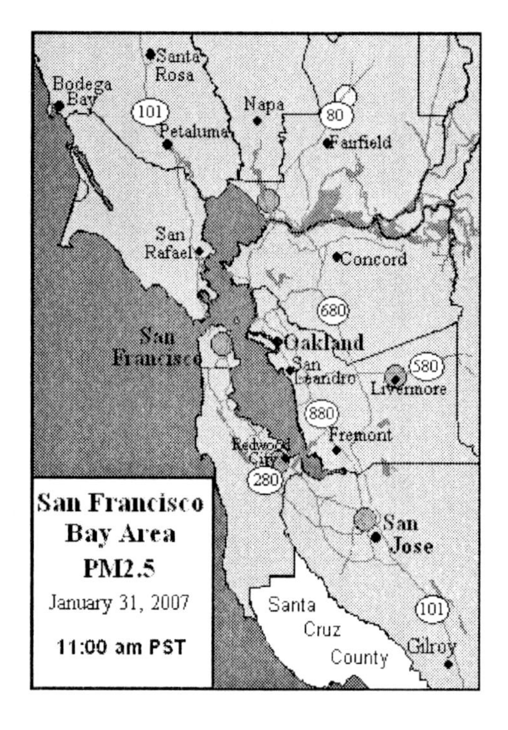

For locations away from these stations, a Gaussian dispersion model extrapolation technique called ASPEN (Assessment System for Population Exposure Nationwide) is used by the EPA. However, there are numerous citizen science projects that have exposed the inaccuracies contained in the ASPEN model. For example, in 1995 the EPA ASPEN model estimated the expected outdoor concentration of a toxic nonflammable carcinogenic liquid used as an industrial and organic solvent and as a dry-cleaning agent called *perchloroethylene* to be less than 2 ppb (parts per **billion**) across the Greenpoint-Williamsburg area of Brooklyn, NY. However, a citizen science community group called the "Watchperson Project" investigated areas of odor complaints, canvassed neighborhoods, and conducted studies documenting (1) 39 of 40 apartments in excess of 100 ppm (parts per **million**), (2) one site with levels at 197,000 ppm, and (3) 24 of 29 apartments above dry cleaners with a eight-day average concentration above 1,000 ppm (Wallace, Groth, Kirrane, Warren, & Halloran, 1995).

Clearly, this sparse sensing strategy and dispersion modeling technique do little to capture the dynamic variability arising from urban micro-climates, daily automobile traffic patterns, human activity, and smaller industries. Are we to believe that the park, subway exit, underground parking lot, building atrium, bus stop, and roadway median are all equivalent environmental places? A geo-logged path of more fine grain ozone measurements across Las Vegas, Nevada begins to expose the detailed dynamic range of ozone across a city (Figure 3) (DAQEM, 2005).

Finally, these measurements only address outdoor air quality. However, every year, indoor air pollution is responsible for the deaths of 2 million people—that's one death every 20 seconds (Ezzati & Kammen, 2002). Indoor air pollution is caused by the burning of solid fuels which over half of the world's population rely on for cooking and heating. More than half of this burden is borne by developing countries. This means that there is an opportunity for not just civic change but the

Figure 3. Geo-logged ozone measurements over Las Vegas, Nevada on 21 August 2005 @ 3:30PM (left) and 22 August 2005 @ 3:00PM (right), measured at 1500 meters. Colors indicate wide variations expressed in parts per billion.

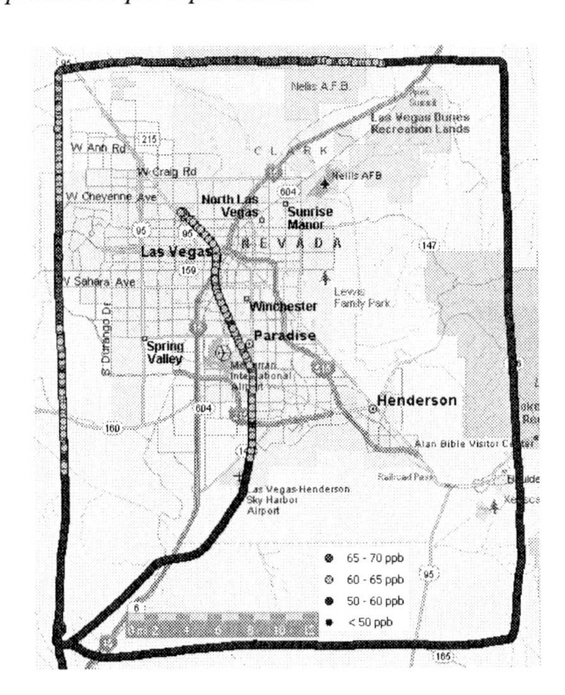

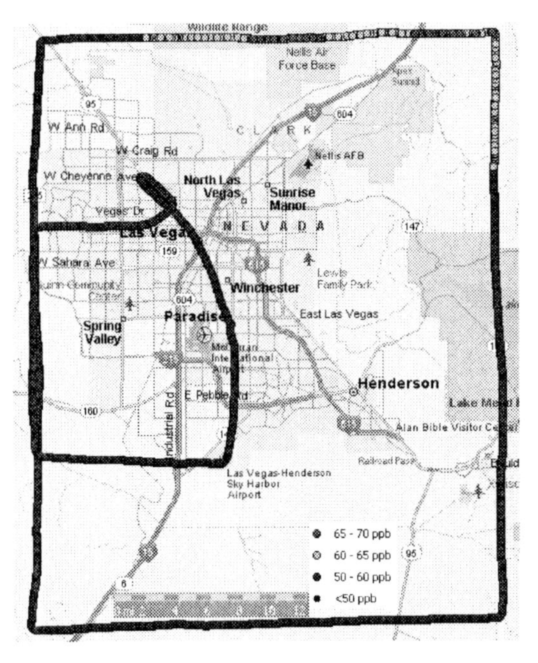

creation of significant life saving technologies for emerging regions and rural villages worldwide. Even in developed nations there is significant indoor air pollution from smoke, radon, paints, carpets, and even toner particles expelled from laser printers (Kay, 2007). However, there are no EPA monitoring stations indoors. Air quality needs to be measured where people go and this includes indoors. Our solution is to integrate air quality sensing with a technology already carried indoors and outdoors by billions of people everyday—their mobile phone.

Environmental Health Related Work

Historically there have been grassroots efforts by communities to address environmental issues, often when conditions become extreme. In 1988 "The Toxic Avengers" (Yahr, June 2001), a group of 15 high school students, operating under the belief that people should have a right to live in a safe environment, successfully canvassed their community about the myriad of risks inherent in having toxic industrial pollutants and a large capacity incinerator in close proximity to residences and schools. In direct response to the citizens, the government eventually took action and many of the polluting sites were cleaned up and the incinerator shutdown.

However, in general the individual citizen has very little direct awareness of the air quality that they encounter daily and almost no public forum to debate strategies for change. Several recent projects have explored citizen measurement of air quality such as Hooker's Pollution e-Sign (Hooker, Gaver, Steed, & Bowers, 2007), Preemptive Media's AIR (Areas Immediate Reading) mobile device (Costa, Schulte, & Singer, http://www.pm-air.net/index.php), Proboscis' Snout (http://socialtapestries.net/snout, 2007), Millecevic's Neighborhood Satellites (Milicevic, 2007), the Everyday Learning Lab's Smoke Rings (Foley-Fisher & Strohecker, 2005), Jeremijenko's Feral Robotic Dogs (Jeremijenko, 2005), EQUATOR's e-science project which included extensive carbon monoxide samples taken by people across several streets in London (Milton & Steed, 2007), and a more recent deployment of such bike based mobile air quality sensing by Cambridge Mobile Urban Sensing (CamMobSens) (Kanjo & Landshoff, 2007).

STUDY 1: PERCEPTIONS OF AIR QUALITY

What does "air quality" mean? How is it measured? Where? How often do people think about their air quality? Where can you find the reported/forecast air quality? We were interested in understanding people's perceptions of air quality and their interest in taking personal measurements. We conducted personal interviews with 12 residents (9M/3F 23-56 years old) who were approached on public streets of a major metropolitan US city using the questions above as well as others. The small sample size prohibits statistically significant data but several insights can be drawn. Mentioning the term "air quality" elicited responses such as pollution, smog, Los Angeles, Athens, soot, pollen, asthma, vehicles, breathing, smells, cleanliness, quality of life, and even global warming. None of the participants had a clear understanding of how and where air quality was measured in their own city and only one knew that reported forecasts could be found in the weather section of the local newspaper. When participants speculated on where air samples were taken, the dominant model was *samples at multiple locations* such as "at least in every district in the city", "hundreds if not more near factories, close to highways, etc", and "all around". Some voiced concern over the management of the data, commenting, "I don't trust the government to collect and report air quality". However, every participant expressed some degree of interest in personally being able to sample air quality, most of them enthusiastically positive responses such as "definitely...what a cool idea", "absolutely", "yes, it has a lot to do with how we breath", "I would try to spread the new across the world", "I want to be part of the solution", "I am concerned and want to be involved and monitor it", "yes, especially if it was useful to other people", "that would be cool ... I'd love to do that", and "definitely but only if it could bring about some global change in policy or action". This led us to further understand the existing air quality system and how we could enable personal sampling.

Ergo: Air Quality On-the-Go

To further study the experience of receiving air quality data on a mobile platform while on-the-go, we designed a public tool called Ergo. Ergo is a simple SMS system that allows anyone with a mobile phone to quickly and easily explore, query, and learn about his or her air quality on-the-go. Ergo uses data from the United States Environmental Protection Agency (EPA) based on fixed metropolitan air quality measurement stations. Sending a text message containing a zip code causes Ergo to deliver current air quality data (usually less than 20 minutes old) and up to three days of forecast for the area. Similar SMS commands to Ergo allow users to request the worst thee polluted locations within the United States that day as well as schedule daily air quality reports to be delivered to their mobile device at any specified frequency. Ergo has delivered nearly 10,000 air quality reports and generated a range of positive feedback including comments from individuals with respiratory problems. For example, several individuals have reported on how the system has improved their lifestyle and provided them with easy access while on-the-go to real-time geographically measured air quality reports.

STUDY 2: AIR QUALITY SENSING FIELD STUDY

What would be the experience of daily living with a personal environmental monitoring mobile device? How can we understand the challenges associated with large, distributed, geo-logged data collection schemes by non-expert users across everyday urban life?

We recruited 7 taxi drivers and 3 students in Accra, Ghana to participate in a two week long study to collect air quality data. We chose Accra because of its poor air quality and common practice of domestic cooking outside using wood, charcoal, and other biofuels. Subjects were modestly compensated even if they did not participate in the full study. Each taxi driver was provided with a dash mounted GPS logger and a tube to hang from their passenger window that contained a carbon monoxide sensor and (a sulfur dioxide sensor or a nitrogen dioxide sensor). At the end of each day, the sensor tube was dropped off at a convenient location where the data was extracted and the sensors charged. Similarly, 3 students were each given a mobile clip sensor pack containing a GPS logger, carbon monoxide sensor and (a sulfur dioxide sensor or a nitrogen dioxide sensor). Similarly, at the end of each day the unit was collected, data extracted, and batteries charged. Both systems are shown in Figure 4. The system was setup to automatically log sensor data every second. Subjects were asked to carry the sensor/GPS loggers as much as possible and during normal everyday activities, including those surrounding work, family, and relaxed social activities. This study allowed us to collect actual geo-logged air quality sensor data by citizens across an urban landscape and influence our design for an integrated air quality sensor into a working mobile phone.

Over the two-week period we collected an impressive and diverse collection of air quality data, reveling dynamic ranges of air quality across neighborhoods and over time. Looking only at CO, for which safe levels are defined as 9 ppm for 8 hours and 35 ppm for 1 hour, our participants logged frequent readings around 30 ppm with many samples ranging up to 75 ppm and some as high as 200 ppm. A heat-map visualizing only the CO values over one 24-hour period captures the dynamic range of air quality across the city as never before seen (Figure 5).

While the rich data sampled from our two-week study in Accra provided hard evidence for the need for citizen science air quality sampling, there were even more revealing and unexpected social effects that resulted. Recall that we provided the participants only with the technology and instructions to carry it during their everyday activities. At the end of each day they dropped of the gear where it was charged and the data extracted. When we conducted exit interviews,

Figure 4. Taxi mounted tube with carbon monoxide, sulfur dioxide, and nitrogen dioxide sensors exposed (left) and packaged (center). Also, the student worn setup containing similar air quality sensors and GPS unit (right)

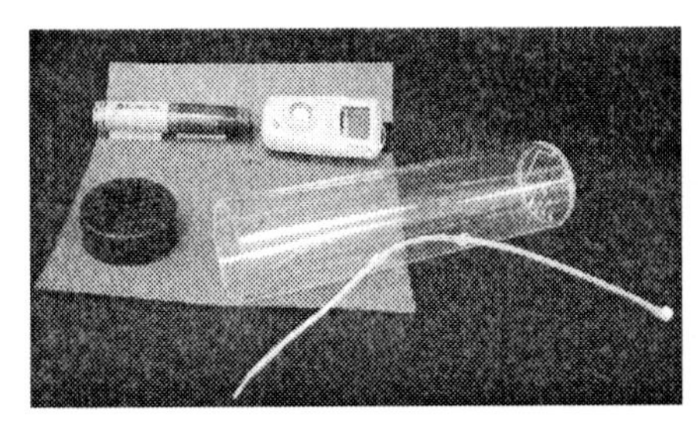

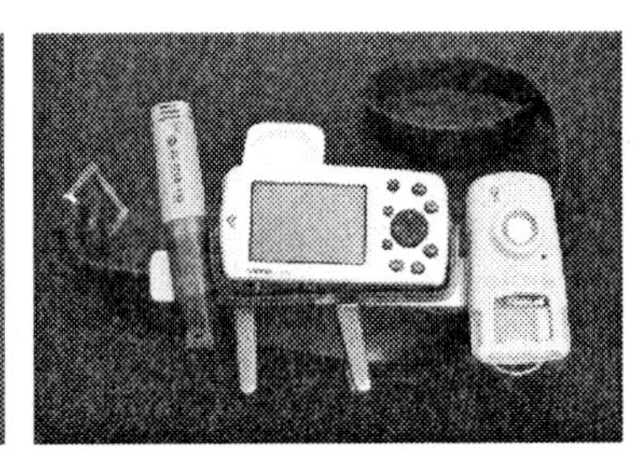

Figure 5. A heat-map visualization of carbon monoxide readings across Accra, Ghana rendered atop Google Earth. Colors represent individual intensity reading of carbon monoxide during a single 24-hour period across the city. Red circles are locations were actual readings were taken

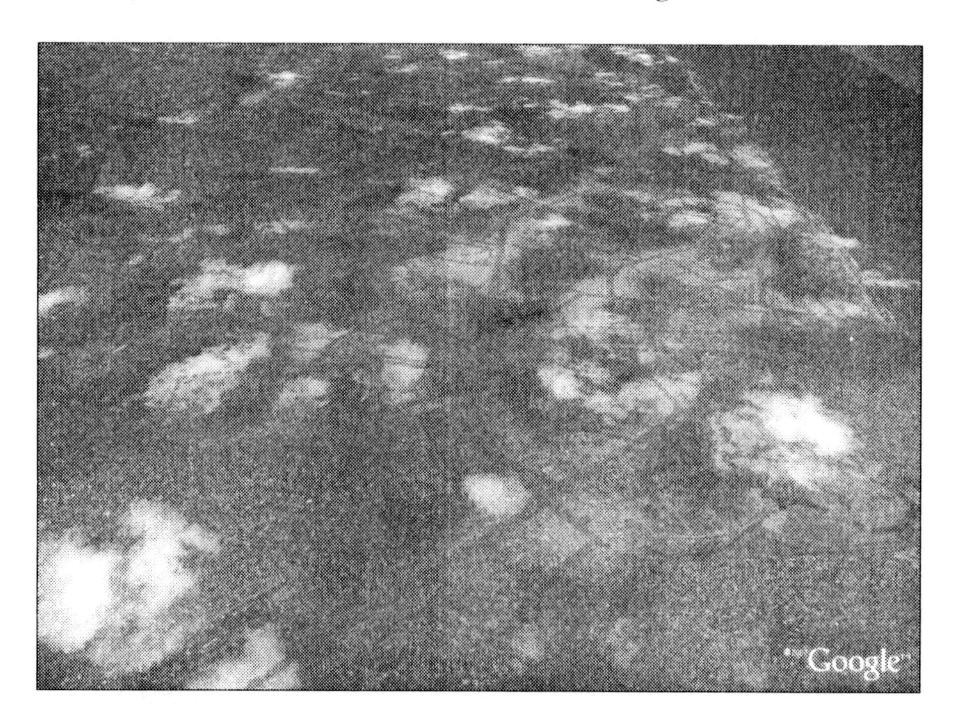

it was clear that, although it was not our intention to provide any humanly readable output on the devices, the participants had begun to look at the numbers and piece together a personal view of the areas of toxic air across Accra. Many discussed how they watched to see how parts of their city changed as they traveled through neighborhoods at different times of day and how they discussed air quality and passed on their new knowledge of hazardous locations to their friends and family. Subjects who thought about air quality once every few months before the study were now thinking about it hourly. Even more interesting, our subjects reported changing their routes through the city, choosing different times to travel downtown, walking further from roadways, being outside later in the day rather than during high pollution times, and even taking their own automobile in for inspection to insure they were not contributing to the problem they were measuring. Several of the participants encountered each other when dropping of their equipment for charging and, although they did not know each other, many of them reported discussing the data they were measuring each day, where they had captured dangerously high readings, alternate routes they had found to be less polluted, and compared graphs from each day's journey.

Most participants reported that the most important thing they learned from the study was that there were extremely unsafe levels of air quality in their city beyond what any government or news agency had reported to them before. Many expressed anger and distrust in their civic leadership for not informing them of such a hazardous condition nor enacting legislation to improve it in the name of public health. Remember, there was no formal networked, real-time sharing mechanism or informed interface for this study and yet spectacular elements of citizen science emerged.

AIR QUALITY SENSING WITH A MOBILE PHONE

Informed by our studies, we designed, built, and integrated various air quality sensors into a mobile phone. We iterated several designs based on various sensors and communications standards such as wired RS-232 and Bluetooth. The resulting 2x4cm hardware combines a carbon monoxide, nitrogen dioxide, and temperature sensor with Bluetooth wireless communication to the mobile phone (Figure 6). The CO/NO_X sensor is small scale, low power (< 350 microamps) and designed to be maintenance-free and stable over long periods. The sensor only needs to be turned on for a few milliseconds while taking a measurement, further reducing overall power consumption. These sensors have a direct response to volume concentration of gas rather than partial pressure and are ideally suited for integration into mobile devices. We chose this sensor combination not only for its small size, low-power, and cost ($60USD), but because they measure two important gases strongly associated with air pollution primarily from automobiles and diesel exhaust.

The sensor board is powered separately and captures measurements when polled by the mobile phone via Bluetooth. It can be attached to the mobile phone, worn, or carried on a purse or backpack. The complete system uses either a Bluetooth mobile phone with assisted GPS (AGPS) or a standard Bluetooth phone without location technology. With a GPS phone, the sensor data is geo-logged at regular intervals with a bias towards taking measurements during voice calls and text messaging usage (ensuring adequate exposure of the carbon monoxide sensor to ambient air samples when the sensor is mounted on the phone). We also wanted to explore the value of using such sensors without any location information using only time. In both conditions the collected data is sent using SMS. SMS was chosen over other data transport mechanisms because of its wide adoption and its use of the carrier's control channel rather than data channel, allowing data to be sent even during voice calls. The format of sampled sensor and corresponding meta-data is in a standardized XML schema. It is shared publicly via a web-based interface with a standardized SQL database backend. This allows the public to query and use the data within the repository.

Figure 6. A wireless Bluetooth based sensor board with a carbon monoxide and nitrogen dioxide sensor (black rectangle in upper left of board) and a graph depicting nitrogen dioxide values from 0-5ppm (top graph) and the measured signal from the sensor (bottom graph)

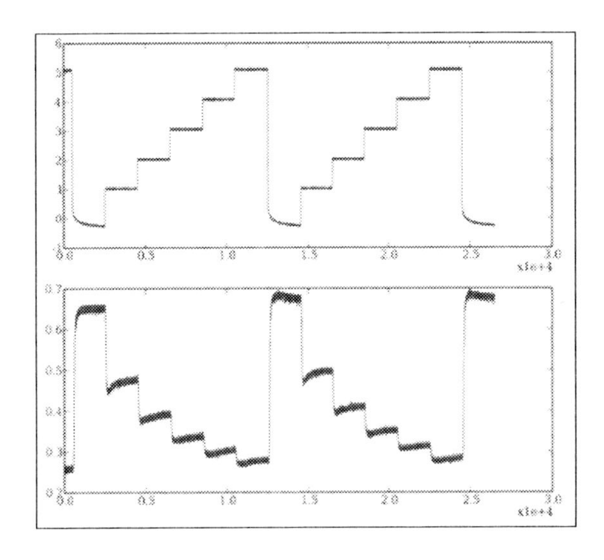

Figure 7. A wristwatch form factor design currently under development integrating air quality sensing and output

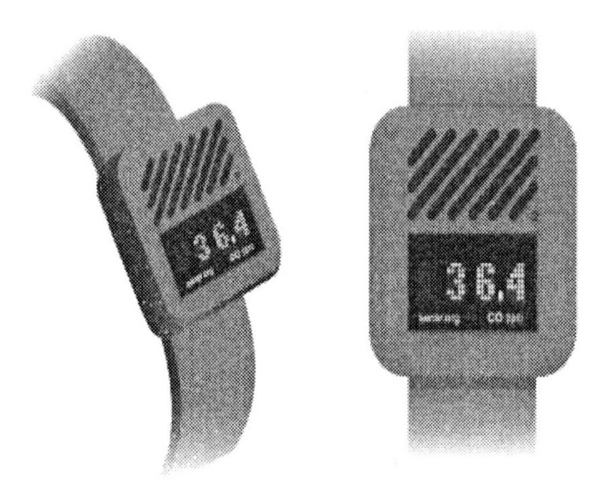

It is not uncommon for people to use Bluetooth headsets (exposed to ambient air) while their mobile phone is in their bag or pocket (not exposed to ambient air). Similarly, we can imagine the use of various Bluetooth based air quality sensors attached to clothing, body, or backpacks for sampling the ambient atmosphere. We are also in the process of developing a wristwatch version of the hardware with a numeric output on a watch face. A sketch of this design concept is shown (Figure 7).

INTERACTING WITH AIR QUALITY MEASUREMENTS

While we have addressed many of the technical challenges associated with the engineering of the overall system from circuit design to the construction of the SMS relay service for handling data to the XML database schema, it is the overall user experience of everyday life capturing, sharing, and comparing air quality data that is fundamental to the success of citizen science. We concentrate our discussion here for the mobile on-the-go experience rather than the online web interaction that is more straightforward. It is the on-the-go experience that promotes curiosity, reflection, learning, and awareness of a user's immediate surroundings and perhaps influences their actions in the moment. In the rest of this section we sketch out several design concepts currently under development for interacting with live air quality data sampled from citizens.

Either prompted by a user's request or triggered automatically by the act of preparing to make a phone call when the device is best exposed to the atmospheric air, the system powers up, captures a CO sample, and displays the numeric value in ppm to the user (Figure 8a). We want to use real values rather than arbitrary ones to preserve the scientific element of the task as well as encourage awareness of air quality based on real numbers. In the same way living with Celsius temperature measurements promotes a clear understanding of the relationship of the values to the feeling of temperature, we want to build a connection of air quality values to user perceptions.

Once the data is uploaded it can be shared with others. An obvious sharing mechanism acts much like a buddy list where you can see a listing of your "sensor buddies" with real-time comparison not of online status but of their current and daily cumulative (*i.e.* dosimeter) air quality exposure (Figure 8b). It also shows up/down trends and a current ranking by exposure level, including the forecast and officially reported levels by the government. A related "sensor buddies" view shows graphs for each person over the course of the day (Figure 8c). This promotes reflection about your exposure compared to others through the day. Why is my exposure so much higher in the morning? Lower at lunch? My office air quality seems worse than my friends? What's my child's exposure today? While your "buddies" could be located across your own city, it is also interesting to compare your values to people in other parts of the world? How does my exposure compare to my friend in Rome? A mother in Shanghai? My parents in Los Angeles?

Figure 8: Mobile interactions and visualizations of air quality (a) taking an air quality measurement, (b) sensor buddies, (c) daily graphed sensor buddies, (d) air journal, (e) my air journal, (f) cleanest route signage, (g) dynamic real-time air quality map, (h) bus stop air quality mashup

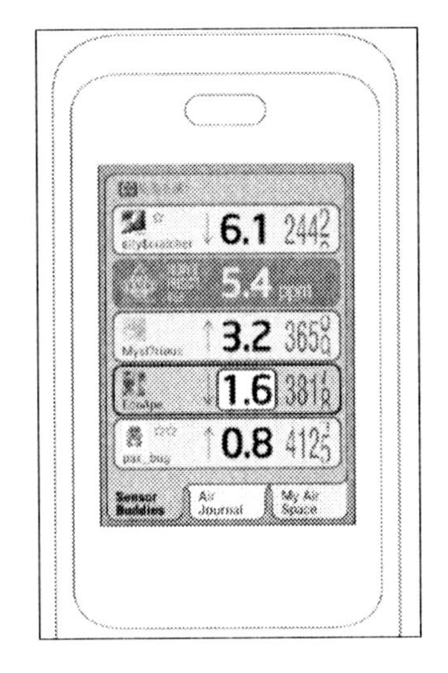

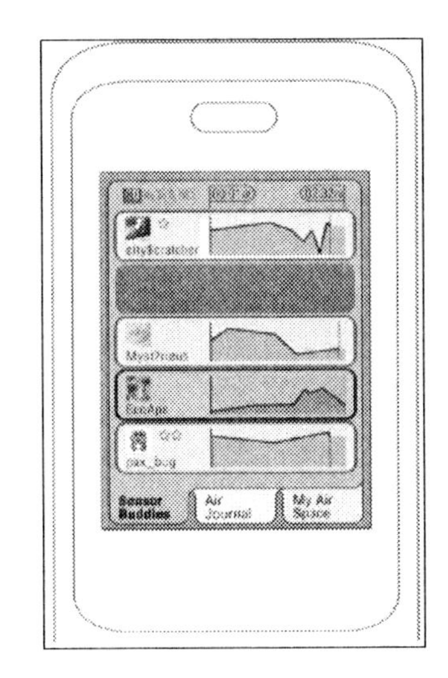

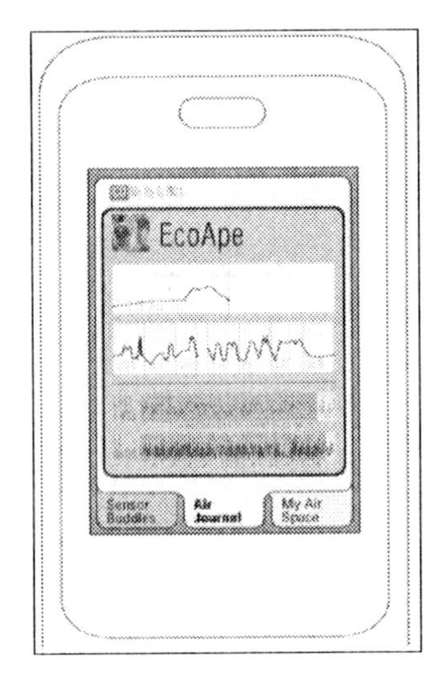

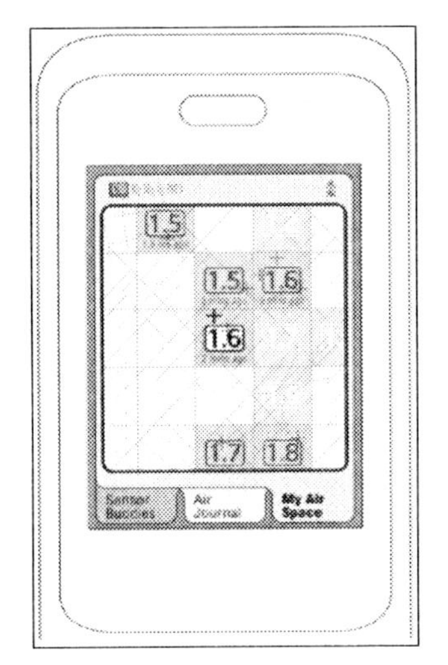

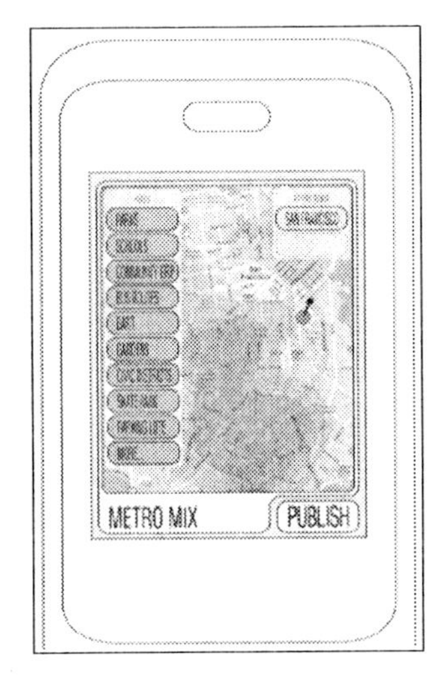

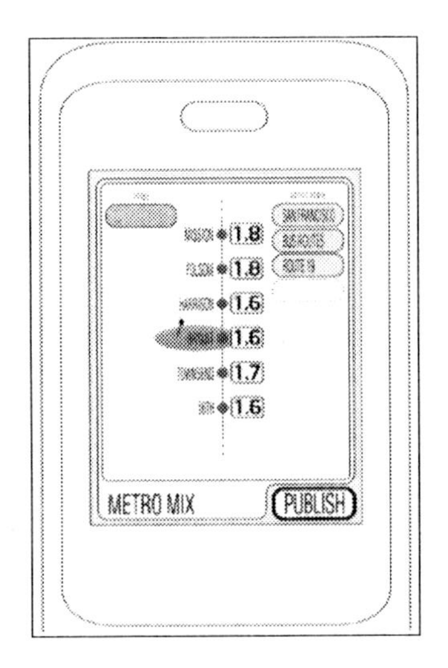

As you start to log more data it becomes interesting to look for trends and longer patterns. The "air journal" mechanism allows a user to look at hourly, daily, weekly, monthly, and yearly cycles (Figure 8d). How does my weekend exposure differ from the weekday? What was a trip to Rome look like? My vacation in Montana? Again the focus is on promoting awareness and reflection to motivate discussions and lifestyle changes within individuals and groups using persuasive technologies.

It is important that individuals are provided feedback that their own measurements are important and making a significant contribution to their community. The "my air space" tab affords a visualization of the location of their own measurements atop a map with data from others. It is a visual reward system that their data is providing vital coverage for a specific neighborhood or subway stop as well as a simple reputation model for publicly showcasing their contributions (Figure 8e).

One of the themes of citizen science is allowing others to remix the public data for other applications. For example, new signage or mobile route planning software could provide, not the quickest or shortest route, but the cleanest (Figure 8f). Similarly, users could not only create visual heat-map style graphs (Figure 8g) but also mashups more specifically tuned to various activities such as creating an air quality mashup of bus stops for a bus line (Figure 8h).

CONCLUSION

We have presented a transformation of the mobile phone from solely communication tool to that of a sensor rich personal measurement instrument that empowers individuals and groups to more easily gain an awareness of their surroundings, engage in grassroots efforts to promote environmental change, and enables an important social paradigm—*citizen science*. Through a series of investigations, interviews, and user field studies around the measurement of air quality by individuals, we designed and built a series of functional air quality hardware solutions integrated with a mobile phone. Finally, we sketched a series of design scenarios around various, on-the-go, mobile interactions with an overall system. We believe that such systems can elevate individuals to have a powerful new voice in society, to act as citizen scientists, and collectively learn and lobby for change within their block, neighborhood, city, and nation.

REFERENCES

AirNOW. (2006). AirNOW: Quality of Air Means Quality of Life.

Akyildiz, I. F., Su, W., Sankarasubramaniam, Y., & Cayirci, E. (2002). A survey on sensor networks. *Communications Magazine, IEEE, 40*(8), 102-114.

Balandina, E., & Trossen, D. (2006). Nokia Remote Sensing Platform Middleware and Demo Application Server: Features and User Interface. *Nokia Research Center/Helsinki.*

Beck, U., Giddens, A., & Lash, S. (1994). *Reflexive modernization: politics, tradition and aesthetics in the modern social order.* Stanford, Calif.: Stanford University Press.

Beer, D., & Burrows, R. (2007). Sociology And, of and in Web 2.0: Some Initial Considerations. *Sociological Research Online, 12.*

Burke, J., Estrin, D., Hansen, M., Parker, A., Ramanathan, N., Reddy, S., et al. (2006). *Participatory Sensing.* Paper presented at the Workshop on World Sensor Web at SenSys.

Campbell, A. T., Eisenman, S. B., Lane, N. D., Miluzzo, E., Peterson, R., Lu, H., et al. (2008). People Power - The Rise of People-Centric Sensing. *IEEE Internet Computing Special Issue on Mesh Networks.*

Corburn, J. (2005). *Street science : community knowledge and environmental health justice*. Cambridge, MA: MIT Press.

Costa, B. d., Schulte, J., & Singer, B. (Artist). (http://www.pm-air.net/index.php). *AIR*

DAQEM. (2005). Regional Ozone Study. *Clark County Department of Air Quality and Environmental Management (DAQEM)*

Dave, B. (2007). Space, sociality, and pervasive computing. *Environment and Planning B: Planning and Design, 34*(3), 381-282.

EPA. (2007). United State Environmental Protection Agency: CO: What does it come from?

EU. (2004). Towards a thematic strategy on the urban environment. In O. J. C. o. 23.04.2004 (Ed.) (Vol. 60): Communication from the Commission to the Council, the European Parliament, the European Economic and Social Committee and the Committee of the Regions -

Ezzati, M., & Kammen, D. M. (2002). The health impacts of exposure to indoor air pollution from solid fuels in developing countries: knowledge, gaps, and data needs. *Environmental Health Perspectives, 110*.

Fall, K. (2003). *A delay-tolerant network architecture for challenged Internets*. Paper presented at the Applications, technologies, architectures, and protocols for computer communications.

Farooq, S. (2006, 27 November). Nonprofit Web Site ParkScan Spearheads Cleanup of SF's Dirtiest, Most Troubled Parks *San Francisco Examiner*.

Foley-Fisher, Z., & Strohecker, C. (2005). *An approach to the presentation of information from multiple sensors*: Media Lab Europe/MLE EL-TR2005-02.

Hooker, B., Gaver, W. W., Steed, A., & Bowers, J. (2007). *The Pollution e-Sign*. Paper presented at the UbiComp Workshop on Ubiquitous Sustainability: Technologies for Green Values.http://socialtapestries.net/snout. (2007). *Snout*.

Jacobs, J. (1961). *The death and life of great American cities*. New York: Random House.

Jeremijenko, N. (Artist). (2005). *Feral Robotic Dogs*

Josef, K., & Hermann, M. (2006). The Transformation of the Web : How Emerging Communities Shape the Information we Consume. *Journal of Universal Computer Science, 12*(2), 187-213.

Justo, P. D. (2004, 9 September). Protests Powered by Cellphone. *New York Times*.

Kanjo, E., & Landshoff, P. (2007). Mobile Phones to Monitor Pollution. *IEEE Distributed Systems Online Urban Computing and Mobile Devices, 8*(7).

Kay, J. (2007, August 1, 2007). Big health risk seen in some laser printers. *San Francisco Chronicle*.

Kindberg, T., Chalmers, M., & Paulos, E. (2007). Urban Computing. Guest Editors' Introduction of special issue *IEEE Pervasive Computing, 6*(3).

Klonoff-Cohen, H., Lam, P. K., & Lewis, A. (2005). Outdoor carbon monoxide, nitrogen dioxide, and sudden infant death syndrome. *Archives of Disease in Childhood 90*, 750-753.

LeBaron, G. (2006). *The 106th Christmas Bird Count*: National Audubon Society.

Lessig, L. (1994). The Creative Commons. *Florida Law Review, 55*(763).

May, M. (2007, 10 February). Creeps beware: Web gives women revenge - Catcall recipients share their stories and men's photos. *San Francisco Chronicle*.

Milicevic, M. (2007). *Imaginary To Dos: Three initiatives for personal environmental explorations*. Paper presented at the UbiComp Workshop on Ubiquitous Sustainability: Technologies for Green Values.

Milton, R., & Steed, A. (2007). Mapping carbon monoxide using GPS tracked sensors. *Environmental Monitoring and Assessment, 124*, 1-19.

Mydans, S. (2001, 5 February). Expecting Praise, Filipinos Are Criticized for Ouster. *New York Times*.

Myers, S. L., & Mydans, S. (2005). Fired Ukraine Premier Sees End of 'Orange Revolution' Unity. *New York Times*, p. 7.

Nath, S., Liu, J., Miller, J., Zhao, F., & Santanche, A. (2006). SensorMap: a web site for sensors world-wide. *Proceedings of the 4th international conference on Embedded networked sensor systems*, 373-374.

O'Reilly, T. (2005). What is Web 2.0: Design Patterns and Business Models for the Next Generation of Software, *O'Reilly Radar Reports*.

Park, J. J., Lee, K. H., Jung, H. J., & Kim, K. H. (2007). Portable device for user's basal body temperature and method for operating the device (Patent #20070191729), *US Patent Office*

Park, K. K., & Hwang, I. D. (2007). Apparatus, method, and medium for measuring body fat using a near infrared signal (Patent #20070185399), *US Patent Office*.

Paulos, E., & Jenkins, T. (2005). *Urban probes: encountering our emerging urban atmospheres*. Paper presented at the ACM SIGCHI, Portland, Oregon, USA.

Paulos, E., Jenkins, T., Joki, A., & Vora, P. (2008). *Objects of Wonderment*. Paper presented at the ACM Designing Interactve Systems (DIS).

Randell, C., Phelps, T., & Rogers, Y. (2003). *Ambient Wood: Demonstration of a Digitally Enhanced Field Trip for Schoolchildren*. Paper presented at the UbiComp Demonstration.

Santanche, A., Nath, S., Liu, J., Priyantha, B., & Zhao, F. (2006). SenseWeb: Browsing the Physical World in Real Time. *IPSN'06: Proceedings of the fifth international conference on Information processing in sensor networks*.

Shklovski, I., & Chang, M. (2006). Guest Editors' Introduction: Urban computing: Navigating space and context. *IEEE Computer 39*(9), 36-37.

UN. (2003). World urbanization prospects: The 2003 Revision. New York: United Nations Dept. of International Economic and Social Affairs.

Viégas, F. B., Wattenberg, M., Ham, F. v., Kriss, J., & McKeon, M. (2007). *Many Eyes: A Site for Visualization at Internet Scale*. Paper presented at the Infovis.

Wallace, D., Groth, E., Kirrane, E., Warren, B., & Halloran, J. (1995). Upstairs, downstairs: Perchloroethylene in the air in aparments above New York City dry cleaners. *Consumers Union of The United States, Inc.*

White, S., Feiner, S., & Kopylec, J. (2006). *Virtual Vouchers: Prototyping a Mobile Augmented Reality User Interface for Botanical Species Identification*. Paper presented at the Proc. 3DUI 2006 (IEEE Symp. on 3D User Interfaces).

WHO. (2006a). *Air quality guidelines. Global update 2005. Particulate matter, ozone, nitrogen dioxide and sulfur dioxide*.

WHO. (2006b). World Health Organization (WHO), Fact sheet N°307.

Winner, L. (1999). Do Artifacts have Politics. In D. MacKenzie & J. Wajcman (Eds.), *The Social Shaping of Technology*: McGraw Hill.

Yahr, H. (June 2001). Toxic Avengers: High School Students in Middletown, New York Wade Ever Deeper into Illegal Dumping Story. *The Independent Film and Video Monthly*.

KEY TERMS

Air Quality Index (AQI): An index for reporting daily air quality. It reports a number, between 0 and 500 to reflect how clean (low AQI) or polluted (high AQI) the air is, and what associated health effects might be of concern to citizens. The AQI focuses on health effects people may experience within a few hours or days after breathing polluted air. AQI is measured for five major air pollutants: ground-level ozone, particle pollution (also known as particulate matter), carbon monoxide, sulfur dioxide, and nitrogen dioxide. For each of these pollutants.

Citizen Science: Citizen science is a term used for projects or ongoing program of scientific work in which individual volunteers or networks of volunteers, many of whom may have no specific scientific training, perform or manage research-related tasks such as observation, measurement or computation. The use of citizen-science networks often allows scientists to accomplish research objectives more feasibly than would otherwise be possible. In addition, these projects aim to promote public engagement with the research, as well as with science in general. Some programs provide materials specifically for use by primary or secondary school students. As such, citizen science is one approach to informal science education. The longest-running citizen science project currently active is probably the Audubon Society's Christmas Bird Count, which started in 1900.

Environmental Protection Agency (EPA): An agency of the federal government of the United States charged with protecting human health and with safeguarding the natural environment: air, water, and land. The are the primary reporting and regulation mechanism for air quality within the United States.

Participatory Urbanism: A research initiative directly focused on the potential for emerging ubiquitous urban and personal mobile technologies to enable citizen action by allowing open measuring, sharing, and remixing of elements of urban living marked by, requiring, or involving participation, especially affording the opportunity for individual citizen participation, sharing, and voice. Participatory urbanism promotes new styles and methods for individual citizens to become proactive in their involvement with their city, neighborhood, and urban self-reflexivity. One such example of Participatory Urbanism is the use of mobile phones to be transformed into environmental sensing platforms that support community action to effect positive societal change.

Toxic Avengers: founded in 1988 by a group of high school students who organized themselves to raise community awareness about environmental pollution in their Brooklyn neighborhood. The name came from a comic book of the same name, whose characters were crusaders against toxic waste." The students were from the El Puente Academy high school and the community organization's program on community health, youth service, and leadership. What began as a science-class project turned into an organization that raised environmental awareness in the community and helped galvanize a community coalition that would be instrumental for taking action against neighborhood environmental hazards. The young people who formed the Toxic Avengers were part of a science class that was doing a unit on understanding the neighborhood environment. The class researched local hazards by gathering readily available information from local, state, and federal environmental agencies on the environmental performance of facilities in the community. The students also searched through newspaper archives to find references to environmental pollution in their neighborhood. They discovered, for example, that the Radiac Corporation—a storage and transfer facility for toxic, flammable, and low-level-radioactive waste located in the neighborhood—was the only facility of its kind in the entire city.

Urban Computing: Urban Computing focuses on our lifestyles and technologies within the context of public urban spaces. Its research challenges differ from those found within the home where technologies readily intermingle across our intimate relations with friends and family members. It diverges from office and work environments where productivity and efficiency often dominate our computing tools. It is also not simply concerned with mobile or social computing. Urban Computing establishes an important new framework for deconstructing and analyzing technology and urban life across five research themes - *people*, *place*, *infrastructure*, *architecture*, and *flow*. The diversity of these important themes promotes rich interdisciplinary research within the field of Urban Computing. Urban Computing leverages such forces as expanding urban populations, rapid adoption of small, powerful, networked, mobile devices, bluetooth radios, tiny ad hoc sensor networks, and the widespread influence of wireless technologies across our growing urban landscapes.

World Health Organization (WHO): The directing and coordinating authority for health within the United Nations system. It is responsible for providing leadership on global health matters, shaping the health research agenda, setting norms and standards, articulating evidence-based policy options, providing technical support to countries and monitoring and assessing health trends. They are the leading international body on air quality health research and standards.

Chapter XXIX
Extreme Informatics:
Toward the De-Saturated City

Mark Shepard
University at Buffalo, USA

ABSTRACT

What happens to urban space given a hypothetical future where all information loses its body, that is, when it is offloaded from the material substrate of the physical city[1] to the personal, portable, or ambient displays of tomorrow's urban information systems? This chapter explores the spatial, technological and social implications of an extreme urban informatics regime. It investigates the total virtualization of the marks, signage, signaling and display systems by which we locate, orient ourselves, and navigate through the city. Taking as a vehicle a series of digitally manipulated photographs of specific locations in New York, this study analyzes the environmental impact of a pervasive evacuation of information–at various sites and scales–from the sidewalks, buildings, streets, intersections, infrastructures and public spaces of a fictional future De-saturated City.

INTRODUCTION

One might argue that urban informatics takes as an implicit goal the expansion of information into ever-greater visual and aural fields of the ordinary urban citizen. For example, consider the current fascination with large-scale "urban screens"—programmable electronic billboards enveloping the facades of prominent urban structures. UN Studio's Galleria West Department Store in Seoul, South Korea, realities:united's SPOTS media façade on Berlin's Potsdamerplatz, Klien Dytham's Uniqlo façade in Tokyo, or Peter Cook's Kunsthaus in Graz, Austria are but a few examples of media-architecture hybrids that demonstrate the transformation of entire architectural surfaces into information displays. Media artists have also explored the building façade as a site for mediated play and social interaction. Berlin-based Chaos Computer Club's "Blinkenlights" (2001) transformed the façade of the Haus des Lehrers on Berlin's Alexanderplatz into a low-resolution game board for playing pong with your mobile phone (www.blinkenlights.de). Christian Moeller's "Nosy" (2006) is a building-scale video installation that employs "a robotic video camera" to capture "the surrounding landscape and people, which are then displayed in bitmap graphics onto three towers covered with white LEDs behind frosted glass panels." (http://www.christian-moeller.com/display.php?project_id=59). Interest in urban screens is also reflected by a number of recent conferences on the subject bringing together researchers and practitioners from the fields of media, architecture, information and communication technology (ICT) to identify issues and strategies for future development.[2]

At a smaller scale, signage displaying precise wait-times for busses, trams or subway trains equipped with GPS or other location-based technologies is becoming common in large cities such as Berlin, Rome, and New York. In 2000, New York City-based Adapt Media of Boston introduced AdRunner: a GPS enabled LED display designed for the roofs of taxi cabs that delivered location-based advertising customized for the neighborhood demographics they move through.[3] In Toronto, 13,000 cabs have been outfitted with touch screen displays that use GPS to target customers as they drive by local businesses (GPS World, 2007). Whether it be building-scale display surfaces or smaller mobile screens distributed throughout the city, the trend appears to be toward finding new and inventive ways for displaying real-time or demographically targeted information in physical urban space.

Visions of the future promulgated by popular media project this trend to the extreme. Take for example a scene from Steven Spielberg's 2002 movie *Minority Report*, which projects the future of location-based advertising in the city of year 2050. In a chase scene depicting renegade cop Jon Anderton's flight through a shopping mall, he is bombarded by advertising tailored to his specific *location*, *time* and *identity* as revealed through a retinal scan. Illuminated advertisements that recall his personal preferences and history of shopping transactions call out for his attention as he darts through a cacophony of sound and image. In developing this scene, Spielberg worked with a combination of advertising "creatives" and researchers from MIT's media lab to extrapolate forward a vision of how advertising might work in the future (Rothkerch, 2002). What's perhaps interesting here is less the imagined technical wizardry at work in, say, performing retinal scans on the fly and correlating the unique patterning of neural cells and capillary structure with a personal transaction history—all in real-time, and for a rapidly moving target. Rather, as with most science fiction, what's striking is how this future scenario carries forward existing paradigms of advertising embedded within the present, and what that might tell us about the *place* of media and information within contemporary culture.

We are certainly living in a time that offers unprecedented access to an increasing amount of information. Yet as computing leaves the space of the screen for that of the sidewalk, as promised by ubiquitous computing, new questions arise for urban informatics that have less to do with the technical challenges involved with increasing access to (or the amount of) available information and more to do with the material, spatial and social relations this shift implies. In the sections that follow, I first explore the role of cultural paradigms in framing the way by which new technologies are conceived and introduced into society. I discuss the role of skeuomorphs–material artifacts that simulate an aspect of a previous time using a technology that has superceded it–as transitional objects that smooth the introduction of "new" technologies into a culture. I suggest that by doing so, technological ideation is at times constrained by conceptual categories that may no longer be relevant to contemporary societies. I then introduce the paradigm of location-based, "personal informatics" (as a subset of ubiquitous computing) that offers an alternative to a broadcast model based on an undifferentiated public (urban screens). Finally, I attempt to project forward this paradigm to an extreme condition where information is *subtracted from* (rather than *added to*) the physical space of the city, in order to examine the spatial, technological and social implications this would imply.

URBAN SCREEN AS SKEUOMORPH

In narrating a history of cybernetics, Hayles (1999) borrows the term "skeuomorph" from anthropology to describe theoretical constructs from one period that serve as transitional objects to the next in the development of cybernetic theory. In archaeology, skeumorphs are material artifacts that simulate an aspect of a previous time using a technology that has superseded it. They are derivative objects that retain structurally necessary elements of the original as ornamentation, stripped of their original function. Skeuomorphs are often deliberately employed to make the "new" look familiar, comfortable and accessible. Examples include the simulated stitching of the vacuum formed vinyl replacing the fabric upholstery of car interiors, the mechanical shutter sound produced by digital cameras, or more abstractly, the metaphor of the "desktop" work-space of the personal computer introduced by Apple with the Macintosh computer in 1984, where the organizational syntax of files and folders serves to orient us within an otherwise unfamiliar space. [4] Hayles notes that these transitional objects serve to smooth the transition from one historical moment to the next. Artifacts (and by extension ways of thinking) of one moment are carried forward into the future by simulating aspects of the past. New technologies in this case are put in service to reproduce the past as simulation to render the unfamiliar in familiar terms.

In a press release announcing plans for a city-wide network of 45 large-scale electronic billboards in Berlin (IBM, 2007), IBM subsidiary Rock Screen Europe invokes the name of Ernst Theodor Amandus Litfass, the German printer and publisher who in 1854 invented the Litfass-Säulen, a free-standing cylindrical column for the posting of advertising. Here, the marketing effort is obviously directed at establishing continuity with a history of "innovation" in Berlin advertising, an important rhetorical device given the fact that each screen will ultimately require the approval of the local governmental district within which it is to be situated. Yet the plan perpetuates a longstanding model of information access and distribution in urban public space, one built on the idea that we need access to *more* information in the physical space of the city, and that this information be uniformly broadcast to an undifferentiated "public." In this respect, it is hardly innovative. The introduction of new technologies has long been linked with the proliferation of advertising through the posting of bills, distribution of flyers and other forms of information in urban public space. Coincident with the introduction of subway trains in Manhattan at the beginning of the 20th century, for example, was the selling of advertising space on the interior surfaces of subway cars. With this came the by now familiar public reaction against the increase of advertising in "public" places, and the counter-argument for the need to subsidize the cost of public services with private revenue streams.

More significantly, Rock Screen Europe's plan depends on an idea of public space that is perhaps no longer fully operative. If public space was once considered the geography of the public sphere (Habermas, 1991)—as a materially grounded condition within which information circulates and informs public opinion—today the two operate within increasingly separate domains. On one level, we are witnessing the displacement of the public sphere to the immaterial nodes and networks of electronic media and information systems. "The public", "publics" and "public opinion" are today formed more through cable and network news channels, Internet blogs and websites than on the sidewalks, streets, cafés, parks or shopping arcades of the contemporary city (Low and Smith, 2005). Online social networking sites such as MySpace and Facebook have replaced the street or the mall as the preferred place to "see, be seen, and connect" for today's youth. Sociable web media such as Flickr enable forms of media sharing and exchange previously unimaginable in physical space. At the same time, the category of the "public" itself has become increasingly differentiated. The idea of a unified body identifiable as a "public" has given way to target markets and increasingly granular demographic categories. Within this context, "one-to-many" broadcast models for information distribution become less successful than advertising targeted to specific consumer profiles compiled from a large array of data sources and transaction histories. Consider for example the success of Google's Ad Sense program, which delivers ads tailored to keywords entered into its search engine or the content displayed on the web page being viewed.

I would argue that the paradigm of large-scale "urban screens" operates as a skeumorph in the evolution of urban informatics. It is based on conceptual categories whose relevance vis-à-vis contemporary societies is questionable. While this paradigm may serve to smooth the transition of integrating digital information systems into urban environments, it does so by reproducing modes of information access and distribution that no longer hold sway. In doing so, it perpetuates design logics regarding "the public" and "public space" that are perhaps less reflective of the way we access, share and distribute information today. In this regard, urban screens serve simply as urban ornament. What other strategies for urban informatics exist that offer alternatives to envisioning the architectural surface in urban environments as confectionary spectacle?

LOCATIVE MEDIA, PERSONAL INFORMATICS, AND PUBLIC AUTHORING

While much has been written regarding the opportunities (and dilemmas) of ubiquitous computing for urban life (Dave, 2007; Ellison et. al, 2007; Galloway, 2004; Greenfield, 2006; Greenfield and Shepard, 2007; Kindberg et. al, 2007; Shklovski and Chang, 2006), few implementations of these ideas exist. The relatively recent emergence of Location-based Services (LBS) for mobile devices, however, is beginning to provide insights to new ways be which information can be accessed, shared and distributed in urban environments. These services deliver information specific to location, time of day/year, and the preferences or profile of an individual, group or network. Common commercial applications include GPS-based navigation systems for mobile phones that direct one to a local restaurant or business meeting, and location-aware city guides that aid visitors to unfamiliar places in finding what they are looking for. But location-based services can also enable ordinary people to tie bits of media and information to specific locations in the physical world, marking up the built environment with personal notations, stories and images. Less common are applications that explore the ability to correlate place, time and identity in ways by which we might not just *read* information in urban space, but also *write* it.

Many early innovations in location-based services came not from industry but from artists and researchers working with positioning technologies such as GPS in unconventional ways. In 2003, Karlis Kalnins coined the term "Locative Media" for work produced by the Locative Media Lab (Galloway and Ward, 2006). A report from the "Locative Media Workshop", held in Latvia in 2003, defined an agenda for location-aware technologies:

Inexpensive receivers for global positioning satellites have given amateurs the means to produce their own cartographic information with military precision. This user-generated cartographic data has recently begun to be shared in a variety of networking machine-searchable environments, which is enabling the development of an 'open source' data pool of human geography. With the arrival of portable, location-aware networked computing devices this "collaborative cartography" will permit users to map their physical environments with geo-annotated, digital data. As opposed to the World Wide Web the focus here is spatially localized, and centered on the individual user; a collaborative cartography of space and mind, places and the connections between them. (RIXC, 2003)

Amsterdam Realtime (2002), a project by the WAAG society in association with Esther Polak, focused on tracing movements through the city of people carrying GPS enabled devices that transmitted their location in real-time to a remote server that in turn projected these movements as an animated "map" in the space of an art gallery (realtime.waag.org). This map represented the city not as a static network of streets, buildings, and open space, but as a series of traces that aggregate over time to represent the city as traversed by different people. Here, the traditional authority attributed to maps and their ability to structure the way we navigate cities is subverted. Rather than a map that informs how one moves through a city, one's movements inform the map.

Other projects include systems for open, public participation in annotating specific places or locations within the city for others to access. "Yellow Arrow" (2004), a global public art project originating in New York, employed simple SMS messaging techniques to enable people to "mark-up" the physical spaces of the city with virtual "tags" or "geo-annotations"– short text messages associated with a specific spot or location (www.yellowarrow.net). Using stickers obtained from the project website, participants affix markers in the public realm with a unique code printed on it. When others encounter a sticker on the street, they send the code printed on it via a text message to a particular phone number. A text message is subsequently received that contains a message left by the person who placed the sticker. In place of the ubiquitous bronze plaque providing "official narratives" affixed to the side of "significant" urban structures or spaces, the project provided for the unofficial annotation of everyday urban places by the ordinary citizen.

"Urban Tapestries" (2004/2006), a project by the London-based social research group Proboscis, further investigated this idea of "public authoring" of urban environments (www.urbantapestries.net). From the website:

Urban Tapestries investigated how, by combining mobile and Internet technologies with geographic information systems, people could 'author' the environment around them; a kind of Mass Observation for the 21st Century. Like the founders of Mass Observation in the 1930s, we were interested creating opportunities for an "anthropology of ourselves"—adopting and adapting new and emerging technologies for creating and sharing everyday knowledge and experience; building up organic, collective memories that trace and embellish different kinds of relationships across places, time and communities.

The project enabled people to create relationships between places and to associate text, images, sounds and videos with them. Using a mobile phone or PDA with wireless connectivity, project participants authored "pockets" consisting of text and media objects related to specific locations in the city. A series of pockets by a single author formed "threads" that connected these locations. Significantly, the system "enabled people not only to personally map their urban spaces, but also read the maps of the neighbours and strangers who share those spaces" (Silverstone and Sujon, 2005).

The project thus focused less on the uniqueness of individual expressions and more on the aggregation of these personal annotations and how they form a collective representation of urban life in a particular place at a particular time.

City Flocks (Belzandic et. al., 2007) builds on the work of Proboscis in exploring how such systems might enable forms of "social navigation" that connect people unfamiliar with a city to the "local knowledge" of its residents. Social navigation is simply the act of asking people (or inferring from the actions of others) directions to something or some place you are looking for. Rather than consulting a map or city guide, one simply asks local residents for directions to a good café, or selects one filled with people, as this may sometimes be a good indicator of quality. City Flocks translates this common social practice into a digital information service: a mobile system enabling local residents to markup particular spots or locations within the physical city with digital annotations that include personal ratings, recommendations or comments. Visitors to the city access this service via their mobile phone, and after entering what they are looking for, receive a list of recommendations along with their calculated average ratings. And if that's not enough guidance, one can even directly call or send a text message to the person who created the original annotation for more information.

Of course, innovations like these (and the many, many others which I haven't mentioned) were not lost on information industry giants. In late November of 2007, Google introduced a version of its popular GoogleMaps application for mobile phones that integrates positioning using GPS and cell tower data with its mapping system to provide turn-by-turn directions to a chosen destination. Google Earth already enables one to markup a map and share it with others. It is likely we will see this service available on mobile phones in the near future. In January 2008, Apple followed suit and announced a partnership with Skyhook Wireless to enable location-based services on the iPhone. Skyhook's technology, based on a similar platform called Placelab developed by Intel Research in Seattle, provides location information by triangulating the signal strength of WiFi access points—an important step for urban applications as GPS signals are blocked by tall buildings in dense urban environments. To be sure, it won't be long before you're walking by a Starbucks and receive on your mobile phone a discount coupon for your latte, or whatever drink you frequently consume there (Greenfield and Shepard, 2007).

Compared to the urban screen paradigm, where information is broadcast to a unified public body, these location-based, personal information systems filter information based on the context within which it is accessed. Factors such as location, time, and personal preference serve to cluster, slice and extract information relevant to the person searching for it. Often accessed via personal, portable, mobile computing devices, this information is displayed not for the public at large but accessed by an individual at a particular place in time.

Toward the De-Saturated City

At a time when we have access to unprecedented amounts of information, I would suggest that the critical issues for urban informatics become less about how to introduce more information into the physical urban realm—an approach represented by urban screens—but rather exploring alternate ways by which information systems in urban environments can be used to sort, parse, slice and dice the overwhelming amount of information that is *already situated* within urban public space. While much has been theorized regarding the technical challenges and social implications of doing so, little attention has been paid to the potential material and spatial consequences of such a shift of focus. What happens to urban space given a hypothetical future where *all* information loses its body, that is, when it is offloaded from the material substrate of the physical city to the personal, portable, or ambient displays of tomorrow's urban information systems?

As discussed above, location-based media and personal informatics influence how we locate and orient ourselves within cities and subsequently navigate through them. Traditionally, architecture and urban design have served to provide the cues by which this occurs. Lynch (1960) attempted to distill a syntax through which a cognitive map of the city is formed over time, through habitual interactions with things like paths, districts, edges, landmarks and nodes. Location-based media and personal informatics suggest a shift from material/tangible cues (streets, squares, rivers, monuments, transportation hubs) to immaterial/ambient ones through which we form our cognitive maps.

If we project forward this approach to the extreme—the offloading of information from the material substrate of urban environments to the personal, portable mobile device –derivative implications for the spatial order of cities emerge. What if urban informatics embraced an agenda for *subtracting from* (rather than *adding to*) the visual field of the city? What if urban informatics worked toward the reduction of information *pollution* in cities, rather than a more compartmentalized concern with captivating the attention of potential consumers or guiding the tourist to the official sights designated as worth visiting?[5]

The set of images reproduced here aims to examine the implications for architecture and urbanism of an alternate urban informatics strategy. Together, they represent a future-fiction that simply assumes technological feasibility. The aim is to raise awareness of the derivative implications of technological development vis-à-vis urban space, and suggest that future development need not only be driven by advertising revenues, the optimization of tourism flows, or making information access generally more efficient. They explore what happens when urban informatics is driven by the visual and spatial impact information systems could have on the spatial order of the physical city itself.

Figures 1-3. "De-Saturated City," digitally altered photographs. Mark Shepard. 2007

On one level the images attempt to catalogue and articulate various sites and scales by which different types of visual information exists within the analog city by methodically removing all the marks, signage and display systems from photographic documentation of existing urban conditions. Through this process, what new sites or situations are uncovered for urban informatics research? What new modalities are implied by which we might access this information? What follows is an attempt to play out a few of the more obvious information displacements from physical to virtual space and an exploration of the corresponding spatial, social and technological implications that arise as a result.

INFORMATION DISPLACEMENTS

Billboards and Electronic Screens

The most obvious displacement is of course that of the large-scale billboard or electronic screen. Times Square in New York is full of them. Displacing these to the screen of a mobile phone, for example, would dramatically reduce the visual overload of this part of the city. Replacing the super-bright LEDs of electronic screens with infrared LEDs only visible to CCD cameras (such as the ones on mobile phones or digital cameras) would be one method by which these displays could remain viewable on demand.[6]

Building Signage

Who needs a sign announcing the name of their store when shoppers are directed turn-by-turn to it via Google Navigator? Clearly stores will miss the casual customer passing by who wasn't looking for them, but people with their mobile device set to "just browsing" could easily receive a more targeted message, that is if they have their bluetooth set to "discoverable." Again, receiving a Starbucks latte discount coupon on your mobile phone while "just passing by" is not that far off.

Traffic Signals, Crosswalk and Street Signage

For the vehicular set, augmented reality (AR) windshield displays could take the place of stop-lights and street markings denoting traffic lanes and crosswalks. Of course in turn, pedestrians would need to make sure they synch their contact-lens displays[7] with the traffic management system to keep in step with the ebbs and flows of automobiles. Signs displaying parking regulations, one-way streets, and turning restrictions could be handled by a future upgrade to your "sat-nav" system. For better or worse, this would also facilitate the issuing of tickets for parking violations.

Taxi Cabs

Why waste time trying to hail a cab and deciphering the occupied/unoccupied/off duty sign mounted on its roof when you can set your mobile device to broadcast your current location, desired destination and pickup time? What's the use of cab-top advertising—no matter how well targeted geo-demographically—when these ads are much more successful when bundled with a service you already use to navigate the city and that already knows quite a bit about your purchasing power, transaction history and personal predilections?

Public Transportation Systems

Bus routes and schedules, train arrivals and departures, updated in real-time, could easily become yet another channel accessed via a mobile device. Finding your way through a transportation hub to your track or gate might prove more challenging. At what point does your route become too granular for mobile navigation systems? Research in wayfinding systems for the blind is already addressing some of these issues (Willis and Helal, 2005).

SPATIAL IMPLICATIONS

The most obvious spatial implication is, of course, a reduction in the visual overload of information encountered in dense urban environments. More specifically one can reasonably suggest that the experience of the city shifts from reading the urban space in terms of linguistic signifiers (names of streets, shops and other attractions) to apprehending its material and spatial signifiers (surface qualities, spatial proportions and scales). The presence (or lack thereof) of *people* occupying space is fore-grounded, as is the particular formal and material articulation of the buildings defining that space. If the words and images saturating the contemporary city offer portals through which we can access other "spaces"—streets named after historical figures, advertisements for travel to remote places—this De-saturated City reflects back its physical boundaries, and highlights the cues we take from their different porosities and material qualities.

Perhaps less obvious are the implications for certain spatial organizations and architectural "programs"—the set of relations between spaces assigned specific functional uses. In an *extreme* informatics regime, one class of sites that would be radically transformed is that of traditional bottlenecks in the flow of people through the city. Ticket counters and subway turnstiles at transportation hubs, for example, become obsolete when near-field communications enable you to wave your mobile phone in front of a RFID reader attached to a computer that subsequently communicates with your bank and debits your fare. While this scenario is perhaps by now a common design cliché within interaction design circles, the implications for the architectural "programming" of these spaces remains under-examined. We have already witnessed how the introduction of the Automated Teller Machine (ATM) has reconfigured the spatial organization of banks, and it is clear we will continue to see similar transformations to come.

TECHNOLOGICAL IMPLICATIONS

Obviously, the complex technological infrastructure required for offloading *all* information from the material substrate of the physical city implies a level of systems integration an order of magnitude far beyond current capabilities. At the same time, the questions surrounding who is responsible for building this infrastructure—or infrastructures—are equally complex. Would this hypothetical future-fiction result from a state-sponsored initiative, or as a private venture such New Songdo, the model "ubiquitous city" currently under development by the Gale Corporation in South Korea (www.new-songdocity.co.kr, see also Hwang in this volume)? If so, who is granted access privileges sufficient to add, modify or delete elements of the infoscape? How is it maintained and updated? By whom? Here, projects that investigate forms collaborative urban markup such as Urban Tapestries and City Flocks might serve as useful references.

Or rather, is this De-saturated City an end-state representing a gradual process of minor displacements, where more limited technological developments render certain forms of urban information vestigial one-by-one, and over time aggregate toward the larger urban condition depicted here? If so, how would the (inevitably) exacerbated digital divide be mitigated? What standards need to be in place to ensure a reasonable level of interoperability between diverse systems and devices?

Above all, perhaps, is the problem of how to handle the inevitable system failures and data losses (Greenfield, 2006). Would we be able to manage these in the same way we do, say, power outages? Do the same techniques and methods developed for graceful system degradation within our current computing paradigm work at the scale of urban infrastructure?

SOCIAL IMPLICATIONS

At the beginning of the 20th century, Georg Simmel, an urban sociologist writing about the transformation of Berlin from an agrarian to an industrial city, notes:

The interpersonal relationships of people in big cities are characterized by a markedly greater emphasis on the use of the eyes than on the ears. This can be attributed chiefly to the institution of public conveyance. Before buses, railways and trams became fully established during the 19C, people were never put in the position of having to stare at one another for minutes or even hours on end without exchanging a word. (Simmel, 1908, p. 486)

As the quote above suggests, the relationship between technological change and the production of new social situations in urban environments is longstanding. In his seminal essay on urban life, *The Metropolis and Mental Life*, Simmel attributes certain cognitive and social shifts in the modern urban dweller to rise of the industrial metropolis. "The psychological basis of the metropolitan type of individuality consists in the intensification of nervous stimulation which results from the swift and uninterrupted change of outer and inner stimuli," he writes. "These are the psychological conditions which the metropolis creates" (Simmel 1997, 175). Among these conditions he includes the rationalization of urban life, the subjection of all social relations to capital, the production of the "blasé" attitude, and the need to increasingly specialize to remain competitive.

Yet if an extreme urban informatics regime might contribute to a reduction of information overload within physical urban environments and a corresponding reduction in the "nervous stimulation" Simmel cites as endemic to the modern metropolis, it would also introduce new psycho-social dilemmas. As information is displaced from physical to virtual environments, our attention now becomes divided not just within our field of vision, but also between two radically different fields of vision. How we will negotiate, say, viewing real-time information streams overlaid on our eyeglasses or contact lenses, while navigating the physical space we are moving through and the social encounters that transpire there remains an unresolved problem. Here, problems familiar to Human-Computer-Interaction (HCI) and Augmented Reality (AR) researchers are compounded. Displacing information from the physical to virtual will inevitably exaggerate a sense of distracted presence if the interface is not well resolved.

Further, if we look at how mobile phones or portable audio devices like the iPod have altered the social space of cities, more dilemmas arise. As Bull (2000) has shown, people use portable audio devices in a variety of ways to manage the contingencies of everyday life. On the one hand, the popularity of the iPod points toward a desire to personalize the experience of the contemporary city with one's own private soundtrack. On the bus, in the park at lunch, while shopping in the deli—the city becomes a film for which you compose the soundtrack. These devices also provide varying degrees of privacy within urban space, affording the speaker/listener certain exceptions to conventions for social interaction within the public domain, absolving them from some responsibility for what is happening around them. Talking on a mobile phone while walking down the sidewalk, text-messaging with a friend while on the bus, or listening to an iPod on the subway are everyday practices for organizing space, time and the boundaries around the body in public.

Whether these practices contribute to a "retreat" of the modern citizen from the public realm (Sennett, 1977; Putnam, 2000), or serve more to connect isolated individuals with social networks based on shared needs, interests or beliefs (Wellman, 2002), is only partly a question of design. We would be remiss if we were not to acknowledge the remarkable ingenuity by which people grapple with new techno-social situations in ways designers and developers of information systems never imagined. If space is a social product (Lefebvre, 1991), then new social practices will produce new spatial conditions. Yet it is practically impossible to predict what new socio-spatial practices will emerge within this extreme informatics regime. If reading a book, magazine or newspaper while commuting via subway became a way to mitigate the social awkwardness Simmel notes above, what new practices will emerge in this future fiction? What unforeseen awkward situations will they arise from?

CONCLUSION

In projecting a future fiction of a De-saturated City under the reign of an extreme urban informatics regime, technological development is cast in relation to larger questions surrounding the spatial, social and technological organization of urban environments. By doing so, alternate trajectories for research and development emerge that look beyond established paths to market, ones that take into account the complexity of the techno-social topographies of living cities.

While these trajectories inevitably raise more questions than they answer, they do so in an attempt to explore how the potential *impact* of information systems on urban space might *drive* future technological development, rather than merely *result* from it. Such an inversion involves integrating speculative/projective methods common to architecture and urban design with more analytical methods common to the design and development of technological systems.

Both urban informatics research as well as architecture and urban design stand to benefit from greater dialog over the opportunities and dilemmas encountered by each with regard to the larger social and cultural concerns surrounding the technological mediation of urban life.

REFERENCES

Bull, M. (2000). Sounding Out the City: Personal Stereos and the Management of Everyday Life. Oxford: Berg.

Dave, B. (Ed.). (2007). Space, sociality, and pervasive computing. Guest editor of a special issue of Environment and Planning B: Planning and Design, 34(3). London: Pion.

Ellison, N., Burrows, R., & Parker, S. (Eds.). (2007). Urban Informatics: Software, Cities and the New Cartographies of Knowing Capitalism. Guest editors of a special issue of Information, Communication & Society, 10(6). London: Routledge.

Galloway, A. and Ward, M. (2006). "Locative Media as Socialising and Spatialising Practice: Learning from Archaeology." Leonardo Electronic Almanac, Vol. 14, Issue 3/4. http://leoalmanac.org/journal/Vol_14/lea_v14_n03-04/ - last accessed February 15, 2008.

Galloway, A. (2004). "Intimations of Everyday Life: Ubiquitous Computing and the City." Cultural Studies, Volume 18, Numbers 2-3, pp. 384-408.

Greenfield, A. (2006). Everyware: The dawning age of ubiquitous computing. Berkeley: New Riders.

GPS World (2007). Wherever you go, ads will follow. GPSworld.com. March 1. http://www.gpsworld.com/gpsworld/The-Business-mdash-March-2007/ArticleStandard/Article/detail/406754 - last accessed February 15, 2008.

Habermas, J. (1991). The Structural Transformation of the Public Sphere, Cambridge: MIT Press.

IBM (2007). Press Release. http://www-05.ibm.com/de/pressroom/presseinfos/2007/03/16_2.html - last accessed February 15, 2008.

Hayles, N. K. (1999). How We Became Posthuman: Virtual Bodies in Cybernetics, Literature, and Informatics, Chicago: University of Chicago Press.

Kindberg, T., Chalmers, M., & Paulos, E. (Eds.). (2007). Urban Computing. Guest editors of a special issue of Pervasive Computing, 6(3). Washington, DC: IEEE.

Lefebvre, H. (1991). The Production of Space, Oxford: Blackwell.

Madge, C. & Harrisson, T.H., (1937). Mass Observation. London: Frederick Muller Ltd.

Putnam, R. D. (2000). Bowling alone. New York: Simon & Schuster.

RIXC (2002). http://locative.x-i.net/report.html - last accessed February 15, 2008.

Rothkerch, I. (2002). Will the future really look like "Minority Report"? Salon.com. July 20. http://dir.salon.com/story/ent/movies/int/2002/07/10/underkoffler_belker/ - last accessed February 15, 2008.

Sennett, R. (1977), The Fall of Public Man, New York: Alfred A. Knopf.

Silverstone, R., & Sujon, Z. (2005). Urban Tapestries: Experimental Ethnography, Technological Identities and Place (No. 7). London: Media@lse, Department of Media and Communications, LSE.

Shklovski, I., & Chang, M. F. (Eds.). (2006). Urban Computing: Navigating Space and Context. Guest editors of a special issue of Computer, 39(9). Washington, DC: IEEE.

Simmel, G. (1908). Soziologie: Untersuchungen über Die Formen Der Vergesellschaftung. Berlin: Duncker & Humblot.

Simmel, G. (1997). Simmel on Culture: Selected Writings, New York: Sage.

Wellman, B. (2002). Little Boxes, Glocalization, and Networked Individualism. In M. Tanabe, P. van den Besselaar & T. Ishida (Eds.), Digital Cities II: Second Kyoto Workshop on Digital Cities (Vol. LNCS 2362, pp. 10-25). Heidelberg, Germany: Springer.

West, N. (2005). Urban Tapestries: The spatial and social on your mobile (Cultural Snapshot No. 10). London: Proboscis, London School of Economics.

Willis, S. & Helal, S. (2005). RFID information grid for blind navigation and wayfinding. Proceedings. Ninth IEEE International Symposium on Wearable Computers, Washington, DC: IEEE.

Worldchanging (2007), "No Logo: São Paulo Bans Outdoor Advertising", worldchanging.com. June 30. http://www.worldchanging.com/archives/006973.html - last accessed February 15, 2008.

KEY TERMS

Geocoding: The process of associating geographic metadata (such as latitude/longitude coordinates) to physical features within natural or constructed environments.

Locative Media: Art and technology practices that employ location-aware technologies in exploring novel or unconventional relations between cartography, place, community, sociality, and technology.

Location-based Services (LBS): A combination of hardware and software technologies that incorporate mobile devices, wireless communication networks, geographic information, and software applications in serving information related to specific locations or routes.

Personal Informatics: Information services, often accessible via a mobile device, that search, sort, mine, correlate or otherwise filter information for a person based on their preferences, transaction logs, location, social networks and other personal data.

Public Authoring: The open, public process of associating virtual media objects and other types of information to actual locations in the city, whereby people are enabled to construct and contribute to this information layer, rather simply consuming pre-authored content.

Skeuomorph: Material artifacts that simulate an aspect of a previous time using a technology that has superseded it. Skeuomorphs are derivative objects that retain structurally necessary elements of the original as ornamentation, stripped of their original function.

Urban Screen: Large, programmable, outdoor electronic billboards displaying media for commercial and advertising purposes, often affixed to buildings or other prominent urban structures.

ENDNOTES

1. By "city", I here refer to dense urban agglomerations such as New York, London, Berlin, Seoul, Tokyo, São Paulo and the like, where the integration of information systems into the material fabric of the city has become commonplace.
2. See for example the Urban Screens Manchester conference (http://www.manchesterurbanscreens.org.uk/) or the Media Architecture conference (http://www.mediaarchitecture.com/), both held in the fall of 2007.
3. The Bureau of Inverse Technology (BIT) has developed BIT Cab, a system that supplements AdRunner content with an overlay of "critical location-based data such as detail of toxic residues at current GPS position of the vehicle and realtime news alerts for street-based public actions" - http://www.bureauit.org/cab/
4. While skeuomorphs sometimes also reintroduce functional value lost in the technological "upgrade"–such as the mechanical shutter sound on digital cameras making someone aware that you are photographing them–there is little that necessitates these functions being performed in the same way as they had previously.
5. Lest anyone think in today's market-driven world this agenda would be doomed to fail, on January 1, 2007, the city of São Paulo, Brazil, enacted a new 'clean city' law banning all advertising: no billboards, no fliers, no neon signs, no electronic panels. See http://www.worldchanging.com/archives/006973.html
6. See for example Osman and Omar Khan's "SEEN: Fruits of our labor" - http://www.fruitsofourlabor.org/
7. Once considered the fantasy of science fiction writers, these displays are getting closer to feasibility. See the research conducted by engineers at the University at Washington: http://www.eurekalert.org/pub_releases/2008-01/uow-clw011708.php

Afterword
Urban Informatics and Social Ontology

Roger J. Burrows
University of York, UK

Is it still the case that one can symptomatically read the early work of the cyberpunk author William Gibson as a form of prefigurative urban theory (Burrows, 1997a; 1997b)? And why would one want to? Having read the various essays in this eclectic, engaging and exciting volume I turned to Gibson in the hope that I might again find buried in his stylistic prose some hint of an analytic insight that might provide a way of satisfactorily articulating the diverse concerns expressed within these pages. Gibson did not let me down. His most recent novel—*Spook Country* (Gibson, 2007)—is, as always, about many things, but at its core it is a novel of ideas about the social and cultural consequences of a whole assemblage of urban informatics technologies—locative technologies in particular. However, although the substantive concerns of this volume and his most recent novel are homologous, it was a passing exchange between two of the main characters about the changed nature of social ontology that made me realise why the study of urban informatics is as important as it is. The exchange occurs on page 103 of the novel.

Hubertus Bigend, a Belgian born Situationist inspired founder of a viral advertising agency called Blue Ant is talking to a woman, Hollis Henry, a former member of an early 1990s cult rock band, but now a freelance journalist supposedly researching an article about locative technology in the art world. They are in a hotel bar. The exchange is as follows:

'The pop star, as we knew her'—and here he bowed slightly, in her direction—'was actually an artefact of preubiquitous media'

'Of-?'

'Of a state in which "mass" media existed, if you will, within the world.'

'As opposed to?'

'Comprising it.'

Now this short exchange has a deep resonance with various analytic materials I have encountered over the last year or so, all concerned with what the cultural theorist Scott Lash (2007a) has recently identified as the emergence of a New 'New Media' Ontology; a situation in which '[w]hat was a medium [...] has become a thing, a product' (Lash, 2007b: 18). For Thrift and French (2002: 309), for instance, 'the technical substrate of Euro-American societies ... has changed decisively as software has come to intervene in nearly all aspects of everyday life'. For these commentators, software now increasingly functions in order to provide a 'new and complex form of automated spatiality... which has important consequences for what we regard as the world's phenomenality'. In much more basic terms, the 'stuff' that makes up the social and urban fabric has changed—it is no longer just about emergent properties that derive from a complex of social associations and interactions. These associations and interactions are now not only *mediated* by software and code, they are becoming *constituted* by it.

The study of urban informatics then, is becoming the study of the emergence of this new social ontology. But this observation is too abstract. What I want to do in this short concluding contribution to this volume is to give some shape and form to this new social ontology through some very recent writings on urban informatics that the various contributory authors here would not have had an opportunity to read at the time they were crafting their chapters (although the chapter by Mark Shepard on 'Extreme Informatics' is ahead of the game here as it does touch on some of this literature). In particular I want to examine a couple of related typologies of different zones of this new social ontology as revealed by recent studies of urban (Crang and Graham, 2007) and domestic (Dodge and Kitchin, 2008) informatics systems.

Crang and Graham (2007), in a far reaching review of the different ways in which software and code mesh with various aspects of the urban environment, develop a three-fold categorisation of different regions of this new social ontology, what they term: *augmented* space; *enacted* space; and *transducted* space.

Augmented space is, in some ways, the most visible but the least interesting. It is based on the recognition that the built environment has long been saturated with information from signage and adverts, but that much of this information is changing from analogue to digital forms. Augmented spaces then are simply physical objects overlain with virtual objects, but virtual objects increasingly able to alter *dynamically*. As Crang and Graham view it, this notion of augmentation simply reflects the observation that new digital media are being *added* to the experiences of urban life without a qualitative alteration in the emergent properties of urban systems. This then is digital information *superimposed* on physical form.

Enacted space is rather different. This refers to environments in which coded devices—RFIDs might be thought of as being paradigmatic—do not just possess additive effects, but come to inhabit 'the most ordinary of things' (Crang and Graham, 2007: 793) and are able to produce more than just enhancements to spaces; rather they relocate agency. This then is the vision of social ontology envisaged by Bill Mitchell (2003) in his popular articulation of the spatially extended cyborg—the cyborg self in the ubiquitously networked city—or what Cuff (2003) has termed the 'cyburg'.

Transducted space is different again, and is concerned not just with the relocation or spatial extension of *human* agency but with the potentialities of *technological* agency *per se* (Dodge and Kitchen, 2005). This is about the productive power of technology to make things happen via reiterative, transformative or recursive practices. This is what Lash (2007c: 71) calls '[p]ower through the algorithm... a society in which power is increasingly in the algorithm'. This is manifest in situations where 'codes offer modes of address—both locating and hailing people and things' (Crang and Graham, 2007: 794) in order to form what Thrift (2004: 177) urges us to think of as a 'technological unconscious'. This then is about processes of automated spatiality (Thrift and French, 2002) as manifest, for example, in the recursive functioning of geodemographic classifications (Parker *et al.*, 2007).

But what is the ontological status of this code? According to Dodge and Kitchin (2004) coded objects summon two different forms of space into existence—what they term *code/space* and *coded space*. They define code/spaces as 'spaces dependent on software to function; that is the relationship is dyadic...[w]ithout software enabled technologies the space would not be produced as intended' (Dodge and Kitchin, 2008: 11). For instance, a web browser unable to connect to a wifi service is simply unable to function. Coded space, on the other hand, is thought of as 'a spatial transduction that is mediated by coded processes, but whose relationship is not dyadic' (Dodge and Kitchin, 2008: 11). Thus some software enabled technologies are able to 'produce particular spatialities, but if they are not present or operative a space is still produced as intended but less efficiently or cost-effectively' (Dodge and Kitchin, 2008: 11). They give the example of a software controlled burglar alarm system failing, but the house remaining (albeit less) safe than if the software had been functioning as intended.

This distinction between different forms of transduced space is useful, but Dodge and Kichin (2008) go further by making another set of distinctions that further refine the role of code within the social ontology of urban informatics. They differentiate first between what they term *unitary coded objects* (UCOs) and contrast these with what they term *logjects*. UCOs are material objects that rely on code to function but do not keep any record of their actions. Logjects, in contrast, are material *objects* that also rely on code to function but, in addition, they possess the ability to make a record—or a *log*—of their actions.

UCOs are of two types: those that function independently of their environment, and those that gather data from their environment and use this in order to function. Examples of the former might include (now) mundane objects such as digital watches, DVD players or USB sticks. The latter group might include (more recent) objects such as digital heating control systems, advanced software saturated washing machines or digital cameras with automatic settings.

Logjects are more interesting, and are likely to become ever more important elements of the social ontology of the advanced information age. The notion is inspired by Bleecker's (2006) more restricted conceptualisation of *blogjects*, but for Dodge and Kitchin (2008: 6) a blogject is just one type of a more general category of logject. For them logjects are coded objects which 'differ from unitary objects in that they not only sense the world but also record their status and usage, and, importantly, can retain these logs even when deactivated and utilise them when reactivated' (Dodge and Kitchin, 2008: 6). Such coded entities can thus monitor and record their own use. More formally they possess the following properties:

- they are uniquely indexical;
- they possess an 'awareness' of their environment and are able to respond to changes in that environment that are meaningful within limits;
- they produce traces and tracks of their own functioning across time and/or space;
- they record that history and can communicate it across a digital network for analysis and use by other agents (other coded objects and/or people);
- they can use the data produced in order to undertake automated, automatic and autonomous decisions and actions in the world without human oversight and thus effect change; and
- they are programmable and thus mutable through the adjustment of settings and software updates (Dodge and Kitchin, 2008: 6-7).

Such logjects are of two different types: *impermeable* and *permeable*. Impermeable logjects are relatively self-contained units such as MP3 players or PDAs. Unlike permeable logjects 'all essential capacities are held locally and primary functionality does not require network connection to operate' (Dodge and Kitchin, 2008: 7). This does not mean that episodic data and software downloading is not necessary—it is. Impermeable logjects can also be connected to wider networks if required and data can be exchanged with other coded devices *but*, and this is the key point, such interactions with wider digital networks and other coded objects/logjects are *not automatic* and they often require considerable work on the part of users, such as syncing an iPod in order to download podcasts that one might subscribe to. Permeable logjects, on the other hand, 'do not function without continuous access to other technologies and networks' (Dodge and Kitchin, 2008: 8). Some of these permeable logjects are spatially fixed, such as satellite television control boxes, whilst some, such as mobile (cell) phones are inherently mobile.

It is with the development and coming ubiquity of such permeable logjects throughout social and urban systems that we can begin to glimpse the possibility of the ontological arrival of a new object likely to be the central concern of (near) future social scientific analyses of urban informatics—the *spime*. Spimes (a neologism of *space* and *time)* are the thought 'invention' of another former cyberpunk author, Bruce Sterling (who—and in order to generate some symmetry to this short afterword—co-authored *The Difference Engine* with William Gibson in 1990). Sterling is interested in what is likely to result when permeable logjects begin to mesh with RFID and similar technologies (Beer, 2007). He suggests that spimes will be:

> *manufactured objects whose informational support is so overwhelmingly extensive and rich that they are regarded as material instantiations of an immaterial system. Spimes begin and end as data. They are designed on screens, fabricated by digital means and precisely tracked through space and time throughout their earthly sojourn.* (Sterling, 2005, p. 11)

Such entities will, of course, be 'eminently data-minable' (Sterling, 2005: 11) to the extent that their value will more often than not be in the extractable information they contain rather than in the physical spime itself. In Sterling-speak: 'in an age of spimes, the object is no longer an object, but an instantiation' (Sterling, 2005: 79). The resonance with my opening quote from Gibson is, I hope, obvious as are some of the potential social consequences and links with ongoing debates about cultures of *inter alia*: surveillance; privacy; visibility; anticipation; risk; mobility; and even perhaps the category of the post-human.

So here we have a new argot for the study of urban informatics. It is not a difficult task to map the concerns of most of the chapters presented in this volume to some combination or other of this variable new 'new media' ontology. In general we can ask of the substantive and conceptual concerns of each:

- Is the concern with *augmented, enacted* and/or *transducted* space?
- Is this an instance of *code/space* and/or *coded space*?
- Is this a spatialization constituted by *unitary coded objects* and/or the increasingly large numbers of both *impermeable* and *permeable logjects* that now populate our homes, neighbourhoods and cities?
- Is this an instance of a prefigurative *spime*?

So, by way of example, the concerns of Andrew Hudson-Smith *et al.* with the 'neogeography of virtual cities' is with a form of transducted space. The concerns of Barbara Crow *et al.* with 'voices from beyond' are analyses of UCOs. The work of

Kostakos and O'Neill on 'cityware' is concerned with logjects, as is that of Kloeckl and Ratti on 'wikicity', whilst the work of Eric Paulos *et al.* on 'citizen science' is explicitly concerned with both UCOs and logjects. Perhaps not surprisingly however analyses of spimes are not so apparent. This will have to be the focus of a follow on volume to this in a year or so, as these (currently) imaginary entities begin to solidify ontologically in all of our social worlds?

REFERENCES

Beer, D. (2007). Thoughtful territories: Imagining the thinking power of things and spaces. *City, 11*(2), 229-238.

Bleecker, J. (2006). *A manifesto for networked objects - cohabiting with pigeons, arphids and aibos in the Internet of things.* Available at: http://www.nearfuturelaboratory.com/index.php?p=185

Burrows, R. (1997a). Cyberpunk as social theory: William Gibson and the sociological imagination. In S. Westwood and J. Williams (eds) *Imagining Cities: Scripts, Signs, and Memories.* London: Routledge, 235-248.

Burrows, R. (1997b). Virtual culture, urban social polarisation and social science fiction. In B. Loader (ed), *The Governance of Cyberspace.* London: Routledge, 38-45.

Crang, M., & Graham, S. (2007). Sentient cities: Ambient intelligence and the politics of urban space. *Information, Commuication, and Society, 10*(6), 789-817.

Cuff, D. (2003). Immanent domain: Pervasive computing and the public realm. *Journal of Architectural Education, 57*(1), 43-49.

Dodge, M., & Kitchin, R. (2005). Code and the transduction of space. *Annals of the Association of American Geographers, 95*(1), 162-180.

Dodge, M. & Kitchin, R. (2008). *Software, Objects and Home Space.* NIRSA Working Paper Series No 35. National Institute for Regional and Spatial Analysis, National University of Ireland, Maynooth.

Gibson, W. (2007). *Spook Country.* London: Viking.

Lash, S. (2007a). New media ontology. Paper Presented to the *Towards a Social Science of Web 2.0 Conference*, York, UK, Sept 2007.

Lash, S. (2007b). Capitalism and metaphysics. *Theory, Culture, & Society, 24*(5), 1-26.

Lash, S. (2007c). Power after hegemony: Cultural studies in mutation. *Theory Culture Society, 24*(3), 55-78.

Michell, W. (2003). *Me++: The cyborg self and the networked city.* Cambridge, MA: MIT Press.

Parker, S., Uprichard, E., & Burrows, R. (2007). Class places and places Classes: Geodemographics and the spatialisation of class. *Information, Communication and Society, 10*(6), 902-921.

Sterling, B. (2005). *Shaping Things.* Cambridge, Massachusetts: The MIT Press.

Thrift, N. (2004b). Remembering the technological unconscious by foregrounding the knowledges of position. *Environment & Planning D: Society and Space, 22*(1), 175-190.

Thrift, N. & S. French (2002). The automatic production of space. *Transactions of the Institute of British Geographers,* 27(4), 309-335.

About the Contributors

Barbara Adkins is education director at the Australasian Cooperative Research Centre for Interaction Design (ACID). Her research brings together insights from sociology, urban and design studies and sociocultural aspects of interaction design. A key focus of her sociological work has been in studies of organisations, inequality, difference and identity, and the sociology of culture, applying the work of Pierre Bourdieu. Her background is based on expertise in ethnographic case study research and qualitative approaches to data collection and analysis. She has applied these theoretical, substantive and methodological insights to the field of housing and urban development, gaining research funding from the Australian Housing and Urban Research Institute and the Office for the Status of Women on issues related to housing vulnerability. More recently she has focused on the relationship between design and disability, examining person-environment relationships experienced by people with cognitive impairment and the development of assistive technologies for this group. She is currently chief investigator on a 3 year project studying new media in a new urban village and a range of ACID projects leading ethnographic research into the social relationships involved in new applications in interaction design.

Mike Ananny is a doctoral candidate and Trudeau Scholar in Stanford University's Department of Communication where he researches technology-supported public communication and, specifically, relationships between journalism and new media design practices. He holds a bachelor of science from the University of Toronto, a master of science from the MIT Media Laboratory. He was a founding member of the researcher staff at Media Lab Europe and has consulted for or worked with LEGO, Mattel and Nortel Networks, helping to translate research concepts and prototypes into new products and services.

Michael Batty is Bartlett Professor of Planning at University College London where he is director of the Centre for Advanced Spatial Analysis (CASA). From 1979 to 1990, he was professor of City and Regional Planning in the University of Cardiff and from 1990 to 1995, director of the National Center for Geographic Information and Analysis in SUNY-Buffalo. His most recent book is *Cities and Complexity* (MIT Press, 2005). He is editor of *Environment and Planning B* and PI of the NCeSS GeoVUE project. He is a fellow of the British Academy and received the CBE in 2004 for 'services to geography'.

Viktor Bedö currently holds a research scholarship from the Deutscher Akademischer Austauschdienst (German Academic Exchange Service) and is hosted by the Hermann von Helmholtz-Zentrum für Kulturtechnik in Berlin, Germany. After graduating in philosophy from the University Vienna, Austria, he was junior researcher at the Institute for Philosophical Research of the Hungarian Academy of Sciences in Budapest from 2003 to 2006. Since 2005 he has been a PhD student at the Doctor School for Philosophy, University of Pécs, Hungary. As a holder of the Eötvös Scholarship he was a visiting researcher at the Berlin-Brandenburg Academy of Sciences and Humanities in the summer term 2007.

Genevieve Bell is the director of User Experience in Intel Corporation's Digital Home Group where she manages an interdisciplinary team of social scientists and designers. A cultural anthropologist by training, her work explores the relationships between new information and communication technologies and every day social practice. Raised in Australia, Bell earned a PhD in cultural anthropology from Stanford University in 1998.

Wayne Beyea is the associate director of Citizen Education and Statewide Coordinator of the Citizen planner Program at the Land policy institute at Michigan State University. Beyea has over 17 years experience in the fields of land use planning, and community and economic development. Beyea holds a BS in urban planning from Michigan State University and a master of public administration from the University of Maine. He is a certified planner through the American Institute of Certified Planners and will complete a juris doctor degree at the Michigan State University College of Law in May, 2008.

Roger Burrows is professor of sociology and co-director of the Social Informatics Research Unit (SIRU) in the Department of Sociology at the University of York in the UK. He has published widely on a number of different topics, but most recently he has concentrated on topics related to cities, social informatics, and health and illness. Between 2005 and 2007 he was the co-ordinator of the UK Economic and Social Research Council (ESRC) e-Society Research Programme—the largest programme of social science research on digital technologies ever funded in the UK.

Francesco Calabrese is postdoctoral associate at the SENSEable City Laboratory of the Massachusetts Institute of Technology, Cambridge, Massachusetts, USA. He received a Laurea (BS and MS) degree in computer engineering, cum laude, in 2004, and a PhD in computer and system engineering at the University of Naples Federico II, Italy, in 2007. During 2006 and 2007, he was also research assistant at the SENSEable City Laboratory of the Massachusetts Institute of Technology, Cambridge, MA, USA, with a Tronchetti Provera fellowship, major recognition awarded by the Italy-MIT Consortium. His research interests include hybrids control systems, embedded control systems, CNC machines, real-time analysis of telecom systems and urban dynamics. Calabrese is a member of the IEEE and the Control Systems Society.

John M. Carroll is Edward M. Frymoyer Chair Professor of information sciences and technology at Pennsylvania State University. His research interests include methods and theory in human-computer interaction, particularly as applied to networking tools for collaborative learning and problem solving, and design of interactive information systems. He has worked in community informatics for the past 15 years. Recent books include *Making Use* (MIT Press, 2000), *HCI in the New Millennium* (Addison-Wesley, 2001), *Usability Engineering* (Morgan-Kaufmann, 2002) and *HCI Models, Theories, and Frameworks* (Morgan-Kaufmann, 2003). He serves on several editorial and advisory boards and is editor-in-chief of the *ACM Transactions on Computer-Human Interactions*. He received the Rigo Award and CHI Lifetime Achievement Award from the Association for Computing Machinery (ACM), the Silver Core Award from International Federation of Information Processing (IFIP), and the Alfred

N. Goldsmith Award from the Institute of Electrical and Electronics Engineers (IEEE). He is a fellow of the ACM, the IEEE, and the Human Factors and Ergonomics Society.

Luis A. Castro is a doctoral student (PhD) at the Manchester Business School of the University of Manchester. His research interests include ubiquitous computing, location aware computing, neural networks and HCI. Castro has a bachelor in information technology from the Institute of Technology in Mexicali, Mexico and a masters in computer science from CICESE, Mexico. Castro participated in a project called "*The Family Newspaper*" aimed to support the relationship maintenance for older Mexican people and their family living abroad using a web-based communication technology. Castro has also participated in international competitions for technology design sponsored by ACM SIGCHI. He is a student member of the ACM.

Jaz Hee-Jeong Choi is a doctoral candidate in the Creative Industries Faculty at Queensland University of Technology. Her research interests are in playful technology, particularly the ways in which various forms of playful interaction are designed, developed, and integrated in an Asian context. Her current research is on the *trans-youth* mobile play culture of South Korea at the intersection of play, culture, technology, people, and urban environment. Her website is located at www.nicemustard.com

Fiorella De Cindio is associate professor at the Computer and Information Science Department of the University of Milano, where she teaches courses on programming languages, distributed systems design and virtual communities. Her research interests includes Petri nets as concurrency theory, programming languages and the applications of the ICTs to support life and work within social and office systems. In 1994, she promoted the Civic Informatics Laboratory (LIC) and set up the Milano Community Network (RCM). She also promoted the Association for Informatics and Civic Networking of Lombardy (A.I.Re.C.) which groups the community networks in the Lombardy Region. Fiorella De Cindio is now president of both.

Barbara Crow is an associate professor and the Graduate Program Director in Communication and Culture at York University in Toronto. Her research interests include digital technologies, feminist theory, social movements, and the political economy of communication.

Joel Dearden is a research assistant in the Centre for Advanced Spatial Analysis (CASA) at University College London. He was previously systems administrator in CASA and is mow working on the porting for 3D and multimedia into *Second Life*, a project focussed on the development of geographic media in virtual worlds. He has worked on the development of graphics games and was trained in computer science at Imperial College London.

Jean-François Doulet, associate professor, has been exploring urban change in China for many years. He is more specifically interested in understanding how increasing mobility is transforming daily life in cities and the production of urban space. His bibliography is largely dedicated to automobile use and innovative mobility solutions in China and abroad. He co-authored (with historian Mathieu Flonneau) *Paris-Pékin, civiliser l'automobile* (2003), a comparative study on the impact of automobile use on urban mobility schemes in Paris and Beijing. In addition, he heads the China Programme of the Paris-based think tank City on the Move Institute (www.city-on-the-move.com). Since 2006, he has been working, as a scientific advisor, in shaping research activities on ICT usage in Chinese cities within Orange Labs Beijing (France Telecom Group). http://www.villeschinoises.com

Ines Di Loreto, degree in philosophy at the University of Milan, is now a PhD student in computer science at the University of Milan. Her research interests include social media and their societal impact. In particular, she investigates the relationship between ICTs and representation of the self, analyzing how representations—and the resulting relationships build through them—are constructed in the age of Web 2.0.

Paul Dourish is professor of Informatics at University of California, Irvine, with courtesy appointments in computer science and in anthropology. His research interests lie at the intersection of computer science and social science, with particular emphasis on human-computer interaction and ubiquitous computing. His empirical and conceptual investigations focus on information and communication technologies as sites of social and cultural production.

Laura Forlano is a visiting fellow at the Information Society Project at Yale Law School and a PhD candidate in communications at Columbia University. Her research interests include organizations, technology (in particular, mobile and wireless technology) and the role of place in communication, collaboration and innovation. Forlano is an adjunct faculty member in the Design and Management department at Parsons and the graduate programs in International Affairs and Media Studies at The New School. She serves as a board member of NYCwireless and the New York City Computer Human Interaction Association. Forlano received a masters in international affairs from Columbia University, a diploma in international relations from The Johns Hopkins University and a Bachelor in Asian Studies from Skidmore College.

Craig H. Ganoe is an instructor in information sciences and technology at Pennsylvania State University. His research interests include multiple-device interactions, computer supported cooperative work and learning, collaboration in community network contexts, and ubiquitous computing. He leads the BRIDGE Tools open source project (bridgetools.sourceforge.net). He is a member of the Association for Computing Machinery (ACM) and currently serves as information director for the *ACM Transactions on Computer-Human Interactions*.

Jens Geelhaar is dean of the Media Faculty at the Bauhaus University of Weimar, Germany. He is professor of interface design and specialises in art and design aspects of interaction in ubiquitous computing environments with multimodal interfaces. He studied at the Universities of Karlsruhe and Heidelberg and at the Art School in Saarbrücken, Germany.

Christine Geith is an assistant provost and executive director of Michigan State University's MSUglobal, the university's entrepreneurial business unit that works with academic partners across the campus and worldwide to develop online institutes, programs and services. Geith's publications and research include costs, benchmarks and business models for online and blended learning. Geith's 20-years of experience in educational technology and online learning includes the role of executive director of e-learning and co-director of the Educational Technology Center at Rochester Institute of Technology. Geith holds an MBA from Rochester Institute of Technology and a PhD from the University of Nebraska-Lincoln.

Victor M. Gonzalez is a lecturer (assistant professor) in human-computer interaction at the Manchester Business School of the University of Manchester. He is also a senior research fellow of Centre for Research on Information Technology and Organizations(CRITO) at the University of California at Irvine, USA. He conducts investigations on technology usage in the home, office and hospital settings and specializes in the use of ethnographic methods (e.g. interviews, participant observation) and participatory design techniques (e.g. scenario-based design). He received a PhD and master's degree in information and computer science from the University of California at Irvine and a master's degree in telecommunications and information systems from the University of Essex, United Kingdom. He is a member of ACM SIGCHI and vice-president of SIG-CHI Mexican Chapter.

Adam Greenfield is an instructor at New York University's Interactive Telecommunication Program, where he teaches urban computing. He is author of the 2006 *Everyware: The dawning age of ubiquitous computing* and co-author, with Mark Shepard, of the recent *Urban Computing and Its Discontents*. He lives and works in New York City with his wife, artist Nurri Kim.

Greg Hearn is research professor in the Creative Industries Faculty at QUT. His work focuses on policy development and research and development for new technologies and services in the creative industries. In 2005 he was an invited member of a working party examining the role of creativity in the innovation economy for the Australian Prime Minister's Science Engineering and Innovation Council. He has authored or co-authored over 20 major research reports and a number of books, including, *Public policy in knowledge-based economies* (2003: Edward Elgar), and T*he knowledge economy handbook* (2005: Edward Elgar).

Sonya Huang is a PhD candidate in urban information systems at the Massachusetts Institute of Technology. Her research interests include the impacts of information and communication technologies on economic geography and the nature of firms and labor, policy implications of georeferenced data, and methods for presenting geospatial data using open-source tools. She received an SB in economics from MIT and joined the SENSEable City Laboratory in 2005.

Andy Hudson-Smith is senior research fellow in CASA in University College London where he leads both the NCeSS GeoVUE and Virtual London projects. His focus is on Web 2.0 technologies in geography and urban planning which he has researched since his PhD. His blog http://www.digitalurban.blogspot.com features daily news and research on the world of urban visualisation. He is the author of *Digital Geography: Geographic Visualisation of Urban Environment* (CASA, UCL, 2008).

R. J. Honicky is a PhD student in the computer science department at UC Berkeley, and a member of the Technology and Infrastructure for Emerging Regions (TIER) research group. His dissertation work focuses on building a distributed scientific instrument by integrating environmental sensors into cell phones. He also studies various low cost wireless and networking technologies. He has worked at Intel Research, Tensilica, Microsoft Research, Network Appliance, Airtouch Cellular (now AT&T) and various startups. He has a master's of science degree in computer science from UC Santa Cruz, and BA in English literature from the University of Michigan.

Ben Hooker is a multimedia designer whose work explores new experiences and aesthetic situations which arise from the intermingling of the phenomenal and intangible worlds of physical materiality and electronic data. His background is computer-based multimedia design. He graduated from the Royal College of Art's computer related design program in 1997, and after this worked for several years as freelance designer for clients such as the BBC, Sony and Philips, while also continuing at the college as a researcher in the Interaction Design Research Studio. Today Hooker divides his time between creative practice, consultancy and teaching. He currently holds a visiting faculty position at Intel's Research Lab in Berkeley, California. Alongside this he continues to work with long-time collaborator Shona Kitchen. Recent projects they have completed include a multi-site electronic installation for San José International Airport and a conceptual housing project for the Vitra Design Museum.

Jong-Sung Hwang is the executive vice president of the National Information Society Agency (NIA), heading the IT Policy Division. Since he joined NIA in 1995, he has developed Korea's national IT strategies, including the Informatization Promotion Basic Plan in 1996, the Cyber Korea 21 in 1999, and the u-Korea Master Plan in 2006. Since 2005, he has been in charge of promoting ubiquitous computing services, including RFID, USN and u-City. Hwang received his PhD in political science from Yonsei University, Korea in 1994. He is serving as an adjunct professor of the Graduate School of Information, Yonsei University.

Toru Ishida (doctor of engineering, Kyoto University) is a professor in the Department of Social Informatics, Kyoto University. Until 1993, he was a research scientist at the NTT Laboratories. He spent some time at Columbia University, Technische Universität München, Universite Pierre et Marie Curie, University of Maryland, Shanghai Jiao Tong University, and Tsinghua University as a visiting scholar/ professor. He is an IEEE fellow from 2002. He has been working on autonomous agents and multiagent systems for twenty years. He also studies social informatics and leads research projects such as digital cities, language grid and intercultural collaboration.

Katrina Jungnickel recently completed her PhD at Incubator for Critical Inquiry into Technology and Ethnography (INCITE) at the Goldsmiths College, University of London. Her doctoral work on volunteer community wireless groups in Australia (http://www.studioincite.com/makingwifi) presents a methodological and theoretical intervention into DIY practice, visual methodology and the sociological accounts of information and communication technologies.

Michael Keane's research interests include creative industries internationalisation and innovation in China; audio-visual industry policy and development in China, South Korea, and Taiwan; and television formats in Asia. He is the author of *Created in China: the Great New Leap Forward* (2007), a study of China's creative economy, and how television, animation, advertising, design, publishing and digital games are reshaping traditional understanding of culture. His most recent co-authored book (with Anthony Fung and Albert Moran) is *New Television, Globalization and the East Asian Cultural Imagination* (2007), a major study of the evolving landscape of television in China, Hong Kong SAR, South Korea, Japan and Taiwan. http://www.cci.edu.au/profile/michael-keane

Helen Klaebe is a senior research fellow at QUT. Her PhD examined new approaches to participatory public history using multi art form storytelling strategies. She is the author of: *Onward Bound: the first 50 years of Outward Bound Australia* (2005); and *Sharing Stories: a social history of Kelvin Grove*

(2006). Klaebe also consults as a public historian, particularly focusing on engaging communities of urban renewal projects, and regularly designs and manages co-creative media workshops for a range of commercial and public sector organisations.

Kristian Kloeckl graduated at the Politecnico di Milano, Italy, with distinction after having attended architecture and industrial design degree courses in Austria, England and Italy. He is currently enrolled in the design science PhD program at the Iuav University of Venice, Italy, and a visiting PhD student at the SENSEable City Laboratory of the Massachusetts Institute of Technology, Cambridge, MA, USA. His research interests have covered areas such as product design in the urban context, medical design and connections design while he is also holding design classes in various institutes. Besides his academic work he has been collaborating with leading design studios in Berlin and Milan before setting up his own design practice in 2003 in Venice.

Satoshi Koizumi is an associate professor at Osaka University. His research interests include computer vision and virtual reality. He received a B.Eng. and M.Eng. in mechanical engineering from Aoyama Gakuin University, and a D.Eng. in computational intelligence and systems science from the Tokyo Institute of Technology.

Vassilis Kostakos is an assistant professor at the Department of Mathematics & Engineering, University of Madeira, and an adjunct assistant professor at the HCI Institute, Carnegie Mellon University. He previously was appointed research associate at the Department of Computer Science, University of Bath. His research interests include mobile and pervasive computing, complex and social networks, security and privacy. Kostakos received a PhD in computer science from the University of Bath.

Kenneth L. Kraemer is research professor of information systems and co-director of the Center for Research on Information Technology and Organizations (CRITO), at the Paul Merage School of Business, University of California, Irvine. His research interests include the social implications of IT, national policies for IT production and use (*Asia's Computer Challenge*, Oxford 1998), and the contributions of IT to productivity and economic development. His recent book is *Globalization of E-Commerce* (Cambridge University Press, 2006). He is engaged new work on the offshoring of knowledge work and who captures the value from innovation radical and incremental innovations.

Michael Longford recently joined the Department of Design at York University in Toronto. His creative work and research activities reside at the intersection of photography, graphic design and digital media. He recently completed a three-year project as the co-principal investigator for the Mobile Digital Commons Network (MDCN), a national research network developing technology and media rich content for mobile devices. He is a founding member of Hexagram: Institute for Research and Creation in Media Arts and Technologies in Montreal, and served for three years as the director for the Advanced Digital Imaging and 3D Rapid Prototyping Group.

Charles McKeown facilitates the establishment of a permanent campus-based land use modeling and outreach initiative at the MSU Land Policy Institute. Along with a multi-disciplinary team he helps develop predictive tools for land use planning and education. Chuck holds two degrees from Michigan State University, a bachelor of science degree in entomology, and a master of science degree in entomology, specializing in ecology. He is also a veteran of the U.S. Marine Corps.

Richard Milton is a research fellow in the Centre for Advanced Spatial Analysis (CASA) at University College London where he works on the NCeSS GeoVUE project. He previously worked on the Equator project where he used GPS tracked sensors to make fine-scale maps of carbon monoxide around the local area. While at CASA he has released the 'GMapCreator' software for creating thematic *Google Maps*, the 'Image Cutter' for publishing large photos on the web and the 'PhotoOverlayCreator' for turning panoramic images into photo overlays for *Google Earth*. He was trained in information systems engineering at Imperial College London.

Colleen Morgan is a research associate at the Australasian CRC for Interaction Design in Brisbane, Australia. Morgan received her bachelor of creative industries and first class communication design honours degree from Queensland University of Technology. Conducting practiced-based research, Morgan designs and implements ICT urban interventions, exploring how they can foster community sentiment and social capital.

Hideyuki Nakanishi is an associate professor in the Department of Adaptive Machine Systems, Osaka University. He received a B.Eng. and M.Eng. in computer science, and a PhD in informatics from Kyoto University in 1996, 1998 and 2001 respectively. He was an assistant professor in the Department of Social Informatics, Kyoto University from 2001 to 2005. At the time he developed a social interaction platform called FreeWalk in the Digital City Project. He has been working on virtual social interaction including virtual casual meetings, agent-mediated communities, and transcendent communication.

Eric Paulos is a senior research scientist at Intel in Berkeley, California where he is the founder and director of the Urban Atmospheres research group—challenged to employ innovative methods to explore urban life and the future fabric of emerging technologies across public urban landscapes. His areas of expertise span a deep body of research territory in urban computing, sustainability, green design, environmental awareness, social telepresence, robotics, physical computing, interaction design, persuasive technologies, and intimate media. Eric is a leading figure in the field of urban computing and is a regular contributor, editorial board member, and reviewer for numerous professional journals and conferences. He received his PhD in electrical engineering and computer science from UC Berkeley where he helped launch a new robotic industry by developing some of the first Internet tele-operated robots including space browsing helium filled blimps and personal roving presence devices (PRoPs).

Cristian Peraboni is a PhD student in computer science at the University of Milan. He took his degree in computer science at the Department of Informatics and Communications of the University of Milan in April 2005 with the thesis *An Ontology of Computer Science to Retrieve Learning Objects*. Now he is involved in the e21 project that is aimed at overcome the hindrances to participation typical of local agenda 21 processes by creating a social environment on a custom-designed, dedicated online-deliberation platform. His research interests include virtual communities, e-Participation, e-Deliberation, Web 2.0, online social networks, folksonomies and knowledge management.

Debra Polson is an academic and independent designer, focused on exploiting unique aspects of digital games and creative social networks. Polson holds an academic position at Queensland University of Technology, Communication Design Department and has recently been seconded to the Australasian CRC for Interaction Design (ACID) as senior research fellow. In this position Polson has led a number of multi-discipline research and development projects, such as CIPHER CITIES, MILK and SCAPE that endeavour to exploit the potentials of computer games and mobile technologies to improve human relationships in environments such as schools, museums and everyday public places.

Nancy Odendaal is currently a senior lecturer in urban and regional development planning at the School of Architecture, Planning and Housing at the University of KwaZulu-Natal in Durban, South Africa. She is also a PhD candidate in the School of Architecture and Planning at Wits University in Johannesburg. Before joining the University full-time in 2001, she worked on the EU-funded Cato Manor Development project in Durban for five years as a planner and manager of the Cato Manor Development Association's Geographic Information System. Prior work experience includes involvement in a number of planning and development projects in Namibia and Swaziland, working in the public and private sectors. In addition to her academic work, she has consulted on a range of projects for municipalities and various national government departments in South Africa.

Eamonn O'Neill is a senior lecturer and director of postgraduate research studies in the Department of Computer Science at the University of Bath. His research interests include mobile and pervasive computing, participatory design, and technological support for creativity. O'Neill received a PhD in computer science from the University of London.

Jenny Preece is professor and dean of the College of Information Studies—Maryland's iSchool—at the University of Maryland. Preece's teaching and research focuses on the intersection between information, community and technology. She is particularly interested in community participation on- and off-line and social computing. She has researched ways to support empathy online, patterns of online participation and what makes technology-supported communities successful. Preece is author of over two hundred articles that include refereed journal and conference proceedings and eight books. Her two most recent books are: *Online Communities: Designing Usability, Supporting Sociability* (2000) and a co-authored text entitled *Interaction Design: Beyond Human-Computer Interaction* (1st Ed. 2002; 2nd Ed. 2007). Both books are published by John Wiley & Sons. She is also a regular conference keynote speaker.

Carlo Ratti is associate professor of the practice of urban technologies at the Massachusetts Institute of Technology. An architect and engineer by education, he directs the SENSEable City Laboratory, a new research initiative that explores how technology is transforming urban design and living. Ratti is also founding partner and director of carlorattiassociati, a rapidly growing architectural practice that was established in Turin, Italy, in 2002. Ratti graduated in structural engineering from the Politecnico di Torino and the Ecole Nationale des Ponts et Chaussées, later specializing in architecture with MPhil and PhD degrees from the University of Cambridge. He is a member of several professional organizations, including the Ordine degli Ingegneri di Torino, the Association des Anciens Eleves de l'Ecole Nationale des Ponts et Chaussees and the UK Architects Registration Board.

Erica Robles is a PhD candidate in the Department of Communication at Stanford University. Her research focuses on intersections between media technologies and the built environment. Posing research questions through diverse methodologies, from controlled laboratory experiments to archival work, interviews, and ethnographic observation she articulates both psychological and cultural components at play within contemporary mediaspaces. She is currently completing a dissertation about the Crystal Cathedral, a pioneering and influential megachurch and media ministry renown for its use of technologies and transparent architectures in the worship space.

Christine Satchell is a senior research fellow at Queensland University of Technology. She is the recipient of an Australian Postdoctoral Fellowship (Industry) supported under the Australian Research Council's Linkage funding scheme (LP0776341). Her project *Swarms in Urban Villages: New Media Design to Augment Social Networks of Residents in Inner-City Developments* informs the design of Web and mobile technology to support social networks of urban residents. She is also an honorary research fellow with the Interaction Design Group at The University of Melbourne. Her research is concerned with understanding the social and cultural nuances of everyday user behaviour in order to inform design. Integral to this is the development of a methodological approach that embeds cultural theory within human computer interaction. A specific focus of her research is the design of mobile artefacts. She is also the developer of the Swarm, a patented mobile phone prototype that allows the user to simultaneously represent multiple digital identities and embed their virtual presence with digital content.

Kim Sawchuk is an associate professor in the department of communication studies at Concordia University in Montreal, where she teaches courses in research methodologies, communications theory, and feminist media studies. She is a founding editor of *Wi: a journal of mobile digital commons network* and is the current editor of the *Canadian Journal of Communication*.

Andres Sevtsuk is a PhD candidate in city design and development and urban information systems at the Massachusetts Institute of Technology. His research interests include accuracy and reliability in using mobile phone data for estimating the distribution of people in cities, the relationship between urban form and movement, and the study of new possibilities in street parking and one-way vehicle rentals using distributed computation and sensing. He received his MS in architecture studies from MIT.

Kai Schubert studied modern and recent history and print media technology at Chemnitz University of Technology in Germany. He wrote his master thesis in Krakow (Poland) to the history of the "Institut für deutsche Ostarbeit" in the Second World War. Currently he is research associate at the University of Siegen.

Dan Shang is sociology researcher and service development manager at Orange Lab Beijing (France Telecom Group); her research focus is on urban mobility and online communities. She is especially interested in how mobile and urban technologies affect the way people use space and reinvent relations between mobility and sociability.

Mark Shepard is an assistant professor at the University at Buffalo, State University of New York, where he holds a joint appointment in the Departments of Architecture and Media Study and co-directs the Center for Virtual Architecture. His research focuses on the implications of mobile and pervasive computing for medial, architectural and urban space. He is co-editor of the *Situated Technologies* pamphlet series, published by the Architectural League of New York. His recent project, the Tactical Sound Garden, has been presented at museums, galleries, conferences and festivals internationally. He holds a MS in advanced architectural design from Columbia University; a MFA in combined media from Hunter College, City University of New York; and a B.Arch from Cornell University.

Carol Strohecker is director of the Center for Design Innovation, an interinstitutional research center of the University of North Carolina and founder of Strohecker Associates, generators of tools, programs and environments for learning. She was principal investigator of the Everyday Learning research group at Media Lab Europe, the European research partner of the MIT Media Lab and worked at Mitsubishi Electric Research Laboratories and in the Human Interface Group of Sun Microsystems. She earned the PhD of media arts and sciences from the Massachusetts Institute of Technology in 1991 and the master of science in visual studies from MIT in 1986. She has served MIT's program in media arts and sciences as a lecturer and as a presidential nominee on the MIT Corporation visiting committee.

Daisuke Tamada is a masters student in Department of Adaptive Machine Systems, Osaka University. He is currently working on social interaction for collaborative geographical content generation. He received a B.Eng. in mechanical engineering from Osaka University.

Tristan Thielmann is an assistant professor of media geography at the University of Siegen, Germany. He studied media management at the University of Siegen, audiovisual media science at the University for Film and Television "Konrad Wolf" in Babelsberg, European media and cultural studies at the University of Bradford and experimental media design at the University of Arts Berlin. He received his PhD in communication studies at the Ludwig Maximilians University Munich. Since 2005 he is senior research fellow at the Collaborative Research Center FK 615 "Media Upheavals", University of Siegen, Germany. Recent publications: *Display I: analog* and *Display II: digital* (edited together with Jens Schröter), Marburg: Schüren 2006 and 2007; *Spatial Turn* (edited together with Jörg Döring), Bielefeld: Transcript 2008. See also http://www.spatialturn.de

Anthony Townsend recently joined the Institute for the Future, an independent non-profit research group based in Palo Alto, California. As a research director, he will contribute to the Institute's long-range technological forecasting programs. Prior to joining the Institute, Anthony enjoyed a brief but productive career in academia, where his research focused on the role of telecommunications in urban development and design. Between 2000 and 2004 he taught courses in geographic information systems, telematics, and urban design in two graduate schools at New York University: the Interactive Telecommunications Program in the Tisch School of the Arts, and the Urban Planning Program in the Wagner Graduate School of Public Service. During this period, he directed several major research projects funded by the National Science Foundation and Department of Homeland Security. Townsend has been a key organizer in the wireless community networking movement since 2001. He is a co-founder and advisory board member of NYCwireless, a non-profit organization that promote community broadband initiatives using unlicensed wireless spectrum. From 2002 to 2004 he was a principal of Emenity, a successful startup company that built and manages public local wireless networks in public spaces in Lower Manhattan. Anthony's work continues to develop an international focus. He has lectured and consulted throughout Asia, Europe and North America. He lived in Korea during the summer of 2004 as a Fulbright scholar, investigating that nation's rapid development of broadband technology. Anthony holds a PhD in urban and regional planning from Massachusetts Institute of Technology, a master's in urban planning from New York University, and a BA from Rutgers University. More information at http://urban.blogs.com/

Michael Veith studied english linguistics, psychologies and informatics (Magister) at the University of Siegen. His master thesis dealt especially with the question of how ethnic minorities in Germany may be better integrated into society by the help of computer supported cooperative project work. These projects at an elementary school are furthermore realized by mixing generations (parents with children) as well as socio-cultural backgrounds.

Amanda Williams is a PhD candidate at UC Irvine's Donald Bren School of Information and Computer Sciences and a member of the Laboratory for Ubiquitous Computing and Interaction. Her research interests include urban computing, mobility, and tangible interfaces. She is currently doing ethnographic field work and system design in Bangkok, focusing on urban mobilities and mobile technology.

Katharine S. Willis is Marie Curie Research Fellow at the Bauhaus University of Weimar, Germany. Her work explores the ways in which we interact with our spatial surroundings, and in particular approaches to understanding how we can create legible environments. These projects investigate wayfinding, identity and the transformative possibilities of mobile and wireless technologies.

Volker Wulf is a professor in information systems and the director of the Media Research Institute at the University of Siegen. At Fraunhofer FIT, he heads the research group User-centred Software-Engineering (USE). His research interests lie primarily in the area of Computer Supported Cooperative Work, Knowledge Management, Human Computer Interaction, and Participatory Design.

Andrea Zeffiro is a doctoral candidate in Communication at Concordia University, Canada. She has worked as a researcher with the Mobile Digital Commons Network and as managing editor of the on-line journal, Wi (www.wi-not.ca). She is senior editorial assistant for the *Canadian Journal of Communication*, (www.cjc-online.ca).

Index

F

G

H

I

K

L

M

T

U

V

W

Y